Gerhard Rosenberger, Annika Schürenberg, Leonard Wienke
Abstract Algebra

Also of Interest

Algebra and Number Theory. A Selection of Highlights
Benjamin Fine, Anja Moldenhauer, Gerhard Rosenberger,
Annika Schürenberg, Leonard Wienke, 2023
ISBN 978-3-11-078998-0, e-ISBN (PDF) 978-3-11-079028-3,
e-ISBN (EPUB) 978-3-11-079039-9

Geometry and Discrete Mathematics. A Selection of Highlights
Benjamin Fine, Anja Moldenhauer, Gerhard Rosenberger,
Annika Schürenberg, Leonard Wienke, 2023
ISBN 978-3-11-074077-6, e-ISBN (PDF) 978-3-11-074078-3,
e-ISBN (EPUB) 978-3-11-074093-6

Abstract Algebra. An Introduction with Applications
Derek J. S. Robinson, 2022
ISBN 978-3-11-068610-4, e-ISBN (PDF) 978-3-11-069116-0,
e-ISBN (EPUB) 978-3-11-069121-4

Elements of Discrete Mathematics. Numbers and Counting, Groups, Graphs, Orders and Lattices
Volker Diekert, Manfred Kufleitner Gerhard Rosenberger, Ulrich Hertrampf, 2023
ISBN 978-3-11-106069-9, e-ISBN (PDF) 978-3-11-106255-6,
e-ISBN (EPUB) 978-3-11-106288-4

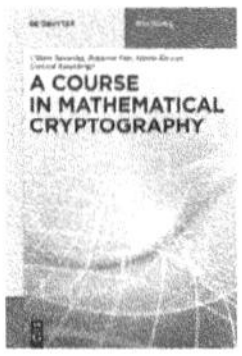

A Course in Mathematical Cryptography
Gilbert Baumslag, Benjamin Fine, Martin Kreuzer, Gerhard Rosenberger, 2015
ISBN 978-3-11-037276-2, e-ISBN (PDF) 978-3-11-037277-9,
e-ISBN (EPUB) 978-3-11-038616-5

Gerhard Rosenberger, Annika Schürenberg,
Leonard Wienke

Abstract Algebra

With Applications to Galois Theory, Algebraic Geometry,
Representation Theory and Cryptography

3rd edition

DE GRUYTER

Mathematics Subject Classification 2020
Primary: 11-01, 12-01, 13-01, 14-01, 16-01, 20-01, 20C15; Secondary: 01-01, 08-01, 94-01

Authors
Prof. Dr. Gerhard Rosenberger
University of Hamburg
Bundesstr. 55
20146 Hamburg
Germany

Dr. Leonard Wienke
University of Bremen
Bibliothekstr. 5
28359 Bremen
Germany

Annika Schürenberg
Grundschule Hoheluft
Wrangelstr. 80
20253 Hamburg
Germany

ISBN 978-3-11-113951-7
e-ISBN (PDF) 978-3-11-114252-4
e-ISBN (EPUB) 978-3-11-114284-5

Library of Congress Control Number: 2024933441

Bibliographic information published by the Deutsche Nationalbibliothek
The Deutsche Nationalbibliothek lists this publication in the Deutsche Nationalbibliografie; detailed bibliographic data are available on the Internet at http://dnb.dnb.de.

Cover image: Comstock / Stockbyte / Getty Images and sakkmesterke / iStock / Getty Images Plus
Typesetting: VTeX UAB, Lithuania

www.degruyter.com

Preface

Traditionally, mathematics has been separated into three main areas: algebra, analysis, and geometry. Of course, there is a great deal of overlap between these areas. In general, algebraic methods and symbolism pervade all of mathematics, and it is essential for anyone learning any advanced mathematics to be familiar with the concepts and methods in abstract algebra.

This is an introductory text on abstract algebra. It grew out of courses given to advanced undergraduate and beginning graduate students in the United States, and to mathematics students and teachers in Germany. We assume that the reader is familiar with calculus and with some linear algebra, primarily matrix algebra and the basic concepts of vector spaces, bases, and dimensions. All other necessary material is introduced and explained in the book. Our expectation is that the material in this text can be completed in a full year's course.

We present the material sequentially, so that polynomials and field extensions precede an in-depth look at advanced topics in group theory and Galois theory. This text follows the new approach of conveying abstract algebra starting with rings and fields, rather than with groups. Our teaching experience shows that examples of groups seem rather abstract and require a certain formal framework and mathematical maturity that would distract a course from its main objectives. The idea is that the integers provide the most natural example of an algebraic structure that students know from school. A student who goes through ring theory first, will attain a solid background in abstract algebra and will be able to move on to more advanced topics.

The centerpiece of our book is the development of Galois theory and its important applications, especially the insolvability of the quintic polynomial. After introducing the basic algebraic structures, groups, rings, and fields, we begin with the theory of polynomials and polynomial equations over fields. We then develop the main ideas of field extensions and adjoining elements to fields. Since the second edition, we include added material on skew field extensions of $\mathbb{C}$ and Frobenius's theorem.

After this, we present the necessary material from group theory needed to complete both the insolvability of the quintic polynomial and solvability by radicals in general. Hence, the middle part of the book, Chapters 9 through 14, are concerned with group theory, including permutation groups, solvable groups, Abelian groups, and group actions. Chapter 14 is somewhat off to the side of the main theme of the book. Here, we give a brief introduction to free groups, group presentations, and combinatorial group theory. In this third edition, we have extended Chapter 14 to include a primer on hyperbolic groups. With the group theory material at hand, we return to Galois theory and study general normal and separable extensions, and the fundamental theorem of Galois theory. Using this approach, we present several major applications of the theory, including solvability by radicals and the insolvability of the quintic, the fundamental theorem of algebra, the construction of regular n-gons, and the famous impossibilities: squaring the circle, doubling the cube, and trisecting an angle.

https://doi.org/10.1515/9783111142524-201

We continue with the theory of modules and prove the fundamental theorem for finitely generated modules over principle ideal domains. We then consider transcendental field extensions and prove Noether's normalization theorem as preparation for algebraic geometry based on Hilbert's basis theorem and the nullstellensatz, and describe several applications. Since the second edition, we include a new chapter on algebras and group representations. We finish in a slightly different direction, giving an introduction to algebraic and noncommutative group-based cryptography. In this third edition, we have devoted a modernized chapter to each of these topics including recent developments and results.

In the bibliography we choose to mention some interesting books and papers which are not used explicitly in our exposition but are very much related to the topics of the present book and could be helpful for additional reading.

We were very pleased with the response to the second edition of this book, and we were very happy to do a third edition. In this third edition, we have added the extensions mentioned above, cleaned up various typos pointed out by readers, and have incorporated their suggestions. Here, we have to give a special thank you to Ahmad Mirzay and O-joung Kwon. We would also like to thank Anja Rosenberger, who helped tremendously with editing and LaTeX, and who made some invaluable suggestions about the contents. Last but not least, we thank De Gruyter for publishing our book.

June 2024

Gerhard Rosenberger
Annika Schürenberg
Leonard Wienke

Contents

1 Groups, Rings and Fields

1.1 Abstract Algebra

Abstract algebra or *modern algebra* can be best described as the theory of *algebraic structures*. Briefly, an *algebraic structure* is a set together with one or more binary operations on it satisfying axioms governing the operations. There are many algebraic structures, but the most commonly studied structures are *groups, rings, fields*, and *vector spaces*. Also, widely used are *modules* and *algebras*. In this first chapter, we will look at some basic preliminaries concerning groups, rings, and fields. We will only briefly touch on groups here; a more extensive treatment will be done later in the book.

Mathematics traditionally has been subdivided into three main areas—*analysis, algebra, and geometry*. These areas overlap in many places so that it is often difficult, for example, to determine whether a topic is one in geometry or in analysis. Algebra and algebraic methods permeate all these disciplines and most of mathematics has been algebraicized; that is, uses the methods and language of algebra. Groups, rings, and fields play a major role in the study of analysis, topology, geometry, and even applied mathematics. We will see these connections in examples throughout the book.

Abstract algebra has its origins in two main areas and questions that arose in these areas—the *theory of numbers* and the *theory of equations*. The theory of numbers deals with the properties of the basic number systems—integers, rationals, and reals, whereas the theory of equations, as the name indicates, deals with solving equations, in particular, polynomial equations. Both are subjects that date back to classical times. A whole section of Euclid's elements is dedicated to number theory. The foundations for the modern study of number theory were laid by Fermat in the 1600s, and then by Gauss in the 1800s. In an attempt to prove Fermat's big theorem, Gauss introduced the complex integers $a + bi$, where a and b are integers and showed that this set has unique factorization. These ideas were extended by Dedekind and Kronecker, who developed a wide ranging theory of algebraic number fields and algebraic integers. A large portion of the terminology used in abstract algebra, such as rings, ideals, and factorization, comes from the study of algebraic number fields. This has evolved into the modern discipline of algebraic number theory.

The second origin of modern *abstract algebra* was the problem of trying to determine a formula for finding the solutions in terms of radicals of a fifth degree polynomial. It was proved first by Ruffini in 1800, and then by Abel that it is impossible to find a formula in terms of radicals for such a solution. Galois in 1820 extended this and showed that such a formula is impossible for any degree five or greater. In proving this, he laid the groundwork for much of the development of modern abstract algebra, especially field theory and finite group theory. Earlier, in 1800, Gauss proved the *fundamental theorem of algebra*, which says that any nonconstant complex polynomial equation must have a solution. One of the goals of this book is to present a comprehensive treatment of Galois theory and a proof of the results mentioned above.

https://doi.org/10.1515/9783111142524-001

The locus of real points (x,y), which satisfy a polynomial equation $f(x,y) = 0$, is called an algebraic plane curve. *Algebraic geometry* deals with the study of algebraic plane curves and extensions to loci in a higher number of variables. Algebraic geometry is intricately tied to abstract algebra and especially commutative algebra. We will touch on this in the book also.

Finally *linear algebra*, although a part of abstract algebra, arose in a somewhat different context. Historically, it grew out of the study of solution sets of systems of linear equations and the study of the geometry of real n-dimensional spaces. It began to be developed formally in the early 1800s with work of Jordan and Gauss, and then later in the century by Cayley, Hamilton, and Sylvester.

1.2 Rings

The primary motivating examples for algebraic structures are the basic number systems: the integers $\mathbb{Z}$, the rational numbers $\mathbb{Q}$, the real numbers $\mathbb{R}$, and the complex numbers $\mathbb{C}$. Each of these has two basic operations, addition and multiplication, and form what is called a *ring*. We formally define this.

Definition 1.2.1. A *ring* is a set R with two binary operations defined on it: addition, denoted by $+$, and multiplication, denoted by $\cdot$, or just by juxtaposition, satisfying the following six axioms:
(1) Addition is commutative: $a + b = b + a$ for each pair a, b in R.
(2) Addition is associative: $a + (b + c) = (a + b) + c$ for $a, b, c \in R$.
(3) There exists an additive identity, denoted by 0, such that $a + 0 = a$ for each $a \in R$.
(4) For each $a \in R$, there exists an additive inverse, denoted by $-a$, such that $a + (-a) = 0$.
(5) Multiplication is associative: $a(bc) = (ab)c$ for $a, b, c \in R$.
(6) Multiplication is left and right distributive over addition: $a(b + c) = ab + ac$, and $(b + c)a = ba + ca$ for $a, b, c \in R$.

The ring R is *commutative* if
(7) Multiplication is commutative: $ab = ba$ for a, b in R.

We call R a *ring with identity* if
(8) There exists a multiplicative identity denoted by 1 such that $a \cdot 1 = a$ and $1 \cdot a = a$ for each a in R.

If R satisfies (1) through (8), then R is a commutative ring with identity.

A set G with one operation, $+$, on it satisfying axioms (1) through (4) is called an *Abelian group*. We will discuss these further later in the chapter.

The numbers systems $\mathbb{Z}$, $\mathbb{Q}$, $\mathbb{R}$, $\mathbb{C}$ are commutative rings with identity.

A ring R with only one element is called *trivial*. A ring R with identity is trivial if and only if $0 = 1$. A *finite ring* is a ring R with only finitely many elements in it. Otherwise, R is an infinite ring. $\mathbb{Z}, \mathbb{Q}, \mathbb{R}, \mathbb{C}$ are all infinite rings. Examples of finite rings are given by the integers modulo n, $\mathbb{Z}_n$, with $n > 1$. The ring $\mathbb{Z}_n$ consists of the elements $0, 1, 2, \ldots, n-1$ with addition and multiplication done modulo n. That is, for example $4 \cdot 3 = 12 = 2$ modulo 5. Hence, in $\mathbb{Z}_5$, we have $4 \cdot 3 = 2$. The rings $\mathbb{Z}_n$ are all finite commutative rings with identity.

To give examples of rings without an identity, consider the set $n\mathbb{Z} = \{nz : z \in \mathbb{Z}\}$ consisting of all multiples of the fixed integer n. It is an easy verification (see exercises) that this forms a ring under the same addition and multiplication as in $\mathbb{Z}$, but that there is no identity for multiplication. Hence, for each $n \in \mathbb{Z}$ with $n > 1$, we get an infinite commutative ring without an identity.

To obtain examples of noncommutative rings, we consider matrices. Let $M(2, \mathbb{Z})$ be the set of (2×2)-matrices with integral entries. Addition of matrices is done componentwise; that is,

$$\begin{pmatrix} a_1 & b_1 \\ c_1 & d_1 \end{pmatrix} + \begin{pmatrix} a_2 & b_2 \\ c_2 & d_2 \end{pmatrix} = \begin{pmatrix} a_1 + a_2 & b_1 + b_2 \\ c_1 + c_2 & d_1 + d_2 \end{pmatrix},$$

whereas multiplication is matrix multiplication

$$\begin{pmatrix} a_1 & b_1 \\ c_1 & d_1 \end{pmatrix} \cdot \begin{pmatrix} a_2 & b_2 \\ c_2 & d_2 \end{pmatrix} = \begin{pmatrix} a_1a_2 + b_1c_2 & a_1b_2 + b_1d_2 \\ c_1a_2 + d_1c_2 & c_1b_2 + d_1d_2 \end{pmatrix}.$$

Then again, it is an easy verification (see exercises) that $M(2, \mathbb{Z})$ forms a ring. Further, since matrix multiplication is noncommutative, this forms a noncommutative ring. However, the identity matrix does form a multiplicative identity for it. $M(2, \mathbb{Z}_n)$ with $n > 1$ provides an example of an infinite noncommutative ring without an identity.

Finally, $M(2, \mathbb{Z}_n)$ for $n > 1$ will give an example of a finite noncommutative ring.

1.3 Integral Domains and Fields

Our basic number systems have the property that if $ab = 0$, then either $a = 0$, or $b = 0$. However, this is not necessarily true in the modular rings. For example, $2 \cdot 3 = 0$ in $\mathbb{Z}_6$.

Definition 1.3.1. A *zero divisor* in a ring R is an element $a \in R$ with $a \neq 0$ such that there exists an element $b \neq 0$ with $ab = 0$. A commutative ring with an identity $1 \neq 0$ and with no zero divisors is called an *integral domain*.

Notice that having no zero divisors is equivalent to the fact that if $ab = 0$ in R, then either $a = 0$, or $b = 0$.

Hence, $\mathbb{Z}, \mathbb{Q}, \mathbb{R}, \mathbb{C}$ are all integral domains, but from the example above, $\mathbb{Z}_6$ is not. In general, we have the following:

Theorem 1.3.2. *$\mathbb{Z}_n$ is an integral domain if and only if n is a prime.*

Proof. First of all, notice that under multiplication modulo n, an element m is 0 if and only if n divides m. We will make this precise shortly. Recall further Euclid's lemma (see Chapter 2), which says that if a prime p divides a product ab, then p divides a, or p divides b.

Now suppose that n is a prime and $ab = 0$ in $\mathbb{Z}_n$. Then n divides ab. From Euclid's lemma it follows that n divides a, or n divides b. In the first case, $a = 0$ in $\mathbb{Z}_n$, whereas in the second, $b = 0$ in $\mathbb{Z}_n$. It follows that there are no zero divisors in $\mathbb{Z}_n$, and since $\mathbb{Z}_n$ is a commutative ring with an identity, it is an integral domain.

Conversely, suppose $\mathbb{Z}_n$ is an integral domain. Suppose that n is not prime. Then $n = ab$ with $1 < a < n, 1 < b < n$. It follows that $ab = 0$ in $\mathbb{Z}_n$ with neither a nor b being zero. Therefore, they are zero divisors, which is a contradiction. Hence, n must be prime. □

In $\mathbb{Q}$, every nonzero element has a multiplicative inverse. This is not true in $\mathbb{Z}$, where only the elements $-1, 1$ have multiplicative inverses within $\mathbb{Z}$.

Definition 1.3.3. A *unit* in a ring R with identity $1 \neq 0$ is an element $a \in R$, which has a multiplicative inverse; that is, an element $b \in R$ such that $ab = ba = 1$. If a is a unit in R, we denote its inverse by a^{-1}. We denote the set of units of R by $R^\star$.

Hence, every nonzero element of $\mathbb{Q}$ and of $\mathbb{R}$ and of $\mathbb{C}$ is a unit, but in $\mathbb{Z}$, the only units are ± 1. In $M(2, \mathbb{R})$, the units are precisely those matrices that have nonzero determinant, whereas in $M(2, \mathbb{Z})$, the units are those integral matrices that have determinant ± 1.

Definition 1.3.4. A *field* K is a commutative ring with an identity $1 \neq 0$, where every nonzero element is a unit.

Hence, a field K always contains at least two elements, a zero element 0 and an identity $1 \neq 0$.

The rationals $\mathbb{Q}$, the reals $\mathbb{R}$, and the complexes $\mathbb{C}$ are all fields. If we relax the commutativity requirement and just require that in the ring R with identity, each nonzero element is a unit, then we get a *skew field* or *division ring*.

Lemma 1.3.5. *If K is a field, then K is an integral domain.*

Proof. Since a field K is already a commutative ring with an identity, we must only show that there are no zero divisors in K.

Suppose that $ab = 0$ with $a \neq 0$. Since K is a field and a is nonzero, it has an inverse a^{-1}. Hence,

$$a^{-1}(ab) = a^{-1}0 = 0 \implies (a^{-1}a)b = 0 \implies b = 0.$$

Therefore, K has no zero divisors and must be an integral domain. □

Recall that $\mathbb{Z}_n$ was an integral domain only when n was a prime. This turns out to also be necessary and sufficient for $\mathbb{Z}_n$ to be a field.

Theorem 1.3.6. *$\mathbb{Z}_n$ is a field if and only if n is a prime.*

Proof. First suppose that $\mathbb{Z}_n$ is a field. Then from Lemma 1.3.5, it is an integral domain. Therefore, from Theorem 1.3.2, n must be a prime.

Conversely, suppose that n is a prime. We must show that $\mathbb{Z}_n$ is a field. Since we already know that $\mathbb{Z}_n$ is an integral domain, we must only show that each nonzero element of $\mathbb{Z}_n$ is a unit. Here, we need some elementary facts from number theory. If a, b are integers, we use the notation $a|b$ to indicate that a divides b.

Recall that given nonzero integers a, b, their *greatest common divisor* or GCD $d > 0$ is a positive integer, which is a common divisor; that is, $d|a$ and $d|b$, and if d_1 is any other common divisor, then $d_1|d$. We denote the greatest common divisor of a, b by either $\gcd(a, b)$ or (a, b). It can be proved that given nonzero integers a, b their GCD exists, is unique and can be characterized as the least positive linear combination of a and b. If the GCD of a and b is 1, then we say that a and b are *relatively prime* or *coprime*. This is equivalent to being able to express 1 as a linear combination of a and b (see Chapter 3 for proofs and more details).

Now let $a \in \mathbb{Z}_n$ with n prime and $a \neq 0$. Since $a \neq 0$, we have that n does not divide a. Since n is prime, it follows that a and n must be relatively prime, $(a, n) = 1$. From the number theoretic remarks above, we then have that there exist x, y with

$$ax + ny = 1.$$

However, in $\mathbb{Z}_n$, the element $ny = 0$. Therefore, in $\mathbb{Z}_n$, we have

$$ax = 1.$$

Therefore, a has a multiplicative inverse in $\mathbb{Z}_n$ and is, hence, a unit. Since a was an arbitrary nonzero element, we conclude that $\mathbb{Z}_n$ is a field. □

The theorem above is actually a special case of a more general result from which Theorem 1.3.6 could also be obtained.

Theorem 1.3.7. *Each finite integral domain is a field.*

Proof. Let K be a finite integral domain. We must show that K is a field. It is clearly sufficient to show that each nonzero element of K is a unit. Let

$$\{0, 1, r_1, \ldots, r_n\}$$

be the elements of K. Let r_i be a fixed nonzero element and multiply each element of K by r_i on the left. Now

$$\text{if } r_i r_j = r_i r_k \quad \text{then } r_i(r_j - r_k) = 0.$$

Since $r_i \neq 0$, it follows that $r_j - r_k = 0$ or $r_j = r_k$. Therefore, all the products $r_i r_j$ are distinct. Hence,

$$R = \{0, 1, r_1, \ldots, r_n\} = r_i R = \{0, r_i, r_i r_1, \ldots, r_i r_n\}.$$

Therefore, the identity element 1 must be in the right-hand list; that is, there is an r_j such that $r_i r_j = 1$. Therefore, r_i has a multiplicative inverse and is, hence, a unit. Therefore, K is a field. □

1.4 Subrings and Ideals

A very important concept in algebra is that of a *substructure* that is a subset having the same structure as the superset.

Definition 1.4.1. A *subring* of a ring R is a nonempty subset S that is also a ring under the same operations as R. If R is a field and S also a field, then it is a *subfield.*

If $S \subset R$, then S satisfies the same basic axioms, associativity, and commutativity of addition, for example. Therefore, S will be a subring if it is nonempty and closed under the operations; that is, closed under addition, multiplication, and taking additive inverses.

Lemma 1.4.2. *A subset S of a ring R is a subring if and only if S is nonempty, and whenever $a, b \in S$, we have $a + b \in S$, $a - b \in S$ and $ab \in S$.*

Example 1.4.3. Show that if $n > 1$, the set $n\mathbb{Z}$ is a subring of $\mathbb{Z}$. Here, clearly $n\mathbb{Z}$ is nonempty. Suppose $a = nz_1$, $b = nz_2$ are two elements of $n\mathbb{Z}$. Then

$$\begin{aligned} a + b &= nz_1 + nz_2 = n(z_1 + z_2) \in n\mathbb{Z} \\ a - b &= nz_1 - nz_2 = n(z_1 - z_2) \in n\mathbb{Z} \\ ab &= nz_1 \cdot nz_2 = n(nz_1 z_2) \in n\mathbb{Z}. \end{aligned}$$

Therefore, $n\mathbb{Z}$ is a subring.

Example 1.4.4. Show that the set of real numbers of the form

$$S = \{u + v\sqrt{2} : u, v \in \mathbb{Q}\}$$

is a subring of $\mathbb{R}$. Here, $1 + \sqrt{2} \in S$; therefore, S is nonempty. Suppose $a = u_1 + v_1\sqrt{2}$, $b = u_2 + v_2\sqrt{2}$ are two element of S. Then

$$\begin{aligned} a + b &= (u_1 + v_1\sqrt{2}) + (u_2 + v_2\sqrt{2}) = u_1 + u_2 + (v_1 + v_2)\sqrt{2} \in S \\ a - b &= (u_1 + v_1\sqrt{2}) - (u_2 + v_2\sqrt{2}) = u_1 - u_2 + (v_1 - v_2)\sqrt{2} \in S \\ a \cdot b &= (u_1 + v_1\sqrt{2}) \cdot (u_2 + v_2\sqrt{2}) = (u_1 u_2 + 2v_1 v_2) + (u_1 v_2 + v_1 u_2)\sqrt{2} \in S. \end{aligned}$$

Therefore, S is a subring.

In fact, S is a field because $\frac{1}{u+v\sqrt{2}} = \frac{u}{u^2-2v^2} - \frac{v\sqrt{2}}{u^2-v^2}$ if $(u,v) \neq (0,0)$. In the following, we are especially interested in special types of subrings called *ideals*.

Definition 1.4.5. Let R be a ring and $I \subset R$. Then I is a (two-sided) *ideal* if the following properties hold:
(1) I is nonempty.
(2) If $a, b \in I$, then $a \pm b \in I$.
(3) If $a \in I$ and r is any element of R, then $ra \in I$, and $ar \in I$.

We denote the fact that I forms an ideal in R by $I \triangleleft R$.

Notice that if $a, b \in I$, then from (3), we have $ab \in I$, and $ba \in I$. Hence, I forms a subring; that is, each ideal is also a subring. The set $\{0\}$ and the whole ring R are trivial ideals of R.

If we assume that in (3), only $ra \in I$, then I is called a left ideal. Analogously, we define a right ideal.

Lemma 1.4.6. *Let R be a commutative ring and $a \in R$. Then the set*

$$\langle a \rangle = aR = \{ar : r \in R\}$$

is an ideal of R.

This ideal is called the *principal ideal generated by a.*

Proof. We must verify the three properties of the definition. Since $a \in R$, we have that aR is nonempty. If $u = ar_1$, $v = ar_2$ are two elements of aR, then

$$u \pm v = ar_1 \pm ar_2 = a(r_1 \pm r_2) \in aR.$$

Therefore, (2) is satisfied.

Finally, let $u = ar_1 \in aR$ and $r \in R$. Then

$$ru = rar_1 = a(rr_1) \in aR, \quad \text{and} \quad ur = ar_1r = a(r_1r) \in aR. \qquad \square$$

Recall that $a \in \langle a \rangle$ if R has an identity.

Notice that if $n \in \mathbb{Z}$, then the principal ideal generated by n is precisely the ring $n\mathbb{Z}$, which we have already examined. Hence, for each $n > 1$, the subring $n\mathbb{Z}$ is actually an ideal. We can show more.

Theorem 1.4.7. *Any subring of $\mathbb{Z}$ is of the form $n\mathbb{Z}$ for some n. Hence, each subring of $\mathbb{Z}$ is actually a principal ideal.*

Proof. Let S be a subring of $\mathbb{Z}$. If $S = \{0\}$, then $S = 0\mathbb{Z}$, so we may assume that S has nonzero elements. Since S is a subring if it has nonzero elements, it must have positive elements (since it has the additive inverse of any element in it).

Let S^+ be the set of positive elements in S. From the remarks above, this is a nonempty set, and so, there must be a least positive element n. We claim that $S = n\mathbb{Z}$.

Let m be a positive element in S. By the division algorithm

$$m = qn + r,$$

where either $r = 0$, or $0 < r < n$ (see Chapter 3). Suppose that $r \neq 0$. Then

$$r = m - qn.$$

Now $m \in S$, and $n \in S$. Since S is a subring, it is closed under addition so that $qn \in S$. But S is a subring, therefore, $m - qn \in S$. It follows that $r \in S$. But this is a contradiction since n was the least positive element in S. Therefore, $r = 0$, and $m = qn$. Hence, each positive element in S is a multiple of n.

Now let m be a negative element of S. Then $-m \in S$, and $-m$ is positive. Hence, $-m = qn$, and thus, $m = (-q)n$. Therefore, every element of S is a multiple of n, and so, $S = n\mathbb{Z}$. It follows that every subring of $\mathbb{Z}$ is of this form and, therefore, every subring of $\mathbb{Z}$ is an ideal. □

We mention that this is true in $\mathbb{Z}$, but not always true. For example, $\mathbb{Z}$ is a subring of $\mathbb{Q}$, but not an ideal. An extension of the proof of Lemma 1.4.6 gives the following. We leave the proof as an exercise.

Lemma 1.4.8. *Let R be a commutative ring and $a_1, \dots, a_n \in R$ be a finite set of elements in R. Then the set*

$$\langle a_1, \dots, a_n \rangle = \{r_1 a_1 + r_2 a_2 + \cdots + r_n a_n : r_i \in R\}$$

is an ideal of R.

This ideal is called the *ideal generated by* $a_1, \dots, a_n$. Recall that $a_1, \dots, a_n$ are in $\langle a_1, \dots, a_n \rangle$ if R has an identity.

Theorem 1.4.9. *Let R be a commutative ring with an identity $1 \neq 0$. Then R is a field if and only if the only ideals in R are $\{0\}$ and R.*

Proof. Suppose that R is a field and $I \lhd R$ is an ideal. We must show that either $I = \{0\}$, or $I = R$. Suppose that $I \neq \{0\}$, then we must show that $I = R$.

Since $I \neq \{0\}$, there exists an element $a \in I$ with $a \neq 0$. Since R is a field, this element a has an inverse a^{-1}. Since I is an ideal, it follows that $a^{-1}a = 1 \in I$. Let $r \in R$, then, since $1 \in I$, we have $r \cdot 1 = r \in I$. Hence, $R \subset I$ and, therefore, $R = I$.

Conversely, suppose that R is a commutative ring with an identity, whose only ideals are $\{0\}$ and R. We must show that R is a field, or equivalently, that every nonzero element of R has a multiplicative inverse.

Let $a \in R$ with $a \neq 0$. Since R is a commutative ring, and $a \neq 0$, the principal ideal aR is a nontrivial ideal in R. Hence, $aR = R$. Therefore, the multiplicative identity $1 \in aR$. It follows that there exists an $r \in R$ with $ar = 1$. Hence, a has a multiplicative inverse, and R must be a field. □

1.5 Factor Rings and Ring Homomorphisms

Given an ideal I in a ring R, we can build a new ring called the *factor ring* or *quotient ring* of R modulo I. The special condition on the subring I, that $rI \subset I$ and $Ir \subset I$ for all $r \in R$, that makes it an ideal, is specifically to allow this construction to be a ring.

Definition 1.5.1. Let I be an ideal in a ring R. Then a *coset* of I is a subset of R of the form

$$r + I = \{r + i : i \in I\}$$

with r a fixed element of R.

Lemma 1.5.2. *Let I be an ideal in a ring R. Then the cosets of I partition R; that is, any two cosets are either coincide or disjoint.*

We leave the proof to the exercises. Now, on the set of all cosets of an ideal, we will build a new ring.

Theorem 1.5.3. *Let I be an ideal in a ring R. Let $R/I = \{r + I : r \in R\}$ be the set of all cosets of I in R. We define addition and multiplication on R/I in the following manner:*

$$(r_1 + I) + (r_2 + I) = (r_1 + r_2) + I$$
$$(r_1 + I) \cdot (r_2 + I) = (r_1 \cdot r_2) + I.$$

Then R/I forms a ring called the factor ring of R modulo I. The zero element of R/I is $0 + I$ and the additive inverse of $r + I$ is $-r + I$. Further, if R is commutative, then R/I is commutative, and if R has an identity, then R/I has an identity $1 + I$.

Proof. The proof that R/I satisfies the ring axioms under the definitions above is straightforward. For example,

$$(r_1 + I) + (r_2 + I) = (r_1 + r_2) + I = (r_2 + r_1) + I = (r_2 + I) + (r_1 + I),$$

and so, addition is commutative. What must be shown is that both addition and multiplication are well defined. That is, if

$$r_1 + I = r_1' + I, \quad \text{and} \quad r_2 + I = r_2' + I$$

then

$$(r_1 + I) + (r_2 + I) = (r_1' + I) + (r_2' + I),$$

and

$$(r_1 + I) \cdot (r_2 + I) = (r_1' + I) \cdot (r_2' + I).$$

Now if $r_1 + I = r_1' + I$, then $r_1 \in r_1' + I$, and so, $r_1 = r_1' + i_1$ for some $i_1 \in I$. Similarly, if $r_2 + I = r_2' + I$, then $r_2 \in r_2' + I$, and so, $r_2 = r_2' + i_2$ for some $i_2 \in I$. Then

$$(r_1 + I) + (r_2 + I) = (r_1' + i_1 + I) + (r_2' + i_2 + I) = (r_1' + I) + (r_2' + I)$$

since $i_1 + I = I$ and $i_2 + I = I$. Similarly,

$$\begin{aligned}(r_1 + I) \cdot (r_2 + I) &= (r_1' + i_1 + I) \cdot (r_2' + i_2 + I) \\ &= r_1' \cdot r_2' + r_1' \cdot i_2 + r_2' \cdot i_1 + r_1' \cdot I + r_2' \cdot I + I \cdot I \\ &= (r_1' \cdot r_2') + I\end{aligned}$$

since all the other products are in the ideal I. This shows that addition and multiplication are well defined. It also shows why the ideal property is necessary. □

As an example, let R be the integers $\mathbb{Z}$. As we have seen, each subring is an ideal and of the form $n\mathbb{Z}$ for some natural number n. The factor ring $\mathbb{Z}/n\mathbb{Z}$ is called the *residue class ring modulo n*, denoted $\mathbb{Z}_n$. Notice that we can take as cosets

$$0 + n\mathbb{Z},\ 1 + n\mathbb{Z},\ \dots,\ (n-1) + n\mathbb{Z}.$$

Addition and multiplication of cosets is then just addition and multiplication modulo n. As we can see, this is just a formalization of the ring $\mathbb{Z}_n$, which we have already looked at. Recall that $\mathbb{Z}_n$ is an integral domain if and only if n is prime and $\mathbb{Z}_n$ is a field for precisely the same n. If $n = 0$, then $\mathbb{Z}/n\mathbb{Z}$ is the same as $\mathbb{Z}$.

We now show that ideals and factor rings are closely related to certain mappings between rings.

Definition 1.5.4. Let R and S be rings. Then a mapping $f : R \to S$ is a *ring homomorphism* if

$$\begin{aligned} f(r_1 + r_2) &= f(r_1) + f(r_2) \quad \text{for any } r_1, r_2 \in R \\ f(r_1 \cdot r_2) &= f(r_1) \cdot f(r_2) \quad \text{for any } r_1, r_2 \in R. \end{aligned}$$

In addition,
(1) f is an *epimorphism* if it is surjective.
(2) f is an *monomorphism* if it is injective.
(3) f is an *isomorphism* if it is bijective; that is, both surjective and injective. In this case, R and S are said to be *isomorphic rings*, which we denote by $R \cong S$.

(4) f is an *endomorphism* if $R = S$; that is, a ring homomorphism from a ring to itself.
(5) f is an *automorphism* if $R = S$ and f is an isomorphism.

Lemma 1.5.5. *Let R and S be rings, and let $f : R \to S$ be a ring homomorphism. Then*
(1) *$f(0) = 0$, where the first and second 0 are the zero elements of R and S, respectively.*
(2) *$f(-r) = -f(r)$ for any $r \in R$.*

Proof. We obtain $f(0) = 0$ from the equation $f(0) = f(0+0) = f(0) + f(0)$. Hence, $0 = f(0) = f(r - r) = f(r + (-r)) = f(r) + f(-r)$; that is, $f(-r) = -f(r)$. □

Definition 1.5.6. Let R and S be rings, and let $f : R \to S$ be a *ring homomorphism*. Then the *kernel* of f is

$$\ker(f) = \{r \in R : f(r) = 0\}.$$

The *image* of f, denoted $\operatorname{im}(f)$, is the range of f within S. That is,

$$\operatorname{im}(f) = \{s \in S : \text{there exists } r \in R \text{ with } f(r) = s\}.$$

Theorem 1.5.7 (Ring isomorphism theorem). *Let R and S be rings, and let*

$$f : R \to S$$

be a ring homomorphism. Then
(1) $\ker(f)$ *is an ideal in R,* $\operatorname{im}(f)$ *is a subring of S, and*

$$R/\ker(f) \cong \operatorname{im}(f).$$

(2) *Conversely, suppose that I is an ideal in a ring R. Then the map $f : R \to R/I$, given by $f(r) = r + I$ for $r \in R$, is a ring homomorphism, whose kernel is I, and whose image is R/I.*

The theorem says that the concepts of ideal of a ring and kernel of a ring homomorphism coincide; that is, each ideal is the kernel of a homomorphism and the kernel of each ring homomorphism is an ideal.

Proof. If $s_1, s_2 \in \operatorname{im}(f)$, then there exist $r_1, r_2 \in R$, such that $f(r_1) = s_1$, and $f(r_2) = s_2$. Then certainly, $\operatorname{im}(f)$ is a subring of S from Definition 1.5.4 and Lemma 1.5.5. Now, let $I = \ker(f)$. We show first that I is an ideal. If $r_1, r_2 \in I$, then $f(r_1) = f(r_2) = 0$. It follows from the homomorphism property that

$$f(r_1 \pm r_2) = f(r_1) \pm f(r_2) = 0 + 0 = 0$$
$$f(r_1 \cdot r_2) = f(r_1) \cdot f(r_2) = 0 \cdot 0 = 0.$$

Therefore, I is a subring.

Now let $i \in I$ and $r \in R$. Then

$$f(r \cdot i) = f(r) \cdot f(i) = f(r) \cdot 0 = 0 \quad \text{and} \quad f(i \cdot r) = f(i) \cdot f(r) = 0 \cdot f(r) = 0$$

and, hence, I is an ideal.

Consider the factor ring R/I. Let $f^* : R/I \to \operatorname{im}(f)$ by $f^*(r + I) = f(r)$. We show that f^* is an isomorphism.

First, we show that it is well defined. Suppose $r_1 + I = r_2 + I$, then $r_1 - r_2 \in I = \ker(f)$. It follows that $f(r_1 - r_2) = 0$, so $f(r_1) = f(r_2)$. Hence, $f^*(r_1 + I) = f^*(r_2 + I)$, and the map f^* is well defined.

Now

$$\begin{aligned} f^*((r_1 + I) + (r_2 + I)) &= f^*((r_1 + r_2) + I) = f(r_1 + r_2) \\ &= f(r_1) + f(r_2) = f^*(r_1 + I) + f^*(r_2 + I), \end{aligned}$$

and

$$\begin{aligned} f^*((r_1 + I) \cdot (r_2 + I)) &= f^*((r_1 \cdot r_2) + I) = f(r_1 \cdot r_2) \\ &= f(r_1) \cdot f(r_2) = f^*(r_1 + I) \cdot f^*(r_2 + I). \end{aligned}$$

Hence, f^* is a homomorphism. We must now show that it is injective and surjective.

Suppose that $f^*(r_1 + I) = f^*(r_2 + I)$. Then $f(r_1) = f(r_2)$ so that $f(r_1 - r_2) = 0$. Hence, $r_1 - r_2 \in \ker(f) = I$. Therefore, $r_1 \in r_2 + I$, and thus, $r_1 + I = r_2 + I$, and the map f^* is injective.

Finally, let $s \in \operatorname{im}(f)$. Then there exists $r \in R$ such that $f(r) = s$. Then $f^*(r + I) = s$, and the map f^* is surjective and, hence, an isomorphism. This proves the first part of the theorem.

To prove the second part, let I be an ideal in R and R/I the factor ring. Consider the map $f : R \to R/I$, given by $f(r) = r + I$. From the definition of addition and multiplication in the factor ring R/I, it is clear that this is a homomorphism. Consider the kernel of f. If $r \in \ker(f)$, then $f(r) = r + I = 0 = 0 + I$. This implies that $r \in I$ and, hence, the kernel of this map is exactly the ideal I, completing the proof. □

Theorem 1.5.7 is called the *ring isomorphism theorem* or the *first ring isomorphism theorem*. We mention that there is an analogous theorem for each algebraic structure, in particular, for groups and vector spaces. We will mention the result for groups in Section 1.8.

1.6 Fields of Fractions

The integers are an integral domain, and the rationals $\mathbb{Q}$ are a field that contains the integers. First, we show that $\mathbb{Q}$ is the smallest field containing $\mathbb{Z}$.

Theorem 1.6.1. *The rationals $\mathbb{Q}$ are the smallest field containing the integers $\mathbb{Z}$. That is, if $\mathbb{Z} \subset K \subset \mathbb{Q}$ with K a subfield of $\mathbb{Q}$, then $K = \mathbb{Q}$.*

Proof. Since $\mathbb{Z} \subset K$, we have $m, n \in K$ for any two integers m, n with $n \neq 0$. Since K is a subfield, it is closed under taking division; that is, taking multiplicative inverses and, hence, the fraction $\frac{m}{n} \in K$. Since each element of $\mathbb{Q}$ is such a fraction, it follows that $\mathbb{Q} \subset K$. Since $K \subset \mathbb{Q}$, it follows that $K = \mathbb{Q}$. □

Notice that to construct the rationals from the integers, we form all fractions $\frac{m}{n}$ with $n \neq 0$, and where $\frac{m_1}{n_1} = \frac{m_2}{n_2}$ if $m_1 n_2 = n_1 m_2$. We then do the standard operations on fractions. If we start with any integral domain D, we can mimic this construction to build a *field of fractions* from D; that is, the smallest field containing D.

Theorem 1.6.2. *Let D be an integral domain. Then there is a field K containing D, called the field of fractions for D, such that each element of K is a fraction from D; that is, an element of the form $d_1 d_2^{-1}$ with $d_1, d_2 \in D$. Further, K is unique up to isomorphism and is the smallest field containing D.*

Proof. The proof is just the mimicking of the construction of the rationals from the integers. Let

$$K' = \{(d_1, d_2) : d_1, d_2 \neq 0, d_1, d_2 \in D\}.$$

Define on K' the equivalence relation

$$(d_1, d_2) = (d_1', d_2') \quad \text{if } d_1 d_2' = d_2 d_1'.$$

Let K be the set of equivalence classes, and define addition and multiplication in the usual manner as for fractions, where the result is the equivalence class:

$$(d_1, d_2) + (d_3, d_4) = (d_1 d_4 + d_2 d_3, d_2 d_4)$$
$$(d_1, d_2) \cdot (d_3, d_4) = (d_1 d_3, d_2 d_4).$$

It is now straightforward to verify the ring axioms for K. The inverse of $(d_1, 1)$ is $(1, d_1)$ for $d_1 \neq 0$ in D. As with $\mathbb{Z}$, we identify the elements of K as fractions $\frac{d_1}{d_2}$. The proof that K is the smallest field containing D is the same as for $\mathbb{Q}$ from $\mathbb{Z}$. □

As examples, we have that $\mathbb{Q}$ is the field of fractions for $\mathbb{Z}$. A familiar, but less common, example is the following:

Let $\mathbb{R}[x]$ be the set of polynomials over the real numbers $\mathbb{R}$. It can be shown that $\mathbb{R}[x]$ forms an integral domain (see Chapter 3). The field of fractions consists of all formal functions $\frac{f(x)}{g(x)}$, where $f(x)$, $g(x)$ are real polynomials with $g(x) \neq 0$. The corresponding field of fractions is called the *field of rational functions* over $\mathbb{R}$ and is denoted $\mathbb{R}(x)$.

1.7 Characteristic and Prime Rings

We saw in the last section that $\mathbb{Q}$ is the smallest field containing the integers. Since any subfield of $\mathbb{Q}$ must contain the identity, it follows that any nontrivial subfield of $\mathbb{Q}$ must contain the integers and, hence, be all of $\mathbb{Q}$. Therefore, $\mathbb{Q}$ has no nontrivial subfields. We say that $\mathbb{Q}$ is a *prime field*.

Definition 1.7.1. A field K is a prime field if K contains no nontrivial subfields.

Lemma 1.7.2. *Let K be any field. Then K contains a prime field K as a subfield.*

Proof. Let K_1, K_2 be subfields of K. If $k_1, k_2 \in K_1 \cap K_2$, then $k_1 \pm k_2 \in K_1$ since K_1 is a subfield, and $k_1 \pm k_2 \in K_2$ since K_2 is a subfield. Therefore, $k_1 \pm k_2 \in K_1 \cap K_2$. Similarly, $k_1 k_2^{-1} \in K_1 \cap K_2$. It follows that $K_1 \cap K_2$ is again a subfield.

Now, let K be the intersection of all subfields of K. From the argument above K is a subfield, and the only nontrivial subfield of K is itself. Hence, K is a prime field. □

Definition 1.7.3. Let R be a commutative ring with an identity $1 \neq 0$. The smallest positive integer n such that $n \cdot 1 = 1 + 1 + \cdots + 1 = 0$ is called the *characteristic* of R. If there is no such n, then R has characteristic 0. We denote the characteristic by $\operatorname{char}(R)$.

First, notice that 0 is the characteristic of $\mathbb{Z}$, $\mathbb{Q}$, $\mathbb{R}$. Further the characteristic of $\mathbb{Z}_n$ is n.

Theorem 1.7.4. *Let R be an integral domain. Then the characteristic of R is either 0 or a prime. In particular, the characteristic of a field is zero or a prime.*

Proof. Suppose that R is an integral domain and $\operatorname{char}(R) = n \neq 0$. Suppose that $n = mk$ with $1 < m < n, 1 < k < n$. Then $n \cdot 1 = 0 = (m \cdot 1)(k \cdot 1)$. Since R is an integral domain, we have no zero divisors and, hence, $m \cdot 1 = 0$, or $k \cdot 1 = 0$. However, this is a contradiction since n is the least positive integer such that $n \cdot 1 = 0$. Therefore, n must be a prime. □

We have seen that every field contains a prime field. We extend this.

Definition 1.7.5. A commutative ring R with an identity $1 \neq 0$ is a *prime ring* if the only subring containing the identity is the whole ring.

Clearly both the integers $\mathbb{Z}$ and the modular integers $\mathbb{Z}_n$ are prime rings. In fact, up to isomorphism, they are the only prime rings.

Theorem 1.7.6. *Let R be a prime ring. Then $\operatorname{char}(R) = 0$ implies $R \cong \mathbb{Z}$, whereas $\operatorname{char}(R) = n > 0$ implies $R \cong \mathbb{Z}_n$.*

Proof. Suppose that $\operatorname{char}(R) = 0$. Let $S = \{r = m \cdot 1 : r \in R, m \in \mathbb{Z}\}$. Then S is a subring of R containing the identity and, hence, $S = R$. However, the map $m \cdot 1 \to m$ gives an isomorphism from S to $\mathbb{Z}$. It follows that R is isomorphic to $\mathbb{Z}$.

If $\operatorname{char}(R) = n > 0$, the proof is identical. Since $n \cdot 1 = 0$, the subring S of R, defined above, is all of R and isomorphic to $\mathbb{Z}_n$. □

Theorem 1.7.6 can be extended to fields with $\mathbb{Q}$, taking the place of $\mathbb{Z}$ and $\mathbb{Z}_p$, with p a prime, taking the place of $\mathbb{Z}_n$.

Theorem 1.7.7. *Let K be a prime field. If K has characteristic 0, then $K \cong \mathbb{Q}$, whereas if K has characteristic p, then $K \cong \mathbb{Z}_p$.*

Proof. The proof is identical to that of Theorem 1.7.6; however, we consider the smallest subfield K_1 of K containing S. □

We mention that there can be infinite fields of characteristic p. Consider, for example, the field of fractions of the polynomial ring $\mathbb{Z}_p[x]$. This is the field of rational functions with coefficients in $\mathbb{Z}_p$.

We give a theorem on fields of characteristic p that will be important much later when we look at Galois theory.

Theorem 1.7.8. *Let K be a field of characteristic p. Then the mapping $\phi : K \to K$, given by $\phi(k) = k^p$, is an injective endomorphism of K. In particular, $(a+b)^p = a^p + b^p$ for any $a, b \in K$.*

This mapping is called the Frobenius homomorphism of K. Further, if K is finite, ϕ is an automorphism.

Proof. We first show that ϕ is a homomorphism. Now

$$\phi(ab) = (ab)^p = a^p b^p = \phi(a)\phi(b).$$

We need a little more work for addition:

$$\phi(a+b) = (a+b)^p = \sum_{i=0}^{p} \binom{p}{i} a^i b^{p-i} = a^p + \sum_{i=1}^{p-1} \binom{p}{i} a^i b^{p-i} + b^p$$

by the binomial expansion, which holds in any commutative ring. However,

$$\binom{p}{i} = \frac{p(p-1)\cdots(p-i+1)}{i\cdot(i-1)\cdots 1},$$

and it is clear that $p \mid \binom{p}{i}$ for $1 \le i \le p-1$. Hence, in K, we have $\binom{p}{i}\cdot 1 = 0$, and so, we have

$$\phi(a+b) = (a+b)^p = a^p + b^p = \phi(a) + \phi(b).$$

Therefore, ϕ is a homomorphism.

Further, ϕ is always injective. To see this, suppose that $\phi(x) = \phi(y)$. Then

$$\phi(x-y) = 0 \implies (x-y)^p = 0.$$

But K is a field, so there are no zero divisors. Therefore, we must have $x - y = 0$, or $x = y$. If K is finite and ϕ is injective, it must also be surjective and, hence, an automorphism of K. □

1.8 Groups

We close this first chapter by introducing some basic definitions and results from group theory that mirror the results, which were presented for rings and fields. We will look at group theory in more detail later in the book. Proofs will be given at that point.

Definition 1.8.1. A *group* G is a set with one binary operation (which we will denote by multiplication) such that
(1) the operation is associative;
(2) there exists an identity for this operation; and
(3) each $g \in G$ has an inverse for this operation.

If, in addition, the operation is commutative, the group G is called an *Abelian group*. The *order* of G is the number of elements in G, denoted by $|G|$. If $|G| < \infty$, G is a *finite group*; otherwise G is an *infinite group*.

Groups most often arise from invertible mappings of a set onto itself. Such mappings are called *permutations*.

Theorem 1.8.2. *The group of all permutations on a set A forms a group called the symmetric group on A, which we denote by S_A. If A has more than 2 elements, then S_A is non-Abelian.*

Definition 1.8.3. Let G_1 and G_2 be groups. Then a mapping $f : G_1 \to G_2$ is a *(group) homomorphism* if

$$f(g_1g_2) = f(g_1)f(g_2) \quad \text{for any } g_1, g_2 \in G_1.$$

As with rings, we have, in addition,
(1) f is an *epimorphism* if it is surjective.
(2) f is an *monomorphism* if it is injective.
(3) f is an *isomorphism* if it is bijective; that is, both surjective and injective. In this case, G_1 and G_2 are said to be *isomorphic groups*, which we denote by $G_1 \cong G_2$.
(4) f is an *endomorphism* if $G_1 = G_2$; that is, a homomorphism from a group to itself.
(5) f is an *automorphism* if $G_1 = G_2$, and f is an isomorphism.

Lemma 1.8.4. *Let G_1 and G_2 be groups, and let $f : G_1 \to G_2$ be a homomorphism. Then*
1. $f(1) = 1$, *where the first* 1 *is the identity element of G_1, and the second is the identity element of G_2.*
2. $f(g^{-1}) = (f(g))^{-1}$ *for any $g \in G_1$.*

If A is a set, $|A|$ denotes the size of A.

Theorem 1.8.5. *If A_1 and A_2 are sets with $|A_1| = |A_2|$, then $S_{A_1} \cong S_{A_2}$. If $|A| = n$ with n finite, we call S_A the symmetric group on n elements, which we denote by S_n. Further, we have $|S_n| = n!$.*

Subgroups are defined in an analogous manner to subrings. Special types of subgroups, called *normal subgroups*, take the place in group theory that ideals play in ring theory.

Definition 1.8.6. A subset H of a group G is a *subgroup* if $H \neq \emptyset$ and H forms a group under the same operation as G. Equivalently, H is a subgroup if $H \neq \emptyset$, and H is closed under the operation and inverses.

Definition 1.8.7. If H is a subgroup of a group G, then a *left coset* of H is a subset of G of the form $gH = \{gh : h \in H\}$. A *right coset* of H is a subset of G of the form $Hg = \{hg : h \in H\}$.

As with rings the cosets of a subgroup partition a group. We call the number of right cosets of a subgroup H in a group G, then *index* of H in G, denoted $|G : H|$. One can prove that the number of right cosets is equal to the number of left cosets. For finite groups, we have the following beautiful result called *Lagrange's theorem*.

Theorem 1.8.8 (Lagrange's theorem). *Let G be a finite group and H a subgroup. Then the order of H divides the order of G. In particular,*

$$|G| = |H||G : H|.$$

Normal subgroups take the place of ideals in group theory.

Definition 1.8.9. A subgroup H of a group G is a *normal subgroup*, denoted $H \triangleleft G$, if every left coset of H is also a right coset; that is, $gH = Hg$ for each $g \in G$. Note that this does not say that g and H commute elementwise, just that the subsets gH and Hg are the same. Equivalently, H is normal if $g^{-1}Hg = H$ for any $g \in G$.

Normal subgroups allow us to construct factor groups, just as ideals allowed us to construct factor rings.

Theorem 1.8.10. *Let H be a normal subgroup of a group G. Let G/H be the set of all cosets of H in G; that is,*

$$G/H = \{gH : g \in G\}.$$

We define multiplication on G/H in the following manner:

$$(g_1H)(g_2H) = g_1g_2H.$$

Then G/H forms a group called the factor group or quotient group of G modulo H. The identity element of G/H is $1H$, and the inverse of gH is $g^{-1}H$. Further, if G is Abelian, then G/H is also Abelian.

Finally, as with rings normal subgroups, factor groups are closely tied to homomorphisms.

Definition 1.8.11. Let G_1 and G_2 be groups, and let $f : G_1 \to G_2$ be a homomorphism. Then the *kernel* of f, denoted $\ker(f)$, is

$$\ker(f) = \{g \in G_1 : f(g) = 1\}.$$

The *image* of f, denoted $\operatorname{im}(f)$, is the range of f within G_2. That is,

$$\operatorname{im}(f) = \{h \in G_2 : \text{there exists } g \in G_1 \text{ with } f(g) = h\}.$$

Theorem 1.8.12 (Group isomorphism theorem). *Let $f : G_1 \to G_2$ be a homomorphism of groups G_1 and G_2. Then*

(1) $\ker(f)$ *is a normal subgroup in G_1.* $\operatorname{im}(f)$ *is a subgroup of G_2, and*

$$G_1/\ker(f) \cong \operatorname{im}(f).$$

(2) *Conversely, suppose that H is a normal subgroup of a group G. Then $f : G \to G/H$, given by $f(g) = gH$ for $g \in G$ is a homomorphism, whose kernel is H and whose image is G/H.*

1.9 Exercises

1. Let $\phi : K \to R$ be a homomorphism from a field K to a ring R. Show that either $\phi(a) = 0$ for all $a \in K$, or ϕ is a monomorphism.
2. Let R be a ring and $M \neq \emptyset$ an arbitrary set. Show that the following are equivalent:
   (i) The ring of all mappings from M to R is a field.
   (ii) M contains only one element and R is a field.
3. Let π be a set of prime numbers. Define

$$\mathbb{Q}_\pi = \left\{\frac{a}{b} : \text{all prime divisors of } b \text{ are in } \pi\right\}.$$

   (i) Show that $\mathbb{Q}_\pi$ is a subring of $\mathbb{Q}$.
   (ii) Let R be a subring of $\mathbb{Q}$ and let $\frac{a}{b} \in R$ with coprime integers a, b. Show that $\frac{1}{b} \in R$.
   (iii) Determine all subrings R of $\mathbb{Q}$.
   (*Hint*: Consider the set of all prime divisors of denominators of reduced elements of R.)
4. Prove Lemma 1.5.2.
5. Let R be a commutative ring with an identity $1 \in R$. Let A, B and C be ideals in R. $A + B := \{a + b : a \in A, b \in B\}$ and $AB := (\{ab : a \in A, b \in B\})$. Show:

(i) $A + B \lhd R, A + B = (A \cup B)$.
(ii) $AB = \{a_1b_1 + \cdots + a_nb_n : n \in \mathbb{N}, a_i \in A, b_i \in B\}, AB \subset A \cap B$.
(iii) $A(B + C) = AB + AC$, $(A + B)C = AB + BC$, $(AB)C = A(BC)$.
(iv) $A = R \Leftrightarrow A \cap R^* \neq \emptyset$.
(v) $a, b \in R \Rightarrow \langle a\rangle + \langle b\rangle = \{xa + yb : x, y \in R\}$.
(vi) $a, b \in R \Rightarrow \langle a\rangle\langle b\rangle = \langle ab\rangle$. Here, $\langle a\rangle = Ra = \{xa : x \in R\}$.

6. Solve the following congruence:

$$3x \equiv 5 \pmod 7.$$

Is this congruence also solvable modulo 17?

7. Show that the set of (2×2)-matrices over a ring R forms a ring.
8. Prove Lemma 1.4.8.
9. Prove that if R is a ring with identity and $S = \{r = m \cdot 1 : r \in R, m \in \mathbb{Z}\}$ then S is a subring of R containing the identity.

2 Maximal and Prime Ideals

In this chapter we use polynomials over integral domains with one or two indeterminates in an elementary fashion. We will consider polynomial rings in detail in later chapters.

2.1 Maximal and Prime Ideals of the Integers

In the first chapter, we defined ideals I in a ring R, and then the factor ring R/I of R modulo the ideal I. We saw, furthermore, that if R is commutative, then R/I is also commutative, and if R has an identity, then so does R/I. This raises further questions concerning the structure of factor rings. In particular, we can ask under what conditions does R/I form an integral domain, and under what conditions does R/I form a field. These questions lead us to define certain special properties of ideals, called prime ideals and maximal ideals.

Let us look back at the integers $\mathbb{Z}$. Recall that each proper ideal in $\mathbb{Z}$ has the form $n\mathbb{Z}$ for some $n > 1$, and the resulting factor ring $\mathbb{Z}/n\mathbb{Z}$ is isomorphic to $\mathbb{Z}_n$. We proved the following result:

Theorem 2.1.1. *The factor ring $\mathbb{Z}_n = \mathbb{Z}/n\mathbb{Z}$ is an integral domain if and only if $n = p$ is a prime. Furthermore, $\mathbb{Z}_n$ is a field again if and only if $n = p$ is a prime.*

Hence, for the integers $\mathbb{Z}$, a factor ring is a field if and only if it is an integral domain. We will see later that this is not true in general. However, what is clear is that special ideals $n\mathbb{Z}$ lead to integral domains and fields when n is a prime. We look at the ideals $p\mathbb{Z}$ with p a prime in two different ways, and then use these in subsequent sections to give the general definitions. We first need a famous result, Euclid's lemma, from number theory. For integers a, b, the notation $a|b$ means that a divides b.

Lemma 2.1.2 (Euclid). *If p is a prime and $p|ab$, then $p|a$ or $p|b$.*

Proof. Recall that the greatest common divisor or GCD of two integers a, b is an integer $d > 0$ such that d is a common divisor of both a and b, and if d_1 is another common divisor of a and b, then $d_1|d$. We express the GCD of a, b by $d = (a, b)$. It is known that for any two integers a, b, their GCD exists and is unique, and is the least positive linear combination of a and b; that is, the least positive integer of the form $ax + by$ for integers x, y. The integers a, b are *relatively prime* if their GCD is 1, $(a, b) = 1$. In this case, 1 is a linear combination of a and b (see Chapter 3 for proofs and more details).

Now suppose $p|ab$, where p is a prime. If p does not divide a, then since the only positive divisors of p are 1 and p, it follows that $(a, p) = 1$. Hence, 1 is expressible as a linear combination of a and p. That is, $ax + py = 1$ for some integers x, y. Multiply through by b, so that

https://doi.org/10.1515/9783111142524-002

$$abx + pby = b.$$

Now $p|ab$, so $p|abx$ and $p|pby$. Therefore, $p|abx + pby$; that is, $p|b$. □

We now recast this lemma in two different ways in terms of the ideal $p\mathbb{Z}$. Notice that $p\mathbb{Z}$ consists precisely of all the multiples of p.

Hence, $p|ab$ is equivalent to $ab \in p\mathbb{Z}$.

Lemma 2.1.3. *If p is a prime and $ab \in p\mathbb{Z}$, then $a \in p\mathbb{Z}$, or $b \in p\mathbb{Z}$.*

This conclusion will be taken as a motivation for the definition of a *prime ideal* in the next section.

Lemma 2.1.4. *If p is a prime and $p\mathbb{Z} \subset n\mathbb{Z}$, then $n = 1$, or $n = p$. That is, every ideal in $\mathbb{Z}$ containing $p\mathbb{Z}$ with p a prime is either all of $\mathbb{Z}$ or $p\mathbb{Z}$.*

Proof. Suppose that $p\mathbb{Z} \subset n\mathbb{Z}$. Then $p \in n\mathbb{Z}$; therefore, p is a multiple of n. Since p is a prime, it follows easily that either $n = 1$, or $n = p$. □

In Section 2.3, the conclusion of this lemma will be taken as a motivation for the definition of a *maximal ideal*.

2.2 Prime Ideals and Integral Domains

Motivated by Lemma 2.1.3, we make the following general definition for commutative rings R with identity:

Definition 2.2.1. Let R be a commutative ring. An ideal P in R with $P \neq R$ is a *prime ideal* if whenever $ab \in P$ with $a, b \in R$, then either $a \in P$, or $b \in P$.

This property of an ideal is precisely what is necessary and sufficient to make the factor ring R/I an integral domain.

Theorem 2.2.2. *Let R be a commutative ring with an identity $1 \neq 0$, and let P be a nontrivial ideal in R. Then P is a prime ideal if and only if the factor ring R/P is an integral domain.*

Proof. Let R be a commutative ring with an identity $1 \neq 0$, and let P be a prime ideal. We show that R/P is an integral domain. From the results in the last chapter, we have that R/P is again a commutative ring with an identity. Therefore, we must show that there are no zero divisors in R/P. Suppose that $(a + I)(b + I) = 0$ in R/P. The zero element in R/P is $0 + P$ and, hence,

$$(a + P)(b + P) = 0 = 0 + P \implies ab + P = 0 + P \implies ab \in P.$$

However, P is a prime ideal; therefore, we must have $a \in P$, or $b \in P$. If $a \in P$, then $a + P = P = 0 + P$ so $a + P = 0$ in R/P. The identical argument works if $b \in P$. Therefore, there are no zero divisors in R/P and, hence, R/P is an integral domain.

Conversely, suppose that R/P is an integral domain. We must show that P is a prime ideal. Suppose that $ab \in P$. Then $(a + P)(b + P) = ab + P = 0 + P$. Hence, in R/P, we have

$$(a + P)(b + P) = 0.$$

However, R/P is an integral domain, so it has no zero divisors. It follows that either $a + P = 0$ and, hence, $a \in P$ or $b + P = 0$, and $b \in P$. Therefore, either $a \in P$, or $b \in P$. Therefore, P is a prime ideal. □

In a commutative ring R, we can define a multiplication of ideals. We then obtain an exact analog of Euclid's lemma. Since R is commutative, each ideal is 2-sided.

Definition 2.2.3. Let R be a commutative ring with an identity $1 \neq 0$, and let A and B be ideals in R. Define

$$AB = \{a_1 b_1 + \cdots + a_n b_n : a_i \in A, b_i \in B, n \in \mathbb{N}\}.$$

That is, AB is the set of finite sums of products ab with $a \in A$ and $b \in B$.

Lemma 2.2.4. *Let R be a commutative ring with an identity $1 \neq 0$, and let A and B be ideals in R. Then AB is an ideal.*

Proof. We must verify that AB is a subring, and that it is closed under multiplication from R. Le $r_1, r_2 \in AB$. Then

$$r_1 = a_1 b_1 + \cdots + a_n b_n \quad \text{for some } a_i \in A,\ b_i \in B,$$

and

$$r_2 = a_1' b_1' + \cdots + a_m' b_m' \quad \text{for some } a_i' \in A,\ b_i' \in B.$$

Then

$$r_1 \pm r_2 = a_1 b_1 + \cdots + a_n b_n \pm a_1' b_1' \pm \cdots \pm a_m' b_m',$$

which is clearly in AB. Furthermore,

$$r_1 \cdot r_2 = a_1 b_1 a_1' b_1' + \cdots + a_n b_n a_m' b_m'.$$

Consider, for example, the first term $a_1 b_1 a_1' b_1'$. Since R is commutative, this is equal to

$$(a_1 a_1')(b_1 b_1').$$

Now $a_1a_1' \in A$ since A is a subring, and $b_1b_1' \in B$ since B is a subring. Hence, this term is in AB. Similarly, for each of the other terms. Therefore, $r_1r_2 \in AB$ and, hence, AB is a subring.

Now let $r \in R$, and consider rr_1. This is then

$$rr_1 = ra_1b_1 + \cdots + ra_nb_n.$$

Now $ra_i \in A$ for each i since A is an ideal. Hence, each summand is in AB, and then $rr_1 \in AB$. Therefore, AB is an ideal. □

Lemma 2.2.5. *Let R be a commutative ring with an identity $1 \neq 0$, and let A and B be ideals in R. If P is a prime ideal in R, then $AB \subset P$ implies that $A \subset P$ or $B \subset P$.*

Proof. Suppose that $AB \subset P$ with P a prime ideal, and suppose that B is not contained in P. We show that $A \subset P$. Since $AB \subset P$, each product $a_ib_j \in P$. Choose a $b \in B$ with $b \notin P$, and let a be an arbitrary element of A. Then $ab \in P$. Since P is a prime ideal, this implies either $a \in P$, or $b \in P$. But by assumption $b \notin P$, so $a \in P$. Since a was arbitrary, we have $A \subset P$. □

2.3 Maximal Ideals and Fields

Now, motivated by Lemma 2.1.4, we define a maximal ideal.

Definition 2.3.1. Let R be a ring and I an ideal in R. Then I is a *maximal ideal* if $I \neq R$, and if J is an ideal in R with $I \subset J$, then $I = J$, or $J = R$.

If R is a commutative ring with an identity this property of an ideal I is precisely what is necessary and sufficient, so that R/I is a field.

Theorem 2.3.2. *Let R be a commutative ring with an identity $1 \neq 0$, and let I be an ideal in R. Then I is a maximal ideal if and only if the factor ring R/I is a field.*

Proof. Suppose that R is a commutative ring with an identity $1 \neq 0$, and let I be an ideal in R. Suppose first that I is a maximal ideal, and we show that the factor ring R/I is a field.

Since R is a commutative ring with an identity, the factor ring R/I is also a commutative ring with an identity. We must show then that each nonzero element of R/I has a multiplicative inverse. Suppose then that $\overline{r} = r + I \in R/I$ is a nonzero element of R/I. It follows that $r \notin I$. Consider the set $\langle r, I\rangle = \{rx + i : x \in R, i \in I\}$. This is also an ideal (see exercises) called the ideal generated by r and I, denoted $\langle r, I\rangle$. Clearly, $I \subset \langle r, I\rangle$, and since $r \notin I$, and $r = r \cdot 1 + 0 \in \langle r, I\rangle$, it follows that $\langle r, I\rangle \neq I$. Since I is a maximal ideal, it follows that $\langle r, I\rangle = R$ the whole ring. Hence, the identity element $1 \in \langle r, I\rangle$, and so, there exist elements $x \in R$ and $i \in I$ such that $1 = rx + i$. But then $1 \in (r + I)(x + I)$, and so, $1 + I = (r + I)(x + I)$. Since $1 + I$ is the multiplicative identity of R/I, it follows that

$x + I$ is the multiplicative inverse of $r + I$ in R/I. Since $r + I$ was an arbitrary nonzero element of R/I, it follows that R/I is a field.

Now suppose that R/I is a field for an ideal I. We show that I must be maximal. Suppose then that I_1 is an ideal with $I \subset I_1$ and $I \neq I_1$. We must show that I_1 is all of R. Since $I \neq I_1$, there exists an $r \in I_1$ with $r \notin I$. Therefore, the element $r + I$ is nonzero in the factor ring R/I, and since R/I is a field, it must have a multiplicative inverse $x + I$. Hence, $(r + I)(x + I) = rx + I = 1 + I$ and, therefore, there is an $i \in I$ with $1 = rx + i$. Since $r \in I_1$, and I_1 is an ideal, we get that $rx \in I_1$. In addition, since $I \subset I_1$, it follows that $rx + i \in I_1$, and so, $1 \in I_1$. If r_1 is an arbitrary element of R, then $r_1 \cdot 1 = r_1 \in I_1$. Hence, $R \subset I_1$, and so, $R = I_1$. Therefore, I is a maximal ideal. □

Recall that a field is already an integral domain. Combining this with the ideas of prime and maximal ideals we obtain:

Theorem 2.3.3. *Let R be a commutative ring with an identity $1 \neq 0$. Then each maximal ideal is a prime ideal.*

Proof. Suppose that R is a commutative ring with an identity and I is a maximal ideal in R. Then from Theorem 2.3.2, we have that the factor ring R/I is a field. But a field is an integral domain, so R/I is an integral domain. Therefore, from Theorem 2.2.2, we have that I must be a prime ideal. □

The converse is not true in general. That is, there are prime ideals that are not maximal. Consider, for example, $R = \mathbb{Z}$ the integers and $I = \{0\}$. Then I is an ideal, and $R/I = \mathbb{Z}/\{0\} \cong \mathbb{Z}$ is an integral domain. Hence, $\{0\}$ is a prime ideal. However, $\mathbb{Z}$ is not a field, so $\{0\}$ is not maximal. Note, however, that in the integers $\mathbb{Z}$, a proper ideal is maximal if and only if it is a prime ideal.

2.4 The Existence of Maximal Ideals

In this section, we prove that in any ring R with an identity, there do exist maximal ideals. Furthermore, given an ideal $I \neq R$, then there exists a maximal ideal I_0 such that $I \subset I_0$. To prove this, we need three important equivalent results from logic and set theory.

First, recall that a *partial order* $\leq$ on a set S is a reflexive, transitive relation on S. That is, $a \leq a$ for all $a \in S$, and if $a \leq b$, $b \leq c$, then $a \leq c$. This is a “partial” order since there may exist elements $a \in S$, where neither $a \leq b$, nor $b \leq a$. If A is any set, then it is clear that containment of subsets is a partial order on the power set $\mathcal{P}(A)$.

If $\leq$ is a partial order on a set M, then a *chain* on M is a subset $K \subset M$ such that $a, b \in K$ implies that $a \leq b$ or $b \leq a$. A chain on M is *bounded* if there exists an $m \in M$ such that $k \leq m$ for all $k \in K$. The element m is called an *upper bound* for K. An element $m_0 \in M$ is *maximal* if whenever $m \in M$ with $m_0 \leq m$, then $m = m_0$. We now state the three important results from logic.

Zorn's lemma. *If each chain of M has an upper bound in M, then there is at least one maximal element in M.*

Axiom of well-ordering. *Each set M can be well-ordered, such that each nonempty subset of M contains a least element.*

Axiom of choice. *Let $\{M_i : i \in I\}$ be a nonempty collection of nonempty sets. Then there is a mapping $f : I \to \bigcup_{i\in I} M_i$ with $f(i) \in M_i$ for all $i \in I$.*

The following can be proved.

Theorem 2.4.1. *Zorn's lemma, the axiom of well-ordering and the axiom of choice are all equivalent.*

We now show the existence of maximal ideals in commutative rings with identity.

Theorem 2.4.2. *Let R be a commutative ring with an identity $1 \neq 0$, and let I be an ideal in R with $I \neq R$. Then there exists a maximal ideal I_0 in R with $I \subset I_0$. In particular, a ring with an identity contains maximal ideals.*

Proof. Let I be an ideal in the commutative ring R. We must show that there exists a maximal ideal I_0 in R with $I \subset I_0$.

Let

$$M = \{X : X \text{ is an ideal with } I \subset X \neq R\}.$$

Then M is partially ordered by containment. We want to show first that each chain in M has a maximal element. If $K = \{X_j : X_j \in M, j \in J\}$ is a chain, let

$$X' = \bigcup_{j\in J} X_j.$$

If $a, b \in X'$, then there exists an $i, j \in J$ with $a \in X_i$, $b \in X_j$. Since K is a chain, either $X_i \subset X_j$ or $X_j \subset X_i$. Without loss of generality, suppose that $X_i \subset X_j$ so that $a, b \in X_j$. Then $a \pm b \in X_j \subset X'$, and $ab \in X_j \subset X'$, since X_j is an ideal. Furthermore, if $r \in R$, then $ra \in X_j \subset X'$, since X_j is an ideal. Therefore, X' is an ideal in R.

Since $X_j \neq R$, it follows that $1 \notin X_j$ for all $j \in J$. Therefore, $1 \notin X'$, and so $X' \neq R$. It follows that under the partial order of containment X' is an upper bound for K.

We now use Zorn's lemma. From the argument above, we have that each chain has a maximal element. Hence, for an ideal I, the set M above has a maximal element. This maximal element I_0 is then a maximal ideal containing I. □

2.5 Principal Ideals and Principal Ideal Domains

Recall again that in the integers $\mathbb{Z}$, each ideal I is of the form $n\mathbb{Z}$ for some integer n. Hence, in $\mathbb{Z}$, each ideal can be generated by a single element.

Lemma 2.5.1. *Let R be a commutative ring and $a_1, \dots, a_n$ be elements of R. Then the set*

$$\langle a_1, \dots, a_n \rangle = \{r_1 a_1 + \cdots + r_n a_n : r_i \in R\}$$

forms an ideal in R called the ideal generated by $a_1, \dots, a_n$.

Proof. The proof is straightforward. Let

$$a = r_1 a_1 + \cdots + r_n a_n, \quad b = s_1 a_1 + \cdots + s_n a_n$$

with $r_1, \dots, r_n, s_1, \dots, s_n$ elements of R, be two elements of $\langle a_1, \dots, a_n \rangle$. Then

$$\begin{aligned} a \pm b &= (r_1 \pm s_1) a_1 + \cdots + (r_n \pm s_n) a_n \in \langle a_1, \dots, a_n \rangle \\ ab &= (r_1 s_1 a_1) a_1 + (r_1 s_2 a_1) a_2 + \cdots + (r_n s_n a_n) a_n \in \langle a_1, \dots, a_n \rangle, \end{aligned}$$

so $\langle a_1, \dots, a_n \rangle$ forms a subring. Furthermore, if $r \in R$, we have

$$ra = (rr_1) a_1 + \cdots + (rr_n) a_n \in \langle a_1, \dots, a_n \rangle,$$

and so $\langle a_1, \dots, a_n \rangle$ is an ideal. □

Definition 2.5.2. Let R be a commutative ring. An ideal $I \subset R$ is a *principal ideal* if it has a single generator. That is,

$$I = \langle a \rangle = aR \quad \text{for some } a \in R.$$

We now restate Theorem 1.4.7 of Chapter 1.

Theorem 2.5.3. *Every nonzero ideal in $\mathbb{Z}$ is a principal ideal.*

Proof. Every ideal I in $\mathbb{Z}$ is of the form $n\mathbb{Z}$. This is the principal ideal generated by n. □

Definition 2.5.4. A *principal ideal domain* or *PID* is an integral domain, in which every ideal is principal.

Corollary 2.5.5. *The integers $\mathbb{Z}$ are a principal ideal domain.*

We mention that the set of polynomials $K[x]$ with coefficients from a field K is also a principal ideal domain. We will return to this in the next chapter.

Not every integral domain is a PID. Consider $K[x,y] = (K[x])[y]$, the set of polynomials over K in two variables x, y (see Chapter 4). Let I consist of all the polynomials with zero constant term.

Lemma 2.5.6. *The set I in $K[x,y]$ as defined above is an ideal, but not a principal ideal.*

Proof. We leave the proof that I forms an ideal to the exercises. To show that it is not a principal ideal, suppose $I = \langle p(x,y) \rangle$. Now the polynomial $q(x) = x$ has zero constant term, so $q(x) \in I$. Hence, $p(x,y)$ cannot be a constant polynomial. In addition, if $p(x,y)$

had any terms with y in them, there would be no way to multiply $p(x,y)$ by a polynomial $h(x,y)$ and obtain just x. Therefore, $p(x,y)$ can contain no terms with y in them. But the same argument, using $s(y) = y$, shows that $p(x,y)$ cannot have any terms with x in them. Therefore, there can be no such $p(x,y)$ generating I, and so, I is not principal, and $K[x,y]$ is not a principal ideal domain. □

2.6 Exercises

1. Consider the set $\langle r, I\rangle = \{rx + i : x \in R, i \in I\}$, where I is an ideal. Prove that this is also an ideal called the ideal generated by r and I, denoted $\langle r, I\rangle$.
2. Let R and S be commutative rings, and let $\phi : R \to S$ be a ring epimorphism. Let M be a maximal ideal in R. Show that $\phi(M)$ is a maximal ideal in S if and only if $\ker(\phi) \subset M$. Is $\phi(M)$ always a prime ideal of S?
3. Let $A_1, \dots, A_t$ be ideals of a commutative ring R. Let P be a prime ideal of R. Show:
   (i) $\bigcap_{i=1}^{t} A_i \subset P$ implies $A_j \subset P$ for at least one index j.
   (ii) $\bigcap_{i=1}^{t} A_i = P$ implies $A_j = P$ for at least one index j.
4. Which of the following ideals A are prime ideals of R? Which are maximal ideals?
   (i) $A = \langle x\rangle, R = \mathbb{Z}[x]$.
   (ii) $A = \langle x^2\rangle, R = \mathbb{Z}[x]$.
   (iii) $A = \langle 1 + \sqrt{5}\rangle, R = \mathbb{Z}[\sqrt{5}] = \{a + b\sqrt{5} : a, b \in \mathbb{Z}\}$.
   (iv) $A = \langle x, y\rangle, R = \mathbb{Q}[x,y]$.
5. Let $w = \frac{1}{2}(1 + \sqrt{-3})$. Show that $\langle 2\rangle$ is a prime ideal and even a maximal ideal of $\mathbb{Z}[w]$, but $\langle 2\rangle$ is neither a prime ideal nor a maximal ideal of $\mathbb{Z}[i]$, $i = \sqrt{-1} \in \mathbb{C}$.
6. Let $R = \{\frac{a}{b} : a, b \in \mathbb{Z}, b \text{ odd}\}$. Show that R is a subring of $\mathbb{Q}$, and that there is only one maximal ideal M in R.
7. Let R be a commutative ring with an identity. Let $x, y \in R$ and $x \neq 0$ not be a zero divisor. Furthermore, let $\langle x\rangle$ be a prime ideal with $\langle x\rangle \subset \langle y\rangle \neq R$. Show that $\langle x\rangle = \langle y\rangle$.
8. Consider $K[x,y]$ the set of polynomials over K in two variables x, y. Let I consist of all the polynomials with zero constant term. Prove that the set I is an ideal.

3 Prime Elements and Unique Factorization Domains

In this chapter we use again polynomials over integral domains with one or two indeterminates in an elementary fashion. We will consider polynomial rings in detail in later chapters.

3.1 The Fundamental Theorem of Arithmetic

The integers $\mathbb{Z}$ have served as much of our motivation for properties of integral domains. In the last chapter, we saw that $\mathbb{Z}$ is a principal ideal domain, and furthermore, that prime ideals $\neq \{0\}$ are maximal. From the viewpoint of the multiplicative structure of $\mathbb{Z}$ and the viewpoint of classical number theory, the most important property of $\mathbb{Z}$ is the *fundamental theorem of arithmetic*. This states that any integer $n \neq 0$ is uniquely expressible as a product of primes, where uniqueness is up to ordering and the introduction of ± 1; that is, units. In this chapter, we show that this property is not unique to the integers, and there are many other integral domains, where this also holds. These are called unique factorization domains, and we will present several examples. First, we review the fundamental theorem of arithmetic, its proof and several other ideas from classical number theory.

Theorem 3.1.1 (Fundamental theorem of arithmetic). *Given any integer $n \neq 0$, there is a factorization*

$$n = cp_1p_2\cdots p_k,$$

where $c = \pm 1$ and $p_1, \ldots, p_k$ are primes. Furthermore, this factorization is unique up to the ordering of the factors.

There are two main ingredients that go into the proof: induction and Euclid's lemma. We presented this in the last chapter. In turn, however, Euclid's lemma depends upon the existence of greatest common divisors and their linear expressibility. Therefore, to begin, we present several basic ideas from number theory.

The starting point for the theory of numbers is *divisibility*.

Definition 3.1.2. If a, b are integers, we say that *a divides b*, or that a is a *factor* or *divisor* of b, if there exists an integer q such that $b = aq$. We denote this by $a|b$. b is then a *multiple* of a. If $b > 1$ is an integer whose only factors are $\pm 1, \pm b$, then b is a *prime*, otherwise, $b > 1$ is *composite*.

The following properties of divisibility are straightforward consequences of the definition.

Lemma 3.1.3. *The following properties hold:*

(1) $a|b \Rightarrow a|bc$ *for any integer c.*

https://doi.org/10.1515/9783111142524-003

(2) *$a|b$ and $b|c$ implies $a|c$.*
(3) *$a|b$ and $a|c$ implies that $a|(bx + cy)$ for any integers x, y.*
(4) *$a|b$ and $b|a$ implies that $a = \pm b$.*
(5) *If $a|b$ and $a > 0$, $b > 0$, then $a \leq b$.*
(6) *$a|b$ if and only if $ca|cb$ for any integer $c \neq 0$.*
(7) *$a|0$ for all $a \in \mathbb{Z}$, and $0|a$ only for $a = 0$.*
(8) *$a| \pm 1$ only for $a = \pm 1$.*
(9) *$a_1|b_1$ and $a_2|b_2$ implies that $a_1a_2|b_1b_2$.*

If b, c, x, y are integers, then an integer $bx + cy$ is called a *linear combination* of b, c. Thus, part (3) of Lemma 3.1.3 says that if a is a *common divisor* of b, c, then a divides any linear combination of b and c.

Furthermore, note that if $b > 1$ is a composite, then there exists $x > 0$ and $y > 0$ such that $b = xy$, and from part (5), we must have $1 < x < b, 1 < y < b$.

In ordinary arithmetic, given a, b, we can always attempt to divide a into b. The next result, called the *division algorithm*, says that if $a > 0$, either a will divide b, or the *remainder* of the division of b by a will be less than a.

Theorem 3.1.4 (Division algorithm). *Given integers a, b with $a > 0$, then there exist unique integers q and r such that $b = qa + r$, where either $r = 0$ or $0 < r < a$.*

One may think of q and r as the *quotient* and *remainder*, respectively, when dividing b by a.

Proof. Given a, b with $a > 0$, consider the set

$$S = \{b - qa \geq 0 : q \in \mathbb{Z}\}.$$

If $b > 0$, then $b + a \geq 0$, and the sum is in S. If $b \leq 0$, then there exists a $q > 0$ with $-qa < b$. Then $b + qa > 0$ and is in S. Therefore, in either case, S is nonempty. Hence, S is a nonempty subset of $\mathbb{N} \cup \{0\}$ and, therefore, has a least element r. If $r \neq 0$, we must show that $0 < r < a$. Suppose $r \geq a$, then $r = a + x$ with $x \geq 0$, and $x < r$ since $a > 0$. Then $b - qa = r = a + x \Rightarrow b - (q + 1)a = x$. This means that $x \in S$. Since $x < r$, this contradicts the minimality of r, which is a contradiction. Therefore, if $r \neq 0$, it follows that $0 < r < a$.

The only thing left is to show the uniqueness of q and r. Suppose $b = q_1a + r_1$ also. By the construction above, r_1 must also be the minimal element of S. Hence, $r_1 \leq r$, and $r \leq r_1$ so $r = r_1$. Now

$$b - qa = b - q_1a \implies (q_1 - q)a = 0,$$

but since $a > 0$, it follows that $q_1 - q = 0$ so that $q = q_1$. □

The next idea that is necessary is the concept of *greatest common divisor*.

Definition 3.1.5. Given nonzero integers a, b, their *greatest common divisor* or GCD $d > 0$ is a positive integer such that it is their common divisor, that is, $d|a$ and $d|b$, and if d_1 is any other common divisor, then $d_1|d$. We denote the greatest common divisor of a, b by either $\gcd(a, b)$ or (a, b).

Certainly, if a, b are nonzero integers with $a > 0$ and $a|b$, then $a = \gcd(a, b)$.

The next result says that given any nonzero integers, they do have a greatest common divisor, and it is unique.

Theorem 3.1.6. *Given nonzero integers a, b, their GCD exists, is unique, and can be characterized as the least positive linear combination of a and b.*

Proof. Given nonzero a, b, consider the set

$$S = \{ax + by > 0 : x, y \in \mathbb{Z}\}.$$

Now, $a^2 + b^2 > 0$, so S is a nonempty subset of $\mathbb{N}$ and, hence, has a least element, $d > 0$. We show that d is the GCD.

First we must show that d is a common divisor. Now $d = ax + by$ and is the least such positive linear combination. By the division algorithm, $a = qd + r$ with $0 \le r < d$. Suppose $r \neq 0$. Then $r = a - qd = a - q(ax + by) = (1 - qx)a - qby > 0$. Hence, r is a positive linear combination of a and b, and therefore in S. But then $r < d$, contradicting the minimality of d in S. It follows that $r = 0$, and so, $a = qd$, and $d|a$. An identical argument shows that $d|b$, and so, d is a common divisor of a and b. Let d_1 be any other common divisor of a and b. Then d_1 divides any linear combination of a and b, and so $d_1|d$. Therefore, d is the GCD of a and b.

Finally, we must show that d is unique. Suppose d_1 is another GCD of a and b. Then $d_1 > 0$, and d_1 is a common divisor of a, b. Then $d_1|d$ since d is a GCD. Identically, $d|d_1$ since d_1 is a GCD. Therefore, $d = \pm d_1$, and then $d = d_1$ since they are both positive. □

If $(a, b) = 1$, then we say that a, b are *relatively prime*. It follows that a and b are relatively prime if and only if 1 is expressible as a linear combination of a and b. We need the following three results:

Lemma 3.1.7. *If $d = (a, b)$, then $a = a_1 d$ and $b = b_1 d$ with $(a_1, b_1) = 1$.*

Proof. If $d = (a, b)$, then $d|a$, and $d|b$. Hence, $a = a_1 d$, and $b = b_1 d$. We have

$$d = ax + by = a_1 dx + b_1 dy.$$

Dividing both sides of the equation by d, we obtain

$$1 = a_1 x + b_1 y.$$

Therefore, $(a_1, b_1) = 1$. □

Lemma 3.1.8. *For any integer c, we have that* $(a, b) = (a, b + ac)$.

Proof. Suppose $(a, b) = d$ and $(a, b + ac) = d_1$. Now d is the least positive linear combination of a and b. Suppose $d = ax + by$. d_1 is a linear combination of $a, b + ac$ so that

$$d_1 = ar + (b + ac)s = a(cs + r) + bs.$$

Hence, d_1 is also a linear combination of a and b; therefore, $d_1 \geq d$. On the other hand, $d_1|a$, and $d_1|(b + ac)$, and so, $d_1|b$. Therefore, $d_1|d$, so $d_1 \leq d$. Combining these, we must have $d_1 = d$. □

The next result, called the *Euclidean algorithm*, provides a technique for both finding the GCD of two integers and expressing the GCD as a linear combination.

Theorem 3.1.9 (Euclidean algorithm). *Given integers b and a* > 0 *with* $a \nmid b$, *the following repeated divisions are formed:*

$$\begin{aligned} b &= q_1 a + r_1, \quad 0 < r_1 < a \\ a &= q_2 r_1 + r_2, \quad 0 < r_2 < r_1 \\ &\vdots \\ r_{n-2} &= q_n r_{n-1} + r_n, \quad 0 < r_n < r_{n-1} \\ r_{n-1} &= q_{n+1} r_n. \end{aligned}$$

The last nonzero remainder r_n *is the GCD of* a, b. *Furthermore,* r_n *can be expressed as a linear combination of a and b by successively eliminating the* r_i*'s in the intermediate equations.*

Proof. In taking the successive divisions as outlined in the statement of the theorem, each remainder r_i gets strictly smaller and still nonnegative. Hence, it must finally end with a zero remainder. Therefore, there is a last nonzero remainder r_n. We must show that this is the GCD.

Now from Lemma 3.1.7, the gcd $(a, b) = (a, b - q_1 a) = (a, r_1) = (r_1, a - q_2 r_1) = (r_1, r_2)$. Continuing in this manner, we have then that $(a, b) = (r_{n-1}, r_n) = r_n$ since r_n divides r_{n-1}. This shows that r_n is the GCD.

To express r_n as a linear combination of a and b, first notice that

$$r_n = r_{n-2} - q_n r_{n-1}.$$

Substituting this in the immediately preceding division, we get

$$r_n = r_{n-2} - q_n(r_{n-3} - q_{n-1} r_{n-2}) = (1 + q_n q_{n-1}) r_{n-2} - q_n r_{n-3}.$$

Doing this successively, we ultimately express r_n as a linear combination of a and b. □

Example 3.1.10. Find the GCD of 270 and 2412, and express it as a linear combination of 270 and 2412.

We apply the Euclidean algorithm

$$\begin{aligned} 2412 &= 8 \cdot 270 + 252 \\ 270 &= 1 \cdot 252 + 18 \\ 252 &= 14 \cdot 18. \end{aligned}$$

Therefore, the last nonzero remainder is 18, which is the GCD. We now must express 18 as a linear combination of 270 and 2412.

From the first equation

$$252 = 2412 - 8 \cdot 270,$$

which gives in the second equation

$$270 = 2412 - 8 \cdot 270 + 18 \implies 18 = -1 \cdot 2412 + 9 \cdot 270,$$

which is the desired linear combination.

The next result that we need is *Euclid's lemma.* We stated and proved this in the last chapter, but we restate it here.

Lemma 3.1.11 (Euclid's lemma). *If p is a prime and $p|ab$, then $p|a$, or $p|b$.*

We can now prove the fundamental theorem of arithmetic. Induction suffices to show that there always exists such a decomposition into prime factors.

Lemma 3.1.12. *Any integer $n > 1$ can be expressed as a product of primes, perhaps with only one factor.*

Proof. The proof is by induction. $n = 2$ is prime. Therefore, it is true at the lowest level. Suppose that any integer $2 \leq k < n$ can be decomposed into prime factors, we must show that n then also has a prime factorization.

If n is prime, then we are done. Suppose then that n is composite. Hence, $n = m_1 m_2$ with $1 < m_1 < n$, $1 < m_2 < n$. By the inductive hypothesis, both m_1 and m_2 can be expressed as products of primes. Therefore, n can, also using the primes from m_1 and m_2, completing the proof. □

Before we continue to the fundamental theorem, we mention that the existence of a prime decomposition, unique or otherwise, can be used to prove that the set of primes is infinite. The proof we give goes back to Euclid and is quite straightforward.

Theorem 3.1.13. *There are infinitely many primes.*

Proof. Suppose that there are only finitely many primes $p_1, \ldots, p_n$. Each of these is positive, so we can form the positive integer

$$N = p_1 p_2 \cdots p_n + 1.$$

From Lemma 3.1.12, N has a prime decomposition. In particular, there is a prime p, which divides N. Then

$$p | (p_1 p_2 \cdots p_n + 1).$$

Since the only primes are assumed $p_1, p_2, \ldots, p_n$, it follows that $p = p_i$ for some $i = 1, \ldots, n$. But then $p | p_1 p_2 \cdots p_i \cdots p_n$ so p cannot divide $p_1 \cdots p_n + 1$, which is a contradiction. Therefore, p is not one of the given primes showing that the list of primes must be endless. □

We can now prove the fundamental theorem of arithmetic.

Proof. We assume that $n \geq 1$. If $n \leq -1$, we use $c = -n$, and the proof is the same. The statement certainly holds for $n = 1$ with $k = 0$. Now suppose $n > 1$. From Lemma 3.1.12, n has a prime decomposition:

$$n = p_1 p_2 \cdots p_m.$$

We must show that this is unique up to the ordering of the factors. Suppose then that n has another such factorization $n = q_1 q_2 \cdots q_k$ with the q_i all prime. We must show that $m = k$, and that, the primes are the same. Now we have

$$n = p_1 p_2 \cdots p_m = q_1 \cdots q_k.$$

Assume that $k \geq m$. From

$$n = p_1 p_2 \cdots p_m = q_1 \cdots q_k,$$

it follows that $p_1 | q_1 q_2 \cdots q_k$. From Lemma 3.1.11 then, we must have that $p_1 | q_i$ for some i. But q_i is prime, and $p_1 > 1$, so it follows that $p_1 = q_i$. Therefore, we can eliminate p_1 and q_i from both sides of the factorization to obtain

$$p_2 \cdots p_m = q_1 \cdots q_{i-1} q_{i+1} \cdots q_k.$$

Continuing in this manner, we can eliminate all the p_i from the left side of the factorization to obtain

$$1 = q_{m+1} \cdots q_k.$$

If $q_{m+1}, \ldots, q_k$ were primes, this would be impossible. Therefore, $m = k$, and each prime p_i was included in the primes $q_1, \ldots, q_m$. Therefore, the factorizations differ only in the order of the factors, proving the theorem. □

3.2 Prime Elements, Units and Irreducibles

We now let R be an arbitrary integral domain and attempt to mimic the divisibility definitions and properties.

Definition 3.2.1. Let R be an integral domain.
(1) Suppose that $a, b \in R$. Then a is a *factor* or *divisor* of b if there exists a $c \in R$ with $b = ac$. We denote this, as in the integers, by $a|b$. If a is a factor of b, then b is called a *multiple* of a.
(2) An element $a \in R$ is a *unit* if a has a multiplicative inverse within R; that is, there exists an element $a^{-1} \in R$ with $aa^{-1} = 1$.
(3) A *prime element* of R is an element $p \neq 0$ such that p is not a unit, and if $p|ab$, then $p|a$ or $p|b$.
(4) An *irreducible element* in R is an element $c \neq 0$ such that c is not a unit, and if $c = ab$, then a or b must be a unit.
(5) a and b in R are *associates* if there exists a unit $e \in R$ with $a = eb$.

Notice that in the integers $\mathbb{Z}$, the units are just ± 1. The set of prime elements coincides with the set of irreducible elements. In $\mathbb{Z}$, these are precisely the set of prime numbers. On the other hand, if K is a field, every nonzero element is a unit. Therefore, in K, there are no prime elements and no irreducible elements.

Recall that the modular rings $\mathbb{Z}_n$ are fields (and integral domains) when n is a prime. In general, if n is not a prime then $\mathbb{Z}_n$ is a commutative ring with an identity, and a unit is still an invertible element. We can characterize the units within $\mathbb{Z}_n$.

Lemma 3.2.2. *$a \in \mathbb{Z}_n$ is a unit if and only if $(a, n) = 1$.*

Proof. Suppose $(a, n) = 1$. Then there exist $x, y \in \mathbb{Z}$ such that $ax + ny = 1$. This implies that $ax \equiv 1 \pmod{n}$, which in turn implies that $ax = 1$ in $\mathbb{Z}_n$ and, therefore, a is a unit.

Conversely, suppose a is a unit in $\mathbb{Z}_n$. Then there is an $x \in \mathbb{Z}_n$ with $ax = 1$. In terms of congruence then

$$ax \equiv 1 \pmod{n} \implies n|(ax - 1) \implies ax - 1 = ny \implies ax - ny = 1.$$

Therefore, 1 is a linear combination of a and n and so $(a, n) = 1$. □

If R is an integral domain, then the set of units within R will form a group.

Lemma 3.2.3. *If R is a commutative ring with an identity, then the set of units in R form an Abelian group under ring multiplication. This is called the unit group of R, denoted $U(R)$.*

Proof. The commutativity and associativity of $U(R)$ follow from the ring properties. The identity of $U(R)$ is the multiplicative identity of R, whereas the ring multiplicative inverse for each unit is the group inverse. We must show that $U(R)$ is closed under ring multiplication. If $a \in R$ is a unit, we denote its multiplicative inverse by a^{-1}. Now suppose $a, b \in U(R)$. Then a^{-1}, b^{-1} exist. It follows that

$$(ab)(b^{-1}a^{-1}) = a(bb^{-1})a^{-1} = aa^{-1} = 1.$$

Hence, ab has an inverse, namely $b^{-1}a^{-1}$ ($= a^{-1}b^{-1}$ in a commutative ring) and, hence, ab is also a unit. Therefore, $U(R)$ is closed under ring multiplication. □

In general, irreducible elements are not prime. Consider for example the subring of the complex numbers (see exercises) given by

$$R = \mathbb{Z}[i\sqrt{5}] = \{x + iy\sqrt{5} : x, y \in \mathbb{Z}\}.$$

This is a subring of the complex numbers $\mathbb{C}$ and, hence, can have no zero divisors. Therefore, R is an integral domain.

For an element $x + iy\sqrt{5} \in R$, define its *norm* by

$$N(x + iy\sqrt{5}) = |x + iy\sqrt{5}| = x^2 + 5y^2.$$

Since $x, y \in \mathbb{Z}$, it is clear that the norm of an element in R is a nonnegative integer. Furthermore, if $a \in R$ with $N(a) = 0$, then $a = 0$.

We have the following result concerning the norm:

Lemma 3.2.4. *Let R and N be as above. Then*
(1) *$N(ab) = N(a)N(b)$ for any elements $a, b \in R$.*
(2) *The units of R are those $a \in R$ with $N(a) = 1$. In R, the only units are ± 1.*

Proof. The fact that the norm is multiplicative is straightforward and left to the exercises. If $a \in R$ is a unit, then there exists a multiplicative inverse $b \in R$ with $ab = 1$. Then $N(ab) = N(a)N(b) = 1$. Since both $N(a)$ and $N(b)$ are nonnegative integers, we must have $N(a) = N(b) = 1$.

Conversely, suppose that $N(a) = 1$. If $a = x + iy\sqrt{5}$, then $x^2 + 5y^2 = 1$. Since $x, y \in \mathbb{Z}$, we must have $y = 0$ and $x^2 = 1$. Then $a = x = \pm 1$. □

Using this lemma we can show that R possesses irreducible elements that are not prime.

Lemma 3.2.5. *Let R be as above. Then $3 = 3 + i0\sqrt{5}$ is an irreducible element in R, but 3 is not prime.*

Proof. Suppose that $3 = ab$ with $a, b \in R$ and a, b nonunits. Then $N(3) = 9 = N(a)N(b)$ with neither $N(a) = 1$, nor $N(b) = 1$. Hence, $N(a) = 3$, and $N(b) = 3$. Let $a = x + iy\sqrt{5}$. It follows that $x^2 + 5y^2 = 3$. Since $x, y \in \mathbb{Z}$, this is impossible. Therefore, one of a or b must be a unit, and 3 is an irreducible element.

We show that 3 is not prime in R. Let $a = 2 + i\sqrt{5}$ and $b = 2 - i\sqrt{5}$. Then $ab = 9$ and, hence, $3|ab$. Suppose $3|a$ so that $a = 3c$ for some $c \in R$. Then

$$9 = N(a) = N(3)N(c) = 9N(c) \implies N(c) = 1.$$

Therefore, c is a unit in R, and from Lemma 3.2.4, we get $c = \pm 1$. Hence, $a = \pm 3$. This is a contradiction, so 3 does not divide a. An identical argument shows that 3 does not divide b. Therefore, 3 is not a prime element in R. □

We now examine the relationship between prime elements and irreducibles.

Theorem 3.2.6. *Let R be an integral domain. Then*
(1) *Each prime element of R is irreducible.*
(2) *$p \in R$ is a prime element if and only if $p \neq 0$, and $\langle p \rangle = pR$ is a prime ideal.*
(3) *$p \in R$ is irreducible if and only if $p \neq 0$, and $\langle p \rangle = pR$ is maximal in the set of all principal ideals of R, which are not equal to R.*

Proof. (1) Suppose that $p \in R$ is a prime element, and $p = ab$. We must show that either a or b must be a unit. Now $p|ab$, so either $p|a$, or $p|b$. Without loss of generality, we may assume that $p|a$, so $a = pr$ for some $r \in R$. Hence, $p = ab = (pr)b = p(rb)$. However, R is an integral domain, so $p - prb = p(1 - rb) = 0$ implies that $1 - rb = 0$ and, hence, $rb = 1$. Therefore, b is a unit and, hence, p is irreducible.

(2) Suppose that p is a prime element. Then $p \neq 0$. Consider the ideal pR, and suppose that $ab \in pR$. Then ab is a multiple of p and, hence, $p|ab$. Since p is prime, it follows that $p|a$ or $p|b$. If $p|a$, then $a \in pR$, whereas if $p|b$, then $b \in pR$. Therefore, pR is a prime ideal.

Conversely, suppose that pR is a prime ideal, and suppose that $p = ab$. Then $ab \in pR$, so $a \in pR$, or $b \in pR$. If $a \in pR$, then $p|a$, and if $b \in pR$, then $p|b$. Therefore, p is prime.

(3) Let p be irreducible, then $p \neq 0$. Suppose that $pR \subset aR$, where $a \in R$. Then $p = ra$ for some $r \in R$. Since p is irreducible, it follows that either a is a unit, or r is a unit. If r is a unit, we have $pR = raR = aR \neq R$ since p is not a unit. If a is a unit, then $aR = R$, and $pR = rR \neq R$. Therefore, pR is maximal in the set of principal ideals not equal to R. Conversely, suppose $p \neq 0$ and pR is a maximal ideal in the set of principal ideals $\neq R$. Let $p = ab$ with a not a unit. We must show that b is a unit. Since $aR \neq R$, and $pR \subset aR$, from the maximality we must have $pR = aR$. Hence, $a = rp$ for some $r \in R$. Then $p = ab = rpb$ and, as before, we must have $rb = 1$ and b a unit. □

Theorem 3.2.7. *Let R be a principle ideal domain. Then we have the following:*
(1) *An element $p \in R$ is irreducible if and only if it is a prime element.*
(2) *A nonzero ideal of R is a maximal ideal if and only if it is a prime ideal.*
(3) *The maximal ideals of R are precisely those ideals pR, where p is a prime element.*

Proof. First note that $\{0\}$ is a prime ideal, but not maximal.

(1) We already know that prime elements are irreducible. To show the converse, suppose that p is irreducible. Since R is a principal ideal domain from Theorem 3.2.6, we

have that pR is a maximal ideal, and each maximal ideal is also a prime ideal. Therefore, from Theorem 3.2.6, we have that p is a prime element.

(2) We already know that each maximal ideal is a prime ideal. To show the converse, suppose that $I \neq \{0\}$ is a prime ideal. Then $I = pR$, where p is a prime element with $p \neq 0$. Therefore, p is irreducible from part (1) and, hence, pR is a maximal ideal from Theorem 3.2.6.

(3) This follows directly from the proof in part (2) and Theorem 3.2.6. □

This Theorem especially explains the following remark at the end of Section 2.3: In the principal ideal domain $\mathbb{Z}$, a proper ideal is maximal if and only if it is a prime ideal.

3.3 Unique Factorization Domains

We now consider integral domains, where there is unique factorization into primes. If R is an integral domain and $a, b \in R$, then we say that a and b are *associates* if there exists a unit $\epsilon \in R$ with $a = \epsilon b$.

Definition 3.3.1. An integral domain D is a *unique factorization domain* or UFD if for each $d \in D$ either $d = 0$, d is a unit, or d has a factorization into primes, which is unique up to ordering and unit factors. This means that if

$$r = p_1 \cdots p_m = q_1 \cdots q_k,$$

then $m = k$, and each p_i is an associate of some q_j.

There are several relationships in integral domains that are equivalent to unique factorization.

Definition 3.3.2. Let R be an integral domain.

(1) R has property (A) if and only if for each nonunit $a \neq 0$ there are irreducible elements $q_1, \ldots, q_r \in R$, satisfying $a = q_1 \cdots q_r$.
(2) R has property (A′) if and only if for each nonunit $a \neq 0$ there are prime elements $p_1, \ldots, p_r \in R$, satisfying $a = p_1 \cdots p_r$.
(3) R has property (B) if and only if whenever $q_1, \ldots, q_r$ and $q'_1, \ldots, q'_s$ are irreducible elements of R with $q_1 \cdots q_r = q'_1 \cdots q'_s$. Then $r = s$, and there is a permutation $\pi \in S_r$ such that for each $i \in \{1, \ldots, r\}$ the elements q_i and $q'_{\pi(i)}$ are associates (uniqueness up to ordering and unit factors).
(4) R has property (C) if and only if each irreducible element of R is a prime element.

Notice that properties (A) and (C) together are equivalent to what we defined as *unique factorization*. Hence, an integral domain satisfying (A) and (C) is a UFD. Next, we show that there are other equivalent formulations.

Theorem 3.3.3. *In an integral domain R, the following are equivalent:*

(1) *R is a UFD.*
(2) *R satisfies properties* (A) *and* (B).
(3) *R satisfies properties* (A) *and* (C).
(4) *R satisfies property* (A′).

Proof. As remarked before, the statement of the theorem by definition (A) and (C) are equivalent to unique factorization. We show here that (2), (3), and (4) are equivalent.

First, we show that (2) implies (3).

Suppose that R satisfies properties (A) and (B). We must show that it also satisfies (C); that is, we must show that if $q \in R$ is irreducible, then q is prime. Suppose that $q \in R$ is irreducible and $q|ab$ with $a, b \in R$. Then we have $ab = cq$ for some $c \in R$. If a is a unit from $ab = cq$, we get that $b = a^{-1}cq$, and $q|b$. The results are identical if b is a unit. Therefore, we may assume that neither a nor b are units.

If $c = 0$, then since R is an integral domain, either $a = 0$, or $b = 0$, and $q|a$, or $q|b$. We may assume then that $c \neq 0$.

If c is a unit, then $q = c^{-1}ab$, and since q is irreducible, either $c^{-1}a$, or b are units. If $c^{-1}a$ is a unit, then a is also a unit. Therefore, if c is a unit, either a or b are units contrary to our assumption.

Therefore, we may assume that $c \neq 0$, and c is not a unit. From (A) we have

$$\begin{aligned} a &= q_1 \cdots q_r \\ b &= q_1' \cdots q_s' \\ c &= q_1'' \cdots q_t'', \end{aligned}$$

where $q_1, \ldots q_r, q_1', \ldots, q_s', q_1'', \ldots q_t''$ are all irreducibles. Hence,

$$q_1 \cdots q_r q_1' \cdots q_s' = q_1'' \cdots q_t'' \cdot q.$$

From (B), q is an associate of some q_i or q_j'. Hence, $q|q_i$ or $q|q_j'$. It follows that $q|a$, or $q|b$ and, therefore, q is a prime element.

That (3) implies (4) is direct.

We show that (4) implies (2). Suppose that R satisfies (A′). We must show that it satisfies both (A) and (B). We show first that (A) follows from (A′) by showing that irreducible elements are prime. Suppose that q is irreducible. Then from (A′), we have

$$q = p_1 \cdots p_r$$

with each p_i prime. It follows, without loss of generality, that $p_2 \cdots p_r$ is a unit, and p_1 is a nonunit and, hence, $p_i|1$ for $i = 2, \ldots, r$. Thus, $q = p_1$, and q is prime. Therefore, (A) holds.

We now show that (B) holds. Let

$$q_1 \cdots q_r = q_1' \cdots q_s',$$

where q_i, q'_j are all irreducibles; hence primes. Then

$$q'_1 | q_1 \cdots q_r,$$

and so, $q'_1 | q_i$ for some i. Without loss of generality, suppose $q'_1 | q_1$. Then $q_1 = aq'_1$. Since q_1 is irreducible, it follows that a is a unit, and q_1 and q'_1 are associates. It follows then that

$$aq_2 \cdots q_r = q'_2 \cdot q'_s$$

since R has no zero divisors. Property (B) holds then by induction, and the theorem is proved. □

Note that in our new terminology, $\mathbb{Z}$ is a UFD. In the next section, we will present other examples of UFD's. However, not every integral domain is a unique factorization domain.

As we defined in the last section, let R be the following subring of $\mathbb{C}$:

$$R = \mathbb{Z}[i\sqrt{5}] = \{x + iy\sqrt{5} : x, y \in \mathbb{Z}\}.$$

R is an integral domain, and we showed, using the norm, that 3 is an irreducible in R. Analogously, we can show that the elements $2 + i\sqrt{5}$, $2 - i\sqrt{5}$ are also irreducibles in R, and furthermore, 3 is not an associate of either $2 + i\sqrt{5}$ or $2 - i\sqrt{5}$. Then

$$9 = 3 \cdot 3 = (2 + i\sqrt{5})(2 - i\sqrt{5})$$

give two different decompositions for an element in terms of irreducible elements. The fact that R is not a UFD also follows from the fact that 3 is an irreducible element, which is not prime.

Unique factorization is tied to the famous solution of Fermat's big theorem. Wiles and Taylor in 1995 proved the following:

Theorem 3.3.4. *The equation $x^p + y^p = z^p$ has no integral solutions with $xyz \neq 0$ for any prime $p \geq 3$.*

Kummer tried to prove this theorem by attempting to factor $x^p = z^p - y^p$. We call the statement of Theorem 3.3.4 in an integral domain R property (F_p). Let $\epsilon = e^{\frac{2\pi i}{p}}$. Then

$$z^p - y^p = \prod_{j=0}^{p-1} (z - \epsilon^j y).$$

View this equation in the ring:

$$R = \mathbb{Z}[\epsilon] = \left\{ \sum_{j=0}^{p-1} a_j \epsilon^j : a_j \in \mathbb{Z} \right\}.$$

Kummer proved that if R is a UFD, then property (F_p) holds. However, independently, from Uchida and Montgomery (1971), R is a UFD only if $p \leq 19$ (see [59]).

3.4 Principal Ideal Domains and Unique Factorization

In this section, we prove that every principal ideal domain (PID) is a unique factorization domain (UFD). We say that an ascending chain of ideals in R

$$I_1 \subset I_2 \subset \cdots \subset I_n \subset \cdots$$

becomes *stationary* if there exists an m such that $I_r = I_m$ for all $r \geq m$.

Theorem 3.4.1. *Let R be an integral domain. If each ascending chain of principal ideals in R becomes stationary, then R satisfies property* (A).

Proof. Suppose that $a \neq 0$ is a not a unit in R. Suppose that a is not a product of irreducible elements. Clearly then, a cannot itself be irreducible. Hence, $a = a_1b_1$ with $a_1, b_1 \in R$, and a_1, b_1 are not units. If both a_1 or b_1 can be expressed as a product of irreducible elements, then so can a. Without loss of generality then, suppose that a_1 is not a product of irreducible elements.

Since $a_1|a$, we have the inclusion of ideals $aR \subseteq a_1R$. If $a_1R = aR$, then $a_1 \in aR$, and $a_1 = ar = a_1b_1r$, which implies that b_1 is a unit contrary to our assumption. Therefore, $aR \neq a_1R$, and the inclusion is proper. By iteration then, we obtain a strictly increasing chain of ideals

$$aR \subset a_1R \subset \cdots \subset a_nR \subset \cdots.$$

From our hypothesis on R, this must become stationary, contradicting the argument above that the inclusion is proper. Therefore, a must be a product of irreducibles. □

Theorem 3.4.2. *Each principal ideal domain R is a unique factorization domain.*

Proof. Suppose that R is a principal ideal domain. R satisfies property (C) by Theorem 3.2.7(1). Therefore, to show that it is a unique factorization domain, we must show that it also satisfies property (A). From the previous theorem, it suffices to show that each ascending chain of principal ideals becomes stationary. Consider such an ascending chain

$$a_1R \subset a_2R \subset \cdots \subset a_nR \subset \cdots.$$

Now let

$$I = \bigcup_{i=1}^{\infty} a_iR.$$

Now I is an ideal in R; hence a principal ideal. Therefore, $I = aR$ for some $a \in R$. Since I is a union, there exists an m such that $a \in a_m R$. Therefore, $I = aR \subset a_m R$ and, hence, $I = a_m R$, and $a_i R \subset a_m R$ for all $i \geq m$. Therefore, the chain becomes stationary and, from Theorem 3.4.1, R satisfies property (A). □

Since we showed that the integers $\mathbb{Z}$ are a PID, we can recover the fundamental theorem of arithmetic from Theorem 3.4.2. We now present another important example of a PID; hence a UFD. In the next chapter, we will look in detail at polynomials with coefficients in an integral domain. Below, we consider polynomials with coefficients in a field, and for the present leave out many of the details.

If K is a field and n is a nonnegative integer, then a *polynomial of degree n over K* is a formal sum of the form

$$P(x) = a_0 + a_1 x + \cdots + a_n x^n$$

with $a_i \in K$ for $i = 0, \ldots, n$, $a_n \neq 0$, and x an indeterminate. A *polynomial* $P(x)$ over K is either a polynomial of some degree or the expression $P(x) = 0$, which is called the *zero polynomial*, and has degree $-\infty$. We denote the degree of $P(x)$ by $\deg P(x)$. A polynomial of zero degree has the form $P(x) = a_0$ and is called a *constant polynomial*, and can be identified with the corresponding element of K. The elements $a_i \in K$ are called the *coefficients of* $P(x)$; a_n is the *leading coefficient*. If $a_n = 1$, $P(x)$ is called a *monic polynomial*. Two nonzero polynomials are equal if and only if they have the same degree and exactly the same coefficients. A polynomial of degree 1 is called a *linear polynomial*, whereas one of degree two is a *quadratic polynomial*.

We denote by $K[x]$ the set of all polynomials over K, and we will show that $K[x]$ becomes a principal ideal domain; hence a unique factorization domain. We first define addition, subtraction, and multiplication on $K[x]$ by algebraic manipulation. That is, suppose $P(x) = a_0 + a_1 x + \cdots + a_n x^n$, $Q(x) = b_0 + b_1 x + \cdots + b_m x^m$, then

$$P(x) \pm Q(x) = (a_0 \pm b_0) + (a_1 \pm b_1)x + \cdots;$$

that is, the coefficient of x^i in $P(x) \pm Q(x)$ is $a_i \pm b_i$, where $a_i = 0$ for $i > n$, and $b_j = 0$ for $j > m$. Multiplication is given by

$$P(x)Q(x) = (a_0 b_0) + (a_1 b_0 + a_0 b_1)x + (a_0 b_2 + a_1 b_1 + a_2 b_0)x^2 + \cdots + (a_n b_m)x^{n+m};$$

that is, the coefficient of x^i in $P(x)Q(x)$ is $(a_0 b_i + a_1 b_{i-1} + \cdots + a_i b_0)$.

Example 3.4.3. Let $P(x) = 3x^2 + 4x - 6$ and $Q(x) = 2x + 7$ be in $\mathbb{Q}[x]$. Then

$$P(x) + Q(x) = 3x^2 + 6x + 1$$

and

$$P(x)Q(x) = (3x^2 + 4x - 6)(2x + 7) = 6x^3 + 29x^2 + 16x - 42.$$

From the definitions, the following degree relationships are clear. The proofs are in the exercises.

Lemma 3.4.4. *Let $0 \neq P(x)$, $0 \neq Q(x)$ in $K[x]$. Then the following hold:*
(1) $\deg P(x)Q(x) = \deg P(x) + \deg Q(x)$.
(2) $\deg(P(x) \pm Q(x)) \leq \max(\deg P(x), \deg Q(x))$ *if* $P(x) \pm Q(x) \neq 0$.

We next obtain the following:

Theorem 3.4.5. *If K is a field, then $K[x]$ forms an integral domain. K can be naturally embedded into $K[x]$ by identifying each element of K with the corresponding constant polynomial. The only units in $K[x]$ are the nonzero elements of K.*

Proof. Verification of the basic ring properties is solely computational and is left to the exercises. Since $\deg P(x)Q(x) = \deg P(x) + \deg Q(x)$, it follows that if neither $P(x) \neq 0$, nor $Q(x) \neq 0$, then $P(x)Q(x) \neq 0$ and, therefore, $K[x]$ is an integral domain.

If $G(x)$ is a unit in $K[x]$, then there exists an $H(x) \in K[x]$ with $G(x)H(x) = 1$. From the degrees, we have $\deg G(x) + \deg H(x) = 0$, and since $\deg G(x) \geq 0$, $\deg H(x) \geq 0$. This is possible only if $\deg G(x) = \deg H(x) = 0$. Therefore, $G(x) \in K$. □

Now that we have $K[x]$ as an integral domain, we proceed to show that $K[x]$ is a principal ideal domain and, hence, there is unique factorization into primes.

We first repeat the definition of a prime in $K[x]$. If $0 \neq f(x)$ has no nontrivial, nonunit factors (it cannot be factorized into polynomials of lower degree), then $f(x)$ is a *prime* in $K[x]$ or a *prime polynomial*. A prime polynomial is also called an *irreducible polynomial*. Clearly, if $\deg g(x) = 1$, then $g(x)$ is irreducible.

The fact that $K[x]$ is a principal ideal domain follows from the division algorithm for polynomials, which is entirely analogous to the division algorithm for integers.

Lemma 3.4.6 (Division algorithm in $K[x]$). *If $0 \neq f(x)$, $0 \neq g(x) \in K[x]$, then there exist unique polynomials $q(x), r(x) \in K[x]$ such that $f(x) = q(x)g(x) + r(x)$, where $r(x) = 0$ or* $\deg r(x) < \deg g(x)$.

(The polynomials $q(x)$ and $r(x)$ are called, respectively, the quotient and remainder.)

We give a formal proof in Chapter 4 on polynomials and polynomial rings. For now we content ourselves here with doing two computations in $\mathbb{Q}[x]$ in the following example.

Example 3.4.7. (1) Let $f(x) = 3x^4 - 6x^2 + 8x - 6$, $g(x) = 2x^2 + 4$. Then

$$\frac{3x^4 - 6x^2 + 8x - 6}{2x^2 + 4} = \frac{3}{2}x^2 - 6 \quad \text{with remainder } 8x + 18.$$

Thus, here, $q(x) = \frac{3}{2}x^2 - 6$, $r(x) = 8x + 18$.

(2) Let $f(x) = 2x^5 + 2x^4 + 6x^3 + 10x^2 + 4x$, $g(x) = x^2 + x$. Then

$$\frac{2x^5 + 2x^4 + 6x^3 + 10x^2 + 4x}{x^2 + x} = 2x^3 + 6x + 4.$$

Thus, here, $q(x) = 2x^3 + 6x + 4$, and $r(x) = 0$.

Theorem 3.4.8. *Let K be a field. Then the polynomial ring $K[x]$ is a principal ideal domain; hence a unique factorization domain.*

Proof. The proof is essentially analogous to the proof in the integers. Let I be an ideal in $K[x]$ with $I \neq K[x]$. Let $f(x)$ be a polynomial in I of minimal degree. We claim that $I = \langle f(x)\rangle$, the principal ideal generated by $f(x)$. Let $g(x) \in I$. We must show that $g(x)$ is a multiple of $f(x)$. By the division algorithm in $K[x]$, we have

$$g(x) = q(x)f(x) + r(x),$$

where $r(x) = 0$, or $\deg(r(x)) < \deg(f(x))$. If $r(x) \neq 0$, then $\deg(r(x)) < \deg(f(x))$. However, $r(x) = g(x) - q(x)f(x) \in I$ since I is an ideal, and $g(x), f(x) \in I$. This is a contradiction since $f(x)$ was assumed to be a polynomial in I of minimal degree. Therefore, $r(x) = 0$ and, hence, $g(x) = q(x)f(x)$ is a multiple of $f(x)$. Therefore, each element of I is a multiple of $f(x)$ and, hence, $I = \langle f(x)\rangle$.

Therefore, $K[x]$ is a principal ideal domain and, from Theorem 3.4.2, a unique factorization domain. □

We proved that in a principal ideal domain, every ascending chain of ideals becomes stationary. In general, a ring R (commutative or not) satisfies the *ascending chain condition* or ACC if every ascending chain of left (or right) ideals in R becomes stationary. A ring satisfying the ACC is called a *Noetherian ring*.

3.5 Euclidean Domains

In analyzing the proof of unique factorization in both $\mathbb{Z}$ and $K[x]$, it is clear that it depends primarily on the division algorithm. In $\mathbb{Z}$, the division algorithm depended on the fact that the positive integers could be ordered, and in $K[x]$, on the fact that the degrees of nonzero polynomials are nonnegative integers and, hence, could be ordered. This basic idea can be generalized in the following way.

Definition 3.5.1. An integral domain D is a *Euclidean domain* if there exists a function N from $D^\star = D \setminus \{0\}$ to the nonnegative integers such that:

(1) $N(r_1) \leq N(r_1 r_2)$ for any $r_1, r_2 \in D^\star$.

(2) For all $r_1, r_2 \in D$ with $r_1 \neq 0$, there exist $q, r \in D$ such that

$$r_2 = q r_1 + r,$$

where either $r = 0$, or $N(r) < N(r_1)$.

The function N is called a *Euclidean norm* on D.

Therefore, Euclidean domains are precisely those integral domains, which allow division algorithms. In the integers $\mathbb{Z}$, define $N(z) = |z|$. Then N is a Euclidean norm on $\mathbb{Z}$ and, hence, $\mathbb{Z}$ is a Euclidean domain. On $K[x]$, define $N(p(x)) = \deg(p(x))$ if $p(x) \neq 0$. Then N is also a Euclidean norm on $K[x]$ so that $K[x]$ is also a Euclidean domain. In any Euclidean domain, we can mimic the proofs of unique factorization in both $\mathbb{Z}$ and $K[x]$ to obtain the following:

Theorem 3.5.2. *Every Euclidean domain is a principal ideal domain; hence a unique factorization domain.*

Before proving this theorem, we must develop some results on the *number theory* of general Euclidean domains. First, some properties of the norm.

Lemma 3.5.3. *If R is a Euclidean domain then the following hold:*
(a) *$N(1)$ is minimal among $\{N(r) : r \in R^\star\}$.*
(b) *$N(u) = N(1)$ if and only if u is a unit.*
(c) *$N(a) = N(b)$ for $a, b \in R^\star$ if a, b are associates.*
(d) *$N(a) < N(ab)$ unless b is a unit.*

Proof. (a) From property (1) of Euclidean norms, we have

$$N(1) \leq N(1 \cdot r) = N(r) \quad \text{for any } r \in R^\star.$$

(b) Suppose u is a unit. Then there exists u^{-1} with $u \cdot u^{-1} = 1$. Then

$$N(u) \leq N(u \cdot u^{-1}) = N(1).$$

From the minimality of $N(1)$, it follows that $N(u) = N(1)$.

Conversely, suppose $N(u) = N(1)$. Apply the division algorithm to get

$$1 = qu + r.$$

If $r \neq 0$, then $N(r) < N(u) = N(1)$, contradicting the minimality of $N(1)$. Therefore, $r = 0$, and $1 = qu$. Then u has a multiplicative inverse and, hence, is a unit.

(c) Suppose $a, b \in R^\star$ are associates. Then $a = ub$ with u a unit. Then

$$N(b) \leq N(ub) = N(a).$$

On the other hand, $b = u^{-1}a$. Therefore,

$$N(a) \leq N(u^{-1}a) = N(b).$$

Since $N(a) \le N(b)$, and $N(b) \le N(a)$, it follows that $N(a) = N(b)$.

(d) Suppose $N(a) = N(ab)$. Apply the division algorithm

$$a = q(ab) + r,$$

where $r = 0$, or $N(r) < N(ab)$. If $r \neq 0$, then

$$r = a - qab = a(1 - qb) \implies N(ab) = N(a) \le N\big(a(1 - qb)\big) = N(r),$$

contradicting that $N(r) < N(ab)$. Hence, $r = 0$, and $a = q(ab) = (qb)a$. Then

$$a = (qb)a = 1 \cdot a \implies qb = 1$$

since there are no zero divisors in an integral domain. Hence, b is a unit.

Since $N(a) \le N(ab)$, it follows that if b is not a unit, we must have $N(a) < N(ab)$. □

We can now prove Theorem 3.5.2.

Proof. Let D be a Euclidean domain. We show that each ideal $I \neq D$ in D is principal. Let $I \neq D$ be an ideal in D. If $I = \{0\}$, then $I = \langle 0 \rangle$, and I is principal. Therefore, we may assume that there are nonzero elements in I. Hence, there are elements $x \in I$ with strictly positive norm. Let a be an element of I of minimal norm. We claim that $I = \langle a \rangle$. Let $b \in I$. We must show that b is a multiple of a. Now by the division algorithm

$$b = qa + r,$$

where either $r = 0$, or $N(r) < N(a)$. As in $\mathbb{Z}$ and $K[x]$, we have a contradiction if $r \neq 0$. In this case, $N(r) < N(a)$, but $r = b - qa \in I$ since I is an ideal, contradicting the minimality of $N(a)$. Therefore, $r = 0$, and $b = qa$ and, hence, $I = \langle a \rangle$. □

As a final example of a Euclidean domain, we consider the *Gaussian integers*

$$\mathbb{Z}[i] = \{a + bi : a, b \in \mathbb{Z}\}.$$

It was first observed by Gauss that this set permits unique factorization. To show this, we need a Euclidean norm on $\mathbb{Z}[i]$.

Definition 3.5.4. If $z = a + bi \in \mathbb{Z}[i]$, then its *norm* $N(z)$ is defined by

$$N(a + bi) = a^2 + b^2.$$

The basic properties of this norm follow directly from the definition (see exercises).

Lemma 3.5.5. *If $\alpha, \beta \in \mathbb{Z}[i]$ then we have the following:*

(1) *$N(\alpha)$ is an integer for all $\alpha \in \mathbb{Z}[i]$.*

(2) *$N(\alpha) \ge 0$ for all $\alpha \in \mathbb{Z}[i]$.*

(3) $N(\alpha) = 0$ *if and only if* $\alpha = 0$.
(4) $N(\alpha) \geq 1$ *for all* $\alpha \neq 0$.
(5) $N(\alpha\beta) = N(\alpha)N(\beta)$*; that is, the norm is multiplicative.*

From the multiplicativity of the norm, we have the following concerning primes and units in $\mathbb{Z}[i]$.

Lemma 3.5.6. (1) $u \in \mathbb{Z}[i]$ *is a unit if and only if* $N(u) = 1$.
(2) *If* $\pi \in \mathbb{Z}[i]$ *and* $N(\pi) = p$, *where* p *is an ordinary prime in* $\mathbb{Z}$, *then* π *is a prime in* $\mathbb{Z}[i]$.

Proof. Certainly u is a unit if and only if $N(u) = N(1)$. But in $\mathbb{Z}[i]$, we have $N(1) = 1$. Therefore, the first part follows.

Suppose next that $\pi \in \mathbb{Z}[i]$ with $N(\pi) = p$ for some $p \in \mathbb{Z}$. Suppose that $\pi = \pi_1\pi_2$. From the multiplicativity of the norm, we have

$$N(\pi) = p = N(\pi_1)N(\pi_2).$$

Since each norm is a positive ordinary integer, and p is a prime, it follows that either $N(\pi_1) = 1$, or $N(\pi_2) = 1$. Hence, either π_1 or π_2 is a unit. Therefore, π is a prime in $\mathbb{Z}[i]$. □

Armed with this norm, we can show that $\mathbb{Z}[i]$ is a Euclidean domain.

Theorem 3.5.7. *The Gaussian integers* $\mathbb{Z}[i]$ *form a Euclidean domain.*

Proof. That $\mathbb{Z}[i]$ forms a commutative ring with an identity can be verified directly and easily. If $\alpha\beta = 0$, then $N(\alpha)N(\beta) = 0$, and since there are no zero divisors in $\mathbb{Z}$, we must have $N(\alpha) = 0$, or $N(\beta) = 0$. But then either $\alpha = 0$, or $\beta = 0$ and, hence, $\mathbb{Z}[i]$ is an integral domain. To complete the proof, we show that the norm N is a Euclidean norm.

From the multiplicativity of the norm, we have, if $\alpha, \beta \neq 0$

$$N(\alpha\beta) = N(\alpha)N(\beta) \geq N(\alpha) \quad \text{since } N(\beta) \geq 1.$$

Therefore, property (1) of Euclidean norms is satisfied. We must now show that the division algorithm holds.

Let $\alpha = a + bi$ and $\beta = c + di$ be Gaussian integers. Recall that the inverse for a nonzero complex number $z = x + iy$ is

$$\frac{1}{z} = \frac{\overline{z}}{|z|^2} = \frac{x - iy}{x^2 + y^2}.$$

Therefore, as a complex number

$$\begin{aligned}\frac{\alpha}{\beta} = \alpha\frac{\overline{\beta}}{|\beta|^2} &= (a + bi)\frac{c - di}{c^2 + d^2} \\ &= \frac{ac + bd}{c^2 + d^2} + \frac{ac - bd}{c^2 + d^2}i = u + iv.\end{aligned}$$

Now since a, b, c, d are integers u, v must be rationals. The set

$$\{u + iv : u, v \in \mathbb{Q}\}$$

is called the set of the *Gaussian rationals.*

If $u, v \in \mathbb{Z}$, then $u + iv \in \mathbb{Z}[i]$, $\alpha = q\beta$ with $q = u + iv$, and we are done. Otherwise, choose ordinary integers m, n satisfying $|u - m| \le \frac{1}{2}$ and $|v - n| \le \frac{1}{2}$, and let $q = m + in$. Then $q \in \mathbb{Z}[i]$. Let $r = \alpha - q\beta$. We must show that $N(r) < N(\beta)$.

Working with complex absolute value, we get

$$|r| = |\alpha - q\beta| = |\beta|\left|\frac{\alpha}{\beta} - q\right|.$$

Now

$$\left|\frac{\alpha}{\beta} - q\right| = \left|(u - m) + i(v - n)\right| = \sqrt{(u - m)^2 + (v - n)^2} \le \sqrt{\left(\frac{1}{2}\right)^2 + \left(\frac{1}{2}\right)^2} < 1.$$

Therefore,

$$|r| < |\beta| \implies |r|^2 < |\beta|^2 \implies N(r) < N(\beta),$$

completing the proof. □

Since $\mathbb{Z}[i]$ forms a Euclidean domain, it follows from our previous results that $\mathbb{Z}[i]$ must be a principal ideal domain; hence a unique factorization domain.

Corollary 3.5.8. *The Gaussian integers are a UFD.*

Since we will now be dealing with many kinds of *integers*, we will refer to the ordinary integers $\mathbb{Z}$ as the *rational integers* and the ordinary primes p as the *rational primes*. It is clear that $\mathbb{Z}$ can be embedded into $\mathbb{Z}[i]$. However, not every rational prime is also prime in $\mathbb{Z}[i]$. The primes in $\mathbb{Z}[i]$ are called the *Gaussian primes*. For example, we can show that both $1 + i$ and $1 - i$ are Gaussian primes; that is, primes in $\mathbb{Z}[i]$. However, $(1 + i)(1 - i) = 2$. Therefore, the rational prime 2 is not a prime in $\mathbb{Z}[i]$. Using the multiplicativity of the Euclidean norm in $\mathbb{Z}[i]$, we can describe all the units and primes in $\mathbb{Z}[i]$.

Theorem 3.5.9. (1) *The only units in $\mathbb{Z}[i]$ are $\pm 1, \pm i$.*

(2) *Suppose π is a Gaussian prime. Then π is one of the following:*
   (a) *a positive rational prime $p \equiv 3 \pmod 4$, or an associate of such a rational prime.*
   (b) *$1 + i$, or an associate of $1 + i$.*
   (c) *$a + bi$, or $a - bi$, where $a > 0$, $b > 0$, a is even, and $N(\pi) = a^2 + b^2 = p$ with p a rational prime congruent to 1 modulo 4, or an associate of $a + bi$, or $a - bi$.*

Proof. (1) Suppose $u = x + iy \in \mathbb{Z}[i]$ is a unit. Then, from Lemma 3.5.6, $N(u) = x^2 + y^2 = 1$, implying that $(x, y) = (0, \pm 1)$ or $(x, y) = (\pm 1, 0)$. Hence, $u = \pm 1$ or $u = \pm i$.

(2) Now suppose that π is a Gaussian prime. Since $N(\pi) = \pi\overline{\pi}$, and $\overline{\pi} \in \mathbb{Z}[i]$, it follows that $\pi | N(\pi)$. $N(\pi)$ is a rational integer, so $N(\pi) = p_1 \cdots p_k$, where the p_i's are rational primes. By Euclid's lemma $\pi | p_i$ for some p_i and, hence, a Gaussian prime must divide at least one rational prime. On the other hand, suppose $\pi | p$ and $\pi | q$, where p, q are different primes. Then $(p, q) = 1$ and, hence, there exist $x, y \in \mathbb{Z}$ such that $1 = px + qy$. It follows that $\pi | 1$ is a contradiction. Therefore, a Gaussian prime divides one and only one rational prime.

Let p be the rational prime that π divides. Then $N(\pi) | N(p) = p^2$. Since $N(\pi)$ is a rational integer, it follows that $N(\pi) = p$, or $N(\pi) = p^2$. If $\pi = a + bi$, then $a^2 + b^2 = p$, or $a^2 + b^2 = p^2$.

If $p = 2$, then $a^2 + b^2 = 2$, or $a^2 + b^2 = 4$. It follows that $\pi = \pm 2, \pm 2i$, or $\pi = 1 + i$, or an associate of $1 + i$. Since $(1 + i)(1 - i) = 2$, and neither $1 + i$, nor $1 - i$ are units, it follows that neither 2, nor any of its associates are primes. Then $\pi = 1 + i$, or an associate of $1 + i$. To see that $1 + i$ is prime supposes $1 + i = \alpha\beta$. Then $N(1 + i) = 2 = N(\alpha)N(\beta)$. It follows that either $N(\alpha) = 1$, or $N(\beta) = 1$, and either α or β is a unit.

If $p \neq 2$, then either $p \equiv 3 \pmod 4$, or $p \equiv 1 \pmod 4$. First suppose $p \equiv 3 \pmod 4$. Then $a^2 + b^2 = p$ would imply (Fermat's two-square theorem, see [53]) $p \equiv 1 \pmod 4$. Therefore, from the remarks above $a^2 + b^2 = p^2$, and $N(\pi) = N(p)$. Since $\pi | p$, we have $\pi = \alpha p$ with $\alpha \in \mathbb{Z}[i]$. From $N(\pi) = N(p)$, we get that $N(\alpha) = 1$, and α is a unit. Therefore, π and p are associates. Hence, in this case, π is an associate of a rational prime congruent to 3 modulo 4.

Finally, suppose $p \equiv 1 \pmod 4$. From the remarks above, either $N(\pi) = p$, or $N(\pi) = p^2$. If $N(\pi) = p^2$, then $a^2 + b^2 = p^2$. Since $p \equiv 1 \pmod 4$, from Fermat's two square theorem, there exist $m, n \in \mathbb{Z}$ with $m^2 + n^2 = p$. Let $u = m + in$, then the norm $N(u) = p$. Since p is a rational prime, it follows that u is a Gaussian prime. Similarly, its conjugate $\overline{u}$ is also a Gaussian prime. Now $u\overline{u} | p^2 = N(\pi)$. Since $\pi | N(\pi)$, it follows that $\pi | u\overline{u}$, and from Euclid's lemma, either $\pi | u$, or $\pi | \overline{u}$. If $\pi | u$, they are associates since both are primes. But this is a contradiction since $N(\pi) \neq N(u)$. The same is true if $\pi | \overline{u}$.

It follows that if $p \equiv 1 \pmod 4$, then $N(\pi) \neq p^2$. Therefore, $N(\pi) = p = a^2 + b^2$. An associate of π has both $a, b > 0$ (see exercises). Furthermore, since $a^2 + b^2 = p$, one of a or b must be even. If a is odd, then b is even; then $i\pi$ is an associate of π with a even, completing the proof. □

Finally, we mention that the methods used in $\mathbb{Z}[i]$ cannot be applied to all quadratic integers. For example, we have seen that there is not unique factorization in $\mathbb{Z}[\sqrt{-5}]$.

3.6 Overview of Integral Domains

Here we present some additional definitions for special types of integral domains.

Definition 3.6.1. (1) A *Dedekind domain* D is an integral domain such that each nonzero proper ideal A ($\{0\} \neq A \neq R$) can be written uniquely as a product of prime ideals

$$A = P_1 \cdots P_r$$

with each P_i being a prime ideal and the factorization being unique up to ordering.

(2) A *Prüfer ring* R is an integral domain such that

$$A \cdot (B \cap C) = AB \cap AC$$

for all ideals A, B, C in R.

Dedekind domains arise naturally in algebraic number theory. It can be proved that the rings of algebraic integers in any algebraic number field are Dedekind domains (see [53]). If R is a Dedekind domain, it is also a Prüfer Ring. If R is a Prüfer ring and a unique factorization domain, then R is a principal ideal domain. In the next chapter, we will prove a Gaussian theorem which states that if R is a UFD, then the polynomial ring $R[x]$ is also a UFD. If K is a field, we have already seen that $K[x]$ is a UFD. Hence, the polynomial ring in several variables $K[x_1, \ldots, x_n]$ is also a UFD. This fact plays an important role in algebraic geometry.

3.7 Exercises

1. Let R be an integral domain, and let $\pi \in R \setminus (U(R) \cup \{0\})$. Show the following:
   (i) If for each $a \in R$ with $\pi \nmid a$, there exist $\lambda, \mu \in R$ with $\lambda\pi + \mu a = 1$, then π is a prime element of R.
   (ii) Give an example for a prime element π in a UFD R, which does not satisfy the conditions of (i).
2. Let R be a UFD, and let $a_1, \ldots, a_t$ be pairwise coprime elements of R. If $a_1 \cdots a_t$ is an m-th power ($m \in \mathbb{N}$), then all factors a_i are associates of an m-th power. Is each a_i necessarily an m-th power?
3. Decide if the unit group of $\mathbb{Z}[k] = \{a + b\sqrt{k} : a, b \in \mathbb{Z}\}$, $k = 3, 5, 7$, is finite or infinite. For which $a \in \mathbb{Z}$ are $(1 - \sqrt{5})$ and $(a + \sqrt{5})$ associates in $\mathbb{Z}[\sqrt{5}]$?
4. Let $k \in \mathbb{Z}$ and $k \neq x^2$ for all $x \in \mathbb{Z}$. Let $\alpha = a + b\sqrt{k}$ and $\beta = c + d\sqrt{k}$ be elements of $\mathbb{Z}[\sqrt{k}]$, and $N(\alpha) = a^2 - kb^2$, $N(\beta) = c^2 - kd^2$. Show the following:
   (i) The equality of the absolute values of $N(\alpha)$ and $N(\beta)$ is necessary for the association of α and β in $\mathbb{Z}[\sqrt{k}]$. Is this constraint also sufficient?
   (ii) Sufficient for the irreducibility of α in $\mathbb{Z}[\sqrt{k}]$ is the irreducibility of $N(\alpha)$ in $\mathbb{Z}$. Is this also necessary?
5. In general irreducible elements are not prime. Consider the set of complex number given by

$$R = \mathbb{Z}[i\sqrt{5}] = \{x + iy\sqrt{5} : x, y \in \mathbb{Z}\}.$$

Show that they form a subring of $\mathbb{C}$.

6. For an element $x + iy\sqrt{5} \in R$ define its *norm* by

$$N(x + iy\sqrt{5}) = |x + iy\sqrt{5}| = x^2 + 5y^2.$$

   Prove that the norm is multiplicative, that is $N(ab) = N(a)N(b)$.
7. Prove Lemma 3.4.4.
8. Prove that the set of polynomials $R[x]$ with coefficients in a ring R forms a ring.
9. Prove the basic properties of the norm of the Gaussian integers. If $\alpha, \beta \in \mathbb{Z}[i]$, then:
   (i) $N(\alpha)$ is an integer for all $\alpha \in \mathbb{Z}[i]$.
   (ii) $N(\alpha) \geq 0$ for all $\alpha \in \mathbb{Z}[i]$.
   (iii) $N(\alpha) = 0$ if and only if $\alpha = 0$.
   (iv) $N(\alpha) \geq 1$ for all $\alpha \neq 0$.
   (v) $N(\alpha\beta) = N(\alpha)N(\beta)$, that is the norm is multiplicative.

4 Polynomials and Polynomial Rings

4.1 Degrees, Reducibility and Roots

In the last chapter, we saw that if K is a field, then the set of polynomials with coefficients in K, which we denoted $K[x]$, forms a unique factorization domain. In this chapter, we take a more detailed look at polynomials over a general ring R. We then prove that if R is a UFD, then the polynomial ring $R[x]$ is also a UFD. We first take a formal look at polynomials.

Let R be a commutative ring with an identity. Consider the set $\tilde{R}$ of functions f from the nonnegative integers $\overline{N} = \mathbb{N} \cup \{0\}$ into R with only a finite number of values nonzero. That is,

$$\tilde{R} = \{f : \overline{N} \to R : f(n) \neq 0 \text{ for only finitely many } n\}.$$

On $\tilde{R}$, we define the following addition and multiplication:

$$(f + g)(n) = f(n) + g(n)$$
$$(f \cdot g)(n) = \sum_{i+j=n} f(i)g(j).$$

If we let $x = (0, 1, 0, \ldots)$ and identify $(r, 0, \ldots)$ with $r \in R$, then

$$x^0 = (1, 0, \ldots) = 1, \quad \text{and} \quad x^{i+1} = x \cdot x^i.$$

Now if $f = (r_0, r_1, r_2, \ldots)$, then f can be written as

$$f = \sum_{i=0}^{\infty} r_i x^i = \sum_{i=0}^{m} r_i x^i$$

for some $m \geq 0$ since $r_i \neq 0$ for only finitely many i. Furthermore, this presentation is unique. We now call x an *indeterminate* over R, and write each element of $\tilde{R}$ as $f(x) = \sum_{i=0}^{m} r_i x^i$ with $f(x) = 0$ or $r_m \neq 0$. We also now write $R[x]$ for $\tilde{R}$. Each element of $R[x]$ is called a *polynomial* over R. The elements $r_0, \ldots, r_m$ are called the *coefficients* of $f(x)$ with r_m the *leading coefficient*. If $r_m \neq 0$, the non-negative integer m is called the *degree* of $f(x)$, which we denote by $\deg f(x)$. We say that $f(x) = 0$ has degree $-\infty$. The uniqueness of the representation of a polynomial implies that two nonzero polynomials are equal if and only if they have the same degree and exactly the same coefficients. A polynomial of degree 1 is called a *linear polynomial*, whereas one of degree two is a *quadratic polynomial*. The set of polynomials of degree 0, together with 0, form a ring isomorphic to R and, hence, can be identified with R, the constant polynomials. Thus, the ring R embeds in the set of polynomials $R[x]$. The following results are straightforward concerning degree:

https://doi.org/10.1515/9783111142524-004

Lemma 4.1.1. *Let $f(x) \neq 0, g(x) \neq 0 \in R[x]$. Then the following hold:*
(a) $\deg f(x)g(x) \leq \deg f(x) + \deg g(x)$.
(b) $\deg(f(x) \pm g(x)) \leq \max(\deg f(x), \deg g(x))$.

If R is an integral domain, then we have equality in (a).

Theorem 4.1.2. *Let R be a commutative ring with an identity. Then the set of polynomials $R[x]$ forms a ring called the ring of polynomials over R. The ring R identified with 0 and the polynomials of degree 0 naturally embeds into $R[x]$. $R[x]$ is commutative. Furthermore, $R[x]$ is uniquely determined by R and x.*

Proof. Set $f(x) = \sum_{i=0}^{n} r_i x^i$ and $g(x) = \sum_{j=0}^{m} s_j x^j$. The ring properties follow directly by computation. The identification of $r \in R$ with the polynomial $r(x) = r$ provides the embedding of R into $R[x]$. From the definition of multiplication in $R[x]$, if R is commutative, then $R[x]$ is commutative. Note that if R has a multiplicative identity $1 \neq 0$, then this is also the multiplicative identity of $R[x]$.

Finally, if S is a ring that contains R and $a \in S$, then

$$R[a] = \left\{\sum_{i\geq 0} r_i a^i : r_i \in R, \text{ and } r_i \neq 0 \text{ for only a finite number of } i\right\}$$

is a homomorphic image of $R[x]$ via the map

$$\sum_{i\geq 0} r_i x^i \mapsto \sum_{i\geq 0} r_i a^i.$$

Hence, $R[x]$ is uniquely determined by R and x. We remark that $R[a]$ must be commutative. □

If R is an integral domain, then irreducible polynomials are defined as irreducibles in the ring $R[x]$. If R is a field, then $f(x)$ is an *irreducible polynomial* if there is no factorization $f(x) = g(x)h(x)$, where $g(x)$ and $h(x)$ are polynomials of lower degree than $f(x)$. Otherwise, $f(x)$ is called *reducible*. In elementary mathematics, polynomials are considered as functions. We recover that idea via the concept of evaluation.

Definition 4.1.3. Let $f(x) = r_0 + r_1 x + \cdots + r_m x^n$ be a polynomial over a commutative ring R with an identity, and let $c \in R$. Then the element

$$f(c) = r_0 + r_1 c + \cdots + r_n c^n \in R$$

is called the *evaluation* of $f(x)$ at c.

Definition 4.1.4. If $f(x) \in R[x]$ and $f(c) = 0$ for $c \in R$, then c is called a *zero* or a root of $f(x)$ in R.

4.2 Polynomial Rings over Fields

We now restate some of the result of the last chapter for $K[x]$, where K is a field. We then consider some consequences of these results to zeros of polynomials.

Theorem 4.2.1. *If K is a field, then $K[x]$ forms an integral domain. K can be naturally embedded into $K[x]$ by identifying each element of K with the corresponding constant polynomial. The only units in $K[x]$ are the nonzero elements of K.*

Proof. Verification of the basic ring properties is solely computational and is left to the exercises. Since $\deg P(x)Q(x) = \deg P(x) + \deg Q(x)$, it follows that if neither $P(x) \neq 0$, nor $Q(x) \neq 0$, then $P(x)Q(x) \neq 0$. Therefore, $K[x]$ is an integral domain.

If $G(x)$ is a unit in $K[x]$, then there exists an $H(x) \in K[x]$ with $G(x)H(x) = 1$.

From the degrees, we have $\deg G(x) + \deg H(x) = 0$, and since $\deg G(x) \geq 0$, $\deg H(x) \geq 0$. This is possible only if $\deg G(x) = \deg H(x) = 0$. Therefore, $G(x) \in K$. □

Now that we have $K[x]$ as an integral domain, we proceed to show that $K[x]$ is a principal ideal domain and, hence, there is unique factorization into primes. We first repeat the definition of a prime in $K[x]$. If $0 \neq f(x)$ has no nontrivial, nonunit factors (it cannot be factorized into polynomials of lower degree), then $f(x)$ is a *prime* in $K[x]$ or a *prime polynomial*. A prime polynomial is also called an *irreducible polynomial* over K. Clearly, if $\deg g(x) = 1$, then $g(x)$ is irreducible.

The fact that $K[x]$ is a principal ideal domain follows from the division algorithm for polynomials, which is entirely analogous to the division algorithm for integers.

Theorem 4.2.2 (Division algorithm in $K[x]$). *If $0 \neq f(x), 0 \neq g(x) \in K[x]$, then there exist unique polynomials $q(x), r(x) \in K[x]$ such that $f(x) = q(x)g(x) + r(x)$, where $r(x) = 0$, or $\deg r(x) < \deg g(x)$. (The polynomials $q(x)$ and $r(x)$ are called respectively the quotient and remainder.)*

Proof. If $\deg f(x) = 0$ and $\deg g(x) \geq 1$, then we just choose $q(x) = 0$, and $r(x) = f(x)$. If $\deg f(x) = 0 = \deg g(x)$, then $f(x) = f \in K$, and $g(x) = g \in K$, and we choose $q(x) = \frac{f}{g}$ and $r(x) = 0$. Hence, Theorem 4.2.2 is proved for $\deg f(x) = 0$, also certainly the uniqueness statement.

Now, let $n > 0$ and Theorem 4.2.2 be proved for all $f(x) \in K[x]$ with $\deg f(x) < n$. Now, given

$$f(x) = a_n x^n + a_{n-1}x^{n-1} + \cdots + a_1 x + a_0, \quad \text{with } a_n \neq 0, \text{ and}$$
$$g(x) = b_m x^m + b_{m-1}x^{m-1} + \cdots + b_1 x + b_0, \quad \text{with } b_m \neq 0, m \geq 0.$$

If $m > n$, then just choose $q(x) = 0$ and $r(x) = f(x)$.

Now, finally, let $0 \leq m \leq n$. We define

$$h(x) = f(x) - \frac{a_n}{b_m} x^{n-m} g(x).$$

We have $\deg h(x) < n$. Hence, by induction assumption, there are $q_1(x)$ and $r(x)$ with $h(x) = q_1(x)g(x) + r(x)$ and $\deg r(x) < \deg g(x)$. Then

$$\begin{aligned} f(x) &= h(x) + \frac{a_n}{b_m}x^{n-m}g(x) \\ &= \left(\frac{a_n}{b_m}x^{n-m} + q_1(x)\right)g(x) + r(x) \\ &= q(x)g(x) + r(x) \quad \text{with } q(x) = \frac{a_n}{b_m}x^{n-m} + q_1(x), \end{aligned}$$

which proves the existence.

We now show the uniqueness. Let

$$\begin{aligned} f(x) &= q_1(x)g(x) + r_1(x) \\ &= q_2(x)g(x) + r_2(x), \end{aligned}$$

with

$$\deg r_1(x) < \deg g(x), \quad \text{and} \quad \deg r_2(x) < \deg g(x).$$

Assume $r_1(x) \neq r_2(x)$. Let $\deg r_1(x) \geq \deg r_2(x)$. We get

$$(q_2(x) - q_1(x))g(x) = r_1(x) - r_2(x),$$

which gives a contradiction because $\deg(r_1(x) - r_2(x)) < \deg g(x)$, and $q_2(x) - q_1(x) \neq 0$ if $r_1(x) \neq r_2(x)$. Therefore, $r_1(x) = r_2(x)$, and furthermore $q_1(x) = q_2(x)$ because $K[x]$ is an integral domain. □

Example 4.2.3. Let $f(x) = 2x^3 + x^2 - 5x + 3$, $g(x) = x^2 + x + 1$. Then

$$\frac{2x^3 + x^2 - 5x + 3}{x^2 + x + 1} = 2x - 1 \quad \text{with remainder } -6x + 4.$$

Hence, $q(x) = 2x - 1$, $r(x) = -6x + 4$, and

$$2x^3 + x^2 - 5x + 3 = (2x - 1)(x^2 + x + 1) + (-6x + 4).$$

Theorem 4.2.4. *Let K be a field. Then the polynomial ring $K[x]$ is a principal ideal domain, and hence a unique factorization domain.*

We now give some consequences relative to zeros of polynomials in $K[x]$.

Theorem 4.2.5. *If $f(x) \in K[x]$ and $c \in K$ with $f(c) = 0$, then*

$$f(x) = (x - c)h(x),$$

where $\deg h(x) < \deg f(x)$.

Proof. Divide $f(x)$ by $x - c$. Then by the division algorithm, we have

$$f(x) = (x - c)h(x) + r(x),$$

where $r(x) = 0$, or $\deg r(x) < \deg(x - c) = 1$. Hence, if $r(x) \neq 0$, then $r(x)$ is a polynomial of degree 0, that is, a constant polynomial, and thus $r(x) = r$ for $r \in K$. Hence, we have

$$f(x) = (x - c)h(x) + r.$$

This implies that

$$0 = f(x) = 0h(c) + r = r$$

and, therefore, $r = 0$, and $f(x) = (x - c)h(x)$. Since $\deg(x - c) = 1$, we must have that $\deg h(x) < \deg f(x)$. □

If $f(x) = (x - c)^k h(x)$ for some $k \geq 1$ with $h(c) \neq 0$, then c is called a zero of order k.

Theorem 4.2.6. *Let $f(x) \in K[x]$ with degree 2 or 3. Then f is irreducible if and only if $f(x)$ does not have a zero in K.*

Proof. Suppose that $f(x)$ is irreducible of degree 2 or 3. If $f(x)$ has a zero c, then from Theorem 4.2.5, we have $f(x) = (x - c)h(x)$ with $h(x)$ of degree 1 or 2. Therefore, $f(x)$ is reducible a contradiction and, hence, $f(x)$ cannot have a zero.

From Theorem 4.2.5, if $f(x)$ has a zero and is of degree greater than 1, then $f(x)$ is reducible.

If $f(x)$ is reducible, then $f(x) = g(x)h(x)$ with $\deg g(x) = 1$ and, hence, $f(x)$ has a zero in K. □

4.3 Polynomial Rings over Integral Domains

Here we consider $R[x]$ where R is an integral domain.

Definition 4.3.1. Let R be an integral domain. Then $a_1, a_2, \ldots, a_n \in R$ are *coprime* if the set of all common divisors of $a_1, a_2, \ldots, a_n$ consists only of units.

Notice, for example, that this concept depends on the ring R. For example, 6 and 9 are not coprime over the integers $\mathbb{Z}$ since $3|6$ and $3|9$ and 3 is not a unit. However, 6 and 9 are coprime over the rationals $\mathbb{Q}$. Here, 3 is a unit.

Definition 4.3.2. Let $f(x) = \sum_{i=0}^{n} r_i x^i \in R[x]$, where R is an integral domain. Then $f(x)$ is a *primitive polynomial* or just *primitive* if $r_0, r_1, \ldots, r_n$ are coprime in R.

Theorem 4.3.3. *Let R be an integral domain. Then the following hold:*

(a) *The units of $R[x]$ are the units of R.*
(b) *If p is a prime element of R, then p is a prime element of $R[x]$.*

Proof. If $r \in R$ is a unit, then since R embeds into $R[x]$, it follows that r is also a unit in $R[x]$. Conversely, suppose that $h(x) \in R[x]$ is a unit. Then there is a $g(x)$ such that $h(x)g(x) = 1$. Hence, $\deg f(x) + \deg g(x) = \deg 1 = 0$. Since degrees are nonnegative integers, it follows that $\deg f(x) = \deg g(x) = 0$ and, hence, $f(x) \in R$.

Now suppose that p is a prime element of R. Then $p \neq 0$, and pR is a prime ideal in R. We must show that $pR[x]$ is a prime ideal in $R[x]$. Consider the map

$$\tau : R[x] \to (R/pR)[x] \quad \text{given by}$$

$$\tau\left(\sum_{i=0}^{n} r_i x^i\right) = \sum_{i=0}^{n} (r_i + pR)x^i.$$

Then τ is an epimorphism with kernel $pR[x]$. Since pR is a prime ideal, we know that R/pR is an integral domain. It follows that $(R/pR)[x]$ is also an integral domain. Hence, $pR[x]$ must be a prime ideal in $R[x]$, and therefore p is also a prime element of $R[x]$. □

Recall that each integral domain R can be embedded into a unique field of fractions K. We can use results on $K[x]$ to deduce some results in $R[x]$.

Lemma 4.3.4. *If K is a field, then each nonzero $f(x) \in K[x]$ is a primitive.*

Proof. Since K is a field, each nonzero element of K is a unit. Therefore, the only common divisors of the coefficients of $f(x)$ are units and, hence, $f(x) \in K[x]$ is primitive. □

Theorem 4.3.5. *Let R be an integral domain. Then each irreducible $f(x) \in R[x]$ of degree > 0 is primitive.*

Proof. Let $f(x)$ be an irreducible polynomial in $R[x]$, and let $r \in R$ be a common divisor of the coefficients of $f(x)$. Then $f(x) = rg(x)$, where $g(x) \in R[x]$.

Then $\deg f(x) = \deg g(x) > 0$, so $g(x) \notin R$. Since the units of $R[x]$ are the units of R, it follows that $g(x)$ is not a unit in $R[x]$. Since $f(x)$ is irreducible, it follows that r must be a unit in $R[x]$ and, hence, r is a unit in R. Therefore, $f(x)$ is primitive. □

Theorem 4.3.6. *Let R be an integral domain and K its field of fractions. If $f(x) \in R[x]$ is primitive and irreducible in $K[x]$, then $f(x)$ is irreducible in $R[x]$.*

Proof. Suppose that $f(x) \in R[x]$ is primitive and irreducible in $K[x]$, and suppose that $f(x) = g(x)h(x)$, where $g(x), h(x) \in R[x] \subset K[x]$. Since $f(x)$ is irreducible in $K[x]$, either $g(x)$ or $h(x)$ must be a unit in $K[x]$. Without loss of generality, suppose that $g(x)$ is a unit in $K[x]$. Then $g(x) = g \in K$. But $g(x) \in R[x]$, and $K \cap R[x] = R$.

Hence, $g \in R$. Then g is a divisor of the coefficients of $f(x)$, and as $f(x)$ is primitive, $g(x)$ must be a unit in R and, therefore, also a unit in $R[x]$. Therefore, $f(x)$ is irreducible in $R[x]$. □

4.4 Polynomial Rings over Unique Factorization Domains

In this section, we prove that if R is a UFD, then the polynomial ring $R[x]$ is also a UFD. We first need the following due to Gauss:

Theorem 4.4.1 (Gauss' lemma). *Let R be a UFD and $f(x)$, $g(x)$ primitive polynomials in $R[x]$. Then their product $f(x)g(x)$ is also primitive.*

Proof. Let R be a UFD and $f(x), g(x)$ primitive polynomials in $R[x]$. Suppose that $f(x)g(x)$ is not primitive. Then there is a prime element $p \in R$ that divides each of the coefficients of $f(x)g(x)$. Then $p|f(x)g(x)$. Since prime elements of R are also prime elements of $R[x]$, it follows that p is also a prime element of $R[x]$ and, hence, $p|f(x)$, or $p|g(x)$. Therefore, either $f(x)$ or $g(x)$ is not primitive, giving a contradiction. □

Theorem 4.4.2. *Let R be a UFD and K its field of fractions.*

(a) *If $g(x) \in K[x]$ is nonzero, then there is a nonzero $a \in K$ such that $ag(x) \in R[x]$ is primitive.*
(b) *Let $f(x), g(x) \in R[x]$ with $g(x)$ primitive and $f(x) = ag(x)$ for some $a \in K$. Then $a \in R$.*
(c) *If $f(x) \in R[x]$ is nonzero, then there is a $b \in R$ and a primitive $g(x) \in R[x]$ such that $f(x) = bg(x)$.*

Proof. (a) Suppose that $g(x) = \sum_{i=0}^{n} a_i x^i$ with $a_i = \frac{r_i}{s_i}$, $r_i, s_i \in R$. Set $s = s_0 s_1 \cdots s_n$. Then $sg(x)$ is a nonzero element of $R[x]$. Let d be a greatest common divisor of the coefficients of $sg(x)$. If we set $a = \frac{s}{d}$, then $ag(x)$ is primitive.

(b) For $a \in K$, there are coprime $r, s \in R$ satisfying $a = \frac{r}{s}$. Suppose that $a \notin R$. Then there is a prime element $p \in R$ dividing s. Since $g(x)$ is primitive, p does not divide all the coefficients of $g(x)$. However, we also have $f(x) = ag(x) = \frac{r}{s}g(x)$. Hence, $sf(x) = rg(x)$, where $p|s$ and p does not divide r. Therefore, p divides all the coefficients of $g(x)$ and, hence, $a \in R$.

(c) From part (a), there is a nonzero $a \in K$ such that $af(x)$ is primitive in $R[x]$. Then $f(x) = a^{-1}(af(x))$. From part (b), we must have $a^{-1} \in R$. Set $g(x) = af(x)$ and $b = a^{-1}$. □

Theorem 4.4.3. *Let R be a UFD and K its field of fractions. Let $f(x) \in R[x]$ be a polynomial of degree ≥ 1.*

(a) *If $f(x)$ is primitive and $f(x)|g(x)$ in $K[x]$, then $f(x)$ divides $g(x)$ also in $R[x]$.*
(b) *If $f(x)$ is irreducible in $R[x]$, then it is also irreducible in $K[x]$.*
(c) *If $f(x)$ is primitive and a prime element of $K[x]$, then $f(x)$ is also a prime element of $R[x]$.*

Proof. (a) Suppose that $g(x) = f(x)h(x)$ with $h(x) \in K[x]$. From Theorem 4.4.2 part (a), there is a nonzero $a \in K$ such that $h_1(x) = ah(x)$ is primitive in $R[x]$. Hence, $g(x) = \frac{1}{a}(f(x)h_1(x))$. From Gauss' lemma $f(x)h_1(x)$ is primitive in $R[x]$. Therefore, from Theorem 4.4.2 part (b), we have $\frac{1}{a} \in R$. It follows that $f(x)|g(x)$ in $R[x]$.

(b) Suppose that $g(x) \in K[x]$ is a factor of $f(x)$. From Theorem 4.4.2 part (a), there is a nonzero $a \in K$ with $g_1(x) = ag(x)$ primitive in $R[x]$. Since a is a unit in K, it follows that

$$g(x)|f(x) \quad \text{in } K[x] \quad \text{implies} \quad g_1(x)|f(x) \quad \text{in } K[x]$$

and, hence, since $g_1(x)$ is primitive

$$g_1(x)|f(x) \quad \text{in } R[x].$$

However, by assumption, $f(x)$ is irreducible in $R[x]$. This implies that either $g_1(x)$ is a unit in R, or $g_1(x)$ is an associate of $f(x)$.

If $g_1(x)$ is a unit, then $g_1 \in K$, and $g_1 = ga$. Hence, $g \in K$; that is, $g = g(x)$ is a unit.

If $g_1(x)$ is an associate of $f(x)$, then $f(x) = bg(x)$, where $b \in K$ since $g_1(x) = ag(x)$ with $a \in K$. Combining these, it follows that $f(x)$ has only trivial factors in $K[x]$, and since—by assumption—$f(x)$ is nonconstant, it follows that $f(x)$ is irreducible in $K[x]$.

(c) Suppose that $f(x)|g(x)h(x)$ with $g(x), h(x) \in R[x]$. Since $f(x)$ is a prime element in $K[x]$, we have that $f(x)|g(x)$ or $f(x)|h(x)$ in $K[x]$. From part (a), we have $f(x)|g(x)$ or $f(x)|h(x)$ in $R[x]$ implying that $f(x)$ is a prime element in $R[x]$. □

We can now state and prove our main result.

Theorem 4.4.4 (Gauss). *Let R be a UFD. Then the polynomial ring $R[x]$ is also a UFD.*

Proof. By induction, on degree, we show that each nonunit $f(x) \in R[x], f(x) \neq 0$, is a product of prime elements. Since R is an integral domain, so is $R[x]$. Therefore, the fact that $R[x]$ is a UFD then follows from Theorem 3.3.3.

If $\deg f(x) = 0$, then $f(x) = f$ is a nonunit in R. Since R is a UFD, f is a product of prime elements in R. However, from Theorem 4.3.3, each prime factor is then also prime in $R[x]$. Therefore, $f(x)$ is a product of prime elements.

Now suppose $n > 0$ and that the claim is true for all polynomials $f(x)$ of degree $< n$. Let $f(x)$ be a polynomial of degree $n > 0$. From Theorem 4.4.2 (c), there is an $a \in R$ and a primitive $h(x) \in R[x]$ satisfying $f(x) = ah(x)$. Since R is a UFD, the element a is a product of prime elements in R, or a is a unit in R. Since the units in $R[x]$ are the units in R, and a prime element in R is also a prime element in $R[x]$, it follows that a is a product of prime elements in $R[x]$, or a is a unit in $R[x]$. Let K be the field of fractions of R. Then $K[x]$ is a UFD. Hence, $h(x)$ is a product of prime elements of $K[x]$.

Let $p(x) \in K[x]$ be a prime divisor of $h(x)$. From Theorem 4.4.2, we can assume by multiplication of field elements that $p(x) \in R[x]$, and $p(x)$ is primitive.

From Theorem 4.4.2 (c), it follows that $p(x)$ is a prime element of $R[x]$. Furthermore, from Theorem 4.4.3 (a), $p(x)$ is a divisor of $h(x)$ in $R[x]$. Therefore,

$$f(x) = ah(x) = ap(x)g(x) \in R[x],$$

where the following hold:
(1) a is a product of prime elements of $R[x]$, or a is a unit in $R[x]$,
(2) $\deg p(x) > 0$, since $p(x)$ is a prime element in $K[x]$,
(3) $p(x)$ is a prime element in $R[x]$, and
(4) $\deg g(x) < \deg f(x)$ since $\deg p(x) > 0$.

By our inductive hypothesis, we have then that $g(x)$ is a product of prime elements in $R[x]$, or $g(x)$ is a unit in $R[x]$. Therefore, the claim holds for $f(x)$, and therefore holds for all $f(x)$ by induction. □

If $R[x]$ is a polynomial ring over R, we can form a polynomial ring in a new indeterminate y over this ring to form $(R[x])[y]$. It is straightforward that $(R[x])[y]$ is isomorphic to $(R[y])[x]$. We denote both of these rings by $R[x,y]$ and consider this as the ring of polynomials in two commuting variables x, y with coefficients in R.

If R is a UFD, then from Theorem 4.4.4, $R[x]$ is also a UFD. Hence, $R[x,y]$ is also a UFD. Inductively then, the ring of polynomials in n commuting variables $R[x_1, x_2, \ldots, x_n]$ is also a UFD.

Here, the ring $R[x_1, \ldots, x_n]$ is inductively given by $R[x_1, \ldots, x_n] = (R[x_1, \ldots, x_{n-1}])[x_n]$ if $n > 2$.

Corollary 4.4.5. *If R is a UFD, then the polynomial ring in n commuting variables $R[x_1, \ldots, x_n]$ is also a UFD.*

We now give a condition for a polynomial in $R[x]$ to have a zero in $K[x]$, where K is the field of fractions of R.

Theorem 4.4.6. *Let R be a UFD and K its field of fractions. Let*

$$f(x) = x^n + r_{n-1}x^{n-1} + \cdots + r_0 \in R[x].$$

Suppose that $\beta \in K$ is a zero of $f(x)$. Then β is in R and is a divisor of r_0.

Proof. Let $\beta = \frac{r}{s}$, where $s \neq 0$, and $r, s \in R$ and r, s are coprime. Now

$$f\left(\frac{r}{s}\right) = 0 = \frac{r^n}{s^n} + r_{n-1}\frac{r^{n-1}}{s^{n-1}} + \cdots + r_0.$$

Hence, it follows that s must divide r^n. Since r and s are coprime, s must be a unit, and then, without loss of generality, we may assume that $s = 1$. Then $\beta \in R$, and

$$r(r^{n-1} + \cdots + r_1) = -r_0,$$

and so $r | a_0$. □

Note that since $\mathbb{Z}$ is a UFD, Gauss' theorem implies that $\mathbb{Z}[x]$ is also a UFD. However, $\mathbb{Z}[x]$ is not a principal ideal domain. For example, the set of integral polynomials with even constant term is an ideal, but not principal. We leave the verification to the exercises. On the other hand, we saw that if K is a field, $K[x]$ is a PID. The question arises as to when $R[x]$ actually is a principal ideal domain. It turns out to be precisely when R is a field.

Theorem 4.4.7. *Let R be a commutative ring with an identity. Then the following are equivalent:*
(a) *R is a field.*
(b) *$R[x]$ is Euclidean.*
(c) *$R[x]$ is a principal ideal domain.*

Proof. From Section 4.2, we know that (a) implies (b), which in turn implies (c). Therefore, we must show that (c) implies (a). Assume then that $R[x]$ is a principal ideal domain. Define the map

$$\tau : R[x] \to R$$

by

$$\tau(f(x)) = f(0).$$

It is easy to see that τ is a ring homomorphism with $R[x]/\ker(\tau) \cong R$. Therefore, $\ker(\tau) \neq R[x]$. Since $R[x]$ is a principal ideal domain, it is an integral domain. It follows that $\ker(\tau)$ must be a prime ideal since the quotient ring is an integral domain. However, since $R[x]$ is a principal ideal domain, prime ideals are maximal ideals; hence, $\ker(\tau)$ is a maximal ideal by Theorem 3.2.7. Therefore, $R \cong R[x]/\ker(\tau)$ is a field. □

We now consider the relationship between irreducibles in $R[x]$ for a general integral domain and irreducibles in $K[x]$, where K is its field of fractions. This is handled by the next result called Eisenstein's criterion.

Theorem 4.4.8 (Eisenstein's criterion). *Let R be an integral domain and K its field of fractions. Let $f(x) = \sum_{i=0}^{n} a_i x^i \in R[x]$ of degree $n > 0$. Let p be a prime element of R satisfying the following:*
(1) *$p | a_i$ for $i = 0, \dots, n-1$.*
(2) *p does not divide a_n.*
(3) *p^2 does not divide a_0.*

Then the following hold:
(a) *If $f(x)$ is primitive, then $f(x)$ is irreducible in $R[x]$.*
(b) *Suppose that R is a UFD. Then $f(x)$ is also irreducible in $K[x]$.*

Proof. (a) Suppose that $f(x) = g(x)h(x)$ with $g(x), h(x) \in R[x]$. Suppose that

$$g(x) = \sum_{i=0}^{k} b_i x^i, \quad b_k \neq 0 \quad \text{and} \quad h(x) = \sum_{j=0}^{l} c_j x^j, \quad c_l \neq 0.$$

Then $a_0 = b_0c_0$. Now $p|a_0$, but p^2 does not divide a_0. This implies that either p does not divide b_0, or p doesn't divide c_0. Without loss of generality, assume that $p|b_0$ and p does not divide c_0.

Since $a_n = b_kc_l$, and p does not divide a_n, it follows that p does not divide b_k. Let b_j be the first coefficient of $g(x)$, which is not divisible by p. Consider

$$a_j = b_jc_0 + \cdots + b_0c_j,$$

where everything after the first term is divisible by p. Since p does not divide both b_j and c_0, it follows that p does not divide b_jc_0. Therefore, p does not divide a_j, which implies that $j = n$. Then from $j \leq k \leq n$, it follows that $k = n$.

Therefore, $\deg g(x) = \deg f(x)$ and, hence, $\deg h(x) = 0$. Thus, $h(x) = h \in R$. Then from $f(x) = hg(x)$ with f primitive, it follows that h is a unit and, therefore, $f(x)$ is irreducible.

(b) Suppose that $f(x) = g(x)h(x)$ with $g(x), h(x) \in R[x]$. The fact that $f(x)$ was primitive was only used in the final part of part (a). Therefore, by the same arguments as in part (a), we may assume—without loss of generality—that $h \in R \subset K$. Therefore, $f(x)$ is irreducible in $K[x]$. □

Following are some examples:

Example 4.4.9. Let $R = \mathbb{Z}$ and p a prime number. Suppose that n, m are integers such that $n \geq 1$ and p does not divide m. Then $x^n \pm pm$ is irreducible in $\mathbb{Z}[x]$ and $\mathbb{Q}[x]$. In particular, $(pm)^{\frac{1}{n}}$ is irrational.

Example 4.4.10. Let $R = \mathbb{Z}$ and p a prime number. Consider the polynomial

$$\Phi_p(x) = \frac{x^p - 1}{x - 1} = x^{p-1} + x^{p-2} + \cdots + 1.$$

Since all the coefficients of $\Phi_p(x)$ are equal to 1, Eisenstein's criterion is not directly applicable. However, the fact that $\Phi_p(x)$ is irreducible implies that for any integer a, the polynomial $\Phi_p(x + a)$ is also irreducible in $\mathbb{Z}[x]$. It follows that

$$\begin{aligned}\Phi_p(x+1) &= \frac{(x+1)^p - 1}{(x+1) - 1} = \frac{x^p + \binom{p}{1}x^{p-1} + \cdots + \binom{p}{p-1}x + 1^p - 1}{x} \\ &= x^{p-1} + \binom{p}{1}x^{p-2} + \cdots + \binom{p}{p-1}.\end{aligned}$$

Now $p|\binom{p}{i}$ for $1 \leq i \leq p - 1$ (see exercises) and, moreover, $\binom{p}{p-1} = p$ is not divisible by p^2. Therefore, we can apply the Eisenstein criterion to conclude that $\Phi_p(x)$ is irreducible in $\mathbb{Z}[x]$ and $\mathbb{Q}[x]$.

Theorem 4.4.11. *Let R be a UFD and K its field of fractions. Let $f(x) = \sum_{i=0}^{n} a_i x^i \in R[x]$ be a polynomial of degree ≥ 1. Let P be a prime ideal in R with $a_n \notin P$. Let $\overline{R} = R/P$, and let $\alpha : R[x] \to \overline{R}[x]$ be defined by*

$$\alpha\left(\sum_{i=0}^{m} r_i x^i\right) = \sum_{i=0}^{m} (r_i + P) x^i.$$

α is an epimorphism. Then if $\alpha(f(x))$ is irreducible in $\overline{R}[x]$, then $f(x)$ is irreducible in $K[x]$.

Proof. By Theorem 4.4.3, there exists an $a \in R$ and a primitive $g(x) \in R[x]$ satisfying $f(x) = ag(x)$. Since $a_n \notin P$, we have that $\alpha(a) \neq 0$. Furthermore, the highest coefficient of $g(x)$ is also not an element of P. If $\alpha(g(x))$ is reducible, then $\alpha(f(x))$ is also reducible. Thus, $\alpha(g(x))$ is irreducible. However, from Theorem 4.4.4, $g(x)$ is irreducible in $K[x]$. Therefore, $f(x) = ag(x)$ is also irreducible in $K[x]$. Therefore, to prove the theorem, it suffices to consider the case where $f(x)$ is primitive in $R[x]$.

Now suppose that $f(x)$ is primitive. We show that $f(x)$ is irreducible in $R[x]$.

Suppose that $f(x) = g(x)h(x)$, $g(x), h(x) \in R[x]$ with $h(x)$, $g(x)$ nonunits in $R[x]$. Since $f(x)$ is primitive, $g, h \notin R$. Therefore, $\deg g(x) < \deg f(x)$, and $\deg h(x) < \deg f(x)$.

Now we have $\alpha(f(x)) = \alpha(g(x))\alpha(h(x))$. Since P is a prime ideal, R/P is an integral domain. Therefore, in $\overline{R}[x]$ we have

$$\deg \alpha(g(x)) + \deg \alpha(h(x)) = \deg \alpha(f(x)) = \deg f(x)$$

since $a_n \notin P$. Since R is a UFD, it has no zero divisors. Therefore,

$$\deg f(x) = \deg g(x) + \deg h(x).$$

Now

$$\deg \alpha(g(x)) \leq \deg g(x)$$
$$\deg \alpha(h(x)) \leq \deg h(x).$$

Therefore, $\deg \alpha(g(x)) = \deg g(x)$, and $\deg \alpha(h(x)) = \deg h(x)$. Therefore, $\alpha(f(x))$ is reducible, and we have a contradiction. □

It is important to note that $\alpha(f(x))$, being reducible, does not imply that $f(x)$ is reducible. For example, $f(x) = x^2 + 1$ is irreducible in $\mathbb{Z}[x]$. However, in $\mathbb{Z}_2[x]$, we have

$$x^2 + 1 = (x + 1)^2$$

and, hence, $f(x)$ is reducible in $\mathbb{Z}_2[x]$.

Example 4.4.12. Let $f(x) = x^5 - x^2 + 1 \in \mathbb{Z}[x]$. Choose $P = 2\mathbb{Z}$ so that

$$\alpha(f(x)) = x^5 + x^2 + 1 \in \mathbb{Z}_2[x].$$

Suppose that in $\mathbb{Z}_2[x]$, we have $a(f(x)) = g(x)h(x)$. Without loss of generality, we may assume that $g(x)$ is of degree 1 or 2.

If $\deg g(x) = 1$, then $a(f(x))$ has a zero c in $\mathbb{Z}_2[x]$. The two possibilities for c are $c = 0$, or $c = 1$. Then the following hold:

$$\text{If } c = 0, \quad \text{then } 0 + 0 + 1 = 1 \neq 0.$$

$$\text{If } c = 1, \quad \text{then } 1 + 1 + 1 = 1 \neq 0.$$

Hence, the degree of $g(x)$ cannot be 1.

Suppose $\deg g(x) = 2$. The polynomials of degree 2 over $\mathbb{Z}_2[x]$ have the form

$$x^2 + x + 1, \quad x^2 + x, \quad x^2 + 1, \quad x^2.$$

The last three, $x^2 + x$, $x^2 + 1$, x^2 all have zeros in $\mathbb{Z}_2[x]$. Therefore, they cannot divide $a(f(x))$. Therefore, $g(x)$ must be $x^2 + x + 1$. Applying the division algorithm, we obtain

$$a(f(x)) = (x^3 + x^2)(x^2 + x + 1) + 1$$

and, therefore, $x^2 + x + 1$ does not divide $a(f(x))$. It follows that $a(f(x))$ is irreducible, and from the previous theorem, $f(x)$ must be irreducible in $\mathbb{Q}[x]$.

4.5 Exercises

1. For which $a, b \in \mathbb{Z}$ does the polynomial $x^2 + 3x + 1$ divide the polynomial

$$x^3 + x^2 + ax + b?$$

2. Let $a + bi \in \mathbb{C}$ be a zero of $f(x) \in \mathbb{R}[x]$. Show that also $a - ib$ is a zero of $f(x)$.
3. Determine all quadratic irreducible polynomials over $\mathbb{R}$.
4. Let R be an integral domain, $I \triangleleft R$ an ideal, and $f \in R[x]$ a monic polynomial. Define $(R/I)[x]$ by the mapping $R[x] \rightarrow (R/I)[x], f = \sum a_i x^i \mapsto \bar{f} = \sum \bar{a}_i x^i$, where $\bar{a} := a + I$. Show, if $(R/I)[x]$ is irreducible, then $f \in R[x]$ is also irreducible.
5. Decide if the following polynomials $f \in R[x]$ are irreducible:
   (i) $f(x) = x^3 + 2x^2 + 3, R = \mathbb{Z}$.
   (ii) $f(x) = x^5 - 2x + 1, R = \mathbb{Q}$.
   (iii) $f(x) = 3x^4 + 7x^2 + 14x + 7, R = \mathbb{Q}$.
   (iv) $f(x) = x^7 + (3 - i)x^2 + (3 + 4i)x + 4 + 2i, R = \mathbb{Z}[i]$.
   (v) $f(x) = x^4 + 3x^3 + 2x^2 + 3x + 4, R = \mathbb{Q}$.
   (vi) $f(x) = 8x^3 - 4x^2 + 2x - 1, R = \mathbb{Z}$.
6. Let R be an integral domain with characteristic 0, let $k \geq 1$ and $\alpha \in R$. In $R[x]$, define the derivatives $f^{(k)}(x), k = 0, 1, 2, \ldots$, of a polynomial $f(x) \in R[x]$ by

$$f^{0}(x) := f(x),$$
$$f^{(k)}(x) := f^{(k-1)'}(x).$$

Show that α is a zero of order k of the polynomial $f(x) \in R[x]$, if $f^{(k-1)}(\alpha) = 0$, but $f^{(k)}(\alpha) \neq 0$.

7. Prove that the set of integral polynomials with even constant term is an ideal, but not principal.
8. Prove that $p | \binom{p}{i}$ for $1 \leq i \leq p-1$.

5 Field Extensions

5.1 Extension Fields and Finite Extensions

Much of algebra in general arose from the theory of equations, specifically polynomial equations. As discovered by Galois and Abel, the solutions of polynomial equations over fields is intimately tied to the theory of field extensions. This theory eventually blossoms into Galois Theory. In this chapter, we discuss the basic material concerning field extensions.

Recall that if L is a field and $K \subset L$ is also a field under the same operations as L, then K is called a *subfield* of L. If we view this situation from the viewpoint of K, we say that L is an *extension field* or *field extension* of K. If K, L are fields with $K \subset L$, we always assume that K is a subfield of L.

Definition 5.1.1. If K, L are fields with $K \subset L$, then we say that L is a *field extension* or *extension field* of K. We denote this by $L|K$.

Note that this is equivalent to having a field monomorphism

$$i : K \to L$$

and then identifying K and $i(K)$.

As examples, we have that $\mathbb{R}$ is an extension field of $\mathbb{Q}$, and $\mathbb{C}$ is an extension field of both $\mathbb{C}$ and $\mathbb{Q}$. If K is any field then the ring of polynomials $K[x]$ over K is an integral domain. Let $K(x)$ be the field of fractions of $K[x]$. This is called the *field of rational functions* over K. Since K can be considered as part of $K[x]$, it follows that $K \subset K(x)$ and, hence, $K(x)$ is an extension field of K.

A crucial concept is that of the degree of a field extension. Recall that a *vector space* V over a field K consists of an Abelian group V together with scalar multiplication from K satisfying the following:

(1) $fv \in V$ if $f \in K, v \in V$.
(2) $f(u + v) = fu + fv$ for $f \in K, u, v \in V$.
(3) $(f + g)v = fv + gv$ for $f, g \in K, v \in V$.
(4) $(fg)v = f(gv)$ for $f, g \in K, v \in V$.
(5) $1v = v$ for $v \in V$.

Notice that if K is a subfield of L, then products of elements of L with elements of K are still in L. Since L is an Abelian group under addition, L can be considered as a vector space over K. Thus, any extension field is a vector space over any of its subfields. Using this, we define the degree $|L : K|$ of an extension $K \subset L$ as the dimension $\dim_K(L)$ of L as a vector space over K. We call L a finite extension of K if $|L : K| < \infty$.

https://doi.org/10.1515/9783111142524-005

Definition 5.1.2. If L is an extension field of K, then the *degree of the extension $L|K$* is defined as the dimension, $\dim_K(L)$, of L, as a vector space over K. We denote the degree by $|L : K|$. The field extension $L|K$ is a *finite extension* if the degree $|L : K|$ is finite.

Lemma 5.1.3. $|\mathbb{C} : \mathbb{R}| = 2$, *but* $|\mathbb{R} : \mathbb{Q}| = \infty$.

Proof. Every complex number can be written uniquely as $a + ib$, where $a, b \in \mathbb{R}$. Hence, the elements $1, i$ constitute a basis for $\mathbb{C}$ over $\mathbb{R}$ and, therefore, the dimension is 2. That is, $|\mathbb{C} : \mathbb{R}| = 2$.

The fact that $|\mathbb{R} : \mathbb{Q}| = \infty$ depends on the existence of *transcendental numbers*. An element $r \in \mathbb{R}$ is *algebraic* (over $\mathbb{Q}$) if it satisfies some nonzero polynomial with coefficients from $\mathbb{Q}$. That is, $P(r) = 0$, where

$$0 \neq P(x) = a_0 + a_1x + \cdots + a_nx^n \quad \text{with } a_i \in \mathbb{Q}.$$

Any $q \in \mathbb{Q}$ is algebraic since if $P(x) = x - q$, then $P(q) = 0$. However, many irrationals are also algebraic. For example, $\sqrt{2}$ is algebraic since $x^2 - 2 = 0$ has $\sqrt{2}$ as a zero. An element $r \in \mathbb{R}$ is *transcendental* if it is not algebraic.

In general, it is very difficult to show that a particular element is transcendental. However, there are uncountably many transcendental elements (see exercises). Specific examples are e and π. We will give a proof of their transcendence in Chapter 20.

Since e is transcendental, for any natural number n, the set $\{1, e, e^2, \dots, e^n\}$ must be independent over $\mathbb{Q}$, for otherwise there would be a polynomial that e would satisfy. Therefore, we have infinitely many independent vectors in $\mathbb{R}$ over $\mathbb{Q}$, which would be impossible if $\mathbb{R}$ had finite degree over $\mathbb{Q}$. □

Lemma 5.1.4. *If K is any field, then* $|K(x) : K| = \infty$.

Proof. For any n, the elements $1, x, x^2, \dots, x^n$ are independent over K. Therefore, as in the proof of Lemma 5.1.3, $K(x)$ must be infinite-dimensional over K. □

If $L|K$ and $L_1|K_1$ are field extensions, then they are *isomorphic* field extensions if there exists a field isomorphism $f : L \to L_1$ such that $f_{|K}$ is an isomorphism from K to K_1.

Suppose that $K \subset L \subset M$ are fields. Below we show that the degrees multiply. In this situation, where $K \subset L \subset M$, we call L an *intermediate field*.

Theorem 5.1.5. *Let K, L, M be fields with $K \subset L \subset M$. Then*

$$|M : K| = |M : L||L : K|.$$

Note that $|M : K| = \infty$ if and only if either $|M : L| = \infty$, or $|L : K| = \infty$.

Proof. Let $\{x_i : i \in I\}$ be a basis for L as a vector space over K, and let $\{y_j : j \in J\}$ be a basis for M as a vector space over L. To prove the result, it is sufficient to show that the set

$$B = \{x_iy_j : i \in I, j \in J\}$$

is a basis for M as a vector space over K. To show this, we must show that B is a linearly independent set over K, and that B spans M.

Suppose that

$$\sum_{i,j} k_{ij} x_i y_j = 0 \quad \text{where } k_{ij} \in K.$$

We can then write this sum as

$$\sum_j \Big(\sum_i k_{ij} x_i \Big) y_j = 0.$$

But $\sum_i k_{ij} x_i \in L$. Since $\{y_j : j \in J\}$ is a basis for M over L, the y_j are independent over L; hence, for each j, we get $\sum_i k_{ij} x_i = 0$. Now since $\{x_i : i \in I\}$ is a basis for L over K, it follows that the x_i are linearly independent, and since for each j we have $\sum_i k_{ij} x_i = 0$, it must be that $k_{ij} = 0$ for all i and for all j. Therefore, the set B is linearly independent over K.

Now suppose that $m \in M$. Then since $\{y_j : j \in J\}$ spans M over L, we have

$$m = \sum_j c_j y_j \quad \text{with } c_j \in L.$$

However, $\{x_i : i \in I\}$ spans L over K, and so for each c_j, we have

$$c_j = \sum_i k_{ij} x_i \quad \text{with } k_{ij} \in K.$$

Combining these two sums, we have

$$m = \sum_{ij} k_{ij} x_i y_j$$

and, hence, B spans M over K. Therefore, B is a basis for M over K, and the result is proved. □

Corollary 5.1.6. (a) *If $|L : K|$ is a prime number, then there exists no proper intermediate field between L and K.*
(b) *If $K \subset L$ and $|L : K| = 1$, then $L = K$.*

Let $L|K$ be a field extension, and suppose that $A \subset L$. Then certainly there are subrings of L containing both A and K, for example L. We denote by $K[A]$ the intersection of all subrings of L containing both K and A. Since the intersection of subrings is a subring, it follows that $K[A]$ is a subring containing both K and A and the smallest such subring. We call $K[A]$ the *ring adjunction* of A to K.

In an analogous manner, we let $K(A)$ be the intersection of all subfields of L containing both K and A. This is then a subfield of L, and the smallest subfield of L containing both K and A. The subfield $K(A)$ is called the *field adjunction* of A to K.

Clearly, $K[A] \subset K(A)$. If $A = \{a_1, \ldots, a_n\}$, then we write

$$K[A] = K[a_1, \ldots, a_n] \quad \text{and} \quad K(A) = K(a_1, \ldots, a_n).$$

Definition 5.1.7. The field extension $L|K$ is *finitely generated* if there exist elements $a_1, \ldots, a_n \in L$ such that $L = K(a_1, \ldots, a_n)$. The extension $L|K$ is a *simple extension* if there is an $a \in L$ with $L = K(a)$. In this case, a is called a *primitive* element of $L|K$.

In Chapter 7, we will look at an alternative way to view the adjunction constructions in terms of polynomials.

5.2 Finite and Algebraic Extensions

We now turn to the relationship between field extensions and the solution of polynomial equations.

Definition 5.2.1. Let $L|K$ be a field extension. An element $a \in L$ is *algebraic* over K if there exists a polynomial $p(x) \in K[x]$ with $p(a) = 0$. L is an *algebraic extension* of K if each element of L is algebraic over K. An element $a \in L$ that is not algebraic over K is called *transcendental.* L is a *transcendental extension* if there are transcendental elements; that is, they are not algebraic over K.

For the remainder of this section, we assume that $L|K$ is a field extension.

Lemma 5.2.2. *Each element of K is algebraic over K.*

Proof. Let $k \in K$. Then k is a zero of the polynomial $p(x) = x - k \in K[x]$. □

We tie now algebraic extensions to finite extensions.

Theorem 5.2.3. *If $L|K$ is a finite extension, then $L|K$ is an algebraic extension.*

Proof. Suppose that $L|K$ is a finite extension and $a \in L$. We must show that a is algebraic over K. Suppose that $|L : K| = n < \infty$, then $\dim_K(L) = n$. It follows that any $n+1$ elements of L are linearly dependent over K.

Now consider the elements $1, a, a^2, \ldots, a^n$ in L. These are $n+1$ distinct elements in L, so they are dependent over K. Hence, there exist $c_0, \ldots, c_n \in K$ not all zero such that

$$c_0 + c_1 a + \cdots + c_n a^n = 0.$$

Let $p(x) = c_0 + c_1 x + \cdots + c_n x^n$. Then $p(x) \in K[x]$, and $p(a) = 0$. Therefore, a is algebraic over K. Since a was arbitrary, it follows that L is an algebraic extension of K. □

From the previous theorem, it follows that every finite extension is algebraic. The converse is not true; that is, there are algebraic extensions that are not finite. We will give examples in Section 5.4.

The following lemma gives some examples of algebraic and transcendental extensions.

Lemma 5.2.4. $\mathbb{C}|\mathbb{R}$ *is algebraic, but* $\mathbb{R}|\mathbb{Q}$ *and* $\mathbb{C}|\mathbb{Q}$ *are transcendental. If* K *is any field, then* $K(x)|K$ *is transcendental.*

Proof. Since $1, i$ constitute a basis for $\mathbb{C}$ over $\mathbb{R}$, we have $|\mathbb{C} : \mathbb{R}| = 2$. Hence, $\mathbb{C}$ is a finite extension of $\mathbb{R}$; therefore, from Theorem 5.2.3, an algebraic extension. More directly, if $\alpha = a + ib \in \mathbb{C}$, then α is a zero of $x^2 - 2ax + (a^2 + b^2) \in \mathbb{R}[x]$.

The existence of transcendental numbers (we will discuss these more fully in Section 5.5) shows that both $\mathbb{R}|\mathbb{Q}$ and $\mathbb{C}|\mathbb{Q}$ are transcendental extensions.

Finally, the element $x \in K(x)$ is not a zero of any polynomial in $K[x]$. Therefore, x is a transcendental element, so the extension $K(x)|K$ is transcendental. □

5.3 Minimal Polynomials and Simple Extensions

If $L|K$ is a field extension and $\alpha \in L$ is algebraic over K, then $p(\alpha) = 0$ for some polynomial $p(x) \in K[x]$. In this section, we consider the smallest such polynomial and tie it to a simple extension of K.

Definition 5.3.1. Suppose that $L|K$ is a field extension and $\alpha \in L$ is algebraic over K. The polynomial $m_\alpha(x) \in K[x]$ is the *minimal polynomial* of α over K if the following hold:
(1) $m_\alpha(x)$ has leading coefficient 1; that is, it is a monic polynomial.
(2) $m_\alpha(\alpha) = 0$.
(3) If $f(x) \in K[x]$ with $f(\alpha) = 0$, then $m_\alpha(x)|f(x)$.

Hence, $m_\alpha(x)$ is the monic polynomial of minimal degree that has α as a zero.

We prove next that every algebraic element has such a minimal polynomial.

Theorem 5.3.2. *Suppose that* $L|K$ *is a field extension and* $\alpha \in L$ *is algebraic over* K. *Then we have:*
(1) *The minimal polynomial* $m_\alpha(x) \in K[x]$ *exists and is irreducible over* K.
(2) $K[\alpha] \cong K(\alpha) \cong K[x]/(m_\alpha(x))$, *where* $(m_\alpha(x))$ *is the principal ideal in* $K[x]$ *generated by* $m_\alpha(x)$.
(3) $|K(\alpha) : K| = \deg(m_\alpha(x))$. *Therefore,* $K(\alpha)|K$ *is a finite extension.*

Proof. (1) Suppose that $\alpha \in L$ is algebraic over K. Let

$$I = \{f(x) \in K[x] : f(\alpha) = 0\}.$$

Since α is algebraic, $I \neq \emptyset$. It is straightforward to show (see exercises) that I is an ideal in $K[x]$. Since K is a field, we have that $K[x]$ is a principal ideal domain.

Hence, there exists $g(x) \in K[x]$ with $I = (g(x))$. Let b be the leading coefficient of $g(x)$. Then $m_a(x) = b^{-1}g(x)$ is a monic polynomial. We claim that $m_a(x)$ is the minimal polynomial of a and that $m_a(x)$ is irreducible. First, it is clear that $I = (g(x)) = (m_a(x))$. If $f(x) \in K[x]$ with $f(a) = 0$, then $f(x) = h(x)m_a(x)$ for some $h(x)$. Therefore, $m_a(x)$ divides any polynomial that has a as a zero. It follows that $m_a(x)$ is the minimal polynomial.

Suppose that $m_a(x) = g_1(x)g_2(x)$. Then since $m_a(a) = 0$, it follows that either $g_1(a) = 0$ or $g_2(a) = 0$. Suppose $g_1(a) = 0$. Then from above, $m_a(x)|g_1(x)$, and since $g_1(x)|m_a(x)$, we must then have that $g_2(x)$ is a unit. Therefore, $m_a(x)$ is irreducible.

(2) Consider the map $\tau : K[x] \to K[a]$ given by

$$\tau\Big(\sum_i k_i x^i\Big) = \sum_i k_i a^i.$$

Then τ is a ring epimorphism (see exercises), and

$$\ker(\tau) = \{f(x) \in K[x] : f(a) = 0\} = (m_a(x))$$

from the argument in the proof of part (1). It follows that

$$K[x]/(m_a(x)) \cong K[a].$$

Since $m_a(x)$ is irreducible, we have $K[x]/(m_a(x))$ is a field and, therefore, $K[a] = K(a)$.

(3) Let $n = \deg(m_a(x))$. We claim that the elements $1, a, \dots, a^{n-1}$ are a basis for $K[a] = K(a)$ over K. First suppose that

$$\sum_{i=1}^{n-1} c_i a^i = 0$$

with not all $c_i = 0$ and $c_i \in K$. Then $h(a) = 0$, where $h(x) = \sum_{i=0}^{n-1} c_i x^i$. But this contradicts the fact that $m_a(x)$ has minimal degree over all polynomials in $K[x]$ that have a as a zero. Therefore, the set $1, a, \dots, a^{n-1}$ is linearly independent over K.

Now let $b \in K[a] \cong K[x]/(m_a(x))$. Then there is a $g(x) \in K[x]$ with $b = g(a)$. By the division algorithm

$$g(x) = h(x)m_a(x) + r(x),$$

where $r(x) = 0$ or $\deg(r(x)) < \deg(m_a(x))$. Now

$$r(a) = g(a) - h(a)m_a(a) = g(a) = b.$$

If $r(x) = 0$, then $b = 0$. If $r(x) \neq 0$, then since $\deg(r(x)) < n$, we have

$$r(x) = c_0 + c_1 x + \dots + c_{n-1}x^{n-1}$$

with $c_i \in K$ and some c_i, but not all might be zero. This implies that

$$b = r(a) = c_0 + c_1 a + \cdots + c_{n-1}a^{n-1}$$

and, hence, b is a linear combination over K of $1, a, \ldots, a^{n-1}$. Hence, $1, a, \ldots, a^{n-1}$ spans $K[a]$ over K and, hence, forms a basis. □

Theorem 5.3.3. *Suppose that $L|K$ is a field extension and $a \in L$ is algebraic over K. Suppose that $f(x) \in K[x]$ is a monic polynomial with $f(a) = 0$. Then $f(x)$ is the minimal polynomial if and only if $f(x)$ is irreducible in $K[x]$.*

Proof. Suppose that $f(x)$ is the minimal polynomial of a. Then $f(x)$ is irreducible from the previous theorem.

Conversely, suppose that $f(x)$ is monic, irreducible and $f(a) = 0$. From the previous theorem $m_a(x)|f(x)$. Since $f(x)$ is irreducible, we have $f(x) = cm_a(x)$ with $c \in K$. However, since both $f(x)$ and $m_a(x)$ are monic, we must have $c = 1$, and $f(x) = m_a(x)$. □

We now show that a finite extension of K is actually finitely generated over K. In addition, it is generated by finitely many algebraic elements.

Theorem 5.3.4. *Let $L|K$ be a field extension. Then the following are equivalent:*
(1) *$L|K$ is a finite extension.*
(2) *$L|K$ is an algebraic extension and there exist $a_1, \ldots, a_n \in L$ with $L = K(a_1, \ldots, a_n)$.*
(3) *There exist algebraic elements $a_1, \ldots, a_n \in L$ such that $L = K(a_1, \ldots, a_n)$.*

Proof. (1) ⇒ (2). We have seen in Theorem 5.2.3 that a finite extension is algebraic. Suppose that $a_1, \ldots, a_n$ are a basis for L over K. Then clearly $L = K(a_1, \ldots, a_n)$.

(2) ⇒ (3). If $L|K$ is an algebraic extension and $L = K(a_1, \ldots, a_n)$, then each a_i is algebraic over K.

(3) ⇒ (1). Suppose that there exist algebraic elements $a_1, \ldots, a_n \in L$ such that $L = K(a_1, \ldots, a_n)$. We show that $L|K$ is a finite extension. We do this by induction on n. If $n = 1$, then $L = K(a)$ for some algebraic element a, and the result follows from Theorem 5.3.2. Suppose now that $n \geq 2$. We assume then that an extension $K(a_1, \ldots, a_{n-1})$ with $a_1, \ldots, a_{n-1}$ algebraic elements is a finite extension. Now suppose that we have $L = K(a_1, \ldots, a_n)$ with $a_1, \ldots, a_n$ algebraic elements.

Then

$$\begin{aligned}&|K(a_1, \ldots, a_n) : K| \\ &\quad = |K(a_1, \ldots, a_{n-1})(a_n) : K(a_1, \ldots, a_{n-1})||K(a_1, \ldots, a_{n-1}) : K|.\end{aligned}$$

The second term $|K(a_1, \ldots, a_{n-1}) : K|$ is finite from the inductive hypothesis. The first term $|K(a_1, \ldots, a_{n-1})(a_n) : K(a_1, \ldots, a_{n-1})|$ is also finite from Theorem 5.3.2 since it is a simple extension of the field $K(a_1, \ldots, a_{n-1})$ by the algebraic element a_n. Therefore, $|K(a_1, \ldots, a_n) : K|$ is finite. □

Theorem 5.3.5. *Suppose that K is a field and R is an integral domain with $K \subset R$. Then R can be viewed as a vector space over K. If $\dim_K(R) < \infty$, then R is a field.*

Proof. Let $r_0 \in R$ with $r_0 \neq 0$. Define the map from R to R given by

$$\tau(r) = rr_0.$$

It is easy to show (see exercises) that this is a linear transformation from R to R, considered as a vector space over K.

Suppose that $\tau(r) = 0$. Then $rr_0 = 0$ and, hence, $r = 0$ since $r_0 \neq 0$ and R is an integral domain. It follows that τ is an injective map. Since R is a finite-dimensional vector space over K, and τ is an injective linear transformation, it follows that τ must also be surjective. This implies that there exists an r_1 with $\tau(r_1) = 1$. Then $r_1 r_0 = 1$ and, hence, r_0 has an inverse within R. Since r_0 was an arbitrary nonzero element of R, it follows that R is a field. □

Theorem 5.3.6. *Suppose that $K \subset L \subset M$ is a chain of field extensions. Then $M|K$ is algebraic if and only if $M|L$ is algebraic, and $L|K$ is algebraic.*

Proof. If $M|K$ is algebraic, then certainly $M|L$ and $L|K$ are algebraic.

Now suppose that $M|L$ and $L|K$ are algebraic. We show that $M|K$ is algebraic. Let $a \in M$. Then since a is algebraic over L, there exist $b_0, b_1, \dots, b_n \in L$ with

$$b_0 + b_1 a + \cdots + b_n a^n = 0.$$

Each b_i is algebraic over K and, hence, $K(b_0, \dots, b_n)$ is finite-dimensional over K. Therefore, $K(b_0, \dots, b_n)(a) = K(b_0, \dots, b_n, a)$ is also finite-dimensional over K. Therefore, $K(b_0, \dots, b_n, a)$ is a finite extension of K and, hence, an algebraic extension K. Since $a \in K(b_0, \dots, b_n, a)$, it follows that a is algebraic over K and, therefore, M is algebraic over K. □

5.4 Algebraic Closures

As before, suppose that $L|K$ is a field extension. Since each element of K is algebraic over K, there are certainly algebraic elements over K within L. Let $\mathcal{A}_K$ denote the set of all elements of L that are algebraic over K. We prove that $\mathcal{A}_K$ is actually a subfield of L. It is called the *algebraic closure* of K within L.

Theorem 5.4.1. *Suppose that $L|K$ is a field extension, and let $\mathcal{A}_K$ denote the set of all elements of L that are algebraic over K. Then $\mathcal{A}_K$ is a subfield of L. $\mathcal{A}_K$ is called the algebraic closure of K in L.*

Proof. Since $K \subset \mathcal{A}_K$, we have that $\mathcal{A}_K \neq \emptyset$. Let $a, b \in \mathcal{A}_K$. Since a, b are both algebraic over K from Theorem 5.3.4, we have that $K(a, b)$ is a finite extension of K. Therefore, $K(a, b)$ is an algebraic extension of K and, hence, each element of $K(a, b)$ is algebraic over K. Now $a, b \in K(a, b)$ if $b \neq 0$, and $K(a, b)$ is a field. Therefore, $a \pm b, ab$, and a/b are all in $K(a, b)$ and, hence, all algebraic over K. Therefore, $a \pm b, ab, a/b$, if $b \neq 0$, are all in $\mathcal{A}_K$. It follows that $\mathcal{A}_K$ is a subfield of L. □

In Section 5.2, we showed that every finite extension is an algebraic extension. We mentioned that the converse is not necessarily true; that is, there are algebraic extensions that are not finite. Here we give an example.

Theorem 5.4.2. *Let $\mathcal{A}$ be the algebraic closure of the rational numbers $\mathbb{Q}$ within the complex numbers $\mathbb{C}$. Then $\mathcal{A}$ is an algebraic extension of $\mathbb{Q}$, but $|\mathcal{A} : \mathbb{Q}| = \infty$.*

Proof. From the previous theorem, $\mathcal{A}$ is an algebraic extension of $\mathbb{Q}$. We show that it cannot be a finite extension.

By Eisenstein's criterion, the rational polynomial $f(x) = x^p + p$ is irreducible over $\mathbb{Q}$ for any prime p. Let a be a zero in $\mathbb{C}$ of $f(x)$. Then $a \in \mathcal{A}$, and $|\mathbb{Q}(a) : \mathbb{Q}| = p$. Therefore, $|\mathcal{A} : \mathbb{Q}| \geq p$ for all primes p. Since there are infinitely many primes, this implies that $|\mathcal{A} : \mathbb{Q}| = \infty$. □

5.5 Algebraic and Transcendental Numbers

In this section, we consider the string of field extensions $\mathbb{Q} \subset \mathbb{R} \subset \mathbb{C}$.

Definition 5.5.1. An *algebraic number* a is an element of $\mathbb{C}$, which is algebraic over $\mathbb{Q}$. Hence, an algebraic number is an $a \in \mathbb{C}$ such that $f(a) = 0$ for some $f(x) \in \mathbb{Q}[x]$. If $a \in \mathbb{C}$ is not algebraic, it is *transcendental*.

We will let $\mathcal{A}$ denote the totality of algebraic numbers within the complex numbers $\mathbb{C}$, and $\mathcal{T}$ the set of transcendentals so that $\mathbb{C} = \mathcal{A} \cup \mathcal{T}$. In the language of the last subsection, $\mathcal{A}$ is the algebraic closure of $\mathbb{Q}$ within $\mathbb{C}$. As in the general case, if $a \in \mathbb{C}$ is algebraic, we will let $m_a(x)$ denote the minimal polynomial of a over $\mathbb{Q}$.

We now examine the sets $\mathcal{A}$ and $\mathcal{T}$ more closely. Since $\mathcal{A}$ is precisely the algebraic closure of $\mathbb{Q}$ in $\mathbb{C}$, we have from our general result that $\mathcal{A}$ actually forms a subfield of $\mathbb{C}$. Furthermore, since the intersection of subfields is again a subfield, it follows that $\mathcal{A}' = \mathcal{A} \cap \mathbb{R}$, the real algebraic numbers form a subfield of the reals.

Theorem 5.5.2. *The set $\mathcal{A}$ of algebraic numbers forms a subfield of $\mathbb{C}$.*
The subset $\mathcal{A}' = \mathcal{A} \cap \mathbb{R}$ of real algebraic numbers forms a subfield of $\mathbb{R}$.

Since each rational is algebraic, it is clear that there are algebraic numbers. Furthermore, there are irrational algebraic numbers, $\sqrt{2}$ for example, since it satisfies the irreducible polynomial $x^2 - 2 = 0$ over $\mathbb{Q}$. On the other hand, we have not examined the question of whether transcendental numbers really exist. To show that any particular complex number is transcendental is, in general, quite difficult. However, it is relatively easy to show that there are uncountably infinitely many transcendentals.

Theorem 5.5.3. *The set $\mathcal{A}$ of algebraic numbers is countably infinite. Therefore, $\mathcal{T}$, the set of transcendental numbers, and $\mathcal{T}' = \mathcal{T} \cap \mathbb{R}$, the real transcendental numbers, are uncountably infinite.*

Proof. Let

$$\mathcal{P}_n = \{f(x) \in \mathbb{Q}[x] : \deg(f(x)) \le n\}.$$

Since if $f(x) \in \mathcal{P}_n, f(x) = q_o + q_1x + \cdots + q_nx^n$ with $q_i \in \mathbb{Q}$, we can identify a polynomial of degree $\le n$ with an $(n+1)$-tuple $(q_0, q_1, \ldots, q_n)$ of rational numbers. Therefore, the set $\mathcal{P}_n$ has the same size as the $(n+1)$-fold Cartesian product of $\mathbb{Q}$:

$$\mathbb{Q}^{n+1} = \mathbb{Q} \times \mathbb{Q} \times \cdots \times \mathbb{Q}.$$

Since a finite Cartesian product of countable sets is still countable, it follows that $\mathcal{P}_n$ is a countable set.

Now let

$$\mathcal{B}_n = \bigcup_{p(x)\in\mathcal{P}_n} \{\text{zeros of } p(x)\};$$

that is, $\mathcal{B}_n$ is the union of all zeros in $\mathbb{C}$ of all rational polynomials of degree $\le n$. Since each such $p(x)$ has a maximum of n zeros, and since $\mathcal{P}_n$ is countable, it follows that $\mathcal{B}_n$ is a countable union of finite sets and, hence, is still countable. Now

$$\mathcal{A} = \bigcup_{n=1}^{\infty} \mathcal{B}_n,$$

so $\mathcal{A}$ is a countable union of countable sets and is, therefore, countable.

Since both $\mathbb{R}$ and $\mathbb{C}$ are uncountably infinite, the second assertions follow directly from the countability of $\mathcal{A}$. If say $\mathcal{T}$ were countable, then $\mathbb{C} = \mathcal{A} \cup \mathcal{T}$ would also be countable, which is a contradiction. □

From Theorem 5.5.3, we know that there exist infinitely many transcendental numbers. Liouville, in 1851, gave the first proof of the existence of transcendentals by exhibiting a few. He gave the following as one example:

Theorem 5.5.4. *The real number*

$$c = \sum_{j=1}^{\infty} \frac{1}{10^{j!}}$$

is transcendental.

Proof. First of all, since $\frac{1}{10^{j!}} < \frac{1}{10^j}$, and $\sum_{j=1}^{\infty} \frac{1}{10^j}$ is a convergent geometric series, it follows from the comparison test that the infinite series defining c converges and defines a real number. Furthermore, since $\sum_{j=1}^{\infty} \frac{1}{10^j} = \frac{1}{9}$, it follows that $c < \frac{1}{9} < 1$. Suppose that c is algebraic so that $g(c) = 0$ for some rational nonzero polynomial $g(x)$. Multiplying through by the least common multiple of all the denominators in $g(x)$, we may suppose that $f(c) = 0$ for some integral polynomial $f(x) = \sum_{j=0}^{n} m_j x^j$. Then c satisfies

$$\sum_{j=0}^{n} m_j c^j = 0$$

for some integers $m_0, \dots, m_n$.

If $0 < x < 1$, then by the triangle inequality

$$|f'(x)| = \left|\sum_{j=1}^{n} j m_j x^{j-1}\right| \le \sum_{j=1}^{n} |j m_j| = B,$$

where B is a real constant depending only on the coefficients of $f(x)$.

Now let

$$c_k = \sum_{j=1}^{k} \frac{1}{10^{j!}}$$

be the k-th partial sum for c. Then

$$|c - c_k| = \sum_{j=k+1}^{\infty} \frac{1}{10^{j!}} < 2 \cdot \frac{1}{10^{(k+1)!}}.$$

Apply the mean value theorem to $f(x)$ at c and c_k to obtain

$$|f(c) - f(c_k)| = |c - c_k||f'(\zeta)|$$

for some ζ with $c_k < \zeta < c < 1$. Now since $0 < \zeta < 1$, we have

$$|c - c_k||f'(\zeta)| < 2B\frac{1}{10^{(k+1)!}}.$$

On the other hand, since $f(x)$ can have at most n zeros, it follows that for all k large enough, we would have $f(c_k) \ne 0$. Since $f(c) = 0$, we have

$$|f(c) - f(c_k)| = |f(c_k)| = \left|\sum_{j=1}^{n} m_j c_k^j\right| > \frac{1}{10^{nk!}}$$

since for each j, $m_j c_k^j$ is a rational number with denominator $10^{jk!}$. However, if k is chosen sufficiently large and n is fixed, we have

$$\frac{1}{10^{nk!}} > \frac{2B}{10^{(k+1)!}},$$

contradicting the equality from the mean value theorem. Therefore, c is transcendental. □

In 1873, Hermite proved that e is transcendental, whereas, in 1882, Lindemann showed that π is transcendental. Schneider, in 1934, showed that a^b is transcendental if $a \neq 0$, a, and b are algebraic and b is irrational. In Chapter 20, we will prove that both e and π are transcendental. An interesting open question is the following:

Is π transcendental over $\mathbb{Q}(e)$?

To close this section, we show that in general if $a \in L$ is transcendental over K, then $K(a)|K$ is isomorphic to the field of rational functions over K.

Theorem 5.5.5. *Suppose that $L|K$ is a field extension and $a \in L$ is transcendental over K. Then $K(a)|K$ is isomorphic to $K(x)|K$. Here the isomorphism $\mu : K(x) \to K(a)$ can be chosen such that $\mu(x) = a$.*

Proof. Define the map $\mu : K(x) \to K(a)$ by

$$\mu\left(\frac{f(x)}{g(x)}\right) = \frac{f(a)}{g(a)}$$

for $f(x), g(x) \in K[x]$ with $g(x) \neq 0$. Then μ is a homomorphism, and $\mu(x) = a$. Since $\mu \neq 0$, it follows that μ is an isomorphism. □

5.6 Exercises

1. Let $a \in \mathbb{C}$ with $a^3 - 2a + 2 = 0$ and $b = a^2 - a$. Compute the minimal polynomial $m_b(x)$ of b over $\mathbb{Q}$ and compute the inverse of b in $\mathbb{Q}(a)$.
2. Determine the algebraic closure of $\mathbb{R}$ in $\mathbb{C}(x)$.
3. Let $a_n := \sqrt[2^n]{2} \in \mathbb{R}$, $n = 1, 2, 3, \dots$ and $A := \{a_n : n \in \mathbb{N}\}$ and $E := \mathbb{Q}(A)$. Show the following:
   (i) $|\mathbb{Q}(a_n) : \mathbb{Q}| = 2^n$.
   (ii) $|E : \mathbb{Q}| = \infty$.
   (iii) $E = \bigcup_{n=1}^{\infty} \mathbb{Q}(a_n)$.
   (iv) E is algebraic over $\mathbb{Q}$.
4. Determine $|E : \mathbb{Q}|$ for
   (i) $E = \mathbb{Q}(\sqrt{2}, \sqrt{-2})$.
   (ii) $E = \mathbb{Q}(\sqrt{3}, \sqrt{3} + \sqrt[3]{3})$.
   (iii) $E = \mathbb{Q}(\frac{1+i}{\sqrt{2}}, \frac{-1+i}{\sqrt{2}})$.
5. Show that $\mathbb{Q}(\sqrt{2}, \sqrt{3}) = \{a + b\sqrt{2} + c\sqrt{3} + d\sqrt{6} : a, b, c, d \in \mathbb{Q}\}$. Determine the degree of $\mathbb{Q}(\sqrt{2}, \sqrt{3})$ over $\mathbb{Q}$. Further show that $\mathbb{Q}(\sqrt{2}, \sqrt{3}) = \mathbb{Q}(\sqrt{2} + \sqrt{3})$.
6. Let K, E be fields and $a \in E$ be transcendental over K.
   Show the following:
   (i) Each element of $K(a)|K$, which is not in K, is transcendental over K.
   (ii) a^n is transcendental over K for each $n > 1$.
   (iii) If $L := K(\frac{a^3}{a+1})$, then a is algebraic over L. Determine the minimal polynomial $m_a(x)$ of a over L.

7. Let K be a field and $a \in K(x) \setminus K$. Show the following:
   (i) x is algebraic over $K(a)$.
   (ii) If L is a field with $K \subset L \subseteq K(x)$ and if $a \in L$, then $|K(x) : L| < \infty$.
   (iii) a is transcendental over K.
8. Suppose that $a \in L$ is algebraic over K. Let

$$I = \{f(x) \in K[x] : f(a) = 0\}.$$

   Since a is algebraic $I \neq \emptyset$. Prove that I is an ideal in $K[x]$.
9. Prove that there are uncountably many transcendental numbers. To do this show that the set $\mathcal{A}$ of algebraic numbers is countable. To do this:
   (i) Show that $\mathbb{Q}_n[x]$, the set of rational polynomials of degree $\leq n$, is countable (finite Cartesian product of countable sets).
   (ii) Let $\mathcal{B}_n = \{\text{Zeros of polynomials in } \mathbb{Q}_n\}$. Show that $\mathcal{B}$ is countable.
   (iii) Show that $\mathcal{A} = \bigcup_{n=1}^{\infty} \mathcal{B}_n$ and conclude that $\mathcal{A}$ is countable.
   (iv) Show that the transcendental numbers are uncountable.
10. Consider the map $\tau : K[x] \to K[a]$ given by

$$\tau\Big(\sum_i k_i x^i\Big) = \sum_i k_i a^i.$$

    Show that τ is a ring epimorphism.
11. Suppose that K is a field and R is an integral domain with $K \subset R$. Then R can be viewed as a vector space over K. Let $r_0 \in R$ with $r_0 \neq 0$. Define the map from R to R given by

$$\tau(r) = rr_0.$$

    Show that this is a linear transformation from R to R, considered as a vector space over K.

6 Field Extensions and Compass and Straightedge Constructions

6.1 Geometric Constructions

Greek mathematicians in the classical period posed the problem of constructing certain geometric figures in the Euclidean plane using only a straightedge and a compass. These are known as *geometric construction problems.*

Recall from elementary geometry that using a straightedge and compass, it is possible to draw a line parallel to a given line segment through a given point, to extend a given line segment, and to erect a perpendicular to a given line at a given point on that line. There were other geometric construction problems that the Greeks could not determine straightedge and compass solutions but, on the other hand, were never able to prove that such constructions were impossible. In particular, there were four famous insolvable (to the Greeks) construction problems. The first is the *squaring of the circle.* This problem is, given a circle, to construct using straightedge and compass a square having an area equal to that of the given circle. The second is the *doubling of the cube.* This problem is, given a cube of given side length, to construct using a straightedge and compass, a side of a cube having double the volume of the original cube. The third problem is the *trisection of an angle.* This problem is to trisect a given angle using only a straightedge and compass. The final problem is the *construction of a regular n-gon.* This problems asks which regular n-gons could be constructed using only straightedge and compass.

By translating each of these problems into the language of field extensions, we can show that each of the first three problems are insolvable in general, and we can give the complete solution to the construction of the regular n-gons.

6.2 Constructible Numbers and Field Extensions

We now translate the geometric construction problems into the language of field extensions. As a first step, we define a *constructible number.*

Definition 6.2.1. Suppose we are given a line segment of unit length. An $\alpha \in \mathbb{R}$ is *constructible* if we can construct a line segment of length $|\alpha|$, in a finite number of steps, from the unit segment using a straightedge and compass.

Our first result is that the set of all constructible numbers forms a subfield of $\mathbb{R}$.

Theorem 6.2.2. *The set $\mathcal{C}$ of all constructible numbers forms a subfield of $\mathbb{R}$. Furthermore, $\mathbb{Q} \subset \mathcal{C}$.*

Proof. Let $\mathcal{C}$ be the set of all constructible numbers. Since the given unit length segment is constructible, we have $1 \in \mathcal{C}$. Therefore, $\mathcal{C} \neq \emptyset$. Thus, to show that it is a field, we must show that it is closed under the field operations.

https://doi.org/10.1515/9783111142524-006

Suppose α, β are constructible. We must show then that $\alpha \pm \beta$, $\alpha\beta$, and α/β for $\beta \neq 0$ are constructible. If $\alpha, \beta > 0$, construct a line segment of length $|\alpha|$. At one end of this line segment, extend it by a segment of length $|\beta|$. This will construct a segment of length $\alpha + \beta$. Similarly, if $\alpha > \beta$, lay off a segment of length $|\beta|$ at the beginning of a segment of length $|\alpha|$. The remaining piece will be $\alpha - \beta$. By considering cases, we can do this in the same manner if either α or β, or both, are negative. These constructions are pictured in Figure 6.1. Therefore, $\alpha \pm \beta$ are constructible.

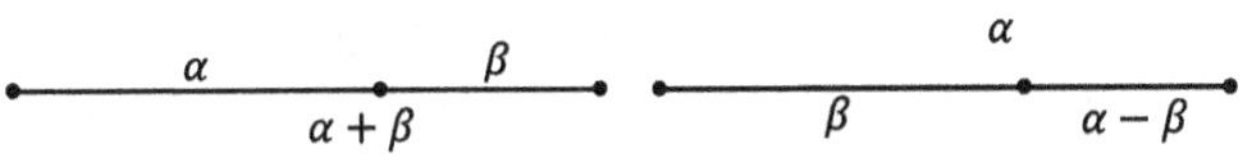

Figure 6.1: Addition of constructible numbers.

In Figure 6.2, we show how to construct $\alpha\beta$. Let the line segment $\overline{OA}$ have length $|\alpha|$. Consider a line L through O not coincident with $\overline{OA}$. Let $\overline{OB}$ have length $|\beta|$ as in the diagram. Let P be on ray $\overline{OB}$ so that $\overline{OP}$ has length 1. Draw $\overline{AP}$ and then find Q on ray $\overline{OA}$ such that $\overline{BQ}$ is parallel to $\overline{AP}$. From similar triangles, we then have

$$\frac{|\overline{OP}|}{|\overline{OB}|} = \frac{|\overline{OA}|}{|\overline{OQ}|} \Rightarrow \frac{1}{|\beta|} = \frac{|\alpha|}{|\overline{OQ}|}.$$

Then $|\overline{OQ}| = |\alpha||\beta|$, and so $\alpha\beta$ is constructible.

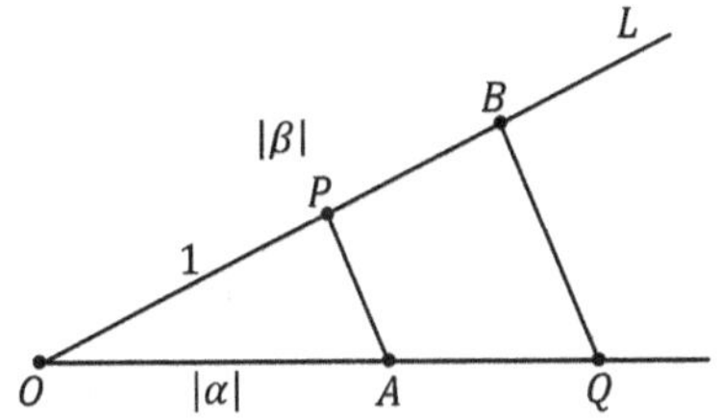

Figure 6.2: Multiplication of constructible numbers.

A similar construction, pictured in Figure 6.3, shows that α/β for $\beta \neq 0$ is constructible. Find $\overline{OA}$, $\overline{OB}$, $\overline{OP}$ as above. Now, connect A to B, and let $\overline{PQ}$ be parallel to $\overline{AB}$. From similar triangles again, we have

$$\frac{1}{|\beta|} = \frac{|\overline{OQ}|}{|\alpha|} \Longrightarrow \frac{|\alpha|}{|\beta|} = |\overline{OQ}|.$$

Hence, α/β is constructible.

Therefore, $\mathcal{C}$ is a subfield of $\mathbb{R}$. Since $\operatorname{char} \mathcal{C} = 0$, it follows that $\mathbb{Q} \subset \mathcal{C}$. □

Let us now consider how a constructible number is found in the plane. Starting at the origin and using the unit length and the constructions above, we can locate any point

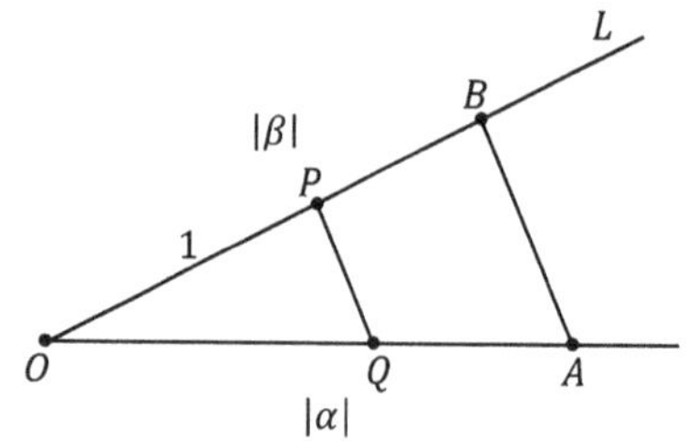

Figure 6.3: Inversion of constructible numbers.

in the plane with rational coordinates. That is, we can construct the point $P = (q_1, q_2)$ with $q_1, q_2 \in \mathbb{Q}$. Using only straightedge and compass, any further point in the plane can be determined in one of the following three ways:

1. The intersection point of two lines, each of which passes through two known points each having rational coordinates.
2. The intersection point of a line passing through two known points having rational coordinates and a circle, whose center has rational coordinates, and whose radius squared is rational.
3. The intersection point of two circles, each of whose centers has rational coordinates, and each of whose radii is the square root of a rational number.

Analytically, the first case involves the solution of a pair of linear equations, each with rational coefficients and, thus, only leads to other rational numbers. In cases two and three, we must solve equations of the form $x^2+y^2+ax+by+c = 0$, with $a, b, c \in \mathbb{Q}$. These will then be quadratic equations over $\mathbb{Q}$ and, thus, the solutions will either be in $\mathbb{Q}$, or in a quadratic extension $\mathbb{Q}(\sqrt{\alpha})$ of $\mathbb{Q}$. Once a real quadratic extension of $\mathbb{Q}$ is found, the process can be iterated. Conversely, using the altitude theorem, if α is constructible, so is $\sqrt{\alpha}$. A much more detailed description of the constructible numbers can be found in [52]. We thus can prove the following theorem:

Theorem 6.2.3. *If γ is constructible with $\gamma \notin \mathbb{Q}$, then there exists a finite number of elements $\alpha_1, \ldots, \alpha_r \in \mathbb{R}$ with $\alpha_r = \gamma$ such that for $i = 1, \ldots, r$, $\mathbb{Q}(\alpha_1, \ldots, \alpha_i)$ is a quadratic extension of $\mathbb{Q}(\alpha_1, \ldots, \alpha_{i-1})$. In particular, $|\mathbb{Q}(\gamma) : \mathbb{Q}| = 2^n$ for some $n \geq 1$.*

Therefore, the constructible numbers are precisely those real numbers that are contained in repeated quadratic extensions of $\mathbb{Q}$. In the next section, we use this idea to show the impossibility of the first three mentioned construction problems.

6.3 Four Classical Construction Problems

We now consider the aforementioned construction problems. Our main technique will be to use Theorem 6.2.3. From this result, we have that if γ is constructible with $\gamma \notin \mathbb{Q}$, then $|\mathbb{Q}(\gamma) : \mathbb{Q}| = 2^n$ for some $n \geq 1$.

6.3.1 Squaring the Circle

Theorem 6.3.1. *It is impossible to square the circle. That is, it is impossible in general, given a circle, to construct using straightedge and compass a square having area equal to that of the given circle.*

Proof. Suppose the given circle has radius 1. It is then constructible and would have an area of π. A corresponding square would then have to have a side of length $\sqrt{\pi}$. To be constructible a number, α must have $|\mathbb{Q}(\alpha) : \mathbb{Q}| = 2^m < \infty$ and, hence, α must be algebraic. However, π is transcendental, so $\sqrt{\pi}$ is also transcendental (see Section 20.4); therefore not constructible. □

6.3.2 The Doubling of the Cube

Theorem 6.3.2. *It is impossible to double the cube. This means that it is impossible in general, given a cube of given side length, to construct using a straightedge and compass, a side of a cube having double the volume of the original cube.*

Proof. Let the given side length be 1, so that the original volume is also 1. To double this, we would have to construct a side of length $2^{1/3}$. However, $|\mathbb{Q}(2^{1/3}) : \mathbb{Q}| = 3$ since the minimal polynomial over $\mathbb{Q}$ is $m_{2^{1/3}}(x) = x^3 - 2$. This is not a power of 2, so $2^{1/3}$ is not constructible. □

6.3.3 The Trisection of an Angle

Theorem 6.3.3. *It is impossible to trisect an angle. This means that it is impossible, in general, to trisect a given angle using only a straightedge and compass.*

Proof. An angle θ is constructible if and only if a segment of length $|\cos\theta|$ is constructible. Since $\cos(\pi/3) = 1/2$, therefore, $\pi/3$ is constructible. We show that it cannot be trisected by straightedge and compass.

The following trigonometric identity holds:

$$\cos(3\theta) = 4\cos^3(\theta) - 3\cos(\theta).$$

Let $\alpha = \cos(\pi/9)$. From the above identity, we have $4\alpha^3 - 3\alpha - \frac{1}{2} = 0$.

The polynomial $4x^3 - 3x - \frac{1}{2}$ is irreducible over $\mathbb{Q}$ and, hence, the minimal polynomial over $\mathbb{Q}$ is $m_\alpha(x) = x^3 - \frac{3}{4}x - \frac{1}{8}$. It follows that $|\mathbb{Q}(\alpha) : \mathbb{Q}| = 3$; hence, α is not constructible. Therefore, the corresponding angle $\pi/9$ is not constructible. Therefore, $\pi/3$ is constructible, but it cannot be trisected. □

6.3.4 Construction of a Regular n-Gon

The final construction problem we consider is the construction of regular n-gons. The algebraic study of the constructibility of regular n-gons was initiated by Gauss in the early part of the nineteenth century.

Notice first that a regular n-gon will be constructible for $n \geq 3$ if and only if the angle $\frac{2\pi}{n}$ is constructible, which is the case if and only if the length $\cos\frac{2\pi}{n}$ is a constructible number. From our techniques, if $\cos\frac{2\pi}{n}$ is a constructible number, then necessarily $|\mathbb{Q}(\cos(\frac{2\pi}{n})) : \mathbb{Q}| = 2^m$ for some m. After we discuss Galois theory, we see that this condition is also sufficient. Therefore, $\cos\frac{2\pi}{n}$ is a constructible number if and only if $|\mathbb{Q}(\cos(\frac{2\pi}{n})) : \mathbb{Q}| = 2^m$ for some m.

The solution of this problem, that is, the determination of when $|\mathbb{Q}(\cos(\frac{2\pi}{n})) : \mathbb{Q}| = 2^m$, involves two concepts from number theory: the *Euler phi-function* and *Fermat primes.*

Definition 6.3.4. For any natural number n, the *Euler phi-function* is defined by

$$\phi(n) = \text{number of integers less than or equal to } n, \text{ and relatively prime to } n.$$

Example 6.3.5. $\phi(6) = 2$ since among 1, 2, 3, 4, 5, 6 only 1, 5 are relatively prime to 6.

It is fairly straightforward to develop a formula for $\phi(n)$. A formula is first determined for primes and for prime powers, and then pasted back together via the fundamental theorem of arithmetic.

Lemma 6.3.6. *For any prime p and $m > 0$,*

$$\phi(p^m) = p^m - p^{m-1} = p^m\left(1 - \frac{1}{p}\right).$$

Proof. If $1 \leq a \leq p$, then either $a = p$, or $(a, p) = 1$. It follows that the positive integers less than or equal to p^m, which are not relatively prime to p^m are precisely the multiples of p; that is, $p, 2p, 3p, \ldots, p^{m-1} \cdot p$. All other positive $a < p^m$ are relatively prime to p^m. Hence, the number relatively prime to p^m is

$$p^m - p^{m-1}.$$

□

Lemma 6.3.7. *If $(a, b) = 1$, then $\phi(ab) = \phi(a)\phi(b)$.*

Proof. Given a natural number n, a *reduced residue system* modulo n is a set of integers $x_1, \ldots, x_k$ such that each x_i is relatively prime to n, $x_i \neq x_j$ modulo n unless $i = j$, and if $(x, n) = 1$ for some integer x, then $x \equiv x_i \pmod{n}$ for some i. Clearly, $\phi(n)$ is the size of a reduced residue system modulo n.

Let $R_a = \{x_1, \ldots, x_{\phi(a)}\}$ be a reduced residue system modulo a, $R_b = \{y_1, \ldots, y_{\phi(b)}\}$ be a reduced residue system modulo b, and let

$$S = \{ay_i + bx_j : i = 1, \ldots, \phi(b), j = 1, \ldots, \phi(a)\}.$$

We claim that S is a reduced residue system modulo ab. Since S has $\phi(a)\phi(b)$ elements, it will follow that $\phi(ab) = \phi(a)\phi(b)$.

To show that S is a reduced residue system modulo ab, we must show three things: first that each $x \in S$ is relatively prime to ab; second that the elements of S are distinct; and, finally, that given any integer n with $(n, ab) = 1$, then $n \equiv s \pmod{ab}$ for some $s \in S$.

Let $x = ay_i + bx_j$. Then since $(x_j, a) = 1$ and $(a, b) = 1$, it follows that $(x, a) = 1$. Analogously, $(x, b) = 1$. Since x is relatively prime to both a and b, we have $(x, ab) = 1$. This shows that each element of S is relatively prime to ab.

Next suppose that

$$ay_i + bx_j \equiv ay_k + bx_l \pmod{ab}.$$

Then

$$ab|(ay_i + bx_j) - (ay_k + bx_l) \implies ay_i \equiv ay_k \pmod{b}.$$

Since $(a, b) = 1$, it follows that $y_i \equiv y_k \pmod{b}$. But then $y_i = y_k$ since R_b is a reduced residue system. Similarly, $x_j = x_l$. This shows that the elements of S are distinct modulo ab.

Finally, suppose $(n, ab) = 1$. Since $(a, b) = 1$, there exist x, y with $ax + by = 1$. Then

$$anx + bny = n.$$

Since $(x, b) = 1$, and $(n, b) = 1$, it follows that $(nx, b) = 1$. Therefore, there is an s_i with $nx = s_i + tb$. In the same manner, $(ny, a) = 1$, and so there is an r_j with $ny = r_j + ua$. Then

$$\begin{aligned} a(s_i + tb) + b(r_j + ua) = n &\implies n = as_i + br_j + (t + u)ab \\ &\implies n \equiv ar_i + bs_j \pmod{ab}, \end{aligned}$$

and we are done. □

We now give the general formula for $\phi(n)$.

Theorem 6.3.8. *Suppose $n = p_1^{e_1} \cdots p_k^{e_k}$, then*

$$\phi(n) = (p_1^{e_1} - p_1^{e_1-1})(p_2^{e_2} - p_2^{e_2-1}) \cdots (p_k^{e_k} - p_k^{e_k-1}).$$

Proof. From the previous lemma, we have

$$\begin{aligned} \phi(n) &= \phi(p_1^{e_1})\phi(p_2^{e_2}) \cdots \phi(p_k^{e_k}) \\ &= (p_1^{e_1} - p_1^{e_1-1})(p_2^{e_2} - p_2^{e_2-1}) \cdots (p_k^{e_k} - p_k^{e_k-1}) \\ &= p_1^{e_1}(1 - 1/p_1) \cdots p_k^{e_k}(1 - 1/p_k) = p_1^{e_1} \cdots p_k^{e_k} \cdot (1 - 1/p_1) \cdots (1 - 1/p_k) \\ &= n \prod_i (1 - 1/p_i). \end{aligned}$$

□

Example 6.3.9. Determine $\phi(126)$. Now

$$126 = 2 \cdot 3^2 \cdot 7 \implies \phi(126) = \phi(2)\phi(3^2)\phi(7) = (1)(3^2 - 3)(6) = 36.$$

Hence, there are 36 units in $\mathbb{Z}_{126}$.

An interesting result with many generalizations in number theory is the following:

Theorem 6.3.10. *For $n > 1$ and for $d \geq 1$*

$$\sum_{d|n} \phi(d) = n.$$

Proof. We first prove the theorem for prime powers and then paste together via the fundamental theorem of arithmetic.

Suppose that $n = p^e$ for p a prime. Then the divisors of n are $1, p, p^2, \ldots, p^e$, so

$$\begin{aligned}\sum_{d|n} \phi(d) &= \phi(1) + \phi(p) + \phi(p^2) + \cdots + \phi(p^e) \\ &= 1 + (p-1) + (p^2 - p) + \cdots + (p^e - p^{e-1}).\end{aligned}$$

Notice that this sum telescopes; that is, $1 + (p-1) = p$, $p + (p^2 - p) = p^2$ and so on. Hence, the sum is just p^e, and the result is proved for n a prime power.

We now do an induction on the number of distinct prime factors of n. The above argument shows that the result is true if n has only one distinct prime factor. Assume that the result is true whenever an integer has less than k distinct prime factors, and suppose $n = p_1^{e_1} \cdots p_k^{e_k}$ has k distinct prime factors. Then $n = p^e c$, where $p = p_1$, $e = e_1$, and c has fewer than k distinct prime factors. By the inductive hypothesis

$$\sum_{d|c} \phi(d) = c.$$

Since $(c, p) = 1$, the divisors of n are all of the form $p^\alpha d_1$, where $d_1 | c$, and $\alpha = 0, 1, \ldots, e$. It follows that

$$\sum_{d|n} \phi(d) = \sum_{d_1|c} \phi(d_1) + \sum_{d_1|c} \phi(pd_1) + \cdots + \sum_{d_1|c} \phi(p^e d_1).$$

Since $(d_1, p^\alpha) = 1$, for any divisor of c, this sum equals

$$\begin{aligned}&\sum_{d_1|c} \phi(d_1) + \sum_{d_1|c} \phi(p)\phi(d_1) + \cdots + \sum_{d_1|c} \phi(p^e)\phi(d_1) \\ &\quad = \sum_{d_1|c} \phi(d_1) + (p-1) \sum_{d_1|c} \phi(d_1) + \cdots + (p^e - p^{e-1}) \sum_{d_1|c} \phi(d_1) \\ &\quad = c + (p-1)c + (p^2 - p)c + \cdots + (p^e - p^{e-1})c.\end{aligned}$$

As in the case of prime powers, this sum telescopes, giving a final result

$$\sum_{d|n} \phi(d) = p^e c = n.$$

□

Example 6.3.11. Consider $n = 10$. The divisors are 1, 2, 5, 10. Then $\phi(1) = 1$, $\phi(2) = 1$, $\phi(5) = 4$, and $\phi(10) = 4$. Then

$$\phi(1) + \phi(2) + \phi(5) + \phi(10) = 1 + 1 + 4 + 4 = 10.$$

We will see later in the book that the Euler phi-function plays an important role in the structure theory of Abelian groups.

We now turn to *Fermat primes*.

Definition 6.3.12. The *Fermat numbers* are the sequence (F_n) of positive integers defined by

$$F_n = 2^{2^n} + 1, \quad n = 0, 1, 2, 3, \ldots.$$

If a particular F_n is prime, it is called a *Fermat prime*.

Fermat believed that all the numbers in this sequence were primes. In fact, F_0, F_1, F_2, F_3, F_4 are all primes, but F_5 is composite and divisible by 641 (see exercises). It is still an open question whether or not there are infinitely many Fermat primes. It has been conjectured that there are only finitely many. On the other hand, if a number of the form $2^n + 1$ is a prime for some integer n, then it must be a Fermat prime.

Theorem 6.3.13. *If $a \geq 2$ and $a^n + 1$ is a prime for some $n \geq 1$, then a is even, and $n = 2^m$ for some nonnegative integer m. In particular, if $p = 2^k + 1$ is a prime for some $k \geq 1$, then $k = 2^n$ for some n, and p is a Fermat prime.*

Proof. If a is odd then $a^n + 1$ is even and, hence, not a prime. Suppose then that a is even and $n = kl$ with k odd and $k \geq 3$. Then

$$\frac{a^{kl} + 1}{a^l + 1} = a^{(k-1)l} - a^{(k-2)l} + \cdots + 1.$$

Therefore, $a^l + 1$ divides $a^{kl} + 1$ if $k \geq 3$. Hence, if $a^n + 1$ is a prime, we must have $n = 2^m$. □

We can now state the solution to the constructibility of regular n-gons.

Theorem 6.3.14. *A regular n-gon is constructible with a straightedge and compass if and only if $n = 2^m p_1 \cdots p_k$, where $p_1, \ldots, p_k$ are distinct Fermat primes.*

For example, before proving the theorem, notice that a regular 20-gon is constructible since $20 = 2^2 \cdot 5$, and 5 is a Fermat prime. On the other hand, a regular 11-gon is not constructible.

Proof. Let $\mu = e^{\frac{2\pi i}{n}}$ be a primitive n-th root of unity. Since

$$e^{\frac{2\pi i}{n}} = \cos\left(\frac{2\pi}{n}\right) + i \sin\left(\frac{2\pi}{n}\right)$$

is easy to compute that (see exercises)

$$\mu + \frac{1}{\mu} = 2\cos\left(\frac{2\pi}{n}\right).$$

Therefore, $\mathbb{Q}(\mu + \frac{1}{\mu}) = \mathbb{Q}(\cos(\frac{2\pi}{n}))$. After we discuss Galois theory in more detail, we will prove that

$$\left|\mathbb{Q}\left(\mu + \frac{1}{\mu}\right) : \mathbb{Q}\right| = \frac{\phi(n)}{2},$$

where $\phi(n)$ is the Euler phi-function. Therefore, $\cos(\frac{2\pi}{n})$ is constructible if and only if $\frac{\phi(n)}{2}$ and, hence, $\phi(n)$ is a power of 2.

Suppose that $n = 2^m p_1^{e_1} \cdots p_k^{e_k}$, all p_i odd primes. Then from Theorem 6.3.8,

$$\phi(n) = 2^{m-1} \cdot (p_1^{e_1} - p_1^{e_1-1})(p_2^{e_2} - p_2^{e_2-1}) \cdots (p_k^{e_k} - p_k^{e_k-1}).$$

If this was a power of 2 each factor must also be a power of 2. Now

$$p_i^{e_i} - p_i^{e_i-1} = p_i^{e_i-1}(p_i - 1).$$

If this is to be a power of 2, we must have $e_i = 1$ and $p_i - 1 = 2^{k_i}$ for some k_i. Therefore, each prime is distinct to the first power, and $p_i = 2^{k_i} + 1$ is a Fermat prime, proving the theorem. □

6.4 Exercises

1. Let ϕ be a given angle. In which of the following cases is the angle ψ constructible from the angle ϕ by compass and straightedge?
   (a) $\phi = \frac{\pi}{13}$ and $\psi = \frac{\pi}{26}$.
   (b) $\phi = \frac{\pi}{33}$ and $\psi = \frac{\pi}{11}$.
   (c) $\phi = \frac{\pi}{7}$ and $\psi = \frac{\pi}{12}$.
2. (The golden section) In the plane, let $\overline{AB}$ be a given segment from A to B with length a. The segment $\overline{AB}$ should be divided such that the proportion of $\overline{AB}$ to the length of the bigger subsegment is equal to the proportion of the length of the bigger subsegment to the length of the smaller subsegment:

$$\frac{a}{b} = \frac{b}{a-b},$$

   where b is the length of the bigger subsegment. Such a division is called division by the golden section. If we write $b = ax$, $0 < x < 1$, then $\frac{1}{x} = \frac{x}{1-x}$, that is, $x^2 = 1 - x$. Do the following:
   (a) Show that $\frac{1}{x} = \frac{1+\sqrt{5}}{2} = \alpha$.

(b) Construct the division of $\overline{AB}$ by the golden section with compass and straightedge.

(c) If we divide the radius $r > 0$ of a circle by the golden section, then the bigger part of the so divided radius is the side of the regular 10-gon with its 10 vertices on the circle.

3. Given a regular 10-gon such that the 10 vertices are on the circle with radius $R > 0$. Show that the length of each side is equal to the bigger part of the radius divided by the golden section. Describe the procedure of the construction of the regular 10-gon and 5-gon.

4. Construct the regular 17-gon with compass and straightedge.
Hint: We have to construct the number $\frac{1}{2}(\omega + \omega^{-1}) = \cos\frac{2\pi}{17}$, where $\omega = e^{\frac{2\pi i}{17}}$. First, construct the positive zero ω_1 of the polynomial $x^2 + x - 4$; we get

$$\omega_1 = \frac{1}{2}(\sqrt{17} - 1) = \omega + \omega^{-1} + \omega^2 + \omega^{-2} + \omega^4 + \omega^{-4} + \omega^8 + \omega^{-8}.$$

Then, construct the positive zero ω_2 of the polynomial $x^2 - \omega_1 x - 1$; we get

$$\omega_2 = \frac{1}{4}\left(\sqrt{17} - 1 + \sqrt{34 - 2\sqrt{17}}\right) = \omega + \omega^{-1} + \omega^4 + \omega^{-4}.$$

From ω_1 and ω_2, construct $\beta = \frac{1}{2}(\omega_2^2 - \omega_1 + \omega_2 - 4)$. Then $\omega_3 = 2\cos\frac{2\pi}{17}$ is the biggest of the two positive zeros of the polynomial $x^2 - \omega_2 x + \beta$.

5. The Fibonacci numbers f_n are defined by $f_0 = 0, f_1 = 1$ and $f_{n+2} = f_{n+1} + f_n$ for $n \in \mathbb{N} \cup \{0\}$. Show the following:
(a) $f_n = \frac{\alpha^n - \beta^n}{\alpha - \beta}$ with $\alpha = \frac{1+\sqrt{5}}{2}, \beta = \frac{1-\sqrt{5}}{2}$.
(b) $\left(\frac{f_{n+1}}{f_n}\right)_{n\in\mathbb{N}}$ converges and $\lim_{n\to\infty}\frac{f_{n+1}}{f_n} = \frac{1+\sqrt{5}}{2} = \alpha$.
(c) $\begin{pmatrix} 0 & 1 \\ 1 & 1 \end{pmatrix}^n = \begin{pmatrix} f_{n-1} & f_n \\ f_n & f_{n+1} \end{pmatrix}, n \in \mathbb{N}$.
(d) $f_1 + f_2 + \cdots + f_n = f_{n+2} - 1, n \geq 1$.
(e) $f_{n-1}f_{n+1} - f_n^2 = (-1)^n, n \in \mathbb{N}$.
(f) $f_1^2 + f_2^2 + \cdots + f_n^2 = f_n f_{n+1}, n \in \mathbb{N}$.

6. Show: The Fermat numbers F_0, F_1, F_2, F_3, F_4 are all prime but F_5 is composite and divisible by 641.

7. Let $\mu = e^{\frac{2\pi i}{n}}$ be a primitive n-th root of unity. Using

$$e^{\frac{2\pi i}{n}} = \cos\left(\frac{2\pi}{n}\right) + i\sin\left(\frac{2\pi}{n}\right),$$

show that

$$\mu + \frac{1}{\mu} = 2\cos\left(\frac{2\pi}{n}\right).$$

7 Kronecker's Theorem and Algebraic Closures

7.1 Kronecker's Theorem

In the last chapter, we proved that if $L|K$ is a field extension, then there exists an intermediate field $K \subset \mathcal{A} \subset L$ such that $\mathcal{A}$ is algebraic over K, and contains all the elements of L that are algebraic over K. We call $\mathcal{A}$ the *algebraic closure* of K within L. In this chapter, we prove that starting with any field K, we can construct an extension field $\overline{K}$ that is algebraic over K and is *algebraically closed*. By this, we mean that there are no algebraic extensions of $\overline{K}$ or, equivalently, that there are no irreducible nonlinear polynomials in $\overline{K}[x]$. In the final section of this chapter, we will give a proof of the famous fundamental theorem of algebra, which in the language of this chapter says that the field $\mathbb{C}$ of complex numbers is algebraically closed. We will present another proof of this important result later in the book after we discuss Galois theory.

First, we need the following crucial result of Kronecker, which says that given a polynomial $f(x)$ in $K[x]$, where K is a field, we can construct an extension field L of K, in which $f(x)$ has a zero a. We say that L has been constructed by *adjoining* a to K. Recall that if $f(x) \in K[x]$ is irreducible, then $f(x)$ can have no zeros in K. We first need the following concept:

Definition 7.1.1. Let $L|K$ and $L'|K$ be field extensions. Then a K-isomorphism is an isomorphism $\tau : L \to L'$, that is, the identity map on K; thus, it fixes each element of K.

Theorem 7.1.2 (Kronecker's theorem). *Let K be a field and $f(x) \in K[x]$. Then there exists a finite extension K' of K, where $f(x)$ has a zero.*

Proof. Suppose that $f(x) \in K[x]$. We know that $f(x)$ factors into irreducible polynomials. Let $p(x)$ be an irreducible factor of $f(x)$. From the material in Chapter 4, we know that since $p(x)$ is irreducible, the principal ideal $\langle p(x)\rangle$ in $K[x]$ is a maximal ideal. To see this, suppose that $g(x) \notin \langle p(x)\rangle$, so that $g(x)$ is not a multiple of $p(x)$. Since $p(x)$ is irreducible, it follows that $(p(x), g(x)) = 1$. Thus, there exist $h(x), k(x) \in K[x]$ with

$$h(x)p(x) + k(x)g(x) = 1.$$

The element on the left is in the ideal $(g(x), p(x))$, so the identity, 1, is in this ideal. Therefore, the whole ring $K[x]$ is in this ideal. Since $g(x)$ was arbitrary, this implies that the principal ideal $\langle p(x)\rangle$ is maximal.

Now let $K' = K[x]/\langle p(x)\rangle$. Since $\langle p(x)\rangle$ is a maximal ideal, it follows that K' is a field. We show that K can be embedded in K', and that $p(x)$ has a zero in K'.

First, consider the map $\alpha : K[x] \to K'$ by $\alpha(f(x)) = f(x) + \langle p(x)\rangle$. This is a homomorphism. Since the identity element $1 \in K$ is not in $\langle p(x)\rangle$, it follows that α restricted to K is nontrivial. Therefore, α restricted to K is a monomorphism since if $\ker(\alpha_{|_K}) \neq K$ then $\ker(\alpha_{|_K}) = \{0\}$. Therefore, K can be embedded into $\alpha(K)$, which is contained in K'. Therefore, K' can be considered as an extension field of K. Consider the element $a =$

https://doi.org/10.1515/9783111142524-007

$x + \langle p(x)\rangle \in K'$. Then $p(a) = p(x) + \langle p(x)\rangle = 0 + \langle p(x)\rangle$ since $p(x) \in \langle p(x)\rangle$. But $0 + \langle p(x)\rangle$ is the zero element 0 of the factor ring $K[x]/\langle p(x)\rangle$. Therefore, in K', we have $p(a) = 0$; hence, $p(x)$ has a zero in K'. Since $p(x)$ divides $f(x)$, we must have $f(a) = 0$ in K' also. Therefore, we have constructed an extension field of K, in which $f(x)$ has a zero. □

In conformity to Chapter 5, we write $K(a)$ for the field adjunction of $a = x + \langle(p(x))\rangle$ to K. We now outline an intuitive construction. From this, we say that the field K is constructed by adjoining the zero (a) to K. We remark that this construction is not a formally correct proof as that given for Theorem 7.1.2.

We can assume that $f(x)$ is irreducible. Suppose that $f(x) = a_0 + a_1x + \cdots + a_nx^n$ with $a_n \neq 0$. Define a to satisfy

$$a_0 + a_1a + \cdots + a_na^n = 0.$$

Now, define $K' = K(a)$ in the following manner. We let

$$K(a) = \{c_0 + c_1a + \cdots + c_{n-1}a^{n-1} : c_i \in K\}.$$

Then on $K(a)$, define addition and subtraction componentwise, and define multiplication by algebraic manipulation, replacing powers of a higher than a^n by using

$$a^n = \frac{-a_0 - a_1a - \cdots - a_{n-1}a^{n-1}}{a_n}.$$

We claim that $K' = K(a)$, then forms a field of finite degree over K. The basic ring properties follow easily by computation (see exercises) using the definitions. We must show then that every nonzero element of $K(a)$ has a multiplicative inverse. Let $g(a) \in K(a)$. Then the corresponding polynomial $g(x) \in K[x]$ is a polynomial of degree $\leq n-1$. Since $f(x)$ is irreducible of degree n, it follows that $f(x)$ and $g(x)$ must be relatively prime; that is, $(f(x), g(x)) = 1$. Hence, there exist $a(x), b(x) \in K[x]$ with

$$a(x)f(x) + b(x)g(x) = 1.$$

Evaluate these polynomials at a to get

$$a(a)f(a) + b(a)g(a) = 1.$$

Since by definition we have $f(a) = 0$, this becomes

$$b(a)g(a) = 1.$$

Now $b(a)$ might have degree higher than $n - 1$ in a. However, using the relation that $f(a) = 0$, we can rewrite $b(a)$ as $\overline{b}(a)$, where $\overline{b}(a)$ now has degree $\leq n-1$ in a and, hence, is in $K(a)$. Therefore,

$$\overline{b}(a)g(a) = 1;$$

hence, $g(a)$ has a multiplicative inverse. It follows that $K(a)$ is a field and, by definition, $f(a) = 0$. The elements $1, a, \dots, a^{n-1}$ form a basis for $K(a)$ over K and, hence,

$$|K(a) : K| = n.$$

Example 7.1.3. Let $f(x) = x^2 + 1 \in \mathbb{R}[x]$. This is irreducible over $\mathbb{R}$. We construct the field, in which this has a zero. Let $K' \cong K[x]/\langle x^2 + 1\rangle$, and let $a \in K'$ with $f(a) = 0$. The extension field $\mathbb{R}(a)$ then has the form

$$K' = \mathbb{R}(a) = \{x + ay : x, y \in \mathbb{R}, a^2 = -1\}.$$

It is clear that this field is $\mathbb{R}$-isomorphic to the complex numbers $\mathbb{C}$; $\mathbb{R}(a) \cong \mathbb{R}(i) \cong \mathbb{C}$.

Theorem 7.1.4. *Let $p(x) \in K[x]$ be an irreducible polynomial, and let $K' = K(a)$ be the extension field of K constructed in Kronecker's theorem, in which $p(x)$ has a zero α. Let L be an extension field of K, and suppose that $a \in L$ is algebraic with minimal polynomial $m_a(x) = p(x)$. Then $K(a)$ is K-isomorphic to $K(a)$.*

Proof. If $L|K$ is a field extension and $a \in L$ with $p(a) = 0$ and if $\deg(p(x)) = n$, then the elements $1, a, \dots, a^{n-1}$ constitute a basis for $K(a)$ over K, and the elements $1, a, \dots, a^{n-1}$ constitute a basis for $K(a)$ over K. The mapping

$$\tau : K(a) \to K(a)$$

defined by $\tau(k) = k$ if $k \in K$ and $\tau(a) = a$, and then extended by linearity, is easily shown to be a K-isomorphism. □

Theorem 7.1.5. *Let K be a field. Then the following are equivalent:*
(1) *Each nonconstant polynomial in $K[x]$ has a zero in K.*
(2) *Each nonconstant polynomial in $K[x]$ factors into linear factors over K. That is, for each $f(x) \in K[x]$, there exist elements $a_1, \dots, a_n, b \in K$ with*

$$f(x) = b(x - a_1) \cdots (x - a_n).$$

(3) *An element of $K[x]$ is irreducible if and only if it is of degree one.*
(4) *If $L|K$ is an algebraic extension, then $L = K$.*

Proof. Suppose that each nonconstant polynomial in $K[x]$ has a zero in K.

Let $f(x) \in K[x]$ with $\deg(f(x)) = n$. Suppose that a_1 is a zero of $f(x)$, then

$$f(x) = (x - a_1)h(x),$$

where the degree of $h(x)$ is $n - 1$. Now $h(x)$ has a zero a_2 in K so that

$$f(x) = (x - a_1)(x - a_2)g(x)$$

with $\deg(g(x)) = n - 2$. Continue in this manner, and $f(x)$ factors completely into linear factors. Hence, (1) implies (2).

Now suppose (2); that is, that each nonconstant polynomial in $K[x]$ factors into linear factors over K. Suppose that $f(x)$ is irreducible. If $\deg(f(x)) > 1$, then $f(x)$ factors into linear factors and, hence, is not irreducible. Therefore, $f(x)$ must be of degree 1, and (2) implies (3).

Now suppose that an element of $K[x]$ is irreducible if and only if it is of degree one, and suppose that $L|K$ is an algebraic extension. Let $a \in L$. Then a is algebraic over K. Its minimal polynomial $m_a(x)$ is monic and irreducible over K and, hence, from (3), is linear. Therefore, $m_a(x) = x - a \in K[x]$. It follows that $a \in K$ and, hence, $K = L$. Therefore, (3) implies (4).

Finally, suppose that whenever $L|K$ is an algebraic extension, then $L = K$. Suppose that $f(x)$ is a nonconstant polynomial in $K[x]$. From Kronecker's theorem, there exists a field extension L, and $a \in L$ with $f(a) = 0$. However, L is an algebraic extension. Therefore, by supposition, $K = L$. Therefore, $a \in K$, and $f(x)$ has a zero in K. Therefore, (4) implies (1), completing the proof. □

In the next section, we will prove that given a field K, we can always find an extension field $\overline{K}$ with the properties of the last theorem.

7.2 Algebraic Closures and Algebraically Closed Fields

A field K is termed *algebraically closed* if K has no algebraic extensions other than K itself. This is equivalent to any one of the conditions of Theorem 7.1.5.

Definition 7.2.1. A field K is *algebraically closed* if every nonconstant polynomial $f(x) \in K[x]$ has a zero in K.

The following theorem is just a restatement of Theorem 7.1.5.

Theorem 7.2.2. *A field K is algebraically closed if and only it satisfies any one of the following conditions:*
(1) *Each nonconstant polynomial in $K[x]$ has a zero in K.*
(2) *Each nonconstant polynomial in $K[x]$ factors into linear factors over K. That is, for each $f(x) \in K[x]$, there exist elements $a_1, \dots, a_n, b \in K$ with*

$$f(x) = b(x - a_1) \cdots (x - a_n).$$

(3) *An element of $K[x]$ is irreducible if and only if it is of degree one.*
(4) *If $L|K$ is an algebraic extension, then $L = K$.*

The prime example of an algebraically closed field is the field $\mathbb{C}$ of complex numbers. The fundamental theorem of algebra says that any nonconstant complex polynomial has a complex zero.

We now show that the algebraic closure of one field within an algebraically closed field is algebraically closed. First, we define a general algebraic closure.

Definition 7.2.3. An extension field $\overline{K}$ of a field K is an *algebraic closure* of K if $\overline{K}$ is algebraically closed and $\overline{K}|K$ is algebraic.

Theorem 7.2.4. *Let K be a field and $L|K$ an extension of K with L algebraically closed. Let $\overline{K} = \mathcal{A}_K$ be the algebraic closure of K within L. Then $\overline{K}$ is an algebraic closure of K.*

Proof. Let $\overline{K} = \mathcal{A}_K$ be the algebraic closure of K within L. We know that $\overline{K}|K$ is algebraic. Therefore, we must show that $\overline{K}$ is algebraically closed.

Let $f(x)$ be a nonconstant polynomial in $\overline{K}[x]$. Then $f(x) \in L[x]$. Since L is algebraically closed, $f(x)$ has a zero a in L. Since $f(a) = 0$ and $f(x) \in \overline{K}[x]$, it follows that a is algebraic over $\overline{K}$. However, $\overline{K}$ is algebraic over K. Therefore, a is also algebraic over K. Hence, $a \in \overline{K}$, and $f(x)$ has a zero in $\overline{K}$. Therefore, $\overline{K}$ is algebraically closed. □

We want to note the distinction between being algebraically closed and being an algebraic closure.

Lemma 7.2.5. *The complex numbers $\mathbb{C}$ are an algebraic closure of $\mathbb{R}$, but not an algebraic closure of $\mathbb{Q}$. An algebraic closure of $\mathbb{Q}$ is $\mathcal{A}$ the field of algebraic numbers within $\mathbb{C}$.*

Proof. $\mathbb{C}$ is algebraically closed (the fundamental theorem of algebra), and since $|\mathbb{C} : \mathbb{R}| = 2$, it is algebraic over $\mathbb{R}$. Therefore, $\mathbb{C}$ is an algebraic closure of $\mathbb{R}$. Although $\mathbb{C}$ is algebraically closed and contains the rational numbers $\mathbb{Q}$, it is not an algebraic closure of $\mathbb{Q}$ since it is not algebraic over $\mathbb{Q}$ as there exist transcendental elements.

On the other hand, $\mathcal{A}$, the field of algebraic numbers within $\mathbb{Q}$, is an algebraic closure of $\mathbb{Q}$ from Theorem 7.2.4. □

We now show that every field has an algebraic closure. To do this, we first show that any field can be embedded into an algebraically closed field.

Theorem 7.2.6. *Let K be a field. Then K can be embedded into an algebraically closed field.*

Proof. We show first that there is an extension field L of K, in which each nonconstant polynomial $f(x) \in K[x]$ has a zero in L.

Assign to each nonconstant $f(x) \in K[x]$ the symbol y_f, and consider

$$R = K[y_f : f(x) \in K[x]],$$

the polynomial ring over K in the variables y_f. Let

$$I = \left\{ \sum_{j=1}^{n} f_j(y_{f_j}) r_j : r_j \in R, f_j(x) \in K[x] \right\}.$$

It is straightforward that I is an ideal in R. Suppose that $I = R$. Then $1 \in I$. Hence, there is a linear combination

$$1 = g_1 f_1(y_{f_1}) + \cdots + g_n f_n(y_{f_n}),$$

where $g_i \in I = R$.

In the n polynomials $g_1, \ldots, g_n$, there are only a finite number of variables, say for example,

$$y_{f_1}, \ldots, y_{f_n}, \ldots, y_{f_m}.$$

Hence,

$$1 = \sum_{i=1}^{n} g_i(y_{f_1}, \ldots, y_{f_m}) f_i(y_{f_i}). \qquad (*)$$

Successive applications of Kronecker's theorem lead us to construct an extension field P of K, in which each f_i has a zero a_i. Substituting a_i for y_{f_i} in $(*)$ above, we get that $1 = 0$ a contradiction. Therefore, $I \neq R$.

Since I is a ideal not equal to the whole ring R, it follows that I is contained in a maximal ideal M of R. Set $L = R/M$. Since M is maximal L is a field. Now $K \cap M = \{0\}$. If not, suppose that $a \in K \cap M$ with $a \neq 0$. Then $a^{-1}a = 1 \in M$, and then $M = R$. Now define $\tau : K \to L$ by $\tau(k) = k + M$. Since $K \cap M = \{0\}$, it follows that $\ker(\tau) = \{0\}$. Therefore, τ is a monomorphism. This allows us to identify K and $\tau(K)$, and shows that K embeds into L.

Now suppose that $f(x)$ is a nonconstant polynomial in $K[x]$. Then

$$f(y_f + M) = f(y_f) + M.$$

However, by the construction $f(y_f) \in M$, so that

$$f(y_f + M) = M = \text{the zero element of } L.$$

Therefore, $y_f + M$ is a zero of $f(x)$.

Therefore, we have constructed a field L, in which every nonconstant polynomial in $K[x]$ has a zero in L.

We now iterate this procedure to form a chain of fields

$$K \subset K_1 (= L) \subset K_2 \subset \cdots$$

such that each nonconstant polynomial of $K_i[x]$ has a zero in K_{i+1}.

Now let $\hat{K} = \bigcup_I K_i$. It is easy to show (see exercises) that $\hat{K}$ is a field. If $f(x)$ is a nonconstant polynomial in $\hat{K}[x]$, then there is some i with $f(x) \in K_i[x]$. Therefore, $f(x)$ has a zero in $K_{i+1}[x] \subset \hat{K}$. Hence, $f(x)$ has a zero in $\hat{K}$, and $\hat{K}$ is algebraically closed. □

Theorem 7.2.7. *Let K be a field. Then K has an algebraic closure.*

Proof. Let $\hat{K}$ be an algebraically closed field containing K, which exists from Theorem 7.2.6. Now let $\overline{K} = \mathcal{A}_{\hat{K}}$ be the set of elements of $\hat{K}$ that are algebraic over K. From Theorem 7.2.4, $\hat{K}$ is an algebraic closure of K. □

The following lemma is straightforward. We leave the proof to the exercises.

Lemma 7.2.8. *Let K, K' be fields and $\phi : K \to K'$ a homomorphism. Then*

$$\tilde{\phi} : K[x] \to K'[x], \quad \text{given by}$$
$$\tilde{\phi}\left(\sum_{i=1}^{n} k_i x^i\right) = \sum_{i=0}^{n} (\phi(k_i))x^i,$$

is also a homomorphism. By convention, we identify ϕ and $\tilde{\phi}$ and write $\phi = \tilde{\phi}$. If ϕ is an isomorphism, then so is $\tilde{\phi}$.

Lemma 7.2.9. *Let K, K' be fields and $\phi : K \to K'$ an isomorphism. Let $f(x) \in K[x]$ be irreducible. Let $K \subset K(a)$ and $K' \subset K'(a')$, where a is a zero of $f(x)$ and a' is a zero of $\phi(f(x))$. Then there is an isomorphism $\psi : K(a) \to K'(a')$ with $\psi_{|K} = \phi$ and $\psi(a) = a'$. Furthermore, ψ is uniquely determined.*

Proof. This is a generalized version of Theorem 7.1.4. If $b \in K(a)$, then from the construction of $K(a)$, there is a polynomial $g(x) \in K[x]$ with $b = g(a)$. Define a map

$$\psi : K(a) \to K'(a')$$

by

$$\psi(b) = \phi(g(x))(a').$$

We show that ψ is an isomorphism.

First, ψ is well defined. Suppose that $b = g(a) = h(a)$ with $h(x) \in K[x]$. Then $(g - h)(a) = 0$. Since $f(x)$ is irreducible, this implies that $f(x) = cm_a(x)$, and since a is a zero of $(g - h)(x)$, then $f(x)|(g - h)(x)$. Then

$$\phi(f(x))|(\phi(g(x)) - \phi(h(x))).$$

Since $\phi(f(x))(a') = 0$, this implies that $\phi(g(x))(a') = \phi(h(x))(a')$; hence, the map ψ is well defined.

It is easy to show that ψ is a homomorphism. Let $b_1 = g_1(a)$, $b_2 = g_2(a)$. Then $b_1 b_1 = g_1 g_2(a)$. Hence,

$$\psi(b_1 b_2) = (\phi(g_1 g_2))(a') = \phi(g_1)(a')\phi(g_2)(a') = \psi(b_1)\psi(b_2).$$

In the same manner, we have $\psi(b_1 + b_2) = \psi(b_1) + \psi(b_2)$. Now suppose that $k \in K$ so that $k \in K[x]$ is a constant polynomial. Then $\psi(k) = (\phi(k))(a') = \phi(k)$. Therefore, ψ restricted to K is precisely ϕ. As ψ is not the zero mapping, it follows that ψ is a monomorphism.

Finally, since $K(a)$ is generated from K and a, and ψ restricted to K is ϕ, it follows that ψ is uniquely determined by ϕ and $\psi(a) = a'$. Hence, ψ is unique. □

Theorem 7.2.10. *Let $L|K$ be an algebraic extension. Suppose that L_1 is an algebraically closed field and ϕ is an isomorphism from K to $K_1 \subset L_1$. Then there exists a monomorphism ψ from L to L_1 with $\psi_{|_K} = \phi$.*

Before we give the proof, we note that the theorem gives the following diagram:

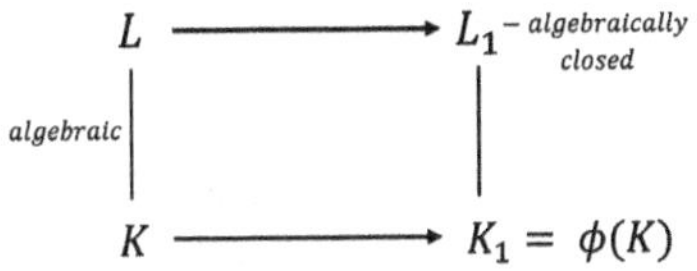

In particular, the theorem can be applied to monomorphisms of a field K within an algebraic closure $\overline{K}$ of K. Specifically, suppose that $K \subset \overline{K}$, where $\overline{K}$ is an algebraic closure of K, and let $\alpha : K \to \overline{K}$ be a monomorphism with $\alpha(K) = K$. Then there exists an automorphism α^* of $\overline{K}$ with $\alpha^*_{|_K} = \alpha$.

Proof of Theorem 7.2.10. Consider the set

$$\mathcal{M} = \{(M, \tau) : M \text{ is a field with } K \subset M \subset L, \\ \text{where there exists a monomorphism } \tau : M \to L_1 \text{ with } \tau_{|_K} = \phi\}.$$

Now the set $\mathcal{M}$ is nonempty since $(K, \phi) \in \mathcal{M}$. Order $\mathcal{M}$ by $(M_1, \tau_1) < (M_2, \tau_2)$ if $M_1 \subset M_2$ and $(\tau_2)_{|_{M_1}} = \tau_1$. Let

$$\mathcal{K} = \{(M_i, \tau_i) : i \in I\}$$

be a chain in $\mathcal{M}$. Let (M, τ) be defined by

$$M = \bigcup_{i \in I} M_i \quad \text{with } \tau(a) = \tau_i(a) \text{ for all } a \in M_i.$$

It is clear that M is an upper bound for the chain $\mathcal{K}$. Since each chain has an upper bound it follows from Zorn's lemma that $\mathcal{M}$ has a maximal element (N, ρ). We show that $N = L$.

Suppose that $N \subsetneq L$. Let $a \in L \setminus N$. Then a is algebraic over N and further algebraic over K, since $L|K$ is algebraic. Let $m_a(x) \in N[x]$ be the minimal polynomial of a relative to N. Since L_1 is algebraically closed, $\rho(m_a(x))$ has a zero $a' \in L_1$. Therefore, there is a monomorphism $\rho' : N(a) \to L_1$ with ρ' restricted to N, the same as ρ. It follows that $(N, \rho) < (N(a), \rho')$ since $a \notin N$. This contradicts the maximality of N. Therefore, $N = L$, completing the proof. □

Combining the previous two theorems, we can now prove that any two algebraic closures of a field K are unique up to K-isomorphism; that is, up to an isomorphism, thus, is the identity on K.

Theorem 7.2.11. *Let L_1 and L_2 be algebraic closures of the field K. Then there is a K-isomorphism $\tau : L \to L_1$. Again by K-isomorphism, we mean that τ is the identity on K.*

Proof. From Theorem 7.2.7, there is a monomorphism $\tau : L_1 \to L_2$ with τ the identity on K. However, since L_1 is algebraically closed, so is $\tau(L_1)$. Then $L_2|\tau(L_1)$ is an algebraic extension. Therefore, since L_2 is algebraically closed, we must have $L_2 = \tau(L_1)$. Therefore, τ is also surjective and, hence, an isomorphism. □

The following corollary is immediate.

Corollary 7.2.12. *Let $L|K$ and $L'|K$ be field extensions with $a \in L$ and $a' \in L'$ algebraic elements over K. Then $K(a)$ is K-isomorphic to $K(a')$ if and only if $|K(a) : K| = |K(a') : K|$, and there is an element $a'' \in K(a')$ with $m_a(x) = m_{a''}(x)$.*

7.3 The Fundamental Theorem of Algebra

The *fundamental theorem of algebra* is one of the most important algebraic results. This says that any nonconstant complex polynomial must have a complex zero. In the language of field extensions, this says that the field of complex numbers $\mathbb{C}$ is algebraically closed. There are many distinct and completely different proofs of this result. In [7], twelve proofs were given covering a wide area of mathematics. In this section we provide an elementary proof of the fundamental theorem of algebra. Before doing this, we briefly mention some of the history surrounding this theorem.

The first mention of the fundamental theorem of algebra, in the form that every polynomial equation of degree n has exactly n zeros, was given by Peter Roth of Nurnberg in 1608. However, its conjecture is generally credited to Girard, who also stated the result in 1629. It was then more clearly stated in 1637 by Descartes, who also distinguished between real and imaginary zeros. The first published proof of the fundamental theorem of algebra was then given by D'Alembert in 1746. However, there were gaps in D'Alembert's proof, and the first fully accepted proof was that given by Gauss in 1797 in his Ph. D. thesis. This was published in 1799. Interestingly enough, in reviewing Gauss' original proof, modern scholars tend to agree that there are as many holes in this proof as in D'Alembert's proof. Gauss, however, published three other proofs with no such holes. He published second and third proofs in 1816, while his final proof, which was essentially another version of the first, was presented in 1849.

First, we need the concept of a *splitting field* for a polynomial.

7.3.1 Splitting Fields

We have just seen that given an irreducible polynomial over a field K, we could always find a field extension, in which this polynomial has a zero. We now push this further to obtain field extensions, where a given polynomial has all its zeros.

Definition 7.3.1. If K is a field and $0 \neq f(x) \in K[x]$, and K' is an extension field of K, then $f(x)$ *splits* in K' (K' may be K), if $f(x)$ factors into linear factors in $K'[x]$. Equivalently, this means that all the zeros of $f(x)$ are in K'.

K' is a *splitting field* for $f(x)$ over K if K' is the smallest extension field of K, in which $f(x)$ splits. (A splitting field for $f(x)$ is the smallest extension field, in which $f(x)$ has all its possible zeros.)

K' is a *splitting field* over K if it is the splitting field for some finite set of polynomials over K.

Theorem 7.3.2. *If K is a field and $0 \neq f(x) \in K[x]$, then there exists a splitting field for $f(x)$ over K.*

Proof. The splitting field is constructed by repeated adjoining of zeros. Suppose, without loss of generality, that $f(x)$ is irreducible of degree n over K. From Theorem 7.1.2, there exists a field K' containing α with $f(\alpha) = 0$. Then $f(x) = (x - \alpha)g(x) \in K'[x]$ with $\deg g(x) = n - 1$. By an inductive argument, $g(x)$ has a splitting field; therefore, so does $f(x)$. □

7.3.2 Permutations and Symmetric Polynomials

To obtain a proof of the fundamental theorem of algebra, we need to go a bit outside of our main discussions of rings and fields and introduce symmetric polynomials. To introduce this concept, we first review some basic ideas from elementary group theory, which we will look at in detail later in the book.

Definition 7.3.3. A *group* G is a set with one binary operation, which we will denote by multiplication, such that the following hold:

(1) The operation is associative; that is, $(g_1g_2)g_3 = g_1(g_2g_3)$ for all $g_1, g_2, g_3 \in G$.
(2) There exists an identity for this operation; that is, an element 1 such that $1g = g$ for each $g \in G$.
(3) Each $g \in G$ has an inverse for this operation; that is, for each g, there exists a g^{-1} with the property that $gg^{-1} = 1$.

If in addition the operation is commutative ($g_1g_2 = g_2g_1$ for all $g_1, g_2 \in G$), the group G is called an *Abelian group*. The *order* of G is the number of elements in G, denoted $|G|$. If $|G| < \infty$, G is a *finite group*. $H \subset G$ is a *subgroup* if H is also a group under the same operation as G. Equivalently, H is a subgroup if $H \neq \emptyset$, and H is closed under the operation and inverses.

Groups most often arise from invertible mappings of a set onto itself. Such mappings are called *permutations.*

Definition 7.3.4. If T is a set, a *permutation* on T is a one-to-one mapping of T onto itself. We denote the set of all permutations on T by S_T.

Theorem 7.3.5. *For any set T, S_T forms a group under composition called the symmetric group on T. If T, T_1 have the same cardinality (size), then $S_T \cong S_{T_1}$. If T is a finite set with $|T| = n$, then S_T is a finite group, and $|S_T| = n!$.*

Proof. If S_T is the set of all permutations on the set T, we must show that composition is an operation on S_T that is associative and has an identity and inverses.

Let $f, g \in S_T$. Then f, g are one-to-one mappings of T onto itself.

Consider $f \circ g : T \to T$. If $f \circ g(t_1) = f \circ g(t_2)$, then $f(g(t_1)) = f(g(t_2))$, and $g(t_1) = g(t_2)$, since f is one-to-one. But then $t_1 = t_2$ since g is one-to-one.

If $t \in T$, there exists $t_1 \in T$ with $f(t_1) = t$ since f is onto. Then there exists $t_2 \in T$ with $g(t_2) = t_1$ since g is onto. Putting these together, $f(g(t_2)) = t$; therefore, $f \circ g$ is onto. Therefore, $f \circ g$ is also a permutation, and composition gives a valid binary operation on S_T.

The identity function $1(t) = t$ for all $t \in T$ will serve as the identity for S_T, whereas the inverse function for each permutation will be the inverse. Such unique inverse functions exist since each permutation is a bijection.

Finally, composition of functions is always associative; therefore, S_T forms a group.

If T, T_1 have the same cardinality, then there exists a bijection $\sigma : T \to T_1$. Define a map $F : S_T \to S_{T_1}$ in the following manner: if $f \in S_T$, let $F(f)$ be the permutation on T_1 given by $F(f)(t_1) = \sigma(f(\sigma^{-1}(t_1)))$. It is straightforward to verify that F is an isomorphism (see the exercises).

Finally, suppose $|T| = n < \infty$. Then $T = \{t_1, \ldots, t_n\}$. Each $f \in S_T$ can be pictured as

$$f = \begin{pmatrix} t_1 & \cdots & t_n \\ f(t_1) & \cdots & f(t_n) \end{pmatrix}.$$

For t_1, there are n choices for $f(t_1)$. For t_2, there are only $n - 1$ choices since f is one-to-one. This continues down to only one choice for t_n. Using the multiplication principle, the number of choices for f and, therefore, the size of S_T is

$$n(n-1)\cdots 1 = n!. \qquad \square$$

For a set with n elements, we denote S_T by S_n called the *symmetric group on n symbols.*

Example 7.3.6. Write down the six elements of S_3, and give the multiplication table for the group.

Name the three elements 1, 2, 3 of T. The six elements of S_3 are then:

$$1 = \begin{pmatrix} 1 & 2 & 3 \\ 1 & 2 & 3 \end{pmatrix}, \quad a = \begin{pmatrix} 1 & 2 & 3 \\ 2 & 3 & 1 \end{pmatrix}, \quad b = \begin{pmatrix} 1 & 2 & 3 \\ 3 & 1 & 2 \end{pmatrix}$$
$$c = \begin{pmatrix} 1 & 2 & 3 \\ 2 & 1 & 3 \end{pmatrix}, \quad d = \begin{pmatrix} 1 & 2 & 3 \\ 3 & 2 & 1 \end{pmatrix}, \quad e = \begin{pmatrix} 1 & 2 & 3 \\ 1 & 3 & 2 \end{pmatrix}.$$

The multiplication table for S_3 can be written down directly by doing the required composition. For example,

$$ac = \begin{pmatrix} 1 & 2 & 3 \\ 2 & 3 & 1 \end{pmatrix}\begin{pmatrix} 1 & 2 & 3 \\ 2 & 1 & 3 \end{pmatrix} = \begin{pmatrix} 1 & 2 & 3 \\ 3 & 2 & 1 \end{pmatrix} = d.$$

To see this, note that $a : 1 \to 2, 2 \to 3, 3 \to 1$; $c : 1 \to 2, 2 \to 1, 3 \to 3$, and so $ac : 1 \to 3, 2 \to 2, 3 \to 1$.

It is somewhat easier to construct the multiplication table if we make some observations. First, $a^2 = b$, and $a^3 = 1$. Next, $c^2 = 1$, $d = ac$, $e = a^2c$ and, finally, $ac = ca^2$.

From these relations, the following multiplication table can be constructed:

	1	a	a^2	c	ac	a^2c
1	1	a	a^2	c	ac	a^2c
a	a	a^2	1	ac	a^2c	c
a^2	a^2	1	a	a^2c	c	ac
c	c	a^2c	ac	1	a^2	a
ac	ac	c	a^2c	a	1	a^2
a^2c	a^2c	ac	c	a^2	a	1

To see this, consider, for example, $(ac)a^2 = a(ca^2) = a(ac) = a^2c$.

More generally, we can say that S_3 has a *presentation* given by

$$S_3 = \langle a, c; a^3 = c^2 = 1, ac = ca^2 \rangle.$$

By this, we mean that S_3 is *generated by* a, c, or that S_3 has *generators* a, c. Thus, the whole group and its multiplication table can be generated by using the *relations* $a^3 = c^2 = 1$, $ac = ca^2$.

An important result, the form of which we will see later in our work on extension fields, is the following:

Lemma 7.3.7. *Let T be a set and $T_1 \subset T$ a subset. Let H be the subset of S_T that fixes each element of T_1; that is, $f \in H$ iff $f(t) = t$ for all $t \in T_1$. Then H is a subgroup.*

Proof. We have $H \neq \emptyset$ since $1 \in H$. Now suppose $h_1, h_2 \in H$. Let $t_1 \in T_1$, and consider $h_1 \circ h_2(t_1) = h_1(h_2(t_1))$. Now $h_2(t_1) = t_1$ since $h_2 \in H$, but then $h_1(t_1) = t_1$ since $h_1 \in H$. Therefore, $h_1 \circ h_2 \in H$, and H is closed under composition. If h_1 fixes t_1, then h_1^{-1} also fixes t_1. Thus, H is also closed under inverses and is, therefore, a subgroup. □

We now apply these ideas of permutations to certain polynomial rings in independent indeterminates over a field. We will look at these in detail in Chapter 11.

Definition 7.3.8. Let $y_1, \dots, y_n$ be (independent) indeterminates over a field K. A polynomial $f(y_1, \dots, y_n) \in K[y_1, \dots, y_n]$ is a *symmetric polynomial* in $y_1, \dots, y_n$ if $f(y_1, \dots, y_n)$ is unchanged by any permutation σ of $\{y_1, \dots, y_n\}$: $f(y_1, \dots, y_n) = f(\sigma(y_1), \dots, \sigma(y_n))$.

If $K \subset K'$ are fields and $\alpha_1, \dots, \alpha_n$ are in K', then we call a polynomial $f(\alpha_1, \dots, \alpha_n)$ with coefficients in K *symmetric* in $\alpha_1, \dots, \alpha_n$ if $f(\alpha_1, \dots, \alpha_n)$ is unchanged by any permutation σ of $\{\alpha_1, \dots, \alpha_n\}$.

Example 7.3.9. Let K be a field and $k_0, k_1 \in K$. Let $h(y_1, y_2) = k_0(y_1 + y_2) + k_1(y_1y_2)$. There are two permutations on $\{y_1, y_2\}$, namely, $\sigma_1 : y_1 \to y_1, y_2 \to y_2$ and $\sigma_2 : y_1 \to y_2, y_2 \to y_1$. Applying either one of these two to $\{y_1, y_2\}$ leaves $h(y_1, y_2)$ invariant. Therefore, $h(y_1, y_2)$ is a symmetric polynomial.

Definition 7.3.10. Let $x, y_1, \dots, y_n$ be indeterminates over a field K (or elements of an extension field K' of K). Form the polynomial $p(x, y_1, \dots, y_n) = (x - y_1) \cdots (x - y_n)$. The *i-th elementary symmetric polynomial* s_i in $y_1, \dots, y_n$ for $i = 1, \dots, n$, is $(-1)^i a_i$, where a_i is the coefficient of x^{n-i} in $p(x, y_1, \dots, y_n)$.

Example 7.3.11. Consider y_1, y_2, y_3. Then

$$\begin{aligned} p(x, y_1, y_2, y_3) &= (x - y_1)(x - y_2)(x - y_3) \\ &= x^3 - (y_1 + y_2 + y_3)x^2 + (y_1y_2 + y_1y_3 + y_2y_3)x - y_1y_2y_3. \end{aligned}$$

Therefore, the three elementary symmetric polynomials in y_1, y_2, y_3 over any field are

(1) $s_1 = y_1 + y_2 + y_3$.
(2) $s_2 = y_1y_2 + y_1y_3 + y_2y_3$.
(3) $s_3 = y_1y_2y_3$.

In general, the pattern of the last example holds for $y_1, \dots, y_n$. That is,

$$\begin{aligned} s_1 &= y_1 + y_2 + \cdots + y_n \\ s_2 &= y_1y_2 + y_1y_3 + \cdots + y_{n-1}y_n \\ s_3 &= y_1y_2y_3 + y_1y_2y_4 + \cdots + y_{n-2}y_{n-1}y_n \\ &\vdots \\ s_n &= y_1 \cdots y_n. \end{aligned}$$

The importance of the elementary symmetric polynomials is that any symmetric polynomial can be built up from the elementary symmetric polynomials. We make this precise in the next theorem called the *fundamental theorem of symmetric polynomials*. We will use this important result several times, and we will give a complete proof in Section 7.4.

Theorem 7.3.12 (Fundamental theorem of symmetric polynomials). *If P is a symmetric polynomial in the indeterminates $y_1, \dots, y_n$ over a field K; that is, $P \in K[y_1, \dots, y_n]$ and P is symmetric, then there exists a unique $g \in K[y_1, \dots, y_n]$ with $f(y_1, \dots, y_n) = g(s_1, \dots, s_n)$. That is, any symmetric polynomial in $y_1, \dots, y_n$ is a polynomial expression in the elementary symmetric polynomials in $y_1, \dots, y_n$.*

From this theorem, we obtain the following two lemmas, which will be crucial in our proof of the fundamental theorem of algebra.

Lemma 7.3.13. *Let $p(x) \in K[x]$, and suppose $p(x)$ has the zeros $\alpha_1, \dots, \alpha_n$ in the splitting field K'. Then the elementary symmetric polynomials in $\alpha_1, \dots, \alpha_n$ are in K.*

Proof. Suppose $p(x) = c_0 + c_1 x + \dots + c_n x^n \in K[x]$. Since $p(x)$ splits in $K'[x]$, with zeros $\alpha_1, \dots, \alpha_n$, we have that, in $K'[x]$,

$$p(x) = c_n(x - \alpha_1) \cdots (x - \alpha_n).$$

The coefficients are then $c_n(-1)^i s_i(\alpha_1, \dots, \alpha_n)$, where the $s_i(\alpha_1, \dots, \alpha_n)$ are the elementary symmetric polynomials in $\alpha_1, \dots, \alpha_n$. However, $p(x) \in K[x]$, so each coefficient is in K. It follows then that for each i, $c_n(-1)^i s_i(\alpha_1, \dots, \alpha_n) \in K$; hence, $s_i(\alpha_1, \dots, \alpha_n) \in K$ since $c_n \in K$. □

Lemma 7.3.14. *Let $p(x) \in K[x]$, and suppose $p(x)$ has the zeros $\alpha_1, \dots, \alpha_n$ in the splitting field K'. Suppose further that $g(x) = g(x, \alpha_1, \dots, \alpha_n) \in K'[x]$. If $g(x)$ is a symmetric polynomial in $\alpha_1, \dots, \alpha_n$, then $g(x) \in K[x]$.*

Proof. If $g(x) = g(x, \alpha_1, \dots, \alpha_n)$ is symmetric in $\alpha_1, \dots, \alpha_n$, then from Theorem 7.3.12, it is a symmetric polynomial in the elementary symmetric polynomials in $\alpha_1, \dots, \alpha_n$. From Lemma 7.3.13, these are in the ground field K, so the coefficients of $g(x)$ are in K. Therefore, $g(x) \in K[x]$. □

We now present a proof of the fundamental theorem of algebra.

Theorem 7.3.15 (Fundamental theorem of algebra). *Any nonconstant complex polynomial has a complex zero. In other words, the complex number field $\mathbb{C}$ is algebraically closed.*

The proof depends on the following sequence of lemmas. The crucial one now is the last, which says that any real polynomial must have a complex zero.

Lemma 7.3.16. *Any odd-degree real polynomial must have a real zero.*

Proof. This is a consequence of the intermediate value theorem from analysis.

Suppose $P(x) \in \mathbb{R}[x]$ with $\deg P(x) = n = 2k + 1$, and suppose the leading coefficient $a_n > 0$ (the proof is almost identical if $a_n < 0$). Then

$$P(x) = a_n x^n + \text{(lower terms)},$$

and n is odd. Then,

(1) $\lim_{x\to\infty} P(x) = \lim_{x\to\infty} a_n x^n = \infty$ since $a_n > 0$.
(2) $\lim_{x\to-\infty} P(x) = \lim_{x\to-\infty} a_n x^n = -\infty$ since $a_n > 0$ and n is odd.

From (1), $P(x)$ gets arbitrarily large positively, so there exists an x_1 with $P(x_1) > 0$. Similarly, from (2) there exists an x_2 with $P(x_2) < 0$.

A real polynomial is a continuous real-valued function for all $x \in \mathbb{R}$. Since $P(x_1)P(x_2) < 0$, it follows from the intermediate value theorem that there exists an x_3, between x_1 and x_2, such that $P(x_3) = 0$. □

Lemma 7.3.17. *Any degree-two complex polynomial must have a complex zero.*

Proof. This is a consequence of the quadratic formula and of the fact that any complex number has a square root.

If $P(x) = ax^2 + bx + c$, $a \neq 0$, then the zeros formally are

$$x_1 = \frac{-b + \sqrt{b^2 - 4ac}}{2a}, \quad x_2 = \frac{-b - \sqrt{b^2 - 4ac}}{2a}.$$

From DeMoivre's theorem, every complex number has a square root; hence, x_1, x_2 exist in $\mathbb{C}$. They of course are the same if $b^2 - 4ac = 0$. □

To go further, we need the concept of the *conjugate of a polynomial* and some straightforward consequences of this idea.

Definition 7.3.18. If $P(x) = a_0 + \cdots + a_n x^n$ is a complex polynomial then its *conjugate* is the polynomial $\overline{P}(x) = \overline{a_0} + \cdots + \overline{a_n} x^n$. That is, the conjugate is the polynomial whose coefficients are the complex conjugates of those of $P(x)$.

Lemma 7.3.19. *For any $P(x) \in \mathbb{C}[x]$, we have the following:*

(1) $\overline{P(z)} = \overline{P}(\overline{z})$ *if $z \in \mathbb{C}$.*
(2) *$P(x)$ is a real polynomial if and only if $P(x) = \overline{P}(x)$.*
(3) *If $P(x)Q(x) = H(x)$, then $\overline{H}(x) = (\overline{P}(x))(\overline{Q}(x))$.*

Proof. (1) Suppose $z \in \mathbb{C}$ and $P(z) = a_0 + \cdots + a_n z^n$. Then

$$\overline{P(z)} = \overline{a_0 + \cdots + a_n z^n} = \overline{a_0} + \overline{a_1}\overline{z} + \cdots + \overline{a_n}\overline{z^n} = \overline{P}(\overline{z}).$$

(2) Suppose $P(x)$ is real, then $a_i = \overline{a_i}$ for all its coefficients; hence, $P(x) = \overline{P}(x)$. Conversely, suppose $P(x) = \overline{P}(x)$. Then $a_i = \overline{a_i}$ for all its coefficients; hence, $a_i \in \mathbb{R}$ for each a_i; therefore, $P(x)$ is a real polynomial.

(3) The proof is a computation and left to the exercises. □

Lemma 7.3.20. *Suppose $G(x) \in \mathbb{C}[x]$. Then $H(x) = G(x)\overline{G}(x) \in \mathbb{R}[x]$.*

Proof. $\overline{H}(x) = \overline{G(x)\overline{G}(x)} = \overline{G}(x)\overline{\overline{G}}(x) = \overline{G}(x)G(x) = G(x)\overline{G}(x) = H(x)$. Therefore, $H(x)$ is a real polynomial. □

Lemma 7.3.21. *If every nonconstant real polynomial has a complex zero, then every nonconstant complex polynomial has a complex zero.*

Proof. Let $P(x) \in \mathbb{C}[x]$, and suppose that every nonconstant real polynomial has at least one complex zero. Let $H(x) = P(x)\overline{P}(x)$. From Lemma 7.3.20, $H(x) \in \mathbb{R}[x]$. By supposition there exists a $z_0 \in \mathbb{C}$ with $H(z_0) = 0$. Then $P(z_0)\overline{P}(z_0) = 0$, and since $\mathbb{C}$ is a field it has no zero divisors.

Hence, either $P(z_0) = 0$, or $\overline{P}(z_0) = 0$. In the first case, z_0 is a zero of $P(x)$. In the second case, $\overline{P}(z_0) = 0$. Then from Lemma 7.3.19, $\overline{P}(z_0) = \overline{\overline{P}(\overline{z_0})} = P(\overline{z_0}) = 0$. Therefore, $\overline{z_0}$ is a zero of $P(x)$. □

Now we come to the crucial lemma.

Lemma 7.3.22. *Any nonconstant real polynomial has a complex zero.*

Proof. Let $f(x) = a_0 + a_1x + \cdots + a_nx^n \in \mathbb{R}[x]$ with $n \geq 1$, $a_n \neq 0$. The proof is an induction on the degree n of $f(x)$.

Suppose $n = 2^m q$, where q is odd. We do the induction on m. If $m = 0$, then $f(x)$ has odd degree, and the theorem is true from Lemma 7.3.16. Assume then that the theorem is true for all degrees $d = 2^k q'$, where $k < m$ and q' is odd. Now assume that the degree of $f(x)$ is $n = 2^m q$.

Suppose K' is the splitting field for $f(x)$ over $\mathbb{R}$, in which the zeros are $\alpha_1, \ldots, \alpha_n$. We show that at least one of these zeros must be in $\mathbb{C}$. (In fact, all are in $\mathbb{C}$, but to prove the lemma, we need only show at least one.)

Let $h \in \mathbb{Z}$, and form the polynomial

$$H(x) = \prod_{i<j} (x - (\alpha_i + \alpha_j + h\alpha_i\alpha_j)).$$

This is in $K'[x]$. In forming $H(x)$, we chose pairs of zeros $\{\alpha_i, \alpha_j\}$, so the number of such pairs is the number of ways of choosing two elements out of $n = 2^m q$ elements. This is given by

$$\frac{(2^m q)(2^m q - 1)}{2} = 2^{m-1} q(2^m q - 1) = 2^{m-1} q'$$

with q' odd. Therefore, the degree of $H(x)$ is $2^{m-1} q'$.

$H(x)$ is a symmetric polynomial in the zeros $\alpha_1, \ldots, \alpha_n$. Since $\alpha_1, \ldots, \alpha_n$ are the zeros of a real polynomial, from Lemma 7.3.14, any polynomial in the splitting field symmetric in these zeros must be a real polynomial.

Therefore, $H(x) \in \mathbb{R}[x]$ with degree $2^{m-1} q'$. By the inductive hypothesis, then, $H(x)$ must have a complex zero. This implies that there exists a pair $\{\alpha_i, \alpha_j\}$ with

$$\alpha_i + \alpha_j + h\alpha_i\alpha_j \in \mathbb{C}.$$

Since h was an arbitrary integer, for any integer h_1, there must exist such a pair $\{\alpha_i, \alpha_j\}$ with

$$\alpha_i + \alpha_j + h_1\alpha_i\alpha_j \in \mathbb{C}.$$

Now let h_1 vary over the integers. Since there are only finitely many such pairs $\{\alpha_i, \alpha_j\}$, it follows that there must be at least two different integers h_1, h_2 such that

$$z_1 = \alpha_i + \alpha_j + h_1\alpha_i\alpha_j \in \mathbb{C}, \quad \text{and} \quad z_2 = \alpha_i + \alpha_j + h_2\alpha_i\alpha_j \in \mathbb{C}.$$

Then $z_1 - z_2 = (h_1 - h_2)\alpha_i\alpha_j \in \mathbb{C}$, and since $h_1, h_2 \in \mathbb{Z} \subset \mathbb{C}$, it follows that $\alpha_i\alpha_j \in \mathbb{C}$. But then $h_1\alpha_i\alpha_j \in \mathbb{C}$, from which it follows that $\alpha_i + \alpha_j \in \mathbb{C}$. Then,

$$p(x) = (x - \alpha_i)(x - \alpha_j) = x^2 - (\alpha_i + \alpha_j)x + \alpha_i\alpha_j \in \mathbb{C}[x].$$

However, $p(x)$ is then a degree-two complex polynomial, and so from Lemma 7.3.17, its zeros are complex. Therefore, $\alpha_i, \alpha_j \in \mathbb{C}$; thus, $f(x)$ has a complex zero. □

It is now easy to give a proof of the fundamental theorem of algebra. From Lemma 7.3.22, every nonconstant real polynomial has a complex zero. From Lemma 7.3.21, if every nonconstant real polynomial has a complex zero, then every nonconstant complex polynomial has a complex zero, proving the fundamental theorem.

Theorem 7.3.23. *If E is a finite-dimensional field extension of $\mathbb{C}$, then $E = \mathbb{C}$.*

Proof. Let $a \in E$. Regard the elements $1, a, a^2, \ldots$. These elements become linearly dependent over $\mathbb{C}$, and we get a nonconstant polynomial over $\mathbb{C}$ with zero a. By the fundamental theorem of algebra, we know that $a \in \mathbb{C}$. □

Corollary 7.3.24. *If E is a finite-dimensional field extension of $\mathbb{R}$, then $E = \mathbb{R}$, or $E = \mathbb{C}$.*

We refer to Section 17.6 where we revisit the fundamental theorem of algebra and provide a Galois theoretic proof.

7.4 The Fundamental Theorem of Symmetric Polynomials

In the proof of the fundamental theorem of algebra that was given in the previous section, we used the fact that any symmetric polynomial in n indeterminates is a polynomial in the elementary symmetric polynomials in these indeterminates. In this section, we give a proof of this theorem.

Let R be an integral domain with $x_1, \ldots, x_n$ (independent) indeterminates over R, and let $R[x_1, \ldots, x_n]$ be the polynomial ring in these indeterminates. Any polynomial $f(x_1, \ldots, x_n) \in R[x_1, \ldots, x_n]$ is composed of a sum of *pieces* of the form $ax_1^{i_1} \cdots x_n^{i_n}$ with $a \in R$. We first put an order on these pieces of a polynomial.

The piece $ax_1^{i_1}\cdots x_n^{i_n}$ with $a \neq 0$ is called *higher* than the piece $bx_1^{j_1}\cdots x_n^{j_n}$ with $b \neq 0$, if the first one of the differences $i_1 - j_1, i_2 - j_2, \ldots, i_n - j_n$ that differs from zero is in fact positive. The highest piece of a polynomial $f(x_1, \ldots, x_n)$ is denoted by $\mathrm{HG}(f)$.

Lemma 7.4.1. *For $f(x_1, \ldots, x_n), g(x_1, \ldots, x_n) \in R[x_1, \ldots, x_n]$, we have*

$$\mathrm{HG}(fg) = \mathrm{HG}(f)\,\mathrm{HG}(g).$$

Proof. We use an induction on n, the number of indeterminates. It is clearly true for $n = 1$, and now assume that the statement holds for all polynomials in k indeterminates with $k < n$ and $n \geq 2$. Order the polynomials via exponents on the first indeterminate x_1 so that

$$\begin{aligned} f(x_1, \ldots, x_n) &= x_1^r \phi_r(x_2, \ldots, x_n) + x_1^{r-1}\phi_{r-1}(x_2, \ldots, x_n) \\ &\quad + \cdots + \phi_0(x_2, \ldots, x_n) \\ g(x_1, \ldots, x_n) &= x_1^s \psi_s(x_2, \ldots, x_n) + x_1^{s-1}\psi_{s-1}(x_2, \ldots, x_n) \\ &\quad + \cdots + \psi_0(x_2, \ldots, x_n). \end{aligned}$$

Then $\mathrm{HG}(fg) = x_1^{r+s}\,\mathrm{HG}(\phi_r \psi_s)$. By the inductive hypothesis

$$\mathrm{HG}(\phi_r \psi_s) = \mathrm{HG}(\phi_r)\,\mathrm{HG}(\psi_s).$$

Hence,

$$\begin{aligned} \mathrm{HG}(fg) &= x_1^{r+s}\,\mathrm{HG}(\phi_r)\,\mathrm{HG}(\psi_s) \\ &= (x_1^r\,\mathrm{HG}(\phi_r))(x_1^s\,\mathrm{HG}(\psi_s)) = \mathrm{HG}(f)\,\mathrm{HG}(g). \end{aligned}$$ □

The elementary symmetric polynomials in n indeterminates $x_1, \ldots, x_n$ are:

$$\begin{aligned} s_1 &= x_1 + x_2 + \cdots + x_n \\ s_2 &= x_1x_2 + x_1x_3 + \cdots + x_{n-1}x_n \\ s_3 &= x_1x_2x_3 + x_1x_2x_4 + \cdots + x_{n-2}x_{n-1}x_n \\ &\vdots \\ s_n &= x_1 \cdots x_n. \end{aligned}$$

These were found by forming the polynomial $p(x, x_1, \ldots, x_n) = (x - x_1)\cdots(x - x_n)$. The i-th elementary symmetric polynomial s_i in $x_1, \ldots, x_n$ is then $(-1)^i a_i$, where a_i is the coefficient of x^{n-i} in $p(x, x_1, \ldots, x_n)$.

In general,

$$s_k = \sum_{i_1 < i_2 < \cdots < i_k, 1 \leq k \leq n} x_{i_1} x_{i_2} \cdots x_{i_k},$$

where the sum is taken over all the $\binom{n}{k}$ different systems of indices $i_1, \dots, i_k$ with $i_1 < i_2 < \cdots < i_k$. Furthermore, a polynomial $s(x_1, \dots, x_n)$ is a *symmetric polynomial* if $s(x_1, \dots, x_n)$ is unchanged by any permutation σ of $\{x_1, \dots, x_n\}$, that is, $s(x_1, \dots, x_n) = s(\sigma(x_1), \dots, \sigma(x_n))$.

Lemma 7.4.2. *In the highest piece $ax_1^{k_1} \cdots x_n^{k_n}$ with $a \neq 0$ of a symmetric polynomial $s(x_1, \dots, x_n)$, we have $k_1 \geq k_2 \geq \cdots \geq k_n$.*

Proof. Assume that $k_i < k_j$ for some $i < j$. As a symmetric polynomial, $s(x_1, \dots, x_n)$ also must then contain the piece $ax_1^{k_1} \cdots x_i^{k_j} \cdots x_j^{k_i} \cdots x_n^{k_n}$, which is higher than $ax_1^{k_1} \cdots x_i^{k_i} \cdots x_j^{k_j} \cdots x_n^{k_n}$, giving a contradiction. □

Lemma 7.4.3. *The product $s_1^{k_1-k_2} s_2^{k_2-k_3} \cdots s_{n-1}^{k_{n-1}-k_n} s_n^{k_n}$ with $k_1 \geq k_2 \geq \cdots \geq k_n$ has the highest piece $x_1^{k_1} x_2^{k_2} \cdots x_n^{k_n}$.*

Proof. From the definition of the elementary symmetric polynomials, we have that

$$\mathrm{HG}(s_k^t) = (x_1 x_2 \cdots x_k)^t, \quad 1 \leq k \leq n,\ t \geq 1.$$

From Lemma 7.3.16,

$$\begin{aligned}
&\mathrm{HG}(s_1^{k_1-k_2} s_2^{k_2-k_3} \cdots s_{n-1}^{k_{n-1}-k_n} s_n^{k_n}) \\
&= x_1^{k_1-k_2} (x_1 x_2)^{k_2-k_3} \cdots (x_1 \cdots x_{n-1}^{k_{n-1}-k_n})(x_1 \cdots x_n)^{k_n} \\
&= x_1^{k_1} x_2^{k_2} \cdots x_n^{k_n}.
\end{aligned}$$

□

Theorem 7.4.4. *Let $s(x_1, \dots, x_n) \in R[x_1, \dots, x_n]$ be a symmetric polynomial. Then $s(x_1, \dots, x_n)$ can be uniquely expressed as a polynomial $f(s_1, \dots, s_n)$ in the elementary symmetric polynomials $s_1, \dots, s_n$ with coefficients from R.*

Proof. We prove the existence of the polynomial f by induction on the size of the highest pieces. If in the highest piece of a symmetric polynomial all exponents are zero, then it is constant, that is, an element of R. Therefore, there is nothing to prove.

Now we assume that each symmetric polynomial with the highest piece smaller than that of $s(x_1, \dots, x_n)$ can be written as a polynomial in the elementary symmetric polynomials. Let $ax_1^{k_1} \cdots x_n^{k_n}$, $a \neq 0$, be the highest piece of $s(x_1, \dots, x_n)$. Let

$$t(x_1, \dots, x_n) = s(x_1, \dots, x_n) - as_1^{k_1-k_2} \cdots s_{n-1}^{k_{n-1}-k_n} s_n^{k_n}.$$

Clearly, $t(x_1, \dots, x_n)$ is another symmetric polynomial, and from Lemma 7.3.19, the highest piece of $t(x_1, \dots, x_n)$ is smaller than that of $s(x_1, \dots, x_n)$. Therefore, $t(x_1, \dots, x_n)$. Hence, $s(x_1, \dots, x_n) = t(x_1, \dots, x_n) + as_1^{k_1-k_2} \cdots s_{n-1}^{k_{n-1}-k_n} s_n^{k_n}$ can be written as a polynomial in $s_1, \dots, s_n$. To prove the uniqueness of this expression, assume that

$$s(x_1, \dots, x_n) = f(s_1, \dots, s_n) = g(s_1, \dots, s_n).$$

Then

$$f(s_1,\ldots,s_n) - g(s_1,\ldots,s_n) = h(s_1,\ldots,s_n) = \phi(x_1,\ldots,x_n)$$

is the zero polynomial in $x_1,\ldots,x_n$. Hence, if we write $h(s_1,\ldots,s_n)$ as a sum of products of powers of the $s_1,\ldots,s_n$, all coefficients disappear because two different products of powers in the $s_1,\ldots,s_n$ have different highest pieces. This follows from the previous set of lemmas. Therefore, f and g are the same, proving the theorem. □

7.5 Skew Field Extensions of $\mathbb{C}$ and the Frobenius Theorem

Let V be a $\mathbb{R}$-vector space with $\dim_{\mathbb{R}}(V) = n < \infty$. We have already seen that as a consequence of the Fundamental theorem of algebra that only for $n = 1$ and $n = 2$, we may provide V with a multiplication such that V becomes a field with respect to the addition in V and this multiplication. Up to isomorphisms, we get $V = \mathbb{R}$ if $n = 1$ and $V = \mathbb{C}$ if $n = 2$.

If we want a suitable multiplication for $n \geq 3$, we have to give up some of the rules of a field. If all the axioms of a field hold except for the commutativity of multiplication, then we have a *skew field* or *division ring*. Hence, a division ring is a noncommutative ring with identity, in which every nonzero element has a multiplicative inverse.

Hamilton described for $n = 4$ a multiplication in V in such a way that V becomes a skew field. In his honor, we talk about the *Hamiltonian skew field*. This skew field is denoted by $\mathbb{H}$ and is called the *quaternions*.

In this section, we want first to describe the skew field $\mathbb{H}$ of Hamilton's quaternions and then to prove that if $n \geq 3$, only for $n = 4$ can we provide V with a multiplication such that V becomes a skew field.

We start with the construction and description of $\mathbb{H}$. Let $\{1, i, j, k\}$ be a basis of V. The addition will be the usual addition in the vector space. We also take scalar multiplication by $\mathbb{R}$. The basis element 1 shall be the unit element for the multiplication (as already mentioned in the case of the complex numbers, this is not a restriction because any nonzero vector in V is a member of a basis). The basis element 1 then should generate the embedding of $\mathbb{R}$.

For i, j, k, we define a multiplication by the following rules of Hamilton:

$$\begin{aligned}
&i^2 = j^2 = k^2 = -1,\\
&ij = k, \quad jk = i, \quad ki = j,\\
&ji = -k, \quad kj = -i, \quad ik = -j.
\end{aligned}$$

For

$$x = x_0 + x_1 i + x_2 j + x_3 k \quad \text{and} \quad y = y_0 + y_1 i + y_2 j + x_3 k,$$

we determine the addition and multiplication in V by following basic algebraic manipulation:

$$\begin{aligned} x+y &:= (x_0+y_0)+(x_1+y_1)i+(x_2+y_2)j+(x_3+y_3)k, \\ x\cdot y &:= (x_0y_0-x_1y_1-x_2y_2-x_3y_3)+(x_0y_1+x_1y_0+x_2y_3-x_3y_2)i \\ &\quad +(x_0y_2-x_1y_3+x_2y_0+x_3y_1)j+(x_0y_3+x_1y_2-x_2y_1+x_3y_0)k. \end{aligned}$$

Together with this addition and multiplication, V becomes a noncommutative ring with unit element 1. For each quaternion

$$x = x_0 + x_1 i + x_2 j + x_3 k,$$

we define the conjugate quaternion by

$$\overline{x} := x_0 - x_1 i - x_2 j - x_3 k.$$

We have the rules

$$\overline{\overline{x}} = x, \quad \overline{x+y} = \overline{x}+\overline{y}, \quad \overline{\lambda x} = \lambda\overline{x}, \quad \lambda \in \mathbb{R}, \quad \text{and} \quad \overline{xy} = \overline{x}\cdot\overline{y}.$$

With help of the conjugation, we may now define the *norm* and the *length* of a quaternion

$$x = x_0 + x_1 i + x_2 j + x_3 k$$

by

$$n(x) = x\overline{x} = \overline{x}x = x_0^2 + x_1^2 + x_2^2 + x_3^2 \quad \text{and} \quad |x| = \sqrt{x_0^2 + x_1^2 + x_2^2 + x_3^2},$$

respectively, in analogy to the complex numbers. If $x \neq 0$, then we get the multiplicative inverse x^{-1} by $x^{-1} = \frac{\overline{x}}{x\overline{x}}$, because

$$xx^{-1} = x\frac{\overline{x}}{x\overline{x}} = 1 = \overline{x}\frac{x}{x\overline{x}}.$$

Hence, together with the addition and multiplication, V becomes a skew field, in which $\mathbb{R}$ can be embedded via $r \mapsto r \cdot 1$ for $r \in \mathbb{R}$.

Theorem 7.5.1. *The set of quaternions $\mathbb{H}$ is a skew field, which contains both the reals and the complexes as subfields. It has dimension 4 as a vector space over $\mathbb{R}$. Furthermore, $rx = xr$ for all $x \in \mathbb{H}$, and all $r \in \mathbb{R}$ (considered as elements of $\mathbb{H}$).*

In $\mathbb{H}$, there is an important multiplicative rule for the norm and the length:

$$n(xy) = n(x)n(y) \quad \text{and} \quad |xy| = |x||y| \quad \text{for } x,y \in \mathbb{H}.$$

This can be shown by an easy calculation.

This result on norms in the quaternions provides the general equation in $\mathbb{R}$ on sums of four squares:

$$\begin{aligned}(x_0^2+x_1^2+x_2^2+x_3^2)(y_0^2+y_1^2+y_2^2+y_3^2) &= (x_0y_0-x_1y_1-x_2y_2-x_3y_3)^2\\ &\quad+(x_0y_1+x_1y_0+x_2y_3-x_3y_2)^2\\ &\quad+(x_0y_2-x_1y_3+x_2y_0+x_3y_1)^2\\ &\quad+(x_0y_3+x_1y_2-x_2y_1+x_3y_0)^2.\end{aligned}$$

This equation is one of the bases for the Theorem of Lagrange.

Theorem 7.5.2 (Theorem of Lagrange). *Each natural number n can be written as a sum*

$$n=a^2+b^2+c^2+d^2$$

of four squares with $a,b,c,d\in\mathbb{Z}$.

Hint: We have only to show that (see [53, Chapter 3.2]) if p is a prime number with $p\equiv 3 \pmod 4$, then $p=a^2+b^2+c^2+d^2$ for some $a,b,c,d\in\mathbb{Z}$. A proof of this can be found for instance in the book [53].

We remark that the skew field $\mathbb{H}$ of the quaternions can be embedded into $M(2,\mathbb{C})$ via

$$1\mapsto\begin{pmatrix}1&0\\0&1\end{pmatrix},\quad i\mapsto\begin{pmatrix}i&0\\0&-i\end{pmatrix},$$

$$j\mapsto\begin{pmatrix}0&1\\-1&0\end{pmatrix},\quad k\mapsto\begin{pmatrix}0&i\\i&0\end{pmatrix}.$$

Using this map, a quaternion $x=x_0+x_1i+x_2j+x_3k$ can be considered as a matrix

$$\begin{pmatrix}x_0+x_1i & x_2+x_3i\\ -x_2+x_3i & x_0-x_1i\end{pmatrix}=\begin{pmatrix}w&z\\-\overline{z}&\overline{w}\end{pmatrix}$$

with $w=x_0+x_1i\in\mathbb{C}$ and $z=x_2+x_3i\in\mathbb{C}$.

We have shown that the quaternions form a skew field of degree 4 over the real numbers. We ask whether there can be other finite degree skew field extensions of $\mathbb{R}$. Let V be a $\mathbb{R}$-vector space of $\dim_{\mathbb{R}}(V)=n<\infty$. For which n, we may provide V with a multiplication such that V with the vector addition and this multiplication becomes a field, or a skew field.

We remark that some nonzero vector in V has to be the unit element 1; therefore, we automatically have an embedding $\mathbb{R}\to V$.

Let $n\ge 2$. Since the irreducible polynomials from $\mathbb{R}[x]$ have degree 1 or 2, then under the existence of such a multiplication, each element $\alpha\in V$, which is not in $\mathbb{R}$ (considered as a subset of V), must be a zero of a quadratic polynomial from $\mathbb{R}[x]$.

We now assume that we have in V a multiplication such that V, together with the addition in V and this multiplication, is a field or a skew field.

If $n = 2$, we get the field $\mathbb{C}$ of the complex numbers.

Now, let $n = 3$. Using analogous thoughts as for the implementation of $\mathbb{C}$, we may construct in two steps a basis $\{1, i, j\}$ of V such that 1 is the unit element of V, and $i^2 = j^2 = -1$. Recall that a two-dimensional subspace of V has to be isomorphic to $\mathbb{C}$ as a subfield of V.

Let $k = ij$. Since $\dim_{\mathbb{R}}(V) = 3$, we must have $k = a_1 + b_1 i + c_1 j$ with $a_1, b_1, c_1 \in \mathbb{R}$. Multiplication from the left with i results in

$$-j = a_1 i - b_1 + c_1 k = a_1 i - b_1 + c_1(a_1 + b_1 i + c_1 j),$$

and since $1, i, j$ are linearly independent, therefore, we get $c_1^2 = -1$, which is impossible in $\mathbb{R}$. Therefore, the case $n = 3$ is not possible.

If $n = 4$, we may construct in V three linearly independent elements $1, i, j$ such that 1 is the unit element of V, and $i^2 = j^2 = -1$. Certainly ij is linearly independent from $1, i$ and j, because otherwise, we get a contradiction as in the case $n = 3$. Also ji is linearly independent from 1, i and j. Now $i + j$ and $i - j$ are both zeros of quadratic polynomials over $\mathbb{R}$; that is, there exists $r_1, s_1, r_2, s_2 \in \mathbb{R}$ with

$$(i + j)^2 + r_1(i + j) + s_1 = 0 \quad \text{and} \quad (i - j)^2 + r_2(i - j) + s_2 = 0.$$

If we add these equations, we see that $r_1 = r_2 = 0$; therefore, we get from the first equation that $ij + ji = c \in \mathbb{R}$. Here, we used that 1, i and j are linearly independent.

Now, we may replace j by $j + \frac{c}{2}i$, which gives

$$i\left(j + \frac{c}{2}i\right) + \left(j + \frac{c}{2}i\right)i = 0.$$

Since the subspace of V generated by 1 and $j + \frac{c}{2}i$ must, as a field, be isomorphic to $\mathbb{C}$, we may normalize $j + \frac{c}{2}i$ to j_1 with $j_1^2 = -1$.

We now define $k = ij_1$. Then automatically

$$k = ij_1 = -j_1 i \quad \text{and} \quad k^2 = -1.$$

So altogether, we may construct a basis $\{1, i, j, k\}$ of V such that 1 is the unit element of V, and $i^2 = j^2 = k^2 = -1$, $k = ij = -ji$. Thereby, V is isomorphic to the skew field $\mathbb{H}$ of the quaternions.

Finally, let $n \geq 5$. Analogously as for the case $n = 4$ and the general observation for the subfield isomorphic to $\mathbb{C}$, we may construct a basis $\{1, i, j, k, l, \ldots\}$ such that

$$i^2 = j^2 = k^2 = -1, \quad k = ij = -ji \quad \text{and} \quad l^2 = -1.$$

Analogously, as in the case $n = 4$, we have that $i + l$ and $i - l$ are both zeros of quadratic polynomials over $\mathbb{R}$.

Therefore, as in the case $n = 4$,

$$il = li = a_2 \in \mathbb{R}.$$

In the same manner, we get

$$jl + lj = b_2 \in \mathbb{R} \quad \text{and} \quad kl + lk = c_2 \in \mathbb{R}.$$

We calculate

$$\begin{aligned} lk = l(ij) &= a_2 j - ilj = a_2 j - i(b_2 - jl) \\ &= a_2 j - b_2 i + ijl = a_2 j - b_2 i + kl \\ &= a_2 j - b_2 i + c_2 - lk. \end{aligned}$$

From this, we get

$$2lk = a_2 j - b_2 i + c_2.$$

Multiplication with k from the right gives

$$-2l = a_2 i + b_2 j + c_2 k,$$

because $jk = i$, and $ik = -j$.

This means that l is linearly dependent of $\{1, i, j, k\}$, which is not the case. This contradiction shows that $n \geq 5$ is not possible.

Altogether, we have proven the following theorem:

Theorem 7.5.3 (Frobenius Theorem). *Let V be an $\mathbb{R}$-vector space, $\dim_{\mathbb{R}}(V) = n < \infty$. Let V be provided in addition with a multiplication, such that V together with the vector addition and the multiplication is a field or a skew field.*

Then $n = 1, 2$ or 4. In particular, if $n = 1$ then V is isomorphic to $\mathbb{R}$, if $n = 2$, then V is isomorphic to $\mathbb{C}$, and if $n = 4$ then V is isomorphic to $\mathbb{H}$.

7.6 Exercises

1. Let $f, g \in K[x]$ be irreducible polynomials of degree 2 over the field K. Let α_1, α_2 (respectively, β_1, β_2) be zeros of f and g. For $1 \leq i, j \leq 2$, let $v_{ij} = \alpha_i + \beta_j$. Show the following:
   (a) $|K(v_{ij}) : K| \in \{1, 2, 3, 4\}$.
   (b) For fixed f, g, there are at most two different degrees in (a).

   (c) Decide which sets of combinations of degrees in (b) (with f, g variable) are possible, and give an example in each case.
2. Let $L|K$ be a field extension; let $v \in L$ and $f(x) \in L[x]$, a polynomial of degree ≥ 1. Let all coefficients of $f(x)$ be algebraic over K. If $f(v) = 0$, then v is algebraic over K.
3. Let $L|K$ be a field extension, and let M be an intermediate field. The extension $M|K$ is algebraic. For $v \in L$, the following are equivalent:
   (a) v is algebraic over M.
   (b) v is algebraic over K.
4. Let $L|K$ be a field extension and $v_1, v_2 \in L$. Then the following are equivalent:
   (a) v_1 and v_2 are algebraic over K.
   (b) $v_1 + v_2$ and $v_1 v_2$ are algebraic over K.
5. Let $L|K$ be a simple field extension. Then there is an extension field L' of L of the form $L' = K(v_1, v_2)$ with the following:
   (a) v_1 and v_2 are transcendental over K.
   (b) The set of all over K algebraic elements of L' is L.
6. In the proof of Theorem 7.1.4, show that the mapping

$$\tau : K(a) \to K(\alpha),$$

   defined by $\tau(k) = k$ if $k \in K$ and $\tau(a) = \alpha$, and then extended by linearity, is a K-isomorphism.
7. Prove Lemma 7.2.8.
8. If T, T_1 are sets with the same cardinality, then there exists a bijection $\sigma : T \to T_1$. Define a map $F : S_T \to S_{T_1}$ in the following manner: if $f \in S_T$, let $F(f)$ be the permutation on T_1 given by $F(f)(t_1) = \sigma(f(\sigma^{-1}(t_1)))$. Prove that F is an isomorphism.
9. Let $P(X), Q(x), H(x) \in \mathbb{C}$. Show that $P(x)Q(x) = H(x)$ implies $\overline{H}(x) = (\overline{P}(x))(\overline{Q}(x))$.
10. Show the multiplicative rule for the norm and the length for the quaternions:

$$n(xy) = n(x)n(y) \quad \text{and} \quad |xy| = |x||y| \quad \text{for } x, y \in \mathbb{H}.$$

11. Determine all irreducible polynomials over $\mathbb{R}$. Factorize $f(x) \in \mathbb{R}[x]$ in irreducible polynomials.

8 Splitting Fields and Normal Extensions

8.1 Splitting Fields

In the last chapter, we introduced *splitting fields* and used this idea to present a proof of the fundamental theorem of algebra. The concept of a splitting field is essential to the Galois theory of equations. Therefore, in this chapter, we look more deeply at this idea.

Definition 8.1.1. Let K be a field and $f(x)$ a nonconstant polynomial in $K[x]$. An extension field L of K is a *splitting field* for $f(x)$ over K if the following hold:
(a) $f(x)$ splits into linear factors in $L[x]$.
(b) $K \subset M \subset L$ and $M \neq L$, resulting in $f(x)$ not splitting into linear factors in $M[x]$.

From part (b) in the definition, the following is clear:

Lemma 8.1.2. *L is a splitting field for $f(x) \in K[x]$ if and only if $f(x)$ splits into linear factors in $L[x]$, and if $f(x) = b(x - a_1)\cdots(x - a_n)$ with $b \in K$, then $L = K(a_1, \ldots, a_n)$.*

Example 8.1.3. The field $\mathbb{C}$ of complex numbers is a splitting field for the polynomial $p(x) = x^2 + 1$ in $\mathbb{R}[x]$. In fact, since $\mathbb{C}$ is algebraically closed, it is a splitting field for any real polynomial $f(x) \in \mathbb{R}[x]$, which has at least one nonreal zero.

The field $\mathbb{Q}(i)$ adjoining i to $\mathbb{Q}$ is a splitting field for $x^2 + 1$ over $\mathbb{Q}[x]$.

The next result was used in the previous chapter. We restate and reprove it here.

Theorem 8.1.4. *Let K be a field. Then each nonconstant polynomial in $K[x]$ has a splitting field.*

Proof. Let $\overline{K}$ be an algebraic closure of K.

Then $f(x)$ splits in $\overline{K}[x]$; that is, $f(x) = b(x - a_1)\cdots(x - a_n)$ with $b \in K$ and $a_i \in \overline{K}$. Let $L = K(a_1, \ldots, a_n)$. Then L is the splitting field for $f(x)$ over K. □

We next show that the splitting field over K of a given polynomial is unique up to K-isomorphism.

Theorem 8.1.5. *Let K, K' be fields and $\phi : K \to K'$ an isomorphism. Let $f(x)$ be a nonconstant polynomial in $K[x]$ and $f'(x) = \phi(f(x))$ its image in $K'[x]$. Suppose that L is a splitting field for $f(x)$ over K, and L' is a splitting field for $f'(x)$ over K'.*
(a) *Suppose that $L' \subset L''$. Then, if $\psi : L \to L''$ is a monomorphism with $\psi_{|_K} = \phi$, then ψ is an isomorphism from L onto L'. Moreover, ψ maps the set of zeros of $f(x)$ in L onto the set of zeros of $f'(x)$ in L'. The map ψ is uniquely determined by the values of the zeros of $f(x)$.*
(b) *If $g(x)$ is an irreducible factor of $f(x)$ in $K[x]$, a is a zero of $g(x)$ in L, and a' is a zero of $g'(x) = \phi(g(x))$ in L', then there is an isomorphism ψ from L to L' with $\psi_{|_K} = \phi$ and $\psi(a) = \psi(a')$.*

https://doi.org/10.1515/9783111142524-008

Before giving the proof of this theorem, we note that the following important result is a direct consequence of it:

Theorem 8.1.6. *A splitting field for $f(x) \in K[x]$ is unique up to K-isomorphism.*

Proof of Theorem 8.1.5. Suppose that $f(x) = b(x - a_1) \cdots (x - a_n) \in L[x]$ and suppose that $f'(x) = b'(x - a_1') \cdots (x - a_n') \in L'[x]$. Then

$$f'(x) = \phi(f(x)) = \psi(f(x)) = (\psi(b))(x - \psi(a_1)) \cdots (x - \psi(a_n)).$$

We have proved that polynomials have unique factorization over fields. Since $L' \subset L''$, it follows that the set of zeros $(\psi(a_1), \dots, \psi(a_n))$ is a permutation of the set of zeros $(a_1', \dots, a_n')$. In particular, this implies that $\psi(a_i) \in L'$; thus,

$$\operatorname{im}(\psi) = L' = K'(a_1, \dots, a_n').$$

Since the image of ψ is $K'(a_1, \dots, a_n') = K'(\psi(a_i), \dots, \psi(a_n))$, it is clear that ψ is uniquely determined by the images $\psi(a_i)$. This proves part (a).

For part (b), embed L' in an algebraic closure L''. Hence, there is a monomorphism

$$\phi' : K(a) \to L''$$

with $\phi'_{|K} = \phi$ and $\phi'(a) = a'$. Hence, there is a monomorphism $\psi : L \to L''$ with $\psi_{|K(a)} = \phi'$. Then from part (a), it follows that $\psi : L \to L'$ is an isomorphism. □

Example 8.1.7. Let $f(x) = x^3 - 7 \in \mathbb{Q}[x]$. This has no zeros in $\mathbb{Q}$, and since it is of degree 3, it follows that it must be irreducible in $\mathbb{Q}[x]$.

Let $\omega = -\frac{1}{2} + \frac{\sqrt{3}}{2}i \in \mathbb{C}$. Then it is easy to show by computation that $\omega^2 = -\frac{1}{2} - \frac{\sqrt{3}}{2}i$, and $\omega^3 = 1$. Therefore, the three zeros of $f(x)$ in $\mathbb{C}$ are as follows:

$$\begin{aligned} a_1 &= 7^{1/3} \\ a_2 &= \omega \cdot 7^{1/3} \\ a_3 &= \omega^2 \cdot 7^{1/3}. \end{aligned}$$

Hence, $L = \mathbb{Q}(a_1, a_2, a_3)$, the splitting field of $f(x)$. Since the minimal polynomial of all three zeros over $\mathbb{Q}$ is the same $f(x)$, it follows that

$$\mathbb{Q}(a_1) \cong \mathbb{Q}(a_2) \cong \mathbb{Q}(a_3).$$

Since $\mathbb{Q}(a_1) \subset \mathbb{R}$ and a_2, a_3 are nonreal, it is clear that $a_2, a_3 \notin \mathbb{Q}(a_1)$. Suppose that $\mathbb{Q}(a_2) = \mathbb{Q}(a_3)$. Then $\omega = a_3 a_2^{-1} \in \mathbb{Q}(a_2)$, and so $7^{1/3} = \omega^{-1} a_2 \in \mathbb{Q}(a_2)$. Hence, $Q(a_1) \subset \mathbb{Q}(a_2)$; therefore, $\mathbb{Q}(a_1) = \mathbb{Q}(a_2)$ since they have the same degree over $\mathbb{Q}$. This contradiction shows that $\mathbb{Q}(a_2)$ and $\mathbb{Q}(a_3)$ are distinct.

By computation, we have $a_3 = a_1^{-1}a_2^2$; hence,

$$L = \mathbb{Q}(a_1, a_2, a_3) = \mathbb{Q}(a_1, a_2) = \mathbb{Q}(7^{1/3}, \omega).$$

Now the degree of L over $\mathbb{Q}$ is

$$|L : \mathbb{Q}| = |Q(7^{1/3}, \omega) : \mathbb{Q}(\omega)||\mathbb{Q}(\omega) : \mathbb{Q}|.$$

Now $|\mathbb{Q}(\omega) : \mathbb{Q}| = 2$ since the minimal polynomial of ω over $\mathbb{Q}$ is $x^2 + x + 1$. Since no zero of $f(x)$ lies in $\mathbb{Q}(\omega)$, and the degree of $f(x)$ is 3, it follows that $f(x)$ is irreducible over $\mathbb{Q}(\omega)$. Therefore, we have that the degree of L over $\mathbb{Q}(\omega)$ is 3. Hence, $|L : \mathbb{Q}| = (2)(3) = 6$.

We now have the following lattice diagram of fields and subfields:

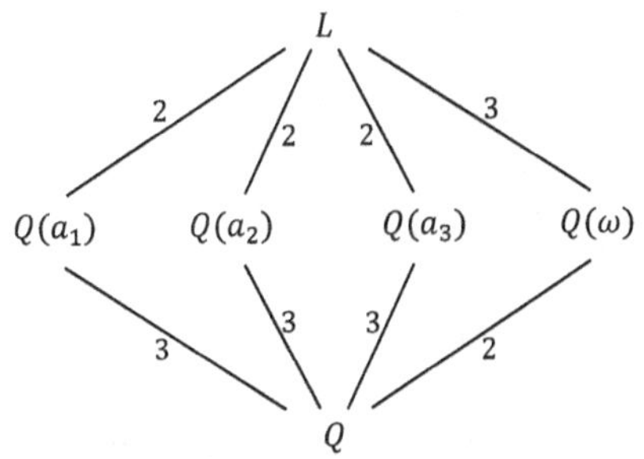

We do not know however if there are any more intermediate fields. There could, for example, be infinitely many. However, as we will see when we do the Galois theory, there are no others.

8.2 Normal Extensions

We now consider algebraic field extensions L of K, which have the property that if $f(x) \in K[x]$ has a zero in L, then $f(x)$ must split in L. In particular, we show that if L is a splitting field of finite degree for some $g(x) \in K[x]$, then L has this property.

Definition 8.2.1. A field extension L of a field K is a *normal extension* if the following hold:

(a) $L|K$ is algebraic.

(b) Each irreducible polynomial $f(x) \in K[x]$ that has a zero in L splits into linear factors in $L[x]$.

Note, in Example 8.1.7, the extension fields $Q(a_i)|\mathbb{Q}$ are not normal extensions. Although $f(x)$ has a zero in $\mathbb{Q}(a_i)$, the polynomial $f(x)$ does not split into linear factors in $\mathbb{Q}(a_i)[x]$.

We now show that $L|K$ is a finite normal extension if and only if L is the splitting field for some $f(x) \in K[x]$.

Theorem 8.2.2. *Let $L|K$ be a finite extension. Then the following are equivalent:*
(a) *$L|K$ is a normal extension.*
(b) *$L|K$ is a splitting field for some $f(x) \in K[x]$.*
(c) *If $L \subset L'$ and $\psi : L \to L'$ is a monomorphism with $\psi_{|K}$, the identity map on K, then ψ is an automorphism of L; that is, $\psi(L) = L$.*

Proof. Suppose that $L|K$ is a finite normal extension. Since $L|K$ is a finite extension, L is algebraic over K, and since of finite degree, we have $L = K(a_1, \ldots, a_n)$ with a_i algebraic over K.

Let $f_i(x) \in K[x]$ be the minimal polynomial of a_i. Since $L|K$ is a normal extension, $f_i(x)$ splits in $L[x]$. This is true for each $i = 1, \ldots, n$. Let $f(x) = f_1(x)f_2(x) \cdots f_n(x)$. Then $f(x)$ splits into linear factors in $L[x]$. Since $K = K(a_1, \ldots, a_n)$, the polynomial $f(x)$ cannot have all its zeros in any intermediate extension between K and L. Therefore, L is the splitting field for $f(x)$. Hence, (a) implies (b).

Now suppose that $L \subset L'$ and $\psi : L \to L'$ is a monomorphism with $\psi_{|K}$ the identity map on K. Then the extension field $\psi(L)$ of K is also a splitting field for $f(x)$ since $\psi_{|K}$ is the identity on K. Hence, ψ maps the zeros of $f(x)$ in $L \subset L'$ onto the zeros of $f(x)$ in $\psi(L) \subset L'$, and thus it follows that $\psi(L) = L$. Hence, (b) implies (c).

Finally, suppose (c). Hence, we assume that if $L \subset L'$ and $\psi : L \to L'$ is a monomorphism with $\psi_{|K}$, the identity map on K, then ψ is an automorphism of L; that is, $\psi(L) = L$.

As before $L|K$ is algebraic since $L|K$ is finite. Suppose that $f(x) \in K[x]$ is irreducible and that $a \in L$ is a zero of $f(x)$. There are algebraic elements $a_1, \ldots, a_n \in L$ with $L = K(a_1, \ldots, a_n)$ since $L|K$ is finite. For $i = 1, \ldots, n$, let $f_i(x) \in K[x]$ be the minimal polynomial of a_i, and let $g(x) = f(x)f_1(x) \cdots f_n(x)$. Let L' be the splitting field of $g(X)$. Clearly, $L \subset L'$. Let $b \in L'$ be a zero of $f(x)$. From Theorem 8.1.5, there is an automorphism ψ of L' with $\psi(a) = b$ and $\psi_{|K}$, the identity on K. Hence, by our assumption, $\psi_{|L}$ is an automorphism of L. It follows that $b \in L$; hence, $f(x)$ splits in $L[x]$. Therefore, (c) implies (a), completing the proof. □

To give simple examples of normal extensions, we have the following:

Lemma 8.2.3. *If L is an extension of K with $|L : K| = 2$, then L is a normal extension of K.*

Proof. Suppose that $|L : K| = 2$. Then $L|K$ is algebraic since it is finite.

Let $f(x) \in K[x]$ be irreducible with leading coefficient 1, and which has a zero in L. Let a be one zero. Then $f(x)$ must be the minimal polynomial of a. However, $\deg(m_a(x)) \leq |L : K| = 2$; hence, $f(x)$ is of degree 1 or 2. Since $f(x)$ has a zero in L, it follows that it must split into linear factors in $L[x]$; therefore, L is a normal extension. □

Later, we will tie this result to group theory when we prove that a subgroup of index 2 must be a normal subgroup.

Example 8.2.4. As a first example of the lemma, consider the polynomial $f(x) = x^2 - 2$. In $\mathbb{R}$, this splits as $(x - \sqrt{2})(x + \sqrt{2})$; hence, the field $\mathbb{Q}(\sqrt{2})$ is the splitting field of $f(x) = x^2 - 2$ over $\mathbb{Q}$. Therefore, $\mathbb{Q}(\sqrt{2})$ is a normal extension of $\mathbb{Q}$.

Example 8.2.5. As a second example, consider the polynomial $x^4 - 2$ in $\mathbb{Q}[x]$. The zeros in $\mathbb{C}$ are

$$2^{1/4},\ 2^{1/4}i,\ 2^{1/4}i^2,\ 2^{1/4}i^3.$$

Hence,

$$L = \mathbb{Q}(2^{1/4}, 2^{1/4}i, 2^{1/4}i^2, 2^{1/4}i^3)$$

is the splitting field of $x^4 - 2$ over $\mathbb{Q}$.

Now

$$L = \mathbb{Q}(2^{1/4}, 2^{1/4}i, 2^{1/4}i^2, 2^{1/4}i^3) = \mathbb{Q}(2^{1/4}, i).$$

Therefore, we have

$$|L : \mathbb{Q}| = |L : \mathbb{Q}(2^{1/4})||\mathbb{Q}(2^{1/4}) : \mathbb{Q}|.$$

Since $x^4 - 2$ is irreducible over $\mathbb{Q}$, we have $|\mathbb{Q}(2^{1/4}) : \mathbb{Q}| = 4$. Since i has degree 2 over any real field, we have $|L : \mathbb{Q}(2^{1/4})| = 2$. Therefore, L is a normal extension of $\mathbb{Q}(2^{1/4})$, and $x^2 - \sqrt{2} \in \mathbb{Q}(\sqrt{2})[x]$ has the splitting field $\mathbb{Q}(2^{1/4})$.

Altogether, we have that $L|\mathbb{Q}(2^{1/4})$, $\mathbb{Q}(2^{1/4})|\mathbb{Q}(2^{1/2})$, $\mathbb{Q}(2^{1/2})|\mathbb{Q}$, and $L|\mathbb{Q}$ are normal extensions. However, $\mathbb{Q}(2^{1/4})|\mathbb{Q}$ is not normal since $2^{1/4}$ is a zero of $x^4 - 2$, but $\mathbb{Q}(2^{1/4})$ does not contain all the zeros of $x^4 - 2$.

Hence, we get the following Figure 8.1.

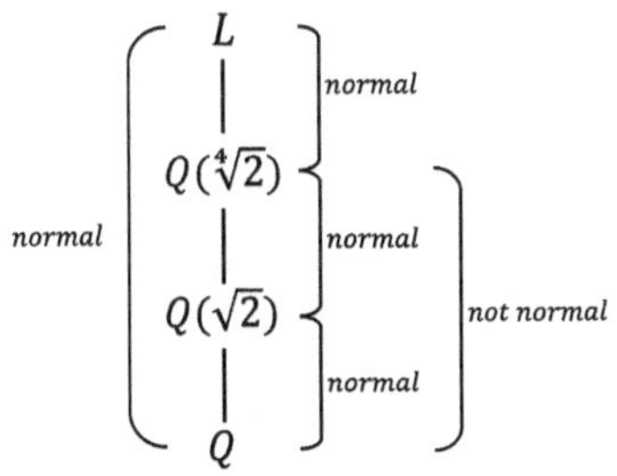

Figure 8.1: Normal extensions.

8.3 Exercises

1. Determine the splitting field of $f(x) \in \mathbb{Q}[x]$ and its degree over $\mathbb{Q}$ in the following cases:
   (a) $f(x) = x^4 - p$, where p is a prime.
   (b) $f(x) = x^p - 2$, where p is a prime.
2. Determine the degree of the splitting field of the polynomial x^4+4 over $\mathbb{Q}$. Determine the splitting field of $x^6 + 4x^4 + 4x^2 + 3$ over $\mathbb{Q}$.
3. For each $a \in \mathbb{Z}$, let $f_a(x) = x^3 - ax^2 + (a-3)x + 1 \in \mathbb{Q}[x]$ be given:
   (a) f_a is irreducible over $\mathbb{Q}$ for each $a \in \mathbb{Z}$.
   (b) If $b \in \mathbb{R}$ is a zero of f_a, then also $(1-b)^{-1}$ and $(b-1)b^{-1}$ are zeros of f_a.
   (c) Determine the splitting field L of $f_a(x)$ over $\mathbb{Q}$ and its degree $|L : \mathbb{Q}|$.
4. Let K be a field and $f(x) \in K[x]$ a polynomial of degree n. Let L be a splitting field of $f(x)$. Show the following:
   (a) If $a_1, \dots, a_n \in L$ are the zeros of f, then $|K(a_1, \dots, a_t) : K| \le n \cdot (n-1) \cdots (n-t+1)$ for each t with $1 \le t \le n$.
   (b) L over K is of degree at most $n!$.
   (c) If $f(x)$ is irreducible over K, then n divides $|L : K|$.

9 Groups, Subgroups and Examples

9.1 Groups, Subgroups and Isomorphisms

Recall from Chapter 1 that the three most commonly studied algebraic structures are groups, rings and fields. We have now looked rather extensively at rings and fields. In this chapter, we consider the basic concepts of group theory. Groups arise in many different areas of mathematics. For example they arise in geometry as groups of congruence motions, and in topology as groups of various types of continuous functions. Later in this book, they will appear in Galois theory as groups of automorphisms of fields. First, we recall the definition of a group given previously in Chapter 1.

Definition 9.1.1. A *group* G is a set with one binary operation, which we will denote by multiplication, such that

(1) The operation is associative; that is, $(g_1g_2)g_3 = g_1(g_2g_3)$ for all $g_1, g_2, g_3 \in G$.
(2) There exists an identity for this operation; that is, an element 1 such that $1g = g$ and $g1 = g$ for each $g \in G$.
(3) Each $g \in G$ has an inverse for this operation; that is, for each g, there exists a g^{-1} with the property that $gg^{-1} = 1$, and $g^{-1}g = 1$.

If, in addition, the operation is commutative; that is, $g_1g_2 = g_2g_1$ for all $g_1, g_2 \in G$, the group G is called an *Abelian group*.

The *order* of G, denoted $|G|$, is the number of elements in the group G. If $|G| < \infty$, G is a *finite group*, otherwise, it is an *infinite group*.

It follows easily from the definition that the identity is unique, and that each element has a unique inverse.

Lemma 9.1.2. *If G is a group, then there is a unique identity. Furthermore, if $g \in G$, its inverse is unique. Finally, if $g_1, g_2 \in G$, then $(g_1g_2)^{-1} = g_2^{-1}g_1^{-1}$.*

Proof. Suppose that 1 and e are both identities for G. Then $1e = e$ since 1 is an identity, and $1e = 1$ since e is an identity. Therefore, $1 = e$, and there is only one identity.

Next suppose that $g \in G$, g_1, and g_2 are inverses for g. Then

$$g_1gg_2 = (g_1g)g_2 = 1g_2 = g_2$$

since $g_1g = 1$. On the other hand,

$$g_1gg_2 = g_1(gg_2) = g_11 = g_1$$

since $gg_2 = 1$. It follows that $g_1 = g_2$, and g has a unique inverse.

Finally, consider

$$(g_1g_2)(g_2^{-1}g_1^{-1}) = g_1(g_2g_2^{-1})g_1^{-1} = g_11g_1^{-1} = g_1g_1^{-1} = 1.$$

https://doi.org/10.1515/9783111142524-009

Therefore, $g_2^{-1}g_1^{-1}$ is an inverse for g_1g_2, and since inverses are unique, it is the inverse of the product. □

Groups most often arise as permutations on a set. We will see this, as well as other specific examples of groups, in the next sections.

Finite groups can be completely described by their *group tables* or multiplication tables. These are sometimes called *Cayley tables.* In general, let $G = \{g_1, \dots, g_n\}$ be a group, then the *multiplication table* of G is

	g_1	g_2	$\cdots$	g_j	$\cdots$	g_n
g_1	$\cdots$					
g_2	$\cdots$					
$\vdots$						
g_i	$\cdots$	$\cdots$	$\cdots$	g_ig_j		
$\vdots$						
g_n	$\cdots$					

The entry in the row of $g_i \in G$ and column of $g_j \in G$ is the product (in that order) g_ig_j in G.

Groups satisfy the *cancellation law for multiplication.*

Lemma 9.1.3. *If G is a group and $a, b, c \in G$ with $ab = ac$ or $ba = ca$, then $b = c$.*

Proof. Suppose that $ab = ac$. Then a has an inverse a^{-1}, so we have

$$a^{-1}(ab) = a^{-1}(ac).$$

From the associativity of the group operation, we then have

$$(a^{-1}a)b = (a^{-1}a)c \implies 1 \cdot b = 1 \cdot c \implies b = c.$$

□

A consequence of Lemma 9.1.3 is that each row and each column in a group table is just a permutation of the group elements. That is, each group element appears exactly once in each row and each column.

A subset $H \subset G$ is a *subgroup* of G if H is also a group under the same operation as G. As for rings and fields, a subset of a group is a subgroup if it is nonempty and closed under both the group operation and inverses.

Lemma 9.1.4.
1. *A subset $H \subset G$ is a subgroup if $H \neq \emptyset$, and H is closed under the operation and inverses. That is, if $a, b \in H$, then $ab \in H$, and $a^{-1}, b^{-1} \in H$.*
2. *A nonempty subset H of a group G is a subgroup if and only if $ab^{-1} \in H$ for all $a, b \in H$. In addition, if G is finite, then H is a subgroup if and only if $ab \in H$ for all $a, b \in H$.*

We leave the proof of this to the exercises.

Let G be a group and $g \in G$; we denote by g^n, $n \in \mathbb{N}$, as with numbers, the product of g taken n times. A negative exponent will indicate the inverse of the positive exponent. As usual, let $g^0 = 1$. Clearly, group exponentiation will satisfy the standard laws of exponents. Now consider the set

$$H = \{1 = g^0, g, g^{-1}, g^2, g^{-2}, \dots\}$$

of all powers of g. We will denote this by $\langle g \rangle$.

Lemma 9.1.5. *If G is a group and $g \in G$, then $\langle g \rangle$ forms a subgroup of G called the cyclic subgroup generated by g. $\langle g \rangle$ is Abelian, even if G is not.*

Proof. If $g \in G$, then $g \in \langle g \rangle$; hence, $\langle g \rangle$ is nonempty. Suppose then that $a = g^n$, $b = g^m$ are elements of $\langle g \rangle$. Then $ab = g^n g^m = g^{n+m} \in \langle g \rangle$, so $\langle g \rangle$ is closed under the group operation. Furthermore, $a^{-1} = (g^n)^{-1} = g^{-n} \in \langle g \rangle$ so $\langle g \rangle$ is closed under inverses. Therefore, $\langle g \rangle$ is a subgroup.

Finally, $ab = g^n g^m = g^{n+m} = g^{m+n} = g^m g^n = ba$; hence, $\langle g \rangle$ is Abelian. □

Suppose that $g \in G$ and $g^m = 1$ for some positive integer m. Then let n be the smallest positive integer such that $g^n = 1$. It follows that the set of elements $\{1, g, g^2, \dots, g^{n-1}\}$ are all distinct, but for any other power g^k, we have $g^k = g^t$ for some $k = 0, 1, \dots, n-1$ (see exercises). The cyclic subgroup generated by g then has order n, and we say that g has *order* n, which we denote by $o(g) = n$. If no such n exists, we say that g has *infinite order*. We will look more deeply at cyclic groups and subgroups in Section 9.5.

We introduce one more concept before looking at examples.

Definition 9.1.6. If G and H are groups, then a mapping $f : G \to H$ is a (group) *homomorphism* if $f(g_1 g_2) = f(g_1)f(g_2)$ for any $g_1, g_2 \in G$. If f is also a bijection, then it is an *isomorphism*.

As with rings and fields, we say that two groups G and H are *isomorphic*, denoted by $G \cong H$, if there exists an isomorphism $f : G \to H$. This means that, abstractly, G and H have exactly the same algebraic structure.

9.2 Examples of Groups

As already mentioned, groups arise in many diverse areas of mathematics. In this section and the next, we present specific examples of groups.

First of all, any ring or field under addition forms an Abelian group. Hence, for example, $(\mathbb{Z}, +)$, $(\mathbb{Q}, +)$, $(\mathbb{R}, +)$, $(\mathbb{C}, +)$, where $\mathbb{Z}, \mathbb{Q}, \mathbb{R}, \mathbb{C}$ are respectively the integers, the rationals, the reals, and the complex numbers; all are infinite Abelian groups. If $\mathbb{Z}_n$ is the modular ring $\mathbb{Z}/n\mathbb{Z}$, then for any natural number n, $(\mathbb{Z}_n, +)$ forms a finite Abelian group. In Abelian groups, the group operation is often denoted by $+$ and the identity element by 0 (zero).

In a field K, the nonzero elements are all invertible and form a group under multiplication. This is called the *multiplicative group* of the field K and is usually denoted by K^*. Since multiplication in a field is commutative, the multiplicative group of a field is an Abelian group. Hence, $\mathbb{Q}^*$, $\mathbb{R}^*$, $\mathbb{C}^*$ are all infinite Abelian groups, whereas if p is a prime, $\mathbb{Z}_p^*$ forms a finite Abelian group. Recall that if p is a prime, then the modular ring $\mathbb{Z}_p$ is a field.

Within $\mathbb{Q}^*$, $\mathbb{R}^*$, $\mathbb{C}^*$, there are certain multiplicative subgroups. Since the positive rationals $\mathbb{Q}_+$ and the positive reals $\mathbb{R}_+$ are closed under multiplication and inverse, they form subgroups of $\mathbb{Q}^*$ and $\mathbb{R}^*$, respectively. In $\mathbb{C}$, if we consider the set of all complex numbers z with $|z| = 1$, these form a multiplicative subgroup. Further within this subgroup, if we consider the set of n-th roots of unity z (that is $z^n = 1$) for a fixed n, this forms a subgroup, this time of finite order.

The multiplicative group of a field is a special case of the unit group of a ring. If R is a ring with identity, recall that a *unit* is an element of R with a multiplicative inverse. Hence, in $\mathbb{Z}$, the only units are ± 1, whereas in any field every nonzero element is a unit.

Lemma 9.2.1. *If R is a ring with identity, then the set of units in R forms a group under multiplication called the unit group of R, and is denoted by $U(R)$. If R is a field, then $U(R) = R^*$.*

Proof. Let R be a ring with identity. Then the identity 1 itself is a unit, so $1 \in U(R)$; hence, $U(R)$ is nonempty. If $e \in R$ is a unit, then it has a multiplicative inverse e^{-1}. Clearly then, the multiplicative inverse has an inverse, namely, e so $e^{-1} \in U(R)$ if e is. Hence, to show $U(R)$ is a group, we must show that it is closed under product.

Let $e_1, e_2 \in U(R)$. Then there exist e_1^{-1}, e_2^{-1}. It follows that $e_2^{-1}e_1^{-1}$ is an inverse for e_1e_2. Hence, e_1e_2 is also a unit, and $U(R)$ is closed under product. Therefore, for any ring R with identity $U(R)$ forms a multiplicative group. □

To present examples of non-Abelian groups, we turn to matrices. If K is a field, we let

$$\mathrm{GL}(n, K) = \{n \times n \text{ matrices over } K \text{ with nonzero determinant}\}$$

and

$$\mathrm{SL}(n, K) = \{n \times n \text{ matrices over } K \text{ with determinant one}\}.$$

Lemma 9.2.2. *If K is a field, then for $n \geq 2$, $\mathrm{GL}(n, K)$ forms a non-Abelian group under matrix multiplication, and $\mathrm{SL}(n, K)$ forms a subgroup.*

$\mathrm{GL}(n, K)$ is called the n-dimensional general linear group over K, whereas $\mathrm{SL}(n, K)$ is called the n-dimensional special linear group over K.

Proof. Recall that for two $n \times n$ matrices A and B with $n \geq 2$ over a field, we have $\det(AB) = \det(A)\det(B)$ where det is the determinant.

Now for any field, the $n \times n$ identity matrix I has determinant 1; hence, $I \in \mathrm{GL}(n, K)$. Since the determinant is multiplicative, the product of two matrices with nonzero determinant has nonzero determinant, so $\mathrm{GL}(n, K)$ is closed under product. Furthermore, over a field K, if A is an invertible matrix, then $\det(A^{-1}) = \frac{1}{\det A}$.

Therefore, if A has nonzero determinant, so does its inverse. It follows that $\mathrm{GL}(n, K)$ has the inverse of any of its elements. Since matrix multiplication is associative, it follows that $\mathrm{GL}(n, K)$ forms a group. It is non-Abelian since in general matrix multiplication is noncommutative. $\mathrm{SL}(n, K)$ forms a subgroup of $\mathrm{GL}(n, K)$ because $\det(A^{-1}) = 1$ if $\det(A) = 1$. □

Groups play an important role in geometry. In any metric geometry, an *isometry* is a mapping that preserves distance. To understand a geometry, one must understand the group of isometries. We look briefly at the Euclidean geometry of the plane $\mathcal{E}^2$.

An *isometry* or *congruence motion* of $\mathcal{E}^2$ is a transformation or bijection T of $\mathcal{E}^2$ that preserves distance; that is, $d(a, b) = d(T(a), T(b))$ for all points $a, b \in \mathcal{E}^2$.

Theorem 9.2.3. *The set of congruence motions of $\mathcal{E}^2$ forms a group called the Euclidean group. We denote the Euclidean group by $\mathcal{E}$.*

Proof. The identity map I is clearly an isometry, and since composition of mappings is associative, we need only to show that the product of isometries is an isometry, and that the inverse of an isometry is an isometry.

Let T, U be isometries. Then $d(a, b) = d(T(a), T(b))$ and $d(a, b) = d(U(a), U(b))$ for any points a, b. Now consider

$$d(TU(a), TU(b)) = d(T(U(a)), T(U(b))) = d(U(a), U(b))$$

since T is an isometry. However,

$$d(U(a), U(b)) = d(a, b)$$

since U is an isometry. Combining these, we have that TU is also an isometry.

Consider T^{-1} and points a, b. Then

$$d(T^{-1}(a), T^{-1}(b)) = d(TT^{-1}(a), TT^{-1}(b))$$

since T is an isometry. But $TT^{-1} = I$; hence,

$$d(T^{-1}(a), T^{-1}(b)) = d(TT^{-1}(a), TT^{-1}(b)) = d(a, b).$$

Therefore, T^{-1} is also an isometry; hence, $\mathcal{E}$ is a group. □

One of the major results concerning $\mathcal{E}$ is the following. We refer to [41], [42], [27], and [35] for a more thorough treatment.

Theorem 9.2.4. *If $T \in \mathcal{E}$, then T is either a translation, rotation, reflection, or glide reflection. The set of translations and rotations forms a subgroup.*

Proof. We outline a brief proof. If T is an isometry and T fixes the origin $(0, 0)$, then T is a linear mapping. It follows that T is a rotation or a reflection. If T does not fix the origin, then there is a translation T_0 such that $T_0 T$ fixes the origin. This gives translations and glide reflections. In the exercises, we expand out more of the proof. □

If D is a geometric figure in $\mathcal{E}^2$, such as a triangle or square, then a *symmetry* of D is a congruence motion $T : \mathcal{E}^2 \to \mathcal{E}^2$ that leaves D in place. However, it may move the individual elements of D. For example, a rotation about the center of a circle is a symmetry of the circle.

Lemma 9.2.5. *If D is a geometric figure in $\mathcal{E}^2$, then the set of symmetries of D forms a subgroup of $\mathcal{E}$ called the symmetry group of D, denoted by* $\mathrm{Sym}(D)$.

Proof. We show that $\mathrm{Sym}(D)$ is a subgroup of $\mathcal{E}$. The identity map I fixes D, that is, $I \in \mathrm{Sym}(D)$, and thus $\mathrm{Sym}(D)$ is nonempty. Let $T, U \in \mathrm{Sym}(D)$. Then T maps D to D, and so does U. It follows directly that so does the composition TU; hence, $TU \in \mathrm{Sym}(D)$. If T maps D to D, then certainly the inverse does. □

Example 9.2.6. Let T be an equilateral triangle. Then there are exactly six symmetries of T (see exercises). These are as follows:
- I is the identity,
- r is a rotation of 120° around the center of T,
- r is a rotation of 240° around the center of T,
- f is a reflection over the perpendicular bisector of one of the sides,
- fr is the composition of f and r, and
- fr^2 is the composition of f and r^2.

The group $\mathrm{Sym}(T)$ is called the *dihedral group* D_3. In the next section, we will see that it is isomorphic to S_3, the symmetric group on 3 symbols.

9.3 Permutation Groups

Groups most often appear as groups of transformations or permutations on a set. In this section, we will take a short look at permutation groups, and then examine them more deeply in Chapter 11. We recall some ideas, first introduced in Chapter 7, in relation to the proof of the fundamental theorem of algebra.

Definition 9.3.1. If A is a set, a *permutation* on A is a one-to-one mapping of A onto itself. We denote the set of all permutations on A by S_A.

Theorem 9.3.2. *For any set A, S_A forms a group under composition, called the symmetric group on A. If $|A| > 2$, then S_A is non-Abelian. Furthermore, if A, B have the same cardinality, then $S_A \cong S_B$.*

Proof. If S_A is the set of all permutations on the set A, we must show that composition is an operation on S_A that is associative, and has an identity and inverses. Let $f, g \in S_A$. Then f, g are one-to-one mappings of A onto itself.

Consider $f \circ g : A \to A$. If $f \circ g(a_1) = f \circ g(a_2)$, then $f(g(a_1)) = f(g(a_2))$, and $g(a_1) = g(a_2)$, since f is one-to-one. But then $a_1 = a_2$ since g is one-to-one.

If $a \in A$, there exists $a_1 \in A$ with $f(a_1) = a$ since f is onto. Then there exists $a_2 \in A$ with $g(a_2) = a_1$ since g is onto. Putting these together, $f(g(a_2)) = a$; therefore, $f \circ g$ is onto. Therefore, $f \circ g$ is also a permutation, and composition gives a valid binary operation on S_A.

The identity function $1(a) = a$ for all $a \in A$ will serve as the identity for S_A, whereas the inverse function for each permutation will be the inverse. Such unique inverse functions exist since each permutation is a bijection.

Finally, composition of functions is always associative; therefore, S_A forms a group.

Suppose that $|A| > 2$. Then A has at least 3 elements. Call them a_1, a_2, a_2. Consider the 2 permutations f and g, which fix (leave unchanged) all of A, except a_1, a_2, a_3 and on these three elements:

$$f(a_1) = a_2, \quad f(a_2) = a_3, \quad f(a_3) = a_1$$
$$g(a_1) = a_2, \quad g(a_2) = a_1, \quad g(a_3) = a_3.$$

Then under composition

$$f(g(a_1)) = a_3, \quad f(g(a_2)) = a_2, \quad f(g(a_3)) = a_1,$$

whereas

$$g(f(a_1)) = a_1, \quad g(f(a_2)) = a_3, \quad g(f(a_3)) = a_2.$$

Therefore, $f \circ g \neq g \circ f$; hence, S_A is not Abelian.

If A, B have the same cardinality, then there exists a bijection $\sigma : A \to B$. Define a map $F : S_A \to S_B$ in the following manner: if $f \in S_A$, let $F(f)$ be the permutation on B, given by $F(f)(b) = \sigma(f(\sigma^{-1}(b)))$. It is straightforward to verify that F is an isomorphism (see the exercises). □

If $A_1 \subset A$, then those permutations on A that map A_1 to A_1 form a subgroup of S_A called the *stabilizer* of A_1, denoted as $\text{stab}(A_1)$. We leave the proof to the exercises.

Lemma 9.3.3. *If $A_1 \subset A$, then* $\text{stab}(A_1) = \{f \in S_A : f : A_1 \to A_1\}$ *forms a subgroup of S_A.*

A *permutation group* is any subgroup of S_A for some set A. We now look at finite permutation groups. Let A be a finite set, say $A = \{a_1, a_2, \ldots, a_n\}$. Then each $f \in S_A$ can be pictured as

$$f = \begin{pmatrix} a_1 & \cdots & a_n \\ f(a_1) & \cdots & f(a_n) \end{pmatrix}.$$

For a_1, there are n choices for $f(a_1)$. For a_2, there are only $n-1$ choices since f is one-to-one. This continues down to only one choice for a_n. Using the multiplication principle, the number of choices for f; therefore, the size of S_A is

$$n(n-1)\cdots 1 = n!.$$

We have thus proved the following theorem.

Theorem 9.3.4. *If $|A| = n$ then $|S_A| = n!$.*

For a set A with n elements, we denote S_A by S_n, called the *symmetric group on n symbols*.

Example 9.3.5. Write down the six elements of S_3 and give the multiplication table for the group.

Name the three elements 1, 2, 3. The six elements of S_3 are then as follows:

$$1 = \begin{pmatrix} 1 & 2 & 3 \\ 1 & 2 & 3 \end{pmatrix}, \quad a = \begin{pmatrix} 1 & 2 & 3 \\ 2 & 3 & 1 \end{pmatrix}, \quad b = \begin{pmatrix} 1 & 2 & 3 \\ 3 & 1 & 2 \end{pmatrix}$$

$$c = \begin{pmatrix} 1 & 2 & 3 \\ 2 & 1 & 3 \end{pmatrix}, \quad d = \begin{pmatrix} 1 & 2 & 3 \\ 3 & 2 & 1 \end{pmatrix}, \quad e = \begin{pmatrix} 1 & 2 & 3 \\ 1 & 3 & 2 \end{pmatrix}.$$

The multiplication table for S_3 can be written down directly by doing the required composition. For example,

$$ac = \begin{pmatrix} 1 & 2 & 3 \\ 2 & 3 & 1 \end{pmatrix}\begin{pmatrix} 1 & 2 & 3 \\ 2 & 1 & 3 \end{pmatrix} = \begin{pmatrix} 1 & 2 & 3 \\ 3 & 2 & 1 \end{pmatrix} = d.$$

To see this, note that $a : 1 \to 2, 2 \to 3, 3 \to 1$; $c : 1 \to 2, 2 \to 1, 3 \to 3$, and so $ac : 1 \to 3, 2 \to 2, 3 \to 1$.

It is somewhat easier to construct the multiplication table if we make some observations. First, $a^2 = b$ and $a^3 = 1$. Next, $c^2 = 1$, $d = ac$, $e = a^2c$ and, finally, $ac = ca^2$.

From these relations, the following multiplication table can be constructed:

	1	a	a^2	c	ac	a^2c
1	1	a	a^2	c	ac	a^2c
a	a	a^2	1	ac	a^2c	c
a^2	a^2	1	a	a^2c	c	ac
c	c	a^2c	ac	1	a^2	a
ac	ac	c	a^2c	a	1	a^2
a^2c	a^2c	ac	c	a^2	a	1

To see this, consider, for example, $(ac)a^2 = a(ca^2) = a(ac) = a^2c$.
More generally, we can say that S_3 has a *presentation* given by

$$S_3 = \langle a, c; a^3 = c^2 = 1, ac = ca^2 \rangle.$$

By this, we mean that S_3 is *generated by* a, c, or that S_3 has *generators* a, c, and the whole group and its multiplication table can be generated by using the *relations* $a^3 = c^2 = 1$, $ac = ca^2$.

A theorem of Cayley actually shows that every group is a permutation group. A group G is a permutation group on the group G itself considered as a set. This result, however, does not give much information about the group.

Theorem 9.3.6 (Cayley's theorem). *Let G be a group. Consider the set of elements of G. Then the group G is a permutation group on the set G; that is, G is a subgroup of S_G.*

Proof. We show that to each $g \in G$, we can associate a permutation of the set G. If $g \in G$, let π_g be the map given by

$$\pi_g : g_1 \to gg_1 \quad \text{for each } g_1 \in G.$$

It is straightforward to show that each π_g is a permutation on G. □

9.4 Cosets and Lagrange's Theorem

In this section, given a group G and a subgroup H, we define an equivalence relation on G. The equivalence classes all have the same size and are called the (left) or (right) cosets of H in G.

Definition 9.4.1. Let G be a group and $H \subset G$ a subgroup. For $a, b \in G$, define $a \sim b$ if $a^{-1}b \in H$.

Lemma 9.4.2. *Let G be a group and $H \subset G$ a subgroup. Then the relation defined above is an equivalence relation on G. The equivalence classes all have the form aH for $a \in G$ and are called the left cosets of H in G. Clearly, G is a disjoint union of its left cosets.*

Proof. Let us show, first of all, that this is an equivalence relation. Now $a \sim a$ since $a^{-1}a = e \in H$. Therefore, the relation is reflexive. Furthermore, $a \sim b$ implies $a^{-1}b \in H$, but since H is a subgroup of G, we have $b^{-1}a = (a^{-1}b)^{-1} \in H$. Thus, $b \sim a$. Therefore, the relation is symmetric. Finally, suppose that $a \sim b$ and $b \sim c$. Then $a^{-1}b \in H$, and $b^{-1}c \in H$. Since H is a subgroup $a^{-1}b \cdot b^{-1}c = a^{-1}c \in H$; hence, $a \sim c$. Therefore, the relation is transitive and, hence, is an equivalence relation.

For $a \in G$, the equivalence class is

$$[a] = \{g \in G : a \sim g\} = \{a \in G : a^{-1}g \in H\}.$$

But then, clearly, $g \in aH$. It follows that the equivalence class for $a \in G$ is precisely the set

$$aH = \{g \in G : g = ah \text{ for some } h \in H\}.$$

These classes, aH, are called *left cosets* of H, and since they are equivalence classes, they partition G. This means that every element of g is in one and only one left coset. In particular, $bH = H = eH$ if and only if $b \in H$. □

If aH is a left coset, then we call the element a a *coset representative*. A complete collection

$$\{a \in G : \{aH\} \text{ is the set of all distinct left cosets of } H\}$$

is called a (left) *transversal* of H in G.

One could define another equivalence relation by defining $a \sim b$ if and only if $ba^{-1} \in H$. Again, this can be shown to be an equivalence relation on G, and the equivalence classes here are sets of the form

$$Ha = \{g \in G : g = ha \text{ for some } h \in H\},$$

called *right cosets* of H. Also, of course, G is the (disjoint) union of distinct right cosets.

It is easy to see that any two left (right) cosets have the same order (number of elements). To demonstrate this, consider the mapping $aH \to bH$ via $ah \mapsto bh$, where $h \in H$. It is not hard to show that this mapping is 1–1 and onto (see exercises). Thus, we have $|aH| = |bH|$. (This is also true for right cosets and can be established in a similar manner.) Letting $b \in H$ in the above discussion, we see $|aH| = |H|$, for any $a \in G$. That is, the size of each left or right coset is exactly the same as the subgroup H.

One can also see that the collection $\{aH\}$ of all distinct left cosets has the same number of elements as the collection $\{Ha\}$ of all distinct right cosets. In other words, the number of left cosets equals the number of right cosets (this number may be infinite). For example, consider the map $f : aH \to Ha^{-1}$. This mapping is well defined; for if $aH = bH$, then $b = ah$, where $h \in H$. Thus, $f(bH) = Hb^{-1} = Hh^{-1}a^{-1} = f(aH)$. It is not hard to show that this mapping is 1–1 and onto (see exercises). Hence, the number of left cosets equals the number of right cosets.

Definition 9.4.3. Let G be a group and $H \subset G$ a subgroup. The number of distinct left cosets, which is the same as the number of distinct right cosets, is called the *index* of H in G, denoted by $[G : H]$.

Now let us consider the case where the group G is finite. Each left coset has the same size as the subgroup H; here, both are finite. Hence, $|aH| = |H|$ for each coset. In addition, the group G is a disjoint union of the left cosets; that is,

$$G = H \cup g_1H \cup \cdots \cup g_nH.$$

Since this is a disjoint union, we have

$$|G| = |H| + |g_1H| + \cdots + |g_nH| = |H| + |H| + \cdots + |H| = |H|[G : H].$$

This establishes the following extremely important theorem:

Theorem 9.4.4 (Lagrange's theorem). *Let G be a group and $H \subset G$ a subgroup. Then*

$$|G| = |H|[G : H].$$

If G is a finite group, this implies that both the order of a subgroup and the index of a subgroup are divisors of the order of the group.

This theorem plays a crucial role in the structure theory of finite groups since it greatly restricts the size of subgroups. For example, in a group of order 10, there can be proper subgroups only of orders 1, 2, and 5.

As an immediate corollary, we have the following result:

Corollary 9.4.5. *The order of any element $g \in G$, where G is a finite group, divides the order of the group. In particular, if $|G| = n$ and $g \in G$, then $o(g)|n$, and $g^n = 1$.*

Proof. Let $g \in G$ and $o(g) = m$. Then m is the size of the cyclic subgroup generated by g; hence divides n from Lagrange's theorem. Then $n = mk$, and so

$$g^n = g^{mk} = (g^m)^k = 1^k = 1.$$ □

Before leaving this section, we consider some results concerning general subsets of a group.

Suppose that G is a group and S is an arbitrary nonempty subset of G, $S \subset G$, and $S \neq \emptyset$. Such a set S is usually called a *complex* of G.

If U and V are two complexes of G, the product UV is defined as follows:

$$UV = \{g_1g_2 \in G : u \in U, v \in V\}.$$

Now suppose that U, V are subgroups of G. When is the complex UV again a subgroup of G?

Theorem 9.4.6. *The product UV of two subgroups U, V of a group G is itself a subgroup if and only if U and V commute; that is, if and only if $UV = VU$.*

Proof. We note first that when we say U and V commute, we do *not* demand that this is so elementwise. In other words, it is not required that $uv = vu$ for all $u \in U$ and all $v \in V$. All that is required is that for any $u \in U$ and $v \in V$ $uv = v_1u_1$ for some elements $u_1 \in U$ and $v_1 \in V$.

Assume that UV is a subgroup of G. Let $u \in U$ and $v \in V$. Then $u \in U \cdot 1 \subset UV$ and $v \in 1 \cdot V \subset UV$. But since UV is assumed itself to be a subgroup, it follows that $vu \in UV$.

Hence, each product $vu \in UV$, and so $VU \subset UV$. In an identical manner, $UV \subset VU$, and so $UV = VU$.

Conversely, suppose that $UV = VU$. Let $g_1 = u_1v_1 \in UV$, $g_2 = u_2v_2 \in UV$. Then

$$g_1g_1 = (u_1v_1)(u_2v_2) = u_1(v_1u_2)v_2 = u_1u_3v_3v_2 = (u_1u_3)(v_3v_2) \in UV$$

since $v_1u_2 = u_3v_3$ for some $u_3 \in U$ and $v_3 \in V$. Furthermore,

$$g_1^{-1} = (u_1v_1)^{-1} = v_1^{-1}u_1^{-1} = u_4v_4.$$

It follows that UV is a subgroup. □

Theorem 9.4.7 (Product formula). *Let U, V be subgroups of G, and let R be a left transversal of the intersection $U \cap V$ in U. Then*

$$UV = \bigcup_{r \in R} rV,$$

where this is a disjoint union.

In particular, if U, V are finite, then

$$|UV| = \frac{|U||V|}{|U \cap V|}.$$

Proof. Since $R \subset U$, we have that

$$\bigcup_{r \in R} rV \subset UV.$$

In the other direction, let $uv \in UV$. Then

$$U = \bigcup_{r \in R} r(U \cap V).$$

It follows that $u = rv'$ with $r \in R$, and $v' \in U \cap V$. Hence,

$$uv = rv'v \in rV.$$

The union of cosets of V is disjoint, so

$$uv \in \bigcup_{r \in R} rV.$$

Therefore, $UV \subset \bigcup_{r \in R} rV$, proving the equality.

Now suppose that $|U|$ and $|V|$ are finite. Then we have

$$|UV| = |R||V| = |U : U \cap V||V| = \frac{|U|}{|U \cap V|}|V| = \frac{|U||V|}{|U \cap V|}.$$ □

We now show that index is multiplicative. Later, we will see how this fact is related to the multiplicativity of the degree of field extensions.

Theorem 9.4.8. *Suppose G is a group and U and V are subgroups with $U \subset V \subset G$. Then if G is the disjoint union*

$$G = \bigcup_{r \in R} rV,$$

R a left transversal of V in G, and V is the disjoint union

$$V = \bigcup_{s \in S} sU,$$

S a left transversal of U in V, then we get a disjoint union for G as

$$G = \bigcup_{r \in R, s \in S} rsU.$$

In particular, if $[G : V]$ and $[V : U]$ are finite, then

$$[G : U] = [G : V][V : U].$$

Proof. Now

$$G = \bigcup_{r \in R} rV = \bigcup_{r \in R}\Big(\bigcup_{s \in S} sU\Big) = \bigcup_{r \in R, s \in S} rsU.$$

Suppose that $r_1 s_1 U = r_2 s_2 U$. Then $r_1 s_1 UV = r_2 s_2 UV$. But $s_1 UV = V$, and $s_2 UV = V$ so $r_1 V = r_2 V$, which implies that $r_1 = r_2$. Then $s_1 U = s_2 U$, which implies that $s_1 = s_2$. Therefore, the union is disjoint.

The index formula now follows directly. □

The next result says that the intersection of subgroups of finite index must again be of finite index.

Theorem 9.4.9 (Poincaré). *Suppose that U, V are subgroups of finite index in G. Then $U \cap V$ is also of finite index. Furthermore,*

$$[G : U \cap V] \le [G : U][G : V].$$

If $[G : U]$, $[G : V]$ are relatively prime then equality holds.

Proof. Let r be the number of left cosets of U in G that are contained in UV. r is finite since the index $[G : U]$ is finite. From Theorem 9.4.7, we then have

$$|V : U \cap V| = r \le [G : U].$$

Then from Theorem 9.4.8,

$$[G : U \cap V] = [G : V][V : U \cap V] \leq [G : V][G : U].$$

Since both $[G : U]$ and $[G : V]$ are finite, so is $[G : U \cap V]$.

Now $[G : U] | [G : U \cap V]$, $[G : V] | [G : U \cap V]$. If $[G : U]$, and $[G : V]$ are relatively prime, then

$$[G : U][G : V] | [G : U \cap V] \implies [G : U][G : V] \leq [G : U \cap V].$$

Therefore, we must have equality. □

Corollary 9.4.10. *Suppose that $[G : U]$ and $[G : V]$ are finite and relatively prime. Then $G = UV$.*

Proof. From Theorem 9.4.9, we have

$$[G : U \cap V] = [G : U][G : V].$$

From Theorem 9.4.8

$$[G : U \cap V] = [G : V][V : U \cap V].$$

Combing these, we have

$$[V : U \cap V] = [G : U].$$

The number of left cosets of U in G that are contained in VU is equal to the number of all left cosets of U in G. It follows then that we must have $G = UV$. □

9.5 Generators and Cyclic Groups

We saw that if G is any group and $g \in G$, then the powers of g generate a subgroup of G, called the cyclic subgroup generated by g. Here, we explore more fully the idea of generating a group or subgroup. We first need the following:

Lemma 9.5.1. *If U and V are subgroups of a group G, then their intersection $U \cap V$ is also a subgroup.*

Proof. Since the identity of G is in both U and V, we have that $U \cap V$ is nonempty. Suppose that $g_1, g_2 \in U \cap V$. Then $g_1, g_2 \in U$; hence, $g_1^{-1}g_2 \in U$ since U is a subgroup. Analogously, $g_1^{-1}g_2 \in V$. Hence, $g^{-1}g_2 \in U \cap V$; therefore, $U \cap V$ is a subgroup. □

Now let S be a subset of a group G. The subset S is certainly contained in at least one subgroup of G, namely G itself. Let $\{U_a\}$ be the collection of all subgroups of G containing S. Then $\bigcap_a U_a$ is again a subgroup of G from Lemma 9.5.1. Furthermore, it is the smallest subgroup of G containing S (see the exercises). We call $\bigcap_\alpha U_\alpha$ the subgroup of G generated by S, and denote it by $\langle S\rangle$, or grp(S). We call the set S a set of *generators* for $\langle S\rangle$.

Definition 9.5.2. A subset M of a group G is a *set of generators* for G if $G = \langle M\rangle$; that is, the smallest subgroup of G containing M is all of G. We say that G is *generated* by M, and that M is a set of *generators* for G.

Notice that any group G has at least one set of generators, namely G itself. If we have $G = \langle M\rangle$ and M is a finite set, then G is called *finitely generated*. Clearly, any finite group is finitely generated. Shortly, we will give an example of a finitely generated infinite group.

Example 9.5.3. The set of all reflections forms a set of generators for the Euclidean group $\mathcal{E}$. Recall that any $T \in \mathcal{E}$ is either a translation, a rotation, a reflection, or a glide reflection. It can be shown (see exercises) that any one of these can be expressed as a product of 3, or fewer reflections.

We now consider the case, where a group G has a single generator.

Definition 9.5.4. A group G is *cyclic* if there exists a $g \in G$ such that $G = \langle g\rangle$.

In this case, $G = \{g^n : n \in \mathbb{Z}\}$; that is, G consists of all the powers of the element g. If there exists an integer m such that $g^m = 1$, then there exists a smallest such positive integer say n. It follows that $g^k = g^l$ if and only if $k \equiv l \pmod n$. In this situation, the distinct powers of g are precisely

$$\{1 = g^0, g, g^2, \ldots, g^{n-1}\}.$$

It follows that $|G| = n$. We then call G a *finite cyclic group*. If no such power exists, then all the powers of G are distinct and G is an infinite cyclic group.

We show next that any two cyclic groups of the same order are isomorphic.

Theorem 9.5.5. (a) *If $G = \langle g\rangle$ is an infinite cyclic group, then $G \cong (\mathbb{Z}, +)$; that is, the integers under addition.*

(b) *If $G = \langle g\rangle$ is a finite cyclic group of order n, then $G \cong (\mathbb{Z}_n, +)$; that is, the integers modulo n under addition.*

It follows that for a given order there is only one cyclic group up to isomorphism.

Proof. Let G be an infinite cyclic group with generator g. Map g onto $1 \in (\mathbb{Z}, +)$. Since g generates G and 1 generates $\mathbb{Z}$ under addition, this can be extended to a homomorphism. It is straightforward to show that this defines an isomorphism.

Now let G be a finite cyclic group of order n with generator g. As above, map g to $1 \in \mathbb{Z}_n$ and extend to a homomorphism. Again it is straightforward to show that this defines an isomorphism.

Now let G and H be two cyclic groups of the same order. If both are infinite, then both are isomorphic to $(\mathbb{Z}, +)$ and, hence, isomorphic to each other. If both are finite of order n, then both are isomorphic to $(\mathbb{Z}_n, +)$ and, hence, isomorphic to each other. □

Theorem 9.5.6. *Let $G = \langle g \rangle$ be a finite cyclic group of order n. Then every subgroup of G is also cyclic. Furthermore, if $d|n$, there exists a unique subgroup of G of order d.*

Proof. Let $G = \langle g \rangle$ be a finite cyclic group of order n, and suppose that H is a subgroup of G. Notice that if $g^m \in H$, then g^{-m} is also in H since H is a subgroup. Hence, H must contain positive powers of the generator g. Let t be the smallest positive power of g such that $g^t \in H$. We claim that $H = \langle g^t \rangle$, the cyclic subgroup of G generated by g^t. Let $h \in H$, then $h = g^m$ for some positive integer $m \geq t$. Divide m by t to get

$$m = qt + r, \quad \text{where } r = 0 \text{ or } 0 < r < t.$$

If $r \neq 0$, then $r = m - qt > 0$. Now $g^m \in H, g^t \in H$ so $g^{-qt} \in H$ for any q since H is a subgroup. It follows that $g^m g^{-qt} = g^{m-qt} \in H$. This implies that $g^r \in H$. However, this is a contradiction since $r < t$ and t is the least positive power in H. It follows that $r = 0$ so $m = qt$. This implies that $g^m = g^{qt} = (g^t)^q$; that is, g^m is a multiple of g^t. Therefore, every element of H is a multiple of g^t; thus, g^t generates H and, hence, H is cyclic.

Now suppose that $d|n$ so that $n = kd$. Let $H = \langle g^k \rangle$; that is, the subgroup of G generated by g^k. We claim that H has order d and that any other subgroup H_1 of G with order d coincides with H. Now $(g^k)^d = g^{kd} = g^n = 1$, so the order of g^k divides d, hence is $\leq d$. Suppose that $(g^k)^{d_1} = g^{kd_1} = 1$ with $d_1 < d$. Then since the order of g is n, we have $n = kd|kd_1$ with $d_1 < d$, which is impossible. Therefore, the order of g^k is d, and $h = \langle g^k \rangle$ is a subgroup of G of order d.

Now let H_1 be a subgroup of G of order d. We must show that $H_1 = H$. Let $h \in H_1$, so $h = g^t$; hence, $g^{td} = 1$. It follows that $n|td$, and so $kd|td$; hence $k|t$. That is, $t = qk$ for some positive integer q. Therefore, $g^t = (g^k)^q \in H$. Therefore, $H_1 \subset H$, and since they are of the same size, $H = H_1$. □

Theorem 9.5.7. *Let $G = \langle g \rangle$ be an infinite cyclic group. Then a subgroup H is of the form $H = \langle g^t \rangle$ for a positive integer t. Furthermore, if t_1, t_2 are positive integers with $t_1 \neq t_2$, then $\langle g^{t_1} \rangle$ and $\langle g^{t_2} \rangle$ are distinct.*

Proof. Let $G = \langle g \rangle$ be an infinite cyclic group and H a subgroup of G. As in the proof of Theorem 9.5.6, H must contain positive powers of the generator g. Let t be the smallest positive power of g such that $g^t \in H$. We claim that $H = \langle g^t \rangle$, the cyclic subgroup of G generated by g^t. Let $h \in H$, then $h = g^m$ for some positive integer $m \geq t$. Divide m by t to get

$$m = qt + r \quad \text{where } r = 0 \text{ or } 0 < r < t.$$

If $r \neq 0$, then $r = m - qt > 0$. Now $g^m \in H, g^t \in H$ so $g^{-qt} \in H$ for any q since H is a subgroup. It follows that $g^m g^{-qt} = g^{m-qt} \in H$. This implies that $g^r \in H$. However, this is a contradiction since $r < t$ and t is the least positive power in H. It follows that $r = 0$, so $m = qt$. This implies that $g^m = g^{qt} = (g^t)^q$; that is, g^m is a multiple of g^t. Therefore, every element of H is a multiple of g^t and, therefore, g^t generates H; hence, $H = \langle g^t \rangle$.

From the proof above in the subgroup $\langle g^t \rangle$, the integer t is the smallest positive power of g in $\langle g^t \rangle$. Therefore, if t_1, t_2 are positive integers with $t_1 \neq t_2$, then $\langle g^{t_1} \rangle$ and $\langle g^{t_2} \rangle$ are distinct. □

Theorem 9.5.8. *Let $G = \langle g \rangle$ be a cyclic group. Then the following hold:*
(a) *If $G = \langle g \rangle$ is finite of order n, then g^k is also a generator if and only if $(k, n) = 1$. That is, the generators of G are precisely those powers g^k, where k is relatively prime to n.*
(b) *If $G = \langle g \rangle$ is infinite, then the only generators are g, g^{-1}.*

Proof. (a) Let $G = \langle g \rangle$ be a finite cyclic group of order n, and suppose that $(k, n) = 1$. Then there exist integers x, y with $kx + ny = 1$. It follows that

$$g = g^{kx+ny} = (g^k)^x (g^n)^y = (g^k)^x$$

since $g^n = 1$. Hence, g is a power of g^k, that implies every element of G is also a power of g^k. Therefore, g^k is also a generator.

Conversely, suppose that g^k is also a generator. Then g is a power of g^k, so there exists an x such that $g = g^{kx}$. It follows that $kx \equiv 1 \pmod{n}$, and so there exists a y such that

$$kx + ny = 1.$$

This then implies that $(k, n) = 1$.

(b) If $G = \langle g \rangle$ is infinite, then any power of g other than g^{-1} generates a proper subgroup. If g is a power of g^n for some n so that $g = g^{nx}$, it follows that $g^{nx-1} = 1$, thus, g has finite order, contradicting that G is infinite cyclic. □

Recall that for positive integers n, the Euler phi-function is defined as follows:

Definition 9.5.9. For any $n > 0$, let

$$\phi(n) = \text{number of integers less than or equal to } n, \text{ and relatively prime to } n.$$

Example 9.5.10. $\phi(6) = 2$ since among 1, 2, 3, 4, 5, 6 only 1, 5 are relatively prime to 6.

Corollary 9.5.11. *If $G = \langle g \rangle$ is finite of order n, then there are $\phi(n)$ generators for G, where ϕ is the Euler phi-function.*

Proof. From Theorem 9.5.8, the generators of G are precisely the powers g^k, where $(k, n) = 1$. The numbers relatively prime to n are counted by the Euler phi-function. □

Recall that in an arbitrary group G, if $g \in G$, then the order of g, denoted $o(g)$, is the order of the cyclic subgroup generated by g. Given two elements $g, h \in G$, in general, there is no relationship between $o(g)$, $o(h)$ and the order of the product gh. However, if they commute, there is a very direct relationship.

Lemma 9.5.12. *Let G be an arbitrary group and $g, h \in G$ both of finite order $o(g)$, $o(h)$. If g and h commute; that is, $gh = hg$, then $o(gh)$ divides $\operatorname{lcm}(o(g), o(h))$. In particular, if G is an Abelian group, then $o(gh) | \operatorname{lcm}(o(g), o(h))$ for all $g, h \in G$ of finite order. Furthermore, if $\langle g\rangle \cap \langle h\rangle = \{1\}$, then $o(gh) = \operatorname{lcm}(o(g), o(h))$.*

Proof. Suppose $o(g) = n$ and $o(h) = m$ are finite. If g, h commute, then for any k, we have $(gh)^k = g^k h^k$. Let $t = \operatorname{lcm}(n, m)$, then $t = k_1 m$, $t = k_2 n$. Hence,

$$(gh)^t = g^t h^t = (g^m)^{k_1} (h^n)^{k_2} = 1.$$

Therefore, the order of gh is finite and divides t. Suppose that $\langle g\rangle \cap \langle h\rangle = \{1\}$; that is, the cyclic subgroup generated by g intersects trivially with the cyclic subgroup generated by h. Let $k = o(gh)$, which we know is finite from the first part of the lemma.

Let $t = \operatorname{lcm}(n, m)$. We then have $(gh)^k = g^k h^k = 1$, which implies that $g^k = h^{-k}$. Since the cyclic subgroups have only trivial intersection, this implies that $g^k = 1$ and $h^k = 1$. But then $n|k$ and $m|k$; hence $t|k$. Since $k|t$ it follows that $k = t$. □

Recall that if m and n are relatively prime, then $\operatorname{lcm}(m, n) = mn$. Furthermore, if the orders of g and h are relatively prime, it follows from Lagrange's theorem that $\langle g\rangle \cap \langle h\rangle = \{1\}$. We then get the following:

Corollary 9.5.13. *If g, h commute and $o(g)$ and $o(h)$ are finite and relatively prime, then $o(gh) = o(g)o(h)$.*

Definition 9.5.14. If G is a finite Abelian group, then the *exponent* of G is the lcm of the orders of all elements of G. That is,

$$\exp(G) = \operatorname{lcm}\{o(g) : g \in G\}.$$

As a consequence of Lemma 9.5.12, we obtain

Lemma 9.5.15. *Let G be a finite Abelian group. Then G contains an element of order $\exp(G)$.*

Proof. Suppose that $\exp(G) = p_1^{e_1} \cdots p_k^{e_k}$ with p_i distinct primes. By the definition of $\exp(G)$, there is a $g_i \in G$ with $o(g_i) = p_i^{e_i} r_i$ with p_i and r_i relatively prime. Let $h_i = g_i^{r_i}$. Then from Lemma 9.5.12, we get $o(h_i) = p_i^{e_i}$. Now let $g = h_1 h_2 \cdots h_k$. From the corollary to Lemma 9.5.12, we have $o(g) = p_1^{e_1} \cdots p_k^{e_k} = \exp(G)$. □

If K is a field then the multiplicative subgroup of nonzero elements of K is an Abelian group $K^\star$. The above results lead to the fact that a finite subgroup of $K^\star$ must actually be cyclic.

Theorem 9.5.16. *Let K be a field. Then any finite subgroup of $K^\star$ is cyclic.*

Proof. Let $A \subset K^\star$ with $|A| = n$. Suppose that $m = \exp(A)$. Consider the polynomial $f(x) = x^m - 1 \in K[x]$. Since the order of each element in A divides m, it follows that

$a^m = 1$ for all $a \in A$; hence, each $a \in A$ is a zero of the polynomial $f(x)$. Hence, $f(x)$ has at least n zeros. Since a polynomial of degree m over a field can have at most m zeros, it follows that $n \leq m$. From Lemma 9.5.15, there is an element $a \in A$ with $o(a) = m$. Since $|A| = n$, it follows that $m|n$; hence, $m \leq n$. Therefore, $m = n$; hence, $A = \langle a \rangle$ showing that A is cyclic. □

We close this section with two other results concerning cyclic groups. The first proves, using group theory, a very interesting number theoretic result concerning the Euler phi-function.

Theorem 9.5.17. *For $n > 1$ and for $d \geq 1$*

$$\sum_{d|n} \phi(d) = n.$$

Proof. Consider a cyclic group G of order n. For each $d|n$, $d \geq 1$, there is a unique cyclic subgroup H of order d. H then has $\phi(d)$ generators. Each element in G generates its own cyclic subgroup H_1, say of order d and, hence, must be included in the $\phi(d)$ generators of H_1. Therefore, $\sum_{d|n} \phi(d)$ is the sum of the numbers of generators of the cyclic subgroups of G. But this must be the whole group; hence, this sum is n. □

We shall make use of the above theorem directly in the following theorem.

Theorem 9.5.18. *If $|G| = n$ and if for each positive d such that $d|n$, G has at most one cyclic subgroup of order d, then G is cyclic (and, consequently, has exactly one cyclic subgroup of order d).*

Proof. For each $d|n$, $d > 0$, let $\psi(d)$ denote the number of elements of G of order d. Then

$$\sum_{d|n} \psi(d) = n.$$

Now suppose that $\psi(d) \neq 0$ for a given $d|n$. Then there exists an $a \in G$ of order d, which generates a cyclic subgroup, $\langle a \rangle$, of order d of G. We claim that all elements of G of order d are in $\langle a \rangle$. Indeed, if $b \in G$ with $o(b) = d$ and $b \notin \langle a \rangle$, then $\langle b \rangle$ is a second cyclic subgroup of order d, distinct from $\langle a \rangle$. This contradicts the hypothesis, so the claim is proved. Thus, if $\psi(d) \neq 0$, then $\psi(d) = \phi(d)$. In general, we have $\psi(d) \leq \phi(d)$, for all positive $d|n$. But $n = \sum_{d|n} \psi(d) \leq \sum_{d|n} \phi(d)$, by the previous theorem. It follows, clearly, from this that $\psi(d) = \phi(d)$ for all $d|n$. In particular, $\psi(n) = \phi(n) \geq 1$. Hence, there exists at least one element of G of order n; hence, G is cyclic. This completes the proof. □

Corollary 9.5.19. *If in a group G of order n, for each $d|n$, the equation $x^d = 1$ has at most d solutions in G, then G is cyclic.*

Proof. The hypothesis clearly implies that G can have at most one cyclic subgroup of order d since all elements of such a subgroup satisfy the equation. So Theorem 9.5.18 applies to give our result. □

If H is a subgroup of a group G then G operates as a group of permutations on the set $\{aH : a \in R\}$ of left cosets of H in G where R is a left transversal of H in G. This we can use to show that a finitely generated group has only finitely many subgroups of a given finite index.

Theorem 9.5.20. *Let G be a finitely generated group. The number of subgroups of index $n < \infty$ is finite.*

Proof. Let H be a subgroup of index n. We choose a left transversal $\{c_1, \dots, c_n\}$ for H in G where $c_1 = 1$ represents H. G permutes the set of cosets c_iH by multiplication from the left. This induces a homomorphism ψ_H from G to S_n as follows. For each $g \in G$ let $\psi_H(g)$ be the permutation which maps i to j if $gc_iH = c_jH$. $\psi_H(g)$ fixes the number 1 if and only if $g \in H$ because $c_1H = H$. Now, let H and L be two different subgroups of index n in G. Then there exists $g \in H$ with $g \notin L$ and $\psi_H(g) \neq \psi_L(g)$, and hence ψ_H and ψ_L are different. Since G is finitely generated there are only finitely many homomorphisms from G to S_n. Therefore the number of subgroups of index $n < \infty$ is finite. □

9.6 Exercises

1. Prove Lemma 9.1.4.
2. Let G be a group and H a nonempty subset. H is a subgroup of G if and only if $ab^{-1} \in H$ for all $a, b \in H$.
3. Suppose that $g \in G$ and $g^m = 1$ for some positive integer m. Let n be the smallest positive integer such that $g^n = 1$.
   Show that the set of elements $\{1, g, g^2, \dots, g^{n-1}\}$ are all distinct but for any other power g^k we have $g^k = g^t$ for some $k = 0, 1, \dots, n-1$.
4. Let G be a group and U_1, U_2 be finite subgroups of G. If $|U_1|$ and $|U_2|$ are relatively prime, then $U_1 \cap U_2 = \{e\}$.
5. Let A, B be subgroups of a finite group G. If $|A| \cdot |B| > |G|$ then $A \cap B \neq \{e\}$.
6. Let G be the set of all real matrices of the form $\begin{pmatrix} a & -b \\ b & a \end{pmatrix}$, where $a^2 + b^2 \neq 0$. Show:
   (a) G is a group.
   (b) For each $n \in \mathbb{N}$ there is at least one element of order n in G.
7. Let p be a prime, and let $G = \mathrm{SL}(2, p) = \mathrm{SL}(2, \mathbb{Z}_p)$. Show: G has at least $2p-2$ elements of order p.
8. Let p be a prime and $a \in \mathbb{Z}$. Show that $a^p \equiv a \pmod{p}$.
9. Here we outline a proof that every planar Euclidean congruence motion is either a rotation, translation, reflection or glide reflection. An isometry in this problem is a planar Euclidean congruence motion. Show:
   (a) If T is an isometry then it is completely determined by its action on a triangle—equivalent to showing that if T fixes three noncollinear points then it must be the identity.

(b) If an isometry T has exactly one fixed point then it must be a rotation with that point as center.
(c) If an isometry T has two fixed points then it fixes the line joining them. Then show that if T is not the identity it must be a reflection through this line.
(d) If an isometry T has no fixed point but preserves orientation then it must be a translation.
(e) If an isometry T has no fixed point but reverses orientation then it must be a glide reflection.

10. Let P_n be a regular n-gon and D_n its group of symmetries. Show that $|D_n| = 2n$. (*Hint*: First show that $|D_n| \leq 2n$ and then exhibit $2n$ distinct symmetries.)
11. If A, B have the same cardinality, then there exists a bijection $\sigma : A \to B$. Define a map $F : S_A \to S_B$ in the following manner: if $f \in S_A$, let $F(f)$ be the permutation on B given by $F(f)(b) = \sigma(f(\sigma^{-1}(b)))$. Show that F is an isomorphism.
12. Prove Lemma 9.3.3.

10 Normal Subgroups, Factor Groups and Direct Products

10.1 Normal Subgroups and Factor Groups

In rings, we saw that there were certain special types of subrings, called *ideals*, which allowed us to define factor rings. The analogous object for groups is called a *normal subgroup*, which we will define and investigate in this section.

Definition 10.1.1. Let G be an arbitrary group and suppose that H_1 and H_2 are subgroups of G. We say that H_2 is *conjugate* to H_1 if there exists an element $a \in G$ such that $H_2 = a^{-1}H_1a$. H_1, H_2 are the called *conjugate subgroups* of G.

Lemma 10.1.2. *Let G be an arbitrary group. Then the relation of conjugacy is an equivalence relation on the set of subgroups of G.*

Proof. We must show that conjugacy is reflexive, symmetric, and transitive. If H is a subgroup of G, then $1^{-1}H1 = H$; hence, H is conjugate to itself and, therefore, the relation is reflexive.

Suppose that H_1 is conjugate to H_2. Then there exists a $g \in G$ with $g^{-1}H_1g = H_2$. This implies that $gH_2g^{-1} = H_1$. However, $(g^{-1})^{-1} = g$; hence, letting $g^{-1} = g_1$, we have $g_1^{-1}H_2g_1 = H_1$. Therefore, H_2 is conjugate to H_1 and conjugacy is symmetric.

Finally, suppose that H_1 is conjugate to H_2 and H_2 is conjugate to H_3. Then there exist $g_1, g_2 \in G$ with $H_2 = g_1^{-1}H_1g_1$ and $H_3 = g_2^{-1}H_2g_2$. Then

$$H_3 = g_2^{-1}g_1^{-1}H_1g_1g_2 = (g_1g_2)^{-1}H_1(g_1g_2).$$

Therefore, H_3 is conjugate to H_1 and conjugacy is transitive. □

Lemma 10.1.3. *Let G be an arbitrary group. Then for $g \in G$, the map $g : a \to g^{-1}ag$ is an automorphism on G.*

Proof. For a fixed $g \in G$, define the map $f : G \to G$ by $f(a) = g^{-1}ag$ for $a \in G$. We must show that this is a homomorphism, and that it is one-to-one and onto.

Let $a_1, a_2 \in G$. Then

$$f(a_1a_2) = g^{-1}a_1a_2g = (g^{-1}a_1g)(g^{-1}a_2g) = f(a_1)f(a_2).$$

Hence, f is a homomorphism.

If $f(a_1) = f(a_2)$, then $g^{-1}a_1g = g^{-1}a_2g$. Clearly, by the cancellation law, we then have $a_1 = a_2$; hence, f is one-to-one.

Finally, let $a \in G$, and let $a_1 = gag^{-1}$. Then $a = g^{-1}a_1g$; hence, $f(a_1) = a$. It follows that f is onto; therefore, f is an automorphism on G. □

https://doi.org/10.1515/9783111142524-010

In general, a subgroup H of a group G may have many different conjugates. However, in certain situations, the only conjugate of a subgroup H is H itself. If this is the case, we say that H is a *normal subgroup*. We will see shortly that this is precisely the analog for groups of the concept of an ideal in rings.

Definition 10.1.4. Let G be an arbitrary group. A subgroup H is a *normal subgroup* of G, which we denote by $H \triangleleft G$, if $g^{-1}Hg = H$ for all $g \in G$.

Since the conjugation map is an isomorphism, it follows that if $g^{-1}Hg \subset H$, then $g^{-1}Hg = H$. Hence, in order to show that a subgroup is normal, we need only show inclusion.

Lemma 10.1.5. *Let N be a subgroup of a group G. Then if $a^{-1}Na \subset N$ for all $a \in G$, then $a^{-1}Na = N$. In particular, $a^{-1}Na \subset N$ for all $a \in G$ implies that N is a normal subgroup.*

Notice that if $g^{-1}Hg = H$, then $Hg = gH$. That is as sets the left coset, gH, is equal to the right coset, Hg. Hence, for each $h_1 \in H$, there is an $h_2 \in H$ with $gh_1 = h_2g$. If $H \triangleleft G$, this is true for all $g \in G$. Furthermore, if H is normal, then for the product of two cosets g_1H and g_2H, we have

$$(g_1H)(g_2H) = g_1(Hg_2)H = g_1g_2(HH) = g_1g_2H.$$

If $(g_1H)(g_2H) = (g_1g_2)H$ for all $g_1, g_2 \in G$, we necessarily have $g^{-1}Hg = H$ for all $g \in G$.

Hence, we have proved the following:

Lemma 10.1.6. *Let H be a subgroup of a group G. Then the following are equivalent:*
(1) *H is a normal subgroup of G.*
(2) *$g^{-1}Hg = H$ for all $g \in G$.*
(3) *$gH = Hg$ for all $g \in G$.*
(4) *$(g_1H)(g_2H) = (g_1g_2)H$ for all $g_1, g_2 \in G$.*

This is precisely the condition needed to construct factor groups. First we give some examples of normal subgroups.

Lemma 10.1.7. *Every subgroup of an Abelian group is normal.*

Proof. Let G be Abelian and H a subgroup of G. Suppose $g \in G$, then $gh = hg$ for all $h \in H$ since G is Abelian. It follows that $gH = Hg$. Since this is true for every $g \in G$, it follows that H is normal. □

Lemma 10.1.8. *Let $H \subset G$ be a subgroup of index 2; that is, $[G : H] = 2$. Then H is normal in G.*

Proof. Suppose that $[G : H] = 2$. We must show that $gH = Hg$ for all $g \in G$. If $g \in H$, clearly then, $H = gH = Hg$. Therefore, we may assume that g is not in H. Then there are only 2 left cosets and 2 right cosets. That is,

$$G = H \cup gH = H \cup Hg.$$

Since the union is a disjoint union, we must have $gH = Hg$; hence, H is normal. □

Lemma 10.1.9. *Let K be any field. Then the group $\mathrm{SL}(n, K)$ is a normal subgroup of $\mathrm{GL}(n, K)$ for any positive integer n.*

Proof. Recall that $\mathrm{GL}(n, K)$ is the group of $n \times n$ matrices over the field K with nonzero determinant, whereas $\mathrm{SL}(n, K)$ is the subgroup of $n \times n$ matrices over the field K with determinant equal to 1. Let $U \in \mathrm{SL}(n, K)$ and $T \in \mathrm{GL}(n, K)$. Consider $T^{-1}UT$. Then

$$\begin{aligned}\det(T^{-1}UT) &= \det(T^{-1})\det(U)\det(T) = \det(U)\det(T^{-1}T)\\ &= \det(U)\det(I) = \det(U) = 1.\end{aligned}$$

Hence, $T^{-1}UT \in \mathrm{SL}(n, K)$ for any $U \in \mathrm{SL}(n, K)$, and any $T \in \mathrm{GL}(n, K)$. It follows that $T^{-1}\,\mathrm{SL}(n, K)T \subset \mathrm{SL}(n, K)$; therefore, $\mathrm{SL}(n, K)$ is normal in $\mathrm{GL}(n, K)$. □

The intersection of normal subgroups is again normal, and the product of normal subgroups is normal.

Lemma 10.1.10. *Let N_1, N_2 be normal subgroups of the group G. Then the following hold:*
(1) *$N_1 \cap N_2$ is a normal subgroup of G.*
(2) *N_1N_2 is a normal subgroup of G.*
(3) *If H is any subgroup of G, then $N_1 \cap H$ is a normal subgroup of H, and $N_1H = HN_1$.*

Proof. We first show (1). Let $n \in N_1 \cap N_2$ and $g \in G$. Then $g^{-1}ng \in N_1$ since N_1 is normal. Similarly, $g^{-1}ng \in N_2$ since N_2 is normal. Hence, $g^{-1}ng \in N_1 \cap N_2$. It follows that $g^{-1}(N_1 \cap N_2)g \subset N_1 \cap N_2$; therefore, $N_1 \cap N_2$ is normal.

We now show (2). Let $n_1 \in N_1$, $n_2 \in N_2$. Since N_1, N_2 are both normal $N_1N_2 = N_2N_1$ as sets, and the complex N_1N_2 forms a subgroup of G. Let $g \in G$ and $n_1n_2 \in N_1N_2$. Then

$$g^{-1}(n_1n_2)g = (g^{-1}n_1g)(g^{-1}n_2g) \in N_1N_2$$

since $g^{-1}n_1g \in N_1$ and $g^{-1}n_2g \in N_2$. Therefore, N_1N_2 is normal in G.

We finally show (3). Let $h \in H$ and $n \in N \cap H$. Then as in part (a), $h^{-1}nh \in N \cap H$; therefore, $N \cap H$ is a normal subgroup of H. If $nh \in N_1H$, $n \in N_1$, $h \in H$, then $nh = hn'$ with some $n' \in N_1$. Hence, $N_1H = HN_1$. □

We now construct *factor groups* or *quotient groups* of a group modulo a normal subgroup.

Definition 10.1.11. Let G be an arbitrary group and H a normal subgroup of G. Let G/H denote the set of distinct left (and hence also right) cosets of H in G. On G/H, define the multiplication $(g_1H)(g_2H) = g_1g_2H$ for any elements g_1H, g_2H in G/H.

Theorem 10.1.12. *Let G be a group and H a normal subgroup of G. Then G/H under the operation defined above forms a group. This group is called the factor group or quotient group of G modulo H. The identity element is the coset $1H = H$, and the inverse of a coset gH is $g^{-1}H$.*

Proof. We first show that the operation on G/N is well defined. Suppose that $a'N = aN$ and $b'N = bN$, then $b' \in bN$, and so $b' = bn_1$. Similarly $a' = an_2$, where $n_1, n_2 \in N$. Therefore,

$$a'b'N = an_2bn_1N = an_2bN$$

since $n_1 \in N$. But $b^{-1}n_2b = n_3 \in N$, since N is normal. Therefore, the right-hand side of the equation can be written as

$$an_2bN = abN.$$

Thus, we have shown that if $N \lhd G$, then $a'b'N = abN$, and the operation on G/N is indeed well defined.

The associative law is true, because coset multiplication as defined above uses the ordinary group operation, which is by definition associative.

The coset N serves as the identity element of G/N. Notice that

$$aN \cdot N = aN^2 = aN,$$

and

$$N \cdot aN = aN^2 = aN.$$

The inverse of aN is $a^{-1}N$ since

$$aNa^{-1}N = aa^{-1}N^2 = N. \qquad \square$$

We emphasize that the elements of G/N are cosets; thus, subsets of G. If $|G| < \infty$, then $|G/N| = [G : N]$, the number of cosets of N in G. It is also to be emphasized that for G/N to be a group, N must be a normal subgroup of G.

In some cases, properties of G are preserved in factor groups.

Lemma 10.1.13. *If G is Abelian, then any factor group of G is also Abelian. If G is cyclic, then any factor group of G is also cyclic.*

Proof. Suppose that G is Abelian and H is a subgroup of G. H is necessarily normal from Lemma 10.1.7 so that we can form the factor group G/H. Let $g_1H, g_2H \in G/H$. Since G is Abelian, we have $g_1g_2 = g_2g_1$. Then in G/H,

$$(g_1H)(g_2H) = (g_1g_2)H = (g_2g_1)H = (g_2H)(g_1H).$$

Therefore, G/H is Abelian.

We leave the proof of the second part to the exercises. □

An extremely important concept has to do with when a group contains no proper normal subgroups other than the identity subgroup $\{1\}$.

Definition 10.1.14. A group $G \neq \{1\}$ is *simple*, provided that $N \triangleleft G$ implies $N = G$ or $N = \{1\}$.

One of the most outstanding problems in group theory has been to give a complete classification of all finite simple groups. In other words, this is the program to discover all finite simple groups, and to prove that there are no more to be found. This was accomplished through the efforts of many mathematicians. The proof of this magnificent result took thousands of pages. We refer the reader to [30] for a complete discussion of this. We give one elementary example:

Lemma 10.1.15. *Any finite group of prime order is simple and cyclic.*

Proof. Suppose that G is a finite group and $|G| = p$, where p is a prime. Let $g \in G$ with $g \neq 1$. Then $\langle g \rangle$ is a nontrivial subgroup of G, so its order divides the order of G by Lagrange's theorem. Since $g \neq 1$, and p is a prime, we must have $|\langle g \rangle| = p$. Therefore, $\langle g \rangle$ is all of G; that is, $G = \langle g \rangle$; hence, G is cyclic.

The argument above shows that G has no nontrivial proper subgroups and, therefore, no nontrivial normal subgroups. Therefore, G is simple. □

In the next chapter, we will examine certain other finite simple groups.

10.2 The Group Isomorphism Theorems

In Chapter 1, we saw that there was a close relationship between ring homomorphisms and factor rings. In particular to each ideal, and consequently to each factor ring, there is a ring homomorphism that has that ideal as its kernel. Conversely, to each ring homomorphism, its kernel is an ideal, and the corresponding factor ring is isomorphic to the image of the homomorphism. This was formalized in Theorem 1.5.7, which we called the *ring isomorphism theorem*. We now look at the group theoretical analog of this result, called the *group isomorphism theorem*. We will then examine some consequences of this result that will be crucial in the Galois theory of fields.

Definition 10.2.1. If G_1 and G_2 be groups and $f : G_1 \to G_2$ is a group homomorphism, then the *kernel* of f, denoted $\ker(f)$, is defined as

$$\ker(f) = \{g \in G_1 : f(g) = 1\}.$$

That is, the kernel is the set of the elements of G_1 that map onto the identity of G_2. The *image* of f, denoted $\mathrm{im}(f)$, is the set of elements of G_2 mapped onto by f from elements of G_1. That is,

$$\mathrm{im}(f) = \{g \in G_2 : f(g_1) = g_2 \text{ for some } g_1 \in G_1\}.$$

Note that if f is a surjection, then $\mathrm{im}(f) = G_2$.

As with ring homomorphisms the kernel measures how far a homomorphism is from being an injection, that is, a one-to-one mapping.

Lemma 10.2.2. *Let G_1 and G_2 be groups and $f : G_1 \to G_2$ a group homomorphism. Then f is injective if and only if* $\ker(f) = \{1\}$.

Proof. Suppose that f is injective. Since $f(1) = 1$, we always have $1 \in \ker(f)$. Suppose that $g \in \ker(f)$. Then $f(g) = f(1)$. Since f is injective, this implies that $g = 1$; hence, $\ker(f) = \{1\}$.

Conversely, suppose that $\ker(f) = \{1\}$ and $f(g_1) = f(g_2)$. Then

$$f(g_1)(f(g_2))^{-1} = 1 \implies f(g_1g_2^{-1}) = 1 \implies g_1g_2^{-1} \in \ker(f).$$

Then since $\ker(f) = \{1\}$, we have $g_1g_2^{-1} = 1$; hence, $g_1 = g_2$. Therefore, f is injective. □

We now state the *group isomorphism theorem*. This is entirely analogous to the ring isomorphism theorem replacing ideals by normal subgroups. We note that this theorem is sometimes called the *first group isomorphism theorem*.

Theorem 10.2.3 (Group isomorphism theorem). (a) *Let G_1 and G_2 be groups and $f : G_1 \to G_2$ a group homomorphism. Then* $\ker(f)$ *is a normal subgroup of G_1,* $\mathrm{im}(f)$ *is a subgroup of G_2, and*

$$G/\ker(f) \cong \mathrm{im}(f).$$

(b) *Conversely, suppose that N is a normal subgroup of a group G. Then there exists a group H and a homomorphism $f : G \to H$ such that* $\ker(f) = N$, *and* $\mathrm{im}(f) = H$.

Proof. We first show (a). Since $1 \in \ker(f)$, the kernel is nonempty. Now suppose that $g_1, g_2 \in \ker(f)$. Then $f(g_1) = f(g_2) = 1$. It follows that $f(g_1g_2^{-1}) = f(g_1)(f(g_2))^{-1} = 1$. Hence, $g_1g_2^{-1} \in \ker(f)$; therefore, $\ker(f)$ is a subgroup of G_1. Furthermore, for $g \in G_1$, we have

$$\begin{aligned} f(g^{-1}g_1g) &= (f(g))^{-1}f(g_1)f(g) \\ &= (f(g))^{-1} \cdot 1 \cdot f(g) = f(g^{-1}g) = f(1) = 1. \end{aligned}$$

Hence, $g^{-1}g_1g \in \ker(f)$ and $\ker(f)$ is a normal subgroup. It is straightforward to show that $\mathrm{im}(f)$ is a subgroup of G_2. Consider the map $\hat{f} : G/\ker(f) \to \mathrm{im}(f)$ defined by

$$\hat{f}(g\ker(f)) = f(g).$$

We show that this is an isomorphism.

Suppose that $g_1 \ker(f) = g_2 \ker(f)$, then $g_1 g_2^{-1} \in \ker(f)$ so that $f(g_1 g_2^{-1}) = 1$. This implies that $f(g_1) = f(g_2)$; hence, the map $\hat{f}$ is well defined. Now,

$$\begin{aligned}\hat{f}(g_1 \ker(f) g_2 \ker(f)) &= \hat{f}(g_1 g_2 \ker(f)) = f(g_1 g_2)\\ &= f(g_1) f(g_2) = \hat{f}(g_1 \ker(f)) \hat{f}(g_2 \ker(f));\end{aligned}$$

therefore, $\hat{f}$ is a homomorphism. Suppose that $\hat{f}(g_1 \ker(f)) = \hat{f}(g_2 \ker(f))$, then it follows that $f(g_1) = f(g_2)$; and hence, $g_1 \ker(f) = g_2 \ker(f)$. It follows that $\hat{f}$ is injective.

Finally, suppose that $h \in \operatorname{im}(f)$. Then there exists a $g \in G_1$ with $f(g) = h$. Then $\hat{f}(g \ker(f)) = h$, and $\hat{f}$ is a surjection onto $\operatorname{im}(f)$. Therefore, $\hat{f}$ is an isomorphism completing the proof of part (a).

Conversely, suppose that N is a normal subgroup of G. Define the map $f : G \to G/N$ by $f(g) = gN$ for $g \in G$. By the definition of the product in the quotient group G/N, it is clear that f is a homomorphism with $\operatorname{im}(f) = G/N$. If $g \in \ker(f)$, then $f(g) = gN = N$ since N is the identity in G/N. However, this implies that $g \in N$; hence, it follows that $\ker(f) = N$, completing the proof. □

There are two related theorems that are called the second isomorphism theorem and the third isomorphism theorem.

Theorem 10.2.4 (Second isomorphism theorem). *Let N be a normal subgroup of a group G and U a subgroup of G. Then $U \cap N$ is normal in U, and*

$$(UN)/N \cong U/(U \cap N).$$

Proof. From Lemma 10.1.10, we know that $U \cap N$ is normal in U. We define the map $\alpha : UN \to U/U \cap N$ by $\alpha(un) = u(U \cap N)$. If $un = u'n'$, then $u'^{-1}u = n'n^{-1} \in U \cap N$. Therefore, $u'(U \cap N) = u(U \cap N)$; hence, the map α is well defined.

Suppose that $un, u'n' \in UN$. Since N is normal in G, we have that $unu'n' \in uu'N$. Hence, $unu'n' = uu'n''$ with $n'' \in N$. Then

$$\alpha(unu'n') = \alpha(uu'n) = uu'(U \cap N).$$

However, $U \cap N$ is normal in U, so

$$uu'(U \cap N) = u(U \cap N)u'(U \cap N) = \alpha(un)\alpha(u'n').$$

Therefore, α is a homomorphism. We have $\operatorname{im}(\alpha) = U/(U \cap N)$ by definition. Suppose that $un \in \ker(\alpha)$. Then $\alpha(un) = U \cap N \subset N$, which implies $u \in N$. Therefore, $\ker(f) = N$. From the group isomorphism theorem, we then have

$$UN/N \cong U/(U \cap N),$$

proving the theorem. □

Theorem 10.2.5 (Third isomorphism theorem). *Let N and M be normal subgroups of a group G with N a subgroup of M. Then M/N is a normal subgroup in G/N, and*

$$(G/N)/(M/N) \cong G/M.$$

Proof. Define the map $\beta : G/N \to G/M$ by

$$\beta(gN) = gM.$$

It is straightforward that β is well defined and a homomorphism. If $gN \in \ker(\beta)$, then $\beta(gN) = gM = M$; hence, $g \in M$. It follows that $\ker(\beta) = M/N$. In particular, this shows that M/N is normal in G/N. From the group isomorphism theorem then,

$$(G/N)/(M/N) \cong G/M. \qquad \square$$

For a normal subgroup N in G, the homomorphism $f : G \to G/N$ provides a one-to-one correspondence between subgroups of G containing N and the subgroups of G/N. This correspondence will play a fundamental role in the study of subfields of a field.

Theorem 10.2.6 (Correspondence Theorem). *Let N be a normal subgroup of a group G, and let f be the corresponding homomorphism $f : G \to G/N$. Then the mapping*

$$\phi : H \to f(H),$$

where H is a subgroup of G containing N provides a one-to-one correspondence between all the subgroups of G/N and the subgroups of G containing N.

Proof. We first show that the mapping ϕ is surjective. Let H_1 be a subgroup of G/N, and let

$$H = \{g \in G : f(g) \in H_1\}.$$

We show that H is a subgroup of G, and that $N \subset H$.

If $g_1, g_2 \in H$, then $f(g_1) \in H_1$, and $f(g_2) \in H_1$. Therefore, $f(g_1)f(g_2) \in H_1$; hence, $f(g_1g_2) \in H_1$. Therefore, $g_1g_2 \in H$. In an identical fashion, $g_1^{-1} \in H$. Therefore, H is a subgroup of G. If $n \in N$, then $f(n) = 1 \in H_1$; hence, $n \in H$. Therefore, $N \subset H$, showing that the map ϕ is surjective.

Suppose that $\phi(H_1) = \phi(H_2)$, where H_1 and H_2 are subgroups of G containing N. This implies that $f(H_1) = f(H_2)$. Let $g_1 \in H_1$. Then $f(g_1) = f(g_2)$ for some $g_2 \in H_2$. Then $g_1g_2^{-1} \in \ker(f) = N \subset H_2$. It follows that $g_1g_2^{-1} \in H_2$ so that $g_1 \in H_2$. Hence, $H_1 \subset H_2$. In a similar fashion, $H_2 \subset H_1$; therefore, $H_1 = H_2$. It follows that ϕ is injective. $\square$

10.3 Direct Products of Groups

In this section, we look at a very important construction, the direct product, which allows us to build new groups out of existing groups. This construction is the analog for groups of the direct sum of rings. As an application of this construction, in the next section, we present a theorem, which completely describes the structure of finite Abelian groups.

Let G_1, G_2 be groups and let G be the Cartesian product of G_1 and G_2. That is,

$$G = G_1 \times G_2 = \{(a, b) : a \in G_1, b \in G_2\}.$$

On G, define

$$(a_1, b_1) \cdot (a_2, b_2) = (a_1 a_2, b_1 b_2).$$

With this operation, it is direct to verify the groups axioms for G; hence, G becomes a group.

Theorem 10.3.1. *Let G_1, G_2 be groups and G the Cartesian product $G_1 \times G_2$ with the operation defined above. Then G forms a group called the direct product of G_1 and G_2. The identity element is $(1, 1)$, and $(g, h)^{-1} = (g^{-1}, h^{-1})$.*

This can be iterated to any finite number of groups (also to an infinite number, that we will not consider here) $G_1, \ldots, G_n$ to form the direct product $G_1 \times G_2 \times \cdots \times G_n$.

Theorem 10.3.2. *For groups G_1 and G_2, we have $G_1 \times G_2 \cong G_2 \times G_1$, and $G_1 \times G_2$ is Abelian if and only if each G_i, $i = 1, 2$, is Abelian.*

Proof. The map $(a, b) \to (b, a)$, where $a \in G_1, b \in G_2$ provides an isomorphism $G_1 \times G_2 \to G_2 \times G_1$.

Suppose that both G_1, G_2 are Abelian. Then if $a_1, a_2 \in G_1$, $b_1, b_2 \in G_2$, we have

$$(a_1, b_1)(a_2, b_2) = (a_1 a_2, b_1 b_2) = (a_2 a_1, b_2 b_1) = (a_2, b_2)(a_1, b_1);$$

hence, $G_1 \times G_2$ is Abelian.

Conversely, suppose $G_1 \times G_2$ is Abelian, and suppose that $a_1, a_2 \in G_1$. Then for the identity $1 \in G_2$, we have

$$(a_1 a_2, 1) = (a_1, 1)(a_2, 1) = (a_2, 1)(a_1, 1) = (a_2 a_1, 1).$$

Therefore, $a_1 a_2 = a_2 a_1$, and G_1 is Abelian. Similarly, G_2 is Abelian. □

We show next that in $G_1 \times G_2$, there are normal subgroups H_1, H_2 with $H_1 \cong G_1$ and $H_2 \cong G_2$.

Theorem 10.3.3. *Let $G = G_1 \times G_2$. Let $H_1 = \{(a,1) : a \in G_1\}$ and $H_2 = \{(1,b) : b \in G_2\}$. Then both H_1 and H_2 are normal subgroups of G with $G = H_1H_2$ and $H_1 \cap H_2 = \{1\}$. Furthermore, $H_1 \cong G_1$, $H_2 \cong G_2$, $G/H_1 \cong G_2$, and $G/H_2 \cong G_1$.*

Proof. Map $G_1 \times G_2$ onto G_2 by $(a,b) \to b$. It is clear that this map is a homomorphism, and that the kernel is $H_1 = \{(a,1) : a \in G_1\}$. This establishes that H_1 is a normal subgroup of G, and that $G/H_1 \cong G_2$. In an identical fashion, we get that $G/H_2 \cong G_1$. The map $(a,1) \to a$ provides the isomorphism from H_1 onto G_1. □

If the factors are finite, it is easy to find the order of $G_1 \times G_2$. The size of the Cartesian product is just the product of the sizes of the factors.

Lemma 10.3.4. *If $|G_1|$ and $|G_2|$ are finite, then $|G_1 \times G_2| = |G_1||G_2|$.*

Now suppose that G is a group with normal subgroups G_1, G_2 such that $G = G_1G_2$ and $G_1 \cap G_2 = \{1\}$. Then we will show that G is isomorphic to the direct product $G_1 \times G_2$. In this case, we say that G is the *internal direct product* of its subgroups, and that G_1, G_2 are *direct factors* of G.

Theorem 10.3.5. *Suppose that G is a group with normal subgroups G_1, G_2 with $G = G_1G_2$, and $G_1 \cap G_2 = \{1\}$. Then G is isomorphic to the direct product $G_1 \times G_2$.*

Proof. Since $G = G_1G_2$, each element of G has the form ab with $a \in G_1$, $b \in G_2$. This representation as ab is unique as $G_1 \cap G_2 = \{1\}$. We first show that each $a \in G_1$ commutes with each $b \in G_2$. Consider the element $aba^{-1}b^{-1}$. Since G_1 is normal $ba^{-1}b^{-1} \in G_1$, which implies that $abab^{-1} \in G_1$. Since G_2 is normal, $aba^{-1} \in G_2$, which implies that $aba^{-1}b^{-1} \in G_2$. Therefore, $aba^{-1}b^{-1} \in G_1 \cap G_2 = \{1\}$; hence, $aba^{-1}b^{1} = 1$, so that $ab = ba$.

Now map G onto $G_1 \times G_2$ by $f(ab) \to (a,b)$. We claim that this is an isomorphism. It is clearly onto. Now

$$\begin{aligned} f((a_1b_1)(a_2b_2)) &= f(a_1a_2b_1b_2) = (a_1a_2, b_1b_2) \\ &= (a_1, b_1)(a_2, b_2) = f((a_1, b_1))(f(a_2, b_2)), \end{aligned}$$

so that f is a homomorphism. The kernel is $G_1 \cap G_2 = \{1\}$, and so f is an isomorphism. □

Although the end resulting groups are isomorphic, we call $G_1 \times G_2$ an external direct product if we started with the groups G_1, G_2 and constructed $G_1 \times G_2$, and we call $G_1 \times G_2$ an internal direct product if we started with a group G having normal subgroups, as in the theorem.

10.4 Finite Abelian Groups

We now use the results of the last section to present a theorem that completely provides the structure of finite Abelian groups. This theorem is a special case of a general result on modules that we will examine in detail in Chapter 19.

Theorem 10.4.1 (Basis theorem for finite Abelian groups). *Let G be a finite Abelian group. Then G is a direct product of cyclic groups of prime power order.*

Before giving the proof, we give two examples showing how this theorem leads to the classification of finite Abelian groups.

Since all cyclic groups of order n are isomorphic to $(\mathbb{Z}_n, +)$, we will denote a cyclic group of order n by $\mathbb{Z}_n$.

Example 10.4.2. Classify all Abelian groups of order 60. Let G be an Abelian group of order 60. From Theorem 10.4.1, G must be a direct product of cyclic groups of prime power order. Now $60 = 2^2 \cdot 3 \cdot 5$, so the only primes involved are 2, 3, and 5. Hence, the cyclic group involved in the direct product decomposition of G have order either 2, 4, 3, or 5 (by Lagrange's theorem, they must be divisors of 60). Therefore, G must be of the form

$$G \cong \mathbb{Z}_4 \times \mathbb{Z}_3 \times \mathbb{Z}_5$$
$$G \cong \mathbb{Z}_2 \times \mathbb{Z}_2 \times \mathbb{Z}_3 \times \mathbb{Z}_5.$$

Hence, up to isomorphism, there are only two Abelian groups of order 60.

Example 10.4.3. Classify all Abelian groups of order 180. Now $180 = 2^2 \cdot 3^2 \cdot 5$, so the only primes involved are 2, 3, and 5. Hence, the cyclic group involved in the direct product decomposition of G have order either 2, 4, 3, 9, or 5 (by Lagrange's theorem, they must be divisors of 180). Therefore, G must be of the form

$$G \cong \mathbb{Z}_4 \times \mathbb{Z}_9 \times \mathbb{Z}_5$$
$$G \cong \mathbb{Z}_2 \times \mathbb{Z}_2 \times \mathbb{Z}_9 \times \mathbb{Z}_5$$
$$G \cong \mathbb{Z}_4 \times \mathbb{Z}_3 \times \mathbb{Z}_3 \times \mathbb{Z}_5$$
$$G \cong \mathbb{Z}_2 \times \mathbb{Z}_2 \times \mathbb{Z}_3 \times \mathbb{Z}_3 \times \mathbb{Z}_5.$$

Hence, up to isomorphism, there are four Abelian groups of order 180.

The proof of Theorem 10.4.1 involves the following lemmas:

Lemma 10.4.4. *Let G be a finite Abelian group, and let $p \mid |G|$, where p is a prime. Then all the elements of G, whose orders are a power of p, form a normal subgroup of G. This subgroup is called the p-primary component of G, which we will denote by G_p.*

Proof. Let p be a prime with $p \mid |G|$, and let a and b be two elements of G of order a power of p. Since G is Abelian, the order of ab is the lcm of the orders, which is again a power of p. Therefore, $ab \in G_p$. The order of a^{-1} is the same as the order of a, so $a^{-1} \in G_p$; therefore, G_p is a subgroup. □

Lemma 10.4.5. *Let G be a finite Abelian group of order n. Suppose that $n = p_1^{e_1} \cdots p_k^{e_k}$ with $p_1, \dots, p_k$ distinct primes. Then*

$$G \cong G_{p_1} \times \cdots \times G_{p_k},$$

where G_{p_i} is the p_i-primary component of G.

Proof. Each G_{p_i} is normal since G is Abelian, and since distinct primes are relatively prime, the intersection of the G_{p_i} is the identity. Therefore, Lemma 10.4.5 will follow by showing that each element of G is a product of elements in the G_{p_1}.

Let $g \in G$. Then the order of g is $p_1^{f_1} \cdots p_k^{f_k}$. We write this as $p_i^{f_i} m$ with $(m, p_i) = 1$. Then g^m has order $p_i^{f_i}$ and, hence, is in G_{p_i}. Now since $p_1, \ldots, p_k$ are relatively prime, there exists $m_1, \ldots, m_k$ with

$$m_1 p_1^{f_1} + \cdots + m_k p_k^{f_k} = 1;$$

hence,

$$g = (g^{p_1^{f_1}})^{m_1} \cdots (g^{p_k^{f_k}})^{m_k}.$$

Therefore, g is a product of elements in the G_{p_i}. □

We next need the concept of a *basis*. Let G be any finitely generated Abelian group (finite or infinite), and let $g_1, \ldots, g_n$ be a set of generators for G. The generators $g_1, \ldots, g_n$ form a *basis* if

$$G = \langle g_1 \rangle \times \cdots \times \langle g_n \rangle;$$

that is, G is the direct product of the cyclic subgroups generated by the g_i. The basis theorem for finite Abelian groups says that any finite Abelian group has a basis. Suppose that G is a finite Abelian group with a basis $g_1, \ldots, g_k$ so that $G = \langle g_1 \rangle \times \cdots \times \langle g_k \rangle$. Since G is finite, each g_i has finite order, say m_i. It follows then, from the fact that G is a direct product, that each $g \in G$ can be expressed as

$$g = g_1^{n_1} \cdots g_k^{n_k}$$

and, furthermore, the integers $n_1, \ldots, n_k$ are unique modulo the order of g_i. Hence, each integer n_i can be chosen in the range $0, 1, \ldots, m_i - 1$, and within this range for the element g, the integer n_i is unique.

From the previous lemma, each finite Abelian group splits into a direct product of its p-primary components for different primes p. Hence, to complete the proof of the basis theorem, we must show that any finite Abelian group of order p^m for some prime p has a basis. We call an Abelian group of order p^m an Abelian p-group.

Consider an Abelian group G of order p^m for a prime p. It is somewhat easier to complete the proof if we consider the group using additive notation. That is, the operation is considered $+$, the identity as 0, and powers are given by multiples. Hence, if an element $g \in G$ has order p^k, then in additive notation, $p^k g = 0$.

A set of elements $g_1, \ldots, g_k$ is then a basis for G if each $g \in G$ can be expressed uniquely as $g = m_1 g_1 + \cdots + m_k g_k$, where the m_i are unique modulo the order of g_i. We say that the $g_1, \ldots, g_k$ are *independent*, and this is equivalent to the fact that whenever $m_1 g_1 + \cdots + m_k g_k = 0$, then $m_i \equiv 0$ modulo the order of g_i. We now prove that any Abelian p-group has a basis.

Lemma 10.4.6. *Let G be a finite Abelian group of prime power order p^n for some prime p. Then G is a direct product of cyclic groups.*

Notice that in the group G, we have $p^n g = 0$ for all $g \in G$ as a consequence of Lagrange's theorem. Furthermore, every element has as its order a power of p. The smallest power of p, say p^r such that $p^r g = 0$ for all $g \in G$, is called the *exponent* of G. Any finite Abelian p-group must have some exponent p^r.

Proof. The proof of this lemma is by induction on the exponent.

The lowest possible exponent is p. So, first, suppose that $pg = 0$ for all $g \in G$. Since G is finite it has a finite system of generators. Let $S = \{g_1, \ldots, g_k\}$ be a minimal set of generators for G. We claim that this is a basis. Since this is a set of generators, to show that it is a basis, we must show that they are independent. Hence, suppose that we have

$$m_1 g_1 + \cdots + m_k g_k = 0 \tag{10.1}$$

for some set of integers m_i. Since the order of each g_i is p, as explained above, we may assume that $0 \le m_i < p$ for $i = 1, \ldots, k$. Suppose that one $m_i \neq 0$.

Then $(m_i, p) = 1$; hence, there exists an x_i with $m_i x_i \equiv 1 \pmod{p}$ (see Chapter 4). Multiplying the equation (10.1) by x_i, we get modulo p,

$$m_1 x_i g_1 + \cdots + g_i + \cdots + m_k x_i g_k = 0,$$

and rearranging

$$g_i = -m_1 x_i g_1 - \cdots - m_k x_k g_k.$$

But then g_i can be expressed in terms of the other g_j; therefore, the set $\{g_1, \ldots, g_k\}$ is not minimal. It follows that $g_1, \ldots, g_k$ constitute a basis, and the lemma is true for the exponent p.

Now suppose that any finite Abelian group of exponent p^{n-1} has a basis, and assume that G has exponent p^n. Consider the set $\overline{G} = pG = \{pg : g \in G\}$. It is straightforward that this forms a subgroup (see exercises). Since $p^n g = 0$ for all $g \in G$, it follows that $p^{n-1} g = 0$ for all $g \in \overline{G}$, and so the exponent of $\overline{G} \le p^{n-1}$. By the inductive hypothesis, $\overline{G}$ has a basis

$$S = \{pg_1, \ldots, pg_k\}.$$

Consider the set $\{g_1, \dots, g_k\}$, and adjoin to this set the set of all elements $h \in G$, satisfying $ph = 0$. Call this set S_1, so that we have

$$S_1 = \{g_1, \dots, g_k, h_1, \dots, h_t\}.$$

We claim that S_1 is a set of generators for G. Let $g \in G$. Then $pg \in \overline{G}$, which has the basis $pg_1, \dots, pg_k$, so that

$$pg = m_1 pg_1 + \cdots + m_k pg_k.$$

This implies that

$$p(g - m_1 g_1 - \cdots - m_k g_k) = 0,$$

so that $g_1 - m_1 g_1 - \cdots - m_k g_k$ must be one of the h_i. Hence,

$$g - m_1 g_1 - \cdots - m_k g_k = h_i, \quad \text{so that } g = m_1 g_1 + \cdots + m_k g_k + h_i,$$

proving the claim.

Now S_1 is finite, so there is a minimal subset of S_1 that is still a generating system for G. Call this S_0, and suppose that S_0, renumbering if necessary, is

$$S_0 = \{g_1, \dots, g_r, h_1, \dots, h_s\} \quad \text{with } ph_i = 0 \text{ for } i = 1, \dots, s.$$

The subgroup generated by $h_1, \dots, h_s$ has exponent p. Therefore, by inductive hypothesis, has a basis. We may assume then that $h_1, \dots, h_s$ is a basis for this subgroup and, hence, is independent. We claim now that $g_1, \dots, g_r, h_1, \dots, h_s$ are independent and, hence, form a basis for G.

Suppose that

$$m_1 g_1 + \cdots + m_r g_r + n_1 h_1 + \cdots + n_s h_s = 0 \tag{10.2}$$

for some integers $m_1, \dots, m_r, h_1, \dots, h_s$. Each m_i, n_i must be divisible by p. Suppose, for example, that an m_i is not. Then $(m_i, p) = 1$, and then $(m_i, p^n) = 1$. This implies that there exists an x_i with $m_i x_i \equiv 1 \pmod{p^n}$. Multiplying through by x_i and rearranging, we then obtain

$$g_i = -m_1 x_i g_1 - \cdots - n_s x_i h_s.$$

Therefore, g_i can be expressed in terms of the remaining elements of S_0, contradicting the minimality of S_0. An identical argument works if an n_i is not divisible by p. Therefore, the relation (10.2) takes the form

$$a_1 pg_1 + \cdots + a_r pg_r + b_1 ph_1 + \cdots + b_s ph_s = 0. \tag{10.3}$$

Each of the terms $ph_i = 0$, so that (10.3) becomes

$$a_1pg_1 + \cdots + a_rpg_r = 0.$$

The $g_1, \ldots, g_r$ are independent and, hence, $a_ip = 0$ for each i; hence, $a_i = 0$. Now (10.2) becomes

$$n_1h_1 + \cdots + n_sh_s = 0.$$

However, $h_1, \ldots, h_s$ are independent, so each $n_i = 0$, completing the claim.

Therefore, the whole group G has a basis proving the lemma by induction. □

For more details see the proof of the general result on modules over principal ideal domains later in the book. There is also an additional elementary proof for the basis theorem for finitely generated Abelian groups.

10.5 Some Properties of Finite Groups

Classification is an extremely important concept in algebra. A large part of the theory is devoted to classifying all structures of a given type, for example all UFD's. In most cases, this is not possible. Since for a given finite n, there are only finitely many group tables, it is theoretically possible to classify all groups of order n. However, even for small n, this becomes impractical. We close the chapter by looking at some further results on finite groups, and then using these to classify all the finite groups up to order 10.

Before stating the classification, we give some further examples of groups that are needed.

Example 10.5.1. In Example 9.2.6, we saw that the symmetry group of an equilateral triangle had 6 elements, and is generated by elements r and f, which satisfy the relations $r^3 = f^2 = 1, f^{-1}rf = r^{-1}$, where r is a rotation of 120° about the center of the triangle, and f is a reflection through an altitude. This was called the dihedral group D_3 of order 6.

This can be generalized to any regular n-gon, $n > 2$. If D is a regular n-gon, then the symmetry group D_n has $2n$ elements, and is called the *dihedral group* of order $2n$. It is generated by elements r and f, which satisfy the relations $r^n = f^2 = 1, f^{-1}rf = r^{n-1}$, where r is a rotation of $\frac{2\pi}{n}$ about the center of the n-gon, and f is a reflection.

Hence, D_4, the symmetries of a square, has order 8 and D_5, the symmetries of a regular pentagon, has order 10.

Example 10.5.2. Let i, j, k be the generators of the quaternions. Then we have

$$i^2 = j^2 = k^2 = -1, \quad (-1)^2 = 1, \quad \text{and} \quad ijk = 1.$$

These elements then form a group of order 8 called the *quaternion group* denoted by Q. Since $ijk = 1$, we have $ij = -ji$, and the generators i and j satisfy the relations $i^4 = j^4 = 1$, $i^2 = j^2$, $ij = i^2ji$.

We now state the main classification, and then prove it in a series of lemmas.

Theorem 10.5.3. *Let G be a finite group.*
(a) *If $|G| = 2$, then $G \cong \mathbb{Z}_2$.*
(b) *If $|G| = 3$, then $G \cong \mathbb{Z}_3$.*
(c) *If $|G| = 4$, then $G \cong \mathbb{Z}_4$, or $G \cong \mathbb{Z}_2 \times \mathbb{Z}_2$.*
(d) *If $|G| = 5$, then $G \cong \mathbb{Z}_5$.*
(e) *If $|G| = 6$, then $G \cong \mathbb{Z}_6 \cong \mathbb{Z}_2 \times \mathbb{Z}_3$, or $G \cong D_3$, the dihedral group with 6 elements. (Note $D_3 \cong S_3$ the symmetric group on 3 symbols.)*
(f) *If $|G| = 7$, then $G \cong \mathbb{Z}_7$.*
(g) *If $|G| = 8$, then $G \cong \mathbb{Z}_8$, or $G \cong \mathbb{Z}_4 \times \mathbb{Z}_2$, or $G \cong \mathbb{Z}_2 \times \mathbb{Z}_2 \times \mathbb{Z}_2$, or $G \cong D_4$, the dihedral group of order 8, or $G \cong Q$, the quaternion group.*
(h) *If $|G| = 9$, then $G \cong \mathbb{Z}_9$, or $G \cong \mathbb{Z}_3 \times \mathbb{Z}_3$.*
(i) *If $|G| = 10$, then $G \cong \mathbb{Z}_{10} \cong \mathbb{Z}_2 \times \mathbb{Z}_5$, or $G \cong D_5$, the dihedral group with 10 elements.*

Recall from Section 10.1, that a finite group of prime order must be cyclic. Hence, in the theorem, the cases $|G| = 2, 3, 5, 7$ are handled. We next consider the case, where G has order p^2, and where p is a prime.

Definition 10.5.4. If G is a group, then its *center* denoted $Z(G)$, is the set of elements in G, which commute with everything in G. That is,

$$Z(G) = \{g \in G : gh = hg \text{ for any } h \in G\}.$$

Lemma 10.5.5. *For any group G the following hold:*
(a) *The center $Z(G)$ is a normal subgroup.*
(b) *$G = Z(G)$ if and only if G is Abelian.*
(c) *If $G/Z(G)$ is cyclic, then G is Abelian.*

Proof. (a) and (b) are direct, and we leave them to the exercises. Consider the case, where $G/Z(G)$ is cyclic. Then each coset of $Z(G)$ has the form $g^m Z(G)$, where $g \in G$. Let $a, b \in G$. Then since a, b are in cosets of the center, we have $a = g^m u$ and $b = g^n v$ with $u, v \in Z(G)$. Then

$$ab = (g^m u)(g^n v) = (g^m g^n)(uv) = (g^n g^m)(vu) = (g^n v)(g^m u) = ba$$

since u, v commute with everything. Therefore, G is Abelian. □

A *p-group* is any finite group of prime power order p^k. We need the following: The proof of this is based on what is called the *class equation*, which we will prove in Chapter 13.

Lemma 10.5.6. *A finite p-group has a nontrivial center of order at least p.*

Lemma 10.5.7. *If $|G| = p^2$ with p a prime, then G is Abelian; hence we have $G \cong \mathbb{Z}_{p^2}$, or $G \cong \mathbb{Z}_p \times \mathbb{Z}_p$.*

Proof. Suppose that $|G| = p^2$. Then from the previous lemma, G has a nontrivial center; hence, $|Z(G)| = p$, or $|Z(G)| = p^2$. If $|Z(G)| = p^2$, then $G = Z(G)$, and G is Abelian. If $|Z(G)| = p$, then $|G/Z(G)| = p$. Since p is a prime this implies that $G/Z(G)$ is cyclic; hence, from Lemma 10.5.5, G is Abelian. □

Lemma 10.5.7 handles the cases $n = 4$ and $n = 9$. Therefore, if $|G| = 4$, we must have $G \cong \mathbb{Z}_4$, or $G \cong \mathbb{Z}_2 \times \mathbb{Z}_2$, and if $|G| = 9$, we must have $G \cong \mathbb{Z}_9$, or $G \cong \mathbb{Z}_3 \times \mathbb{Z}_3$.

This leaves $n = 6, 8, 10$. We next handle the cases 6 and 10.

Lemma 10.5.8. *If G is any group, where every nontrivial element has order 2, then G is Abelian.*

Proof. Suppose that $g^2 = 1$ for all $g \in G$. This implies that $g = g^{-1}$ for all $g \in G$. Let a, b be arbitrary elements of G. Then

$$(ab)^2 = 1 \implies abab = 1 \implies ab = b^{-1}a^{-1} = ba.$$

Therefore, a, b commute, and G is Abelian. □

Lemma 10.5.9. *If $|G| = 6$, then $G \cong \mathbb{Z}_6$, or $G \cong D_3$.*

Proof. Since $6 = 2 \cdot 3$, if G was Abelian, then $G \cong \mathbb{Z}_2 \times \mathbb{Z}_3$. Notice that if an Abelian group has an element of order m and an element of order n with $(n, m) = 1$, then it has an element of order mn. Therefore, for 6 if G is Abelian, there is an element of order 6; hence, $G \cong \mathbb{Z}_2 \times \mathbb{Z}_3 \cong \mathbb{Z}_6$.

Now suppose that G is non-Abelian. The nontrivial elements of G have orders 2, 3, or 6. If there is an element of order 6, then G is cyclic, and hence Abelian. If every element has order 2, then G is Abelian. Therefore, there is an element of order 3, say $g \in G$. The cyclic subgroup $\langle g \rangle = \{1, g, g^2\}$ then has index 2 in G and is, therefore, normal. Let $h \in G$ with $h \notin \langle g \rangle$. Since g, g^2 both generate $\langle g \rangle$, we must have $\langle g \rangle \cap \langle h \rangle = \{1\}$. If h also had order 3, then $|\langle g, h \rangle| = \frac{|\langle g \rangle||\langle h \rangle|}{|\langle g \rangle \cap \langle h \rangle|} = 9$, which is impossible. Therefore, h must have order 2. Since $\langle g \rangle$ is normal, we have $h^{-1}gh = g^t$ for $t = 1, 2$. If $h^{-1}gh = g$, then g, h commute, and the group G is Abelian. Therefore, $h^{-1}gh = g^2 = g^{-1}$. It follows that g, h generate a subgroup of G, satisfying

$$g^3 = h^2 = 1, \quad h^1gh = g^{-1}.$$

This defines a subgroup of order 6 isomorphic to D_3 and, hence, must be all of G. □

Lemma 10.5.10. *If $|G| = 10$, then $G \cong \mathbb{Z}_{10}$, or $G \cong D_5$.*

Proof. The proof is almost identical to that for $n = 6$. Since $10 = 2 \cdot 5$, if G were Abelian, $G \cong \mathbb{Z}_2 \times \mathbb{Z}_5 = \mathbb{Z}_{10}$.

Now suppose that G is non-Abelian. As for $n = 6$, G must contain a normal cyclic subgroup of order 5, say $\langle g\rangle = \{1, g, g^2, g^3, g^4\}$. If $h \notin \langle g\rangle$, then exactly as for $n = 6$, it follows that h must have order 2, and $h^{-1}gh = g^t$ for $t = 1, 2, 3, 4$. If $h^{-1}gh = g$, then g, h commute, and G is Abelian. Notice that $h^{-1} = h$. Suppose that $h^{-1}gh = hgh = g^2$. Then

$$(hgh)^3 = (g^2)^3 = g^6 = g \implies g = h^2gh^2 = hg^2h = g^4 \implies g = 1,$$

which is a contradiction. Similarly, $hgh = g^3$ leads to a contradiction. Therefore, $h^{-1}gh = g^4 = g^{-1}$, and g, h generate a subgroup of order 10, satisfying

$$g^5 = h^2 = 1; \quad h^{-1}gh = g^{-1}.$$

Therefore, this is all of G, and is isomorphic to D_5. □

This leaves the case $n = 8$, the most difficult. If $|G| = 8$, and G is Abelian, then clearly, $G \cong \mathbb{Z}_8$, or $G \cong \mathbb{Z}_4\times\mathbb{Z}_2$, or $G \cong \mathbb{Z}_2\times\mathbb{Z}_2\times\mathbb{Z}_2$. The proof of Theorem 10.5.3 is then completed with the following:

Lemma 10.5.11. *If G is a non-Abelian group of order 8, then $G \cong D_4$, or $G \cong Q$.*

Proof. The nontrivial elements of G have orders 2, 4, or 8. If there is an element of order 8, then G is cyclic, and hence Abelian, whereas if every element has order 2, then G is Abelian. Hence, we may assume that G has an element of order 4, say g. Then $\langle g\rangle$ has index 2 and is a normal subgroup. First, suppose that G has an element $h \notin \langle g\rangle$ of order 2. Then

$$h^{-1}gh = g^t \quad \text{for some } t = 1, 2, 3.$$

If $h^{-1}gh = g$, then as in the cases 6 and 10, $\langle g, h\rangle$ defines an Abelian subgroup of order 8; hence, G is Abelian. If $h^{-1}gh = g^2$, then

$$(h^{-1}gh)^2 = (g^2)^2 = g^4 = 1 \implies g = h^{-2}gh^2 = h^{-1}g^2h = g^4 \implies g^3 = 1,$$

contradicting the fact that g has order 4. Therefore, $h^{-1}gh = g^3 = g^{-1}$. It follows that g, h define a subgroup of order 8, isomorphic to D_4. Since $|G| = 8$, this must be all of G and $G \cong D_4$.

Therefore, we may now assume that every element $h \in G$ with $h \notin \langle g\rangle$ has order 4. Let h be such an element. Then h^2 has order 2, so $h^2 \in \langle g\rangle$, which implies that $h^2 = g^2$. This further implies that g^2 is central; that is, commutes with everything. Identifying g with i, h with j, and g^2 with -1, we get that G is isomorphic to Q, completing Lemma 10.5.11 and the proof of Theorem 10.5.3. □

In principle, this type of analysis can be used to determine the structure of any finite group, although it quickly becomes impractical. A major tool in this classification is the following important result known as the Sylow theorem, which we just state. We will prove this theorem in Chapter 13. If $|G| = p^m n$ with p a prime and $(n, p) = 1$, then a subgroup of G of order p^m is called a *p-Sylow subgroup*. It is not clear at first that a group will contain p-Sylow subgroups.

Theorem 10.5.12 (Sylow theorem). *Let $|G| = p^m n$ with p a prime and $(n, p) = 1$.*
(a) *G contains a p-Sylow subgroup.*
(b) *All p-Sylow subgroups of G are conjugate.*
(c) *Any p-subgroup of G is contained in a p-Sylow subgroup.*
(d) *The number of p-Sylow subgroups of G is of the form $1 + pk$ and divides n.*

10.6 Automorphisms of a Group

Let G be a group. A homomorphism $f : G \to G$ is called an automorphism of G if f is bijective. Let $\mathrm{Aut}(G)$ be the set of all automorphisms of G.

Theorem 10.6.1. *$\mathrm{Aut}(G)$ is a group.*

Proof. The identity map 1 is the identity of $\mathrm{Aut}(G)$.
Let $f, g \in \mathrm{Aut}(G)$.
Then certainly $fg \in \mathrm{Aut}(g)$. Now

$$\begin{aligned} f^{-1}(ab) &= f^{-1}(ff^{-1}(a)ff^{-1}(b)) \\ &= f^{-1}(f(f^{-1}(a)f^{-1}(b))) \\ &= f^{-1}(a)f^{-1}(b) \end{aligned}$$

for $a, b \in G$, because $f \in \mathrm{Aut}(G)$.
Hence, $f^{-1} \in \mathrm{Aut}(G)$. □

A special automorphism of G is as follows: Let $a \in G$, and

$$i_a : G \to G, \quad i_a(x) = axa^{-1}.$$

By Lemma 10.1.3, we have that $i_a \in \mathrm{Aut}(G)$.

Definition 10.6.2. i_a is called an *inner automorphism of G by a*.
Let $\mathrm{Inn}(G)$ be the set of all inner automorphisms of G.

Theorem 10.6.3. *The map $\varphi : G \to \mathrm{Aut}(G)$, $a \mapsto i_a$, is an epimorphism; that is, a surjective homomorphism.*

Proof. Certainly $\varphi(G) = \mathrm{Inn}(G)$. We have the following:

$$\begin{aligned} \varphi(a)\varphi(b)(x) &= i_a(i_b(x)) = i_a(bxb^{-1}) \\ &= abxb^{-1}a^{-1} = (ab)x(ab)^{-1} \\ &= i_{ab}(x) = \varphi(ab)(x), \end{aligned}$$

that is, $\varphi(ab) = \varphi(a)\varphi(b)$. □

Theorem 10.6.4. Inn(G) *is a normal subgroup of* Aut(G); *that is,* Inn$(G) \lhd$ Aut(G).

Proof. From Theorem 10.6.3, Inn(G) is a homomorphic image $\varphi(G)$ of G. Therefore, Inn$(G) <$ Aut(G). Let $f \in$ Aut(G). Then

$$\begin{aligned} fi_a f^{-1}(x) &= f(af^{-1}(x)a^{-1}) = f(a)ff^{-1}(x)f(a^{-1}) \\ &= f(a)x(f(a))^{-1} = i_{f(a)}(x), \end{aligned}$$

that is, $fi_a f^{-1} = i_{f(a)} \in$ Inn(G). □

We now consider the kernel $\ker(\varphi)$ of the map $\varphi : G \to$ Aut(G), $a \mapsto i_a$.

We have

$$\begin{aligned} \ker(\varphi) &= \{a \in G : i_a(x) = x \text{ for all } x \in G\} \\ &= \{a \in G : axa^{-1} = x \text{ for all } x \in G\}. \end{aligned}$$

Hence, $\ker(\varphi) = Z(G)$, the center of G. Now, from Theorem 10.2.3, we get the following:

Theorem 10.6.5. *For a group G we have* Inn$(G) \cong G/Z(G)$.

Let G be a group and $f \in$ Aut(G). If $a \in G$ has order n, then $f(a)$ also has order n; if $a \in G$ has infinite order then $f(a)$ also has infinite order.

Example 10.6.6. Let $V \cong \mathbb{Z}_2 \times \mathbb{Z}_2$; that is, V has four elements $1, a, b$ and ab with $a^2 = b^2 = (ab)^2 = 1$.

V is often called the Klein four group. An automorphism of V permutes the three elements a, b and ab of order 2, and each permutation of $\{a, b, ab\}$ defines an automorphism of V. Hence, Aut$(V) \cong S_3$.

Example 10.6.7. We have $S_3 \cong$ Inn$(S_3) =$ Aut(S_3). By Theorem 10.6.5, $S_3 \cong$ Inn(S_3), because $Z(S_3) = \{1\}$. Now, let $f \in$ Aut(S_3). Analogously, as in Example 10.6.6, the automorphism f permutes the three transpositions $(1,2)$, $(1,3)$, and $(2,3)$.

This gives $|$ Aut$(S_3)| \le |S_3| = 6$, because S_3 is generated by these transpositions. From $S_3 \cong$ Inn$(S_3) \lhd$ Aut(S_3), we have $|$ Aut$(S_3)| \ge 6$.

Hence, Aut$(S_3) \cong$ Inn$(S_3) \cong S_3$.

Example 10.6.8. Let $G_n = \langle g \rangle \cong (\mathbb{Z}_n, +)$, $n \in \mathbb{N}$, be a cyclic group of order n.

If $f \in$ Aut(G_n), then $G_n = \langle f(g) \rangle = \langle g^k \rangle$, and $(k, n) = 1$ by Theorem 9.5.8. Hence, Aut$(G_n) \cong \mathbb{Z}_n^\star$, the group of units of the ring $\mathbb{Z}_n = \mathbb{Z}/n\mathbb{Z}$.

In particular, $|$ Aut$(G_n)| = \varphi(n)$. If $n = p$ a prime number, then Aut$(G_p) \cong \mathbb{Z}_p^\star$ is cyclic by Theorem 9.5.16.

In general, Aut(G_n) is not cyclic. If, for instance, $n = 8$, then $\varphi(8) = 4$. The four automorphisms of G_8 are given by $f_1(g) = g$, $f_2(g) = g^3$, $f_3(g) = g^5$, and $f_4(g) = g^7$.

We have $f_i^2(g) = g$ for $i = 1, 2, 3, 4$. Hence, Aut$(G_8) \cong \mathbb{Z}_2 \times \mathbb{Z}_2$. We remark that certainly Aut$(\mathbb{Z}, +) = \mathbb{Z}_2$, because $f(1) = 1$ or $f(1) = -1$ for $f \in$ Aut$(\mathbb{Z}, +)$.

10.7 Exercises

1. Prove that if G is cyclic, then any factor group of G is also cyclic.
2. Prove that for any group G, the center $Z(G)$ is a normal subgroup, and $G = Z(G)$ if and only if G is Abelian.
3. Let U_1 and U_2 be subgroups of a group G. Let $x, y \in G$. Show the following:
   (i) If $xU_1 = yU_2$, then $U_1 = U_2$.
   (ii) An example that $xU_1 = U_2x$ does not imply $U_1 = U_2$.
4. Let U, V be subgroups of a group G. Let $x, y \in G$. If $UxV \cap UyV \neq \emptyset$, then $UxV = UyV$.
5. Let N be a cyclic normal subgroup of the group G. Then all subgroups of N are normal subgroups of G. Give an example to show that the statement is not correct if N is not cyclic.
6. Let N_1 and N_2 be normal subgroups of G. Show the following:
   (i) If all elements in N_1 and N_2 have finite order, then also the elements of N_1N_2.
   (ii) Let $e_1, e_2 \in \mathbb{N}$. If $n_i^{e_i} = 1$ for all $n_i \in N_i$ $(i = 1, 2)$, then $x^{e_1e_2} = 1$ for all $x \in N_1N_2$.
7. Find groups N_1, N_2 and G with $N_1 \lhd N_2 \lhd G$, but N_1 is not a normal subgroup of G.
8. Let G be a group generated by a and b and let $bab^{-1} = a^r$ and $a^n = 1$ for suitable $r \in \mathbb{Z}$, $n \in \mathbb{N}$. Show the following:
   (i) The subgroup $A := \langle a \rangle$ is a normal subgroup of G.
   (ii) $G/A = \langle bA \rangle$.
   (iii) $G = \{b^j a^i : i, j \in \mathbb{Z}\}$.
9. Prove that any group of order 24 cannot be simple.
10. Let G be a group with subgroups G_1, G_2. Then the following are equivalent:
   (i) $G \cong G_1 \times G_2$.
   (ii) $G_1 \lhd G$, $G_2 \lhd G$, $G = G_1G_2$, and $G_1 \cap G_2 = \{1\}$.
   (iii) Every $g \in G$ has a unique expression $g = g_1g_2$, where $g_1 \in G_1$, $g_2 \in G_2$, and $g_1g_2 = g_2g_1$ for each $g_1 \in G_1$, $g_2 \in G_2$.
11. Suppose that G is a finite group with normal subgroups G_1, G_2 such that $(|G_1|, |G_2|) = 1$. If $|G| = |G_1||G_2|$, then $G \cong G_1 \times G_2$.
12. Let G be a group with normal subgroups G_1 and G_2 such that $G = G_1G_2$. Then

$$G/(G_1 \cap G_2) \cong G_1/(G_1 \cap G_2) \times G_2/(G_1 \cap G_2).$$

11 Symmetric and Alternating Groups

11.1 Symmetric Groups and Cycle Decomposition

Groups most often appear as groups of transformations or permutations on a set. In Galois Theory, groups will appear as permutation groups on the zeros of a polynomial. In Section 9.3, we introduced permutation groups and the symmetric group S_n. In this chapter, we look more carefully at the structure of S_n, and for each n introduce a very important normal subgroup, A_n of S_n, called the *alternating group* on n symbols.

Recall that if A is a set, a *permutation* on A is a one-to-one mapping of A onto itself. The set S_A of all permutations on A forms a group under composition called the *symmetric group* on A. If $|A| > 2$, then S_A is non-Abelian. Furthermore, if A, B have the same cardinality, then $S_A \cong S_B$.

If $|A| = n$, then $|S_A| = n!$ and, in this case, we denote S_A by S_n, called the *symmetric group on n symbols*. For example, $|S_3| = 6$. In Example 9.3.5, we showed that the six elements of S_3 can be given by the following:

$$1 = \begin{pmatrix} 1 & 2 & 3 \\ 1 & 2 & 3 \end{pmatrix}, \quad a = \begin{pmatrix} 1 & 2 & 3 \\ 2 & 3 & 1 \end{pmatrix}, \quad b = \begin{pmatrix} 1 & 2 & 3 \\ 3 & 1 & 2 \end{pmatrix}$$

$$c = \begin{pmatrix} 1 & 2 & 3 \\ 2 & 1 & 3 \end{pmatrix}, \quad d = \begin{pmatrix} 1 & 2 & 3 \\ 3 & 2 & 1 \end{pmatrix}, \quad e = \begin{pmatrix} 1 & 2 & 3 \\ 1 & 3 & 2 \end{pmatrix}.$$

In addition, we saw that S_3 has a *presentation* given by

$$S_3 = \langle a, c; a^3 = c^2 = 1, ac = ca^2 \rangle.$$

By this, we mean that S_3 is *generated by* a, c, or that S_3 has *generators* a, c, and the whole group and its multiplication table can be generated by using the *relations* $a^3 = c^2 = 1, ac = ca^2$.

In general, a *permutation* group is any subgroup of S_A for a set A.

For the remainder of this chapter, we will only consider finite symmetric groups S_n and always consider the set A as $A = \{1, 2, 3, \dots, n\}$.

Definition 11.1.1. Suppose that f is a permutation of $A = \{1, 2, \dots, n\}$, which has the following effect on the elements of A: There exists an element $a_1 \in A$ with $f(a_1) = a_2$, $f(a_2) = a_3, \dots, f(a_{k-1}) = a_k, f(a_k) = a_1$, and f leaves all other elements (if there are any) of A fixed; that is, $f(a_j) = a_j$ for $a_j \neq a_i$, $i = 1, 2, \dots, k$. Such a permutation f is called a *cycle* or a *k-cycle*.

We use the following notation for a k-cycle, f, as given above:

$$f = (a_1, a_2, \dots, a_k).$$

https://doi.org/10.1515/9783111142524-011

The cycle notation is read from left to right. It says f takes a_1 into a_2, a_2 into a_3, et cetera, and finally a_k, the last symbol, into a_1, the first symbol. Moreover, f leaves all the other elements not appearing in the representation above fixed.

Note that one can write the same cycle in many ways using this type of notation; for example, $f = (a_2, a_3, \dots, a_k, a_1)$. In fact, any cyclic rearrangement of the symbols gives the same cycle. The integer k is the *length* of the cycle. Note we allow a cycle to have length 1, that is, $f = (a_1)$, for instance. This is just the identity map. For this reason, we will usually designate the identity of S_n by (1), or just 1. (Of course, it also could be written as (a_i), where $a_i \in A$.)

If f and g are two cycles, they are called *disjoint cycles* if the elements moved by one are left fixed by the other; that is, their representations contain different elements of the set A (their representations are disjoint as sets).

Lemma 11.1.2. *If f and g are disjoint cycles, then they must commute; that is, $fg = gf$.*

Proof. Since the cycles f and g are disjoint, each element moved by f is fixed by g, and vice versa. First, suppose $f(a_i) \neq a_i$. This implies that $g(a_i) = a_i$, and $f^2(a_i) \neq f(a_i)$. But since $f^2(a_i) \neq f(a_i)$, $g(f(a_i)) = f(a_i)$. Thus, $(fg)(a_i) = f(g(a_i)) = f(a_i)$, whereas $(gf)(a_i) = g(f(a_i)) = f(a_i)$. Similarly, if $g(a_j) \neq a_j$, then $(fg)(a_j) = (gf)(a_j)$. Finally, if $f(a_k) = a_k$ and $g(a_k) = a_k$, clearly then, $(fg)(a_k) = a_k = (gf)(a_k)$. Thus, $gf = fg$. □

Before proceeding further with the theory, let us consider a specific example. Let $A = \{1, 2, \dots, 8\}$, and let

$$f = \begin{pmatrix} 1 & 2 & 3 & 4 & 5 & 6 & 7 & 8 \\ 2 & 4 & 6 & 5 & 1 & 7 & 3 & 8 \end{pmatrix}.$$

We pick an arbitrary number from the set A, say 1. Then $f(1) = 2$, $f(2) = 4$, $f(4) = 5$, $f(5) = 1$. Now select an element from A not in the set $\{1, 2, 4, 5\}$, say 3. Then $f(3) = 6$, $f(6) = 7$, $f(7) = 3$.

Next select any element of A that does not occur in the set $\{1, 2, 4, 5\} \cup \{3, 6, 7\}$. The only element left is 8, and $f(8) = 8$. It is clear that we can now write the permutation f as a product of cycles:

$$f = (1, 2, 4, 5)(3, 6, 7)(8),$$

where the order of the cycles is immaterial since they are disjoint and, therefore, commute. It is customary to omit such cycles as (8) and write f simply as

$$f = (1, 2, 4, 5)(3, 6, 7)$$

with the understanding that the elements of A not appearing are left fixed by f.

It is not difficult to generalize what was done here for a specific example, and show that any permutation f can be written uniquely, except for order, as a product of disjoint cycles. Thus, let f be a permutation on the set $A = \{1, 2, \ldots, n\}$, and let $a_1 \in A$. Let $f(a_1) = a_2, f^2(a_1) = f(a_2) = a_3$, et cetera, and continue until a repetition is obtained. We claim that this first occurs for a_1; that is, the first repetition is, say

$$f^k(a_1) = f(a_k) = a_{k+1} = a_1.$$

For suppose the first repetition occurs at the k-th iterate of f and

$$f^k(a_1) = f(a_k) = a_{k+1},$$

and $a_{k+1} = a_j$, where $j < k$. Then

$$f^k(a_1) = f^{j-1}(a_1),$$

and so $f^{k-j+1}(a_1) = a_1$. However, $k - j + 1 < k$ if $j \neq 1$, and we assumed that the first repetition occurred for k. Thus, $j = 1$, and so f does cyclically permute the set $\{a_1, a_2, \ldots, a_k\}$. If $k < n$, then there exists $b_1 \in A$ such that $b_1 \notin \{a_1, a_2, \ldots, a_k\}$, and we may proceed similarly with b_1. We continue in this manner until all the elements of A are accounted for. It is then seen that f can be written in the form

$$f = (a_1, \ldots, a_k)(b_1, \ldots, b_\ell)(c_1, \ldots, c_m) \cdots (h_1, \ldots, h_t).$$

Note that all powers $f^i(a_1)$ belong to the set

$$\{a_1 = f^0(a_1) = f^k(a_1),\ a_2 = f^1(a_1), \ldots, a_k = f^{k-1}(a_1)\};$$

all powers $f^i(b_1)$ belong to the set

$$\{b_1 = f^0(b_1) = f^\ell(b_1),\ b_2 = f^1(b_1), \ldots, b_\ell = f^{\ell-1}(b_1)\};$$

and so on. Here, by definition, b_1 is the smallest element in $\{1, 2, \ldots, n\}$, which does not belong to $\{a_1 = f^0(a_1) = f^k(a_1),\ a_2 = f^1(a_1), \ldots, a_k = f^{k-1}(a_1)\}$; c_1 is the smallest element in $\{1, 2, \ldots, n\}$, which does not belong to

$$\begin{aligned}&\{a_1 = f^0(a_1) = f^k(a_1),\ a_2 = f^1(a_1),\ \ldots,\ a_k = f^{k-1}(a_1)\}\\ &\cup \{b_1 = f^0(b_1) = f^\ell(b_1),\ b_2 = f^1(b_1),\ \ldots,\ b_\ell = f^{\ell-1}(b_1)\}.\end{aligned}$$

Therefore, by construction, all the cycles are disjoint.

From this, it follows that $k + \ell + m + \cdots + t = n$. It is clear that this factorization is unique, except for the order of the factors, since it tells explicitly what effect f has on each element of A.

In summary, we have proven the following result.

Theorem 11.1.3. *Every permutation of S_n can be written uniquely as a product of disjoint cycles (up to order).*

Example 11.1.4. The elements of S_3 can be written in cycle notation as $1 = (1), (1,2), (1,3), (2,3), (1,2,3), (1,3,2)$. This is the largest symmetric group, which consists entirely of cycles.

In S_4, for example, the element $(1,2)(3,4)$ is not a cycle, but a product of cycles. Suppose we multiply two elements of S_3, say $(1,2)$ and $(1,3)$. In forming the product or composition here, we read from right to left. Thus, to compute $(1,2)(1,3)$: We note the permutation $(1,3)$ takes 1 into 3, and then the permutation $(1,2)$ takes 3 into 3. Therefore, the composite $(1,2)(1,3)$ takes 1 into 3. Continuing the permutation, $(1,3)$ takes 3 into 1, and then the permutation $(1,2)$ takes 1 into 2. Therefore, the composite $(1,2)(1,3)$ takes 3 into 2. Finally, $(1,3)$ takes 2 into 2, and then $(1,2)$ takes 2 into 1. So $(1,2)(1,3)$ takes 2 into 1. Thus, we see $(1,2)(1,3) = (1,3,2)$.

As another example of this *cycle multiplication* consider $(1,2)(2,4,5)(1,3)(1,2,5)$ in S_5:

Reading from right to left $1 \mapsto 2 \mapsto 2 \mapsto 4 \mapsto 4$ so $1 \mapsto 4$. Now $4 \mapsto 4 \mapsto 4 \mapsto 5 \mapsto 5$ so $4 \mapsto 5$. Next $5 \mapsto 1 \mapsto 3 \mapsto 3 \mapsto 3$ so $5 \mapsto 3$. Then $3 \mapsto 3 \mapsto 1 \mapsto 1 \mapsto 2$ so $3 \mapsto 2$. Finally, $2 \mapsto 5 \mapsto 5 \mapsto 2 \mapsto 1$, so $2 \mapsto 1$. Since all the elements of $A = \{1,2,3,4,5\}$ have been accounted for, we have $(1,2)(2,4,5)(1,3)(1,2,5) = (1,4,5,3,2)$.

Let $f \in S_n$. If f is a cycle of length 2, that is, $f = (a_1, a_2)$, where $a_1, a_2 \in A$, then f is called a *transposition*. Any cycle can be written as a product of transpositions, namely,

$$(a_1, \ldots, a_k) = (a_1, a_k)(a_1, a_{k-1}) \cdots (a_1, a_2).$$

From Theorem 11.1.3, any permutation can be written in terms of cycles, but from the above, any cycle can be written as a product of transpositions. Thus, we have the following result:

Theorem 11.1.5. *Let $f \in S_n$ be any permutation. Then f can be written as a product of transpositions.*

11.2 Parity and the Alternating Groups

If f is a permutation with a cycle decomposition

$$(a_1, \ldots, a_k)(b_1, \ldots, b_j) \cdots (m_1, \ldots, m_t),$$

then f can be written as a product of

$$W(f) = (k-1) + (j-1) + \cdots + (t-1)$$

transpositions. The number $W(f)$ is uniquely associated with the permutation f since f is uniquely represented (up to order) as a product of disjoint cycles. However, there is nothing unique about the number of transpositions occurring in an arbitrary representation of f as a product of transpositions. For example, in S_3,

$$(1,3,2) = (1,2)(1,3) = (1,2)(1,3)(1,2)(1,2),$$

since $(1,2)(1,2) = (1)$, the identity permutation of S_3.

Although the number of transpositions is not unique in the representation of a permutation f as a product of transpositions, we will show that the parity (evenness or oddness) of that number *is* unique. Moreover, this depends solely on the number $W(f)$ uniquely associated with the representation of f. More explicitly, we have the following result:

Theorem 11.2.1. *If f is a permutation written as a product of disjoint cycles, and if $W(f)$ is the associated integer given above, then if $W(f)$ is even (odd), any representation of f, as a product of transpositions, must contain an even (odd) number of transpositions.*

Proof. We first observe the following:

$$(a,b)(b,c_1,\dots,c_t)(a,b_1,\dots,b_k) = (a,b_1,\dots,b_k,b,c_1,\dots,c_t),$$
$$(a,b)(a,b_1,\dots,b_k,b,c_1,\dots,c_t) = (a,b_1,\dots,b_k)(b,c_1,\dots,c_t).$$

Suppose now that f is represented as a product of disjoint cycles, where we include all the 1-cycles of elements of A, which f fixes, if any. If a and b occur in the same cycle in this representation for f,

$$f = \cdots (a,b_1,\dots,b_k,b,c_1,\dots,c_t)\cdots,$$

then, in the computation of $W(f)$, this cycle contributes $k+t+1$. Now consider $(a,b)f$. Since the cycles are disjoint and disjoint cycles commute,

$$(a,b)f = \cdots (a,b)(a,b_1,\dots,b_k,b,c_1,\dots,c_t)\cdots$$

since neither a nor b can occur in any factor of f other than

$$(a,b_1,\dots,b_k,b,c_1,\dots,c_t).$$

So that (a,b) cancels out, and we find that

$$(a,b)f = \cdots (b,c_1,\dots,c_t)(a,b_1,\dots,b_k)\cdots.$$

Since $W((b,c_1,\dots,c_t)(a,b_1,\dots,b_k)) = k+t$, but $W(a,b_1,\dots,b_k,b,c_1,\dots,c_t) = k+t+1$, we have $W((a,b)f) = W(f) - 1$.

A similar analysis shows that in the case, where a and b occur in different cycles in the representation of f, then $W((a,b)f) = W(f) + 1$. Combining both cases, we have

$$W((a,b)f) = W(f) \pm 1.$$

Now let f be written as a product of m transpositions, say

$$f = (a_1, b_1)(a_2, b_2) \cdots (a_m, b_m).$$

Then

$$(a_m, b_m) \cdots (a_2, b_2)(a_1, b_1)f = 1.$$

Iterating this, together with the fact that $W(1) = 0$, shows that

$$W(f)(\pm 1)(\pm 1)(\pm 1) \cdots (\pm 1) = 0,$$

where there are m terms of the form ± 1. Thus,

$$W(f) = (\pm 1)(\pm 1) \cdots (\pm 1),$$

m times.

Note, if exactly p are $+$ and $q = m - p$ are $-$, then $m = p + q$, and $W(f) = p - q$. Hence, $m \equiv W(f) \pmod 2$. Thus, $W(f)$ is even if and only if m is even, and this completes the proof. □

It now makes sense to state the following definition since we know that the parity is indeed unique:

Definition 11.2.2. A permutation $f \in S_n$ is said to be *even* if it can be written as a product of an even number of transpositions. Similarly, f is called *odd* if it can be written as a product of an odd number of transpositions.

Definition 11.2.3. For $n \geq 2$ we define the *sign function* $\operatorname{sgn} : S_n \to (\mathbb{Z}_2, +)$ by setting $\operatorname{sgn}(\pi) = 0$ if π is an even permutation and $\operatorname{sgn}(\pi) = 1$ if π is an odd permutation.

We note that if f and g are even permutations, then so are fg and f^{-1} and also the identity permutation is even. Furthermore, if f is even and g is odd, it is clear that fg is odd. From this it is straightforward to establish the following:

Lemma 11.2.4. *The map* sgn *is a homomorphism from S_n, for $n \geq 2$, onto $(\mathbb{Z}_2, +)$.*

We now let

$$A_n = \{\pi \in S_n : \operatorname{sgn}(\pi) = 0\}.$$

That is, A_n is precisely the set of even permutations in S_n.

Theorem 11.2.5. *For each $n \in \mathbb{N}$, $n \geq 2$, the set A_n forms a normal subgroup of index 2 in S_n, called the alternating group on n symbols. Furthermore, $|A_n| = \frac{n!}{2}$.*

Proof. By Lemma 11.2.4 sgn : $S_n \to (\mathbb{Z}_2, +)$ is a homomorphism. Then $\ker(\text{sgn}) = A_n$; therefore, A_n is a normal subgroup of S_n. Since $\text{im}(\text{sgn}) = \mathbb{Z}_2$, we have $|\text{im}(\text{sgn})| = 2$, hence, $|S_n/A_n| = 2$. Therefore, $[S_n : A_n] = 2$. Since $|S_n| = n!$, then $|A_n| = \frac{n!}{2}$ follows from Lagrange's theorem. □

11.3 The Conjugation in S_n

Recall that in a group G, two elements $x, y \in G$ are *conjugates* if there exists a $g \in G$ with $g^{-1}xg = y$. Conjugacy is an equivalence relation on G. In the symmetric groups S_n, it is easy to determine if two elements are conjugates. We say that two permutations in S_n have the *same cycle structure* if they have the same number of cycles and the lengths are the same. Hence, for example in S_8 the permutations

$$\pi_1 = (1, 3, 6, 7)(2, 5) \quad \text{and} \quad \pi_2 = (2, 3, 5, 6)(1, 8)$$

have the same cycle structure. In particular, if π_1, π_2 are two permutations in S_n, then π_1, π_2 are conjugates if and only if they have the same cycle structure. Therefore, in S_8, the permutations

$$\pi_1 = (1, 3, 6, 7)(2, 5) \quad \text{and} \quad \pi_2 = (2, 3, 5, 6)(1, 8)$$

are conjugates.

Lemma 11.3.1. *Let*

$$\pi = (a_{11}, a_{12}, \ldots, a_{1k_1}) \cdots (a_{s1}, a_{s2}, \ldots, a_{sk_s})$$

be the cycle decomposition of $\pi \in S_n$. Let $\tau \in S_n$, and denote the image of a_{ij} under τ by a_{ij}^{τ}. Then

$$\tau\pi\tau^{-1} = (a_{11}^{\tau}, a_{12}^{\tau}, \ldots, a_{1k_1}^{\tau}) \cdots (a_{s1}^{\tau}, a_{s2}^{\tau}, \ldots, a_{sk_s}^{\tau}).$$

Proof. (a) Consider a_{11}, then operating on the left like functions, we have

$$\tau\pi\tau^{-1}(a_{11}^{\tau}) = \tau\pi(a_{11}) = \tau(a_{12}) = a_{12}^{\tau}.$$

The same computation then follows for all the symbols a_{ij}, proving the lemma. □

Theorem 11.3.2. *Two permutations $\pi_1, \pi_2 \in S_n$ are conjugates if and only if they are of the same cycle structure.*

Proof. Suppose that $\pi_2 = \tau\pi_1\tau^{-1}$. Then, from Lemma 11.3.1, we have that π_1 and π_2 are of the same cycle structure.

Conversely, suppose that π_1 and π_2 are of the same cycle structure. Let

$$\pi_1 = (a_{11}, a_{12}, \ldots, a_{1k_1}) \cdots (a_{s1}, a_{s2}, \ldots, a_{sk_s})$$
$$\pi_2 = (b_{11}, b_{12}, \ldots, b_{1k_1}) \cdots (b_{s1}, b_{s2}, \ldots, b_{sk_s}),$$

where we place the cycles of the same length under each other. Let τ be the permutation in S_n that maps each symbol in π_1 to the digit below it in π_2. Then, from Lemma 11.3.1, we have $\tau\pi_1\tau^{-1} = \pi_2$; hence, π_1 and π_2 are conjugate. □

11.4 The Simplicity of A_n

A *simple group* is a group G with no nontrivial proper normal subgroups. Up to this point, the only examples we have of simple groups are cyclic groups of prime order. In this section, we prove that if $n \geq 5$, each alternating group A_n is a simple group.

Theorem 11.4.1. *For each $n \geq 3$ each $\pi \in A_n$ is a product of cycles of length* 3.

Proof. Let $\pi \in A_n$. Since π is a product of an even number of transpositions to prove the theorem, it suffices to show that if τ_1, τ_2 are transpositions, then $\tau_1\tau_2$ is a product of 3-cycles.

The statement holds certainly for $n = 3$. Now, let $n \geq 4$.

Suppose that a, b, c, d are different digits in $\{1, \ldots, n\}$. There are three cases to consider. First:

$$\text{Case (1):} \quad (a,b)(a,b) = 1 = (1,2,3)^0;$$

hence, it is true here.

Next:

$$\text{Case (2):} \quad (a,b)(b,c) = (c,a,b);$$

hence, it is also true here.

Finally:

$$\text{Case (3):} \quad (a,b)(c,d) = (a,b)(b,c)(b,c)(c,d) = (c,a,b)(c,d,b)$$

since $(b,c)(b,c) = 1$. Therefore, it is also true here, proving the theorem. □

Now our main result:

Theorem 11.4.2. *For $n \geq 5$, the alternating group A_n is a simple non-Abelian group.*

Proof. Suppose that N is a nontrivial normal subgroup of A_n with $n \geq 5$. We show that $N = A_n$; hence, A_n is simple.

We claim first that N must contain a 3-cycle. Let $1 \neq \pi \in N$, then π is not a transposition since $\pi \in A_n$. Therefore, π moves at least 3 digits. If π moves exactly 3 digits, then it is a 3-cycle, and we are done. Suppose then that π moves at least 4 digits. Let $\pi = \tau_1 \cdots \tau_r$ with τ_i disjoint cycles.

Case (1): There is a $\tau_i = (\ldots, a, b, c, d)$. Set $\sigma = (a, b, c) \in A_n$. Then

$$\pi\sigma\pi^{-1} = \tau_i\sigma\tau_i^{-1} = (b, c, d).$$

However, from Lemma 11.3.1, $(b, c, d) = (a^{\tau_i}, b^{\tau_i}, c^{\tau_i})$. Furthermore, since $\pi \in N$ and N is normal, we have

$$\pi(\sigma\pi^{-1}\sigma^{-1}) = (b, c, d)(a, c, b) = (a, d, b).$$

Therefore, in this case, N contains a 3-cycle.

Case (2): There is a τ_i, which is a 3-cycle. Then

$$\pi = (a, b, c)(d, e, \ldots).$$

Now, set $\sigma = (a, b, d) \in A_n$, and then

$$\pi\sigma\pi^{-1} = (b, c, e) = (a^\pi, b^\pi, d^\pi),$$

and

$$\sigma^{-1}\pi\sigma\pi^{-1} = (a, d, b)(b, c, e) = (b, c, e, d, a) \in N.$$

Now, use Case (1). Therefore, in this case, N has a 3-cycle.

In the final case, π is a disjoint product of transpositions.

Case (3): $\pi = (a, b)(c, d) \cdots$. Since $n \geq 5$, there exists an $e \neq a, b, c, d$. We now set $\sigma = (a, c, e) \in A_n$. Then $\pi\sigma\pi^{-1} = (b, d, e_1)$ with $e_1 = e^\pi \neq b, d$. However, we have $(a^\pi, c^\pi, e^\pi) = (b, d, e_1)$. Let $\gamma = (\sigma^{-1}\pi\sigma)\pi^{-1}$. This is in N since N is normal. If $e = e_1$, then $\gamma = (e, c, a)(b, d, e) = (a, e, b, d, c)$, and we can use Case (1) to get that N contains a 3-cycle. If $e \neq e_1$, then $\gamma = (e, c, a)(b, d, e_1) \in N$, and then we can use Case (2) to obtain that N contains a 3-cycle.

These three cases show that N must contain a 3-cycle.

If N is normal in A_n, then from the argument above, N contains a 3-cycle τ. However, from Theorem 11.3.2, any two 3-cycles in S_n are conjugate. Hence, τ is conjugate to any other 3-cycle in S_n. Since N is normal in A_n and $\tau \in N$, each of these conjugates must also be in N. Therefore, N contains all 3-cycles in S_n. From Theorem 11.4.1, each element of A_n is a product of 3-cycles. It follows then that each element of A_n is in N. However, since $N \subset A_n$, this is only possible if $N = A_n$, completing the proof. □

Theorem 11.4.3. *Let $n \in \mathbb{N}$ and $U \subset S_n$ a subgroup. Let $\tau = (1, 2)$ be a transposition and $\alpha = (1, 2, a_3, \ldots, a_n)$ an n-cycle with $\alpha, \tau \in U$. Then $U = S_n$.*

Proof. Let

$$\pi = \begin{pmatrix} 1 & 2 & a_3 & \cdots & a_n \\ 1 & 2 & 3 & \cdots & n \end{pmatrix}.$$

Then, from Lemma 11.3.1, we have

$$\pi\alpha\pi^{-1} = (1, 2, \ldots, n).$$

Furthermore, $\pi(1, 2)\pi^{-1} = (1, 2)$. Hence, $U_1 = \pi U\pi^{-1}$ contains $(1, 2)$ and $(1, 2, \ldots, n)$.

Now we have

$$(1, 2, \ldots, n)(1, 2)(1, 2, \ldots, n)^{-1} = (2, 3) \in U_1.$$

Analogously,

$$(1, 2, \ldots, n)(2, 3)(1, 2, \ldots, n)^{-1} = (3, 4) \in U_1,$$

and so on until

$$(1, 2, \ldots, n)(n-2, n-1)(1, 2, \ldots, n)^{-1} = (n-1, n) \in U_1.$$

Hence, the transpositions $(1, 2), (2, 3), \ldots, (n-1, n) \in U_1$. Moreover,

$$(1, 2)(2, 3)(1, 2) = (1, 3) \in U_1.$$

In an identical fashion, each $(1, k) \in U_1$. Then for any digits s, t, we have

$$(1, s)(1, t)(1, s) = (s, t) \in U_1.$$

Therefore, U_1 contains all the transpositions of S_n; hence, $U_1 = S_n$. Since $U = \pi U_1\pi^{-1}$, we must also have $U = S_n$. □

We end this chapter with the following corollary.

Corollary 11.4.4. *Let p be a prime number and $U \subset S_p$ a subgroup. Let τ be a transposition and α be a p-cycle with $\alpha, \tau \in U$. Then $U = S_p$.*

Proof. Suppose, without loss of generality, that $\tau = (1, 2)$. Since $\alpha, \ldots, \alpha^{p-1}$ are p-cycles with no fixed points (recall that p is a prime number), there exists an i with $\alpha^i(1) = 2$. Without loss of generality, we may assume that $\alpha = (1, 2, a_3, \ldots, a_p)$. Now the result follows from Theorem 11.4.3. □

11.5 Exercises

1. Show that for $n \geq 3$, the group A_n is generated by $\{(1,2,k) : k \geq 3\}$.
2. Let $\sigma = (k_1, \ldots, k_s) \in S_n$ be a permutation. Show that the order of σ is the least common multiple of $k_1, \ldots, k_s$. Compute the order of $\tau = \begin{pmatrix} 1 & 2 & 3 & 4 & 5 & 6 & 7 \\ 2 & 6 & 5 & 1 & 3 & 4 & 7 \end{pmatrix} \in S_7$.
3. Let $G = S_4$.
   (i) Determine a noncyclic subgroup H of order 4 of G.
   (ii) Show that H is normal.
   (iii) Show that $f(g)(h) := ghg^{-1}$ defines an epimorphism $f : G \to \operatorname{Aut}(H)$ for $g \in G$ and $h \in H$. Determine its kernel.
4. Show that all subgroups of order 6 of S_4 are conjugate.
5. Let $\sigma_1 = (1,2)(3,4)$ and $\sigma_2 = (1,3)(2,4) \in S_4$. Determine $\tau \in S_4$ such that $\tau\sigma_1\tau^{-1} = \sigma_2$.
6. Let $\sigma = (a_1, \ldots, a_k) \in S_n$. Describe σ^{-1}.

12 Solvable Groups

12.1 Solvability and Solvable Groups

The original motivation for Galois theory grew out of a famous problem in the theory of equations. This problem was to determine the solvability or insolvability of a polynomial equation of degree 5 or higher in terms of a formula involving the coefficients of the polynomial and only using algebraic operations and radicals. This question arose out of the well-known quadratic formula.

The ability to solve quadratic equations and, in essence, the quadratic formula was known to the Babylonians some 3600 years ago. With the discovery of imaginary numbers, the quadratic formula then says that any second degree polynomial over $\mathbb{C}$ can be solved by radicals in terms of the coefficients. In the sixteenth century, the Italian mathematician, Niccolo Tartaglia, discovered a similar formula in terms of radicals to solve cubic equations. This *cubic formula* is now known erroneously as *Cardano's formula* in honor of Cardano, who first published it in 1545. An earlier special version of this formula was discovered by Scipione del Ferro. Cardano's student, Ferrari, extended the formula to solutions by radicals for fourth degree polynomials. The combination of these formulas says that polynomial equations of degree four or less over the complex numbers can be solved by radicals.

From Cardano's work until the very early nineteenth century, attempts were made to find similar formulas for degree five polynomials. In 1805, Ruffini proved that fifth degree polynomial equations are insolvable by radicals in general. Therefore, there exists no comparable formula for degree 5. Abel (in 1825–1826) and Galois (in 1831) extended Ruffini's result and proved the insolubility by radicals for all degrees five or greater. In doing this, Galois developed a general theory of field extensions and its relationship to group theory. This has come to be known as *Galois theory* and is really the main focus of this book.

The solution of the insolvability of the quintic and higher polynomials involved a translation of the problem into a group theory setting. For a polynomial equation to be solvable by radicals, its corresponding Galois group (a concept we will introduce in Chapter 16) must be a *solvable group*. This is a group with a certain defined structure. In this chapter, we introduce and discuss this class of groups.

A *normal series* for a group G is a finite chain of subgroups beginning with G and ending with the identity subgroup $\{1\}$

$$G = G_0 \supset G_1 \supset G_2 \supset \cdots \supset G_{n-1} \supset G_n = \{1\},$$

in which each G_{i+1} is a proper normal subgroup of G_i. The factor groups G_i/G_{i+1} are called the *factors* of the series, and n is the length of the series.

https://doi.org/10.1515/9783111142524-012

Definition 12.1.1. A group G is *solvable* if it has a normal series with Abelian factors; that is, G_i/G_{i+1} is Abelian for all $i = 0, 1, \ldots, n-1$. Such a normal series is called a *solvable series*.

If G is an Abelian group, then $G = G_0 \supset \{1\}$ provides a solvable series. Hence, any Abelian group is solvable. Furthermore, the symmetric group S_3 on 3-symbols is also solvable, however, non-Abelian. Consider the series

$$S_3 \supset A_3 \supset \{1\}.$$

Since $|S_3| = 6$, we have $|A_3| = 3$; hence, A_3 is cyclic and therefore Abelian. Furthermore, $|S_3/A_3| = 2$; hence, the factor group S_3/A_3 is also cyclic, thus Abelian. Therefore, the series above gives a solvable series for S_3.

Lemma 12.1.2. *If G is a finite solvable group, then G has a normal series with cyclic factors.*

Proof. If G is a finite solvable group, then by definition, it has a normal series with Abelian factors. Hence, to prove the lemma, it suffices to show that a finite Abelian group has a normal series with cyclic factors. Let A be a nontrivial finite Abelian group. We do an induction on the order of A. If $|A| = 2$, then A itself is cyclic, and the result follows. Suppose that $|A| > 2$. Choose an $1 \neq a \in A$. Let $N = \langle a \rangle$ so that N is cyclic. Then we have the normal series $A \supset N \supset \{1\}$ with A/N Abelian. Moreover, A/N has order less than A, so A/N has a normal series with cyclic factors, and the result follows. □

Solvability is preserved under subgroups and factor groups.

Theorem 12.1.3. *Let G be a solvable group. Then the following hold:*
(1) *Any subgroup H of G is also solvable.*
(2) *Any factor group G/N of G is also solvable.*

Proof. (1) Let G be a solvable group, and suppose that

$$G = G_0 \supset G_1 \supset \cdots \supset G_r = \{1\}$$

is a solvable series for G. Hence, G_{i+1} is a normal subgroup of G_i for each i, and the factor group G_i/G_{i+1} is Abelian.

Now let H be a subgroup of G, and consider the chain of subgroups

$$H = H \cap G_0 \supset H \cap G_1 \supset \cdots \supset H \cap G_r = \{1\}.$$

Since G_{i+1} is normal in G_i, we know that $H \cap G_{i+1}$ is normal in $H \cap G_i$; this gives a finite normal series for H. Furthermore, from the second isomorphism theorem, we have

$$\begin{aligned}(H \cap G_i)/(H \cap G_{i+1}) &= (H \cap G_i)/((H \cap G_i) \cap G_{i+1}) \\ &\cong (H \cap G_i)G_{i+1}/G_{i+1} \subset G_i/G_{i+1}\end{aligned}$$

for each i. However, G_i/G_{i+1} is Abelian, so each factor in the normal series for H is Abelian. Therefore, the above series is a solvable series for H; hence, H is also solvable.

(2) Let N be a normal subgroup of G. Then from (1) N is also solvable. As above, let

$$G = G_0 \supset G_1 \supset \cdots \supset G_r = \{1\}$$

be a solvable series for G. Consider the chain of subgroups

$$G/N = G_0N/N \supset G_1N/N \supset \cdots \supset G_rN/N = N/N = \{1\}.$$

Let $m \in G_{i-1}$, $n \in N$. Then since N is normal in G,

$$\begin{aligned}(mn)^{-1}G_iN(mn) &= n^{-1}m^{-1}G_imnN = n^{-1}G_inN \\ &= n^{-1}NG_i = NG_i = G_iN.\end{aligned}$$

It follows that $G_{i+1}N$ is normal in G_iN for each i; therefore, the series for G/N is a normal series.

Again, from the isomorphism theorems,

$$\begin{aligned}(G_iN/N)/(G_{i+1}N/N) &\cong G_i/(G_i \cap G_{i+1}N) \\ &\cong (G_i/G_{i+1})/((G_i \cap G_{i+1}N)/G_{i+1}).\end{aligned}$$

However, the last group $(G_i/G_{i+1})/((G_i \cap G_{i+1}N)/G_{i+1})$ is a factor group of the group G_i/G_{i+1}, which is Abelian. Hence, this last group is also Abelian; therefore, each factor in the normal series for G/N is Abelian. Hence, this series is a solvable series, and G/N is solvable. □

The following is a type of converse of the above theorem:

Theorem 12.1.4. *Let G be a group and N a normal subgroup of G. If both N and G/N are solvable, then G is solvable.*

Proof. Suppose that

$$\begin{aligned}N &= N_0 \supset N_1 \supset \cdots \supset N_r = \{1\} \\ G/N &= G_0/N \supset G_1/N \supset \cdots \supset G_s/N = N/N = \{1\}\end{aligned}$$

are solvable series for N and G/N, respectively. Then

$$G = G_0 \supset G_1 \supset \cdots \supset G_s = N \supset N_1 \supset \cdots \supset N_r = \{1\}$$

gives a normal series for G. Furthermore, from the isomorphism theorems again,

$$G_i/G_{i+1} \cong (G_i/N)/(G_{i+1}/N);$$

hence, each factor is Abelian. Therefore, this is a solvable series for G; hence, G is solvable. □

This theorem allows us to prove that solvability is preserved under direct products.

Corollary 12.1.5. *Let G and H be solvable groups. Then their direct product $G \times H$ is also solvable.*

Proof. Suppose that G and H are solvable groups and $K = G \times H$. Recall from Chapter 10 that G can be considered as a normal subgroup of K with $K/G \cong H$. Therefore, G is a solvable subgroup of K, and K/G is a solvable quotient. It follows then, from Theorem 12.1.4, that K is solvable. □

We saw that the symmetric group S_3 is solvable. However, the following theorem shows that the symmetric group S_n is not solvable for $n \geq 5$. This result will be crucial to the proof of the insolvability of the quintic and higher polynomials.

Theorem 12.1.6. *For $n \geq 5$, the symmetric group S_n is not solvable.*

Proof. For $n \geq 5$, we saw that the alternating group A_n is simple. Furthermore, A_n is non-Abelian. Hence, A_n cannot have a nontrivial normal series, and so no solvable series. Therefore, A_n is not solvable. If S_n were solvable for $n \geq 5$, then from Theorem 12.1.3, A_n would also be solvable. Therefore, S_n must also be nonsolvable for $n \geq 5$. □

In general, for a simple, solvable group we have the following:

Lemma 12.1.7. *If a group G is both simple and solvable, then G is cyclic of prime order.*

Proof. Suppose that G is a nontrivial simple, solvable group. Since G is simple, the only normal series for G is $G = G_0 \supset \{1\}$. Since G is solvable, the factors are Abelian; hence, G is Abelian. Again, since G is simple, G must be cyclic. If G were infinite, then $G \cong (\mathbb{Z}, +)$. However, then $2\mathbb{Z}$ is a proper normal subgroup, a contradiction. Therefore, G must be finite cyclic. If the order were not prime, then for each proper divisor of the order, there would be a nontrivial proper normal subgroup. Therefore, G must be of prime order. □

In general, a finite p-group is solvable.

Theorem 12.1.8. *A finite p-group G is solvable.*

Proof. Suppose that $|G| = p^n$. We do this by induction on n. If $n = 1$, then $|G| = p$, and G is cyclic, hence Abelian and therefore solvable. Suppose that $n > 1$. Then as used previously G has a nontrivial center $Z(G)$. If $Z(G) = G$, then G is Abelian; hence solvable. If $Z(G) \neq G$, then $Z(G)$ is a finite p-group of order less than p^n. From our inductive hypothesis, $Z(G)$ must be solvable. Furthermore, $G/Z(G)$ is then also a finite p-group of order less than p^n, so it is also solvable. Hence, $Z(G)$ and $G/Z(G)$ are both solvable. Therefore, from Theorem 12.1.4, G is solvable. □

12.2 The Derived Series

Let G be a group, and let $a, b \in G$. The product $aba^{-1}b^{-1}$ is called the *commutator* of a and b. We write $[a, b] = aba^{-1}b^{-1}$.

Clearly, $[a, b] = 1$ if and only if a and b commute.

Definition 12.2.1. Let G' be the subgroup of G, which is generated by the set of all commutators

$$G' = gp(\{[x, y] : x, y \in G\}).$$

G' is called the *commutator* or *(derived) subgroup* of G. We sometimes write $G' = [G, G]$.

Theorem 12.2.2. *For any group G, the commutator subgroup G' is a normal subgroup of G, and G/G' is Abelian. Furthermore, if H is a normal subgroup of G, then G/H is Abelian if and only if $G' \subset H$.*

Proof. The commutator subgroup G' consists of all finite products of commutators and inverses of commutators. However,

$$[a, b]^{-1} = (aba^{-1}b^{-1})^{-1} = bab^{-1}a^{-1} = [b, a],$$

and so the inverse of a commutator is once again a commutator. It then follows that G' is precisely the set of all finite products of commutators; that is, G' is the set of all elements of the form

$$h_1 h_2 \cdots h_n,$$

where each h_i is a commutator of elements of G.

If $h = [a, b]$ for $a, b \in G$, then for $x \in G$, $xhx^{-1} = [xax^{-1}, xbx^{-1}]$ is again a commutator of elements of G. Now from our previous comments, an arbitrary element of G' has the form $h_1 h_2 \cdots h_n$, where each h_i is a commutator.

Thus, $x(h_1 h_2 \cdots h_n)x^{-1} = (xh_1x^{-1})(xh_2x^{-1}) \cdots (xh_nx^{-1})$ and, since by the above each xh_ix^{-1} is a commutator, $x(h_1 h_2 \cdots h_n)x^{-1} \in G'$. It follows that G' is a normal subgroup of G.

Consider the factor group G/G'. Let aG' and bG' be any two elements of G/G'. Then

$$\begin{aligned}[aG', bG'] &= aG' \cdot bG' \cdot (aG')^{-1} \cdot (bG')^{-1} \\ &= aG' \cdot bG' \cdot a^{-1}G' \cdot b^{-1}G' = aba^{-1}b^{-1}G' = G'\end{aligned}$$

since $[a, b] \in G'$. In other words, any two elements of G/G' commute; therefore, G/G' is Abelian.

Now let N be a normal subgroup of G with G/N Abelian. Let $a, b \in G$, then aN and bN commute since G/N is Abelian. Therefore,

$$[aN, bN] = aNbNa^{-1}Nb^{-1}N = aba^{-1}b^{-1}N = N.$$

It follows that $[a, b] \in N$. Therefore, all commutators of elements in G lie in N; thus, $G' \subset N$. □

From the second part of Theorem 12.2.2, we see that G' is the minimal normal subgroup of G such that G/G' is Abelian. We call $G/G' = G_{\text{ab}}$ the *Abelianization* of G.

We consider next the following inductively defined sequence of subgroups of an arbitrary group G called the *derived series*:

Definition 12.2.3. For an arbitrary group G, define $G^{(0)} = G$ and $G^{(1)} = G'$, and then, inductively, $G^{(n+1)} = (G^{(n)})'$. That is, $G^{(n+1)}$ is the commutator subgroup or derived group of $G^{(n)}$. The chain of subgroups

$$G = G^{(0)} \supset G^{(1)} \supset \cdots \supset G^{(n)} \supset \cdots$$

is called the *derived series* for G.

Notice that since $G^{(i+1)}$ is the commutator subgroup of $G^{(i)}$, we have $G^{(i)}/G^{(i+1)}$ is Abelian. If the derived series was finite, then G would have a normal series with Abelian factors; hence would be solvable. The converse is also true and characterizes solvable groups in terms of the derived series.

Theorem 12.2.4. *A group G is solvable if and only if its derived series is finite. That is, there exists an n such that $G^{(n)} = \{1\}$.*

Proof. If $G^{(n)} = \{1\}$ for some n, then as explained above, the derived series provides a solvable series for G; hence, G is solvable. Conversely, suppose that G is solvable, and let

$$G = G_0 \supset G_1 \supset \cdots \supset G_r = \{1\}$$

be a solvable series for G. We claim first that $G_i \supset G^{(i)}$ for all i. We do this by induction on r. If $r = 0$, then $G = G_0 = G^{(0)}$. Suppose that $G_i \supset G^{(i)}$. Then $G_i' \supset (G^{(i)})' = G^{(i+1)}$. Since G_i/G_{i+1} is Abelian, it follows, from Theorem 12.2.2, that $G_{i+1} \supset G_i'$. Therefore, $G_{i+1} \supset G^{(i+1)}$, establishing the claim. Now if G is solvable, from the claim, we have that $G_r \supset G^{(r)}$. However, $G_r = \{1\}$; therefore, $G^{(r)} = \{1\}$, proving the theorem. □

The length of the derived series is called the *solvability length* of a solvable group G. The class of solvable groups of *class c* consists of those solvable groups of solvability length c, or less.

12.3 Composition Series and the Jordan–Hölder Theorem

The concept of a normal series is extremely important in the structure theory of groups. This is especially true for finite groups. If

$$G = G_0 \supset G_1 \supset \cdots \supset G_s = \{1\} \quad \text{and} \quad G = H_0 \supset H_1 \supset \cdots \supset H_t = \{1\}$$

are two normal series for the group G, then the second is a *refinement* of the first if all the terms of the second occur in the first series. Furthermore, two normal series are called *equivalent* or (*isomorphic*) if there exists a 1–1 correspondence between the factors (hence the length must be the same) of the two series such that the corresponding factors are isomorphic.

Theorem 12.3.1 (Schreier's theorem). *Any two normal series for a group G have equivalent refinements.*

Proof. Consider two normal series for G:

$$G = G_0 \supset G_1 \supset \cdots \supset G_{s-1} \supset G_s = \{1\},$$
$$G = H_0 \supset H_1 \supset \cdots \supset H_{t-1} \supset H_t = \{1\}.$$

Now define

$$G_{ij} = (G_i \cap H_j)G_{i+1}, \quad j = 0, 1, 2, \ldots, t,$$
$$H_{ji} = (G_i \cap H_j)H_{j+1}, \quad i = 0, 1, 2, \ldots, s.$$

Then we have

$$G = G_{00} \supset G_{01} \supset \cdots \supset G_{0t} = G_1$$
$$= G_{10} \supset \cdots \supset G_{1t} = G_2 \supset \cdots \supset G_{st} = \{e\},$$

and

$$G = H_{00} \supset H_{01} \supset \cdots \supset H_{0s} = H_1$$
$$= H_{10} \supset \cdots \supset H_{1s} = H_2 \supset \cdots \supset H_{ts} = \{e\}.$$

Now, applying the third isomorphism theorem to the groups G_i, H_j, G_{i+1}, H_{j+1}, we have that $G_{i(j+1)} = (G_i \cap H_{j+1})G_{i+1}$ is a normal subgroup of $G_{ij} = (G_i \cap H_j)G_{i+1}$, and also that $H_{j(i+1)} = (G_{i+1} \cap H_j)H_{j+1}$ is a normal subgroup of $H_{ji} = (G_i \cap H_j)H_{j+1}$. Furthermore,

$$G_{ij}/G_{i(j+1)} \cong H_{ji}/H_{j(i+1)}.$$

Thus, the above two are normal series, which are refinements of the two given series, and they are equivalent. □

A proper normal subgroup N of a group G is called *maximal* in G, if there does not exist any normal subgroup $N \subset M \subset G$ with all inclusions proper. This is the group theoretic analog of a maximal ideal. An alternative characterization is the following: N is a maximal normal subgroup of G if and only if G/N is simple.

A normal series, where each factor is simple can have no refinements.

Definition 12.3.2. A *composition series* for a group G is a normal series, where all the inclusions are proper and such that G_{i+1} is maximal in G_i. Equivalently, a normal series, where each factor is simple.

It is possible that an arbitrary group does not have a composition series, or even if it does have one, a subgroup of it may not have one. Of course, a finite group does have a composition series.

In the case in which a group G does have a composition series, the following important theorem, called the Jordan–Hölder theorem, provides a type of unique factorization.

Theorem 12.3.3 (Jordan–Hölder theorem). *If a group G has a composition series, then any two composition series are equivalent; that is, the composition factors are unique.*

Proof. Suppose we are given two composition series. Applying Theorem 12.3.1, we get that the two composition series have equivalent refinements. But the only refinement of a composition series is one obtained by introducing repetitions. If in the 1–1 correspondence between the factors of these refinements, the paired factors equal to $\{e\}$ are disregarded; that is, if we drop the repetitions, clearly, we get that the original composition series are equivalent. □

We remarked in Chapter 10 that the simple groups are important, because they play a role in finite group theory somewhat analogous to that of the primes in number theory. In particular, an arbitrary finite group G can be broken down into simple components. These uniquely determined simple components are, according to the Jordan–Hölder theorem, the factors of a composition series for G.

12.4 Exercises

1. Let K be a field and

$$G = \left\{ \begin{pmatrix} a & x & y \\ 0 & b & z \\ 0 & 0 & c \end{pmatrix} : a, b, c, x, y, z \in K, abc \neq 0 \right\}.$$

   Show that G is solvable.
2. A group G is called *polycyclic* if it has a normal series with cyclic factors. Show:
   (i) Each subgroup and each factor group of a polycyclic group is polycyclic.
   (ii) In a polycyclic group, each normal series has the same number of infinite cyclic factors.
3. Let G be a group. Show the following:
   (i) If G is finite and solvable, then G is polycyclic.

(ii) If G is polycyclic, then G is finitely generated.
(iii) The group $(\mathbb{Q}, +)$ is solvable, but not polycyclic.

4. Let N_1 and N_2 be normal subgroups of G. Show the following:
   (i) If N_1 and N_2 are solvable, then also N_1N_2 is a solvable normal subgroup of G.
   (ii) Is (i) still true, if we replace "solvable" by "Abelian"?
5. Let $N_1, \dots, N_t$ be normal subgroups of a group G. If all factor groups G/N_i are solvable, then also $G/(N_1 \cap \dots \cap N_t)$ is solvable.

13 Group Actions and the Sylow Theorems

13.1 Group Actions

A *group action* of a group G on a set A is a homomorphism from G into S_A, the symmetric group on A. We say that G *acts* on A. Hence, G acts on A if to each $g \in G$ corresponds a permutation

$$\pi_g : A \to A$$

such that

(1) $\pi_{g_1}(\pi_{g_2}(a)) = \pi_{g_1 g_2}(a)$ for all $g_1, g_2 \in G$ and for all $a \in A$,

(2) $\pi_1(a) = a$ for all $a \in A$.

For the remainder of this chapter, if $g \in G$ and $a \in A$, we will write ga for $\pi_g(a)$. Group actions are an extremely important idea, and we use this idea in the present chapter to prove several fundamental results in group theory. If G acts on the set A, then we say that two elements $a_1, a_2 \in A$ are *congruent* under G if there exists a $g \in G$ with $ga_1 = a_2$. The set

$$G_a = \{a_1 \in A : a_1 = ga \text{ for some } g \in G\}$$

is called the *orbit* of a. It consists of elements congruent to a under G.

Lemma 13.1.1. *If G acts on A, then congruence under G is an equivalence relation on A.*

Proof. Any element $a \in A$ is congruent to itself via the identity map; hence, the relation is reflexive. If $a_1 \sim a_2$ so that $ga_1 = a_2$ for some $g \in G$, then $g^{-1}a_2 = a_1$, and so $a_2 \sim a_1$, and the relation is symmetric. Finally, if $g_1 a_1 = a_2$ and $g_2 a_2 = a_3$, then $g_2 g_1 a_1 = a_3$, and the relation is transitive. □

Recall that the equivalence classes under an equivalence relation partition a set. For a given $a \in A$, its equivalence class under this relation is precisely its orbit G_a, as defined above.

Corollary 13.1.2. *If G acts on the set A, then the orbits under G partition the set A.*

We say that G *acts transitively* on A if any two elements of A are congruent under G. That is, the action is transitive if for any $a_1, a_2 \in A$ there is some $g \in G$ such that $ga_1 = a_2$. If $a \in A$, the *stabilizer* of a consists of those $g \in G$ that *fix* a. Hence,

$$\mathrm{Stab}_G(a) = \{g \in G : ga = a\}.$$

The following lemma is easily proved and left to the exercises.

Lemma 13.1.3. *If G acts on A, then for any $a \in A$, the stabilizer $\mathrm{Stab}_G(a)$ is a subgroup of G.*

https://doi.org/10.1515/9783111142524-013

We now prove the crucial theorem concerning group actions.

Theorem 13.1.4. *Suppose that G acts on A and* $a \in A$. *Let* G_a *be the orbit of a under G and* $\mathrm{Stab}_G(a)$ *its stabilizer. Then*

$$|G : \mathrm{Stab}_G(a)| = |G_a|.$$

That is, the size of the orbit of a is the index of its stabilizer in G.

Proof. Suppose that $g_1, g_2 \in G$ with $g_1 \mathrm{Stab}_G(a) = g_2 \mathrm{Stab}_G(a)$; that is, they define the same left coset of the stabilizer. Then $g_2^{-1} g_1 \in \mathrm{Stab}_G(a)$. This implies that $g_2^{-1} g_1 a = a$ so that $g_2 a = g_1 a$. Hence, any two elements in the same left coset of the stabilizer produce the same image of a in G_a. Conversely, if $g_1 a = g_2 a$, then g_1, g_2 define the same left coset of $\mathrm{Stab}_G(a)$. This shows that there is a one-to-one correspondence between left cosets of $\mathrm{Stab}_G(a)$ and elements of G_a. It follows that the size of G_a is precisely the index of the stabilizer. □

We will use this theorem repeatedly with different group actions to obtain important group theoretic results.

13.2 Conjugacy Classes and the Class Equation

In Section 10.5, we introduced the *center* of a group

$$Z(G) = \{g \in G : gg_1 = g_1 g \text{ for all } g_1 \in G\},$$

and showed that it is a normal subgroup of G. We use this normal subgroup in conjunction with what we call the *class equation* to show that any finite p-group has a nontrivial center. In this section, we use group actions to derive the class equation and prove the result for finite p-groups.

Recall that if G is a group, then two elements $g_1, g_2 \in G$ are *conjugate* if there exists a $g \in G$ with $g^{-1} g_1 g = g_2$. We saw that conjugacy is an equivalence relation on G. For The equivalence class of $g \in G$ is called its *conjugacy class*, which we will denote by $\mathrm{Cl}(g)$. Thus,

$$\mathrm{Cl}(g) = \{g_1 \in G : g_1 \text{ is conjugate to } g\}.$$

If $g \in G$, then its *centralizer* $C_G(g)$ is the set of elements in G that commute with g:

$$C_G(g) = \{g_1 \in G : gg_1 = g_1 g\}.$$

Theorem 13.2.1. *Let G be a finite group and* $g \in G$. *Then the centralizer of g is a subgroup of G, and*

$$|G : C_G(g)| = |\mathrm{Cl}(g)|.$$

That is, the index of the centralizer of g is the size of its conjugacy class.

In particular, for a finite group the size of each conjugacy class divides the order of the group.

Proof. Let the group G act on itself by conjugation. That is, $g(g_1) = g^{-1}g_1g$. It is easy to show that this is an action on the set G (see exercises). The orbit of $g \in G$ under this action is precisely its conjugacy class $\mathrm{Cl}(g)$, and the stabilizer is its centralizer $C_G(g)$. The statements in the theorem then follow directly from Theorem 13.1.4. □

For any group G, since conjugacy is an equivalence relation, the conjugacy classes partition G. Hence,

$$G = \dot{\bigcup_{g \in G}} \mathrm{Cl}(g),$$

where this union is taken over the distinct conjugacy classes. It follows that

$$|G| = \sum_{g \in G} |\mathrm{Cl}(g)|,$$

where this sum is taken over distinct conjugacy classes.

If $\mathrm{Cl}(g) = \{g\}$; that is, the conjugacy class of g is g alone, then $C_G(g) = G$ so that g commutes with all of G. Therefore, in this case, $g \in Z(G)$. This is true for every element of the center; therefore,

$$G = Z(G) \cup \dot{\bigcup_{g \notin Z(G)}} \mathrm{Cl}(g),$$

where again the second union is taken over the distinct conjugacy classes $\mathrm{Cl}(g)$ with $g \notin Z(G)$. The size of G is then the sum of these disjoint pieces, so

$$|G| = |Z(G)| + \sum_{g \notin Z(G)} |\mathrm{Cl}(g)|,$$

where the sum is taken over the distinct conjugacy classes $\mathrm{Cl}(g)$ with $g \notin Z(G)$. However, from Theorem 13.2.1, $|\mathrm{Cl}(g)| = |G : C_G(g)|$, so the equation above becomes

$$|G| = |Z(G)| + \sum_{g \notin Z(G)} |G : C_G(g)|,$$

where the sum is taken over the distinct indices $|G : C_G(g)|$ with $g \notin Z(G)$. This is known as the *class equation*.

Theorem 13.2.2 (Class equation). *Let G be a finite group. Then*

$$|G| = |Z(G)| + \sum_{g \notin Z(G)} |G : C_G(g)|,$$

where the sum is taken over the distinct centralizers.

As a first application, we prove the result that finite p-groups have nontrivial centers (see Lemma 10.5.6).

Theorem 13.2.3. *Let G be a finite p-group. Then G has a nontrivial center.*

Proof. Let G be a finite p-group so that $|G| = p^n$ for some n, and consider the class equation

$$|G| = |Z(G)| + \sum_{g \notin Z(G)} |G : C_G(g)|,$$

where the sum is taken over the distinct centralizers. Since $|G : C_G(g)|$ divides $|G|$ for each $g \in G$, we must have that $p \mid |G : C_G(g)|$ for each $g \in G$. Furthermore, $p \mid |G|$. Therefore, p must divide $|Z(G)|$; hence, $|Z(G)| = p^m$ for some $m \geq 1$. Therefore, $Z(G)$ is nontrivial. □

The idea of conjugacy and the centralizer of an element can be extended to subgroups. If H_1, H_2 are subgroups of a group G, then H_1, H_2 are conjugate if there exists a $g \in G$ such that $g^{-1}H_1g = H_2$. As for elements, conjugacy is an equivalence relation on the set of subgroups of G.

If $H \subset G$ is a subgroup, then its *conjugacy class* consists of all the subgroups of G conjugate to it. The *normalizer* of H is

$$N_G(H) = \{g \in G : g^{-1}Hg = H\}.$$

As for elements, let G act on the set of subgroups of G by conjugation. That is, for $g \in G$, the map is given by $H \mapsto g^{-1}Hg$. For $H \subset G$, the stabilizer under this action is precisely the normalizer. Hence, exactly as for elements, we obtain the following theorem:

Theorem 13.2.4. *Let G be a group and $H \subset G$ a subgroup. Then the normalizer $N_G(H)$ of H is a subgroup of G, H is normal in $N_G(H)$, and*

$$|G : N_G(H)| = \text{number of conjugates of } H \text{ in } G.$$

13.3 The Sylow Theorems

If G is a finite group and $H \subset G$ is a subgroup, then Lagrange's theorem guarantees that the order of H divides the order of G. However, the converse of Lagrange's theorem is false. That is, if G is a finite group of order n and if $d \mid n$, then G need not contain a subgroup of order d. If d is a prime p or a power of a prime p^e, however, then we shall see that G must contain subgroups of that order. In particular, we shall see that if p^d is the highest power of p that divides n, then all subgroups of that order are actually conjugate, and we shall finally get a formula concerning the number of such subgroups. These theorems constitute the Sylow theorems, which we will examine in this section. First, we give an example, where the converse of Lagrange's theorem is false.

Lemma 13.3.1. *The alternating group on* 4 *symbols* A_4 *has order* 12, *but has no subgroup of order* 6.

Proof. Suppose that there exists a subgroup $U \subset A_4$ with $|U| = 6$. Then $|A_4 : U| = 2$ since $|A_4| = 12$; hence, U is normal in A_4.

Now id, $(1,2)(3,4)$, $(1,3)(2,4)$, $(1,4)(2,3)$ are in A_4. These each have order 2 and commute, so they form a normal subgroup $V \subset A_4$ of order 4. This subgroup V is isomorphic to $\mathbb{Z}_2 \times \mathbb{Z}_2$. Then

$$12 = |A_4| \geq |VU| = \frac{|V||U|}{|V \cap U|} = \frac{4 \cdot 6}{|V \cap U|}.$$

It follows that $V \cap U \neq \{1\}$, and since U is normal, we have that $V \cap U$ is also normal in A_4.

Now $(1,2)(3,4) \in V$, and by renaming the entries in V, if necessary, we may assume that it is also in U, so that $(1,2)(3,4) \in V \cap U$. Since $(1,2,3) \in A_4$, we have

$$(3,2,1)(1,2)(3,4)(1,2,3) = (1,3)(2,4) \in V \cap U,$$

and then

$$(3,2,1)(1,4)(2,3)(1,2,3) = (1,2)(3,4) \in V \cap U.$$

But then $V \subset V \cap U$, and so $V \subset U$. But this is impossible since $|V| = 4$, which does not divide $|U| = 6$. □

Definition 13.3.2. Let G be a finite group with $|G| = n$, and let p be a prime such that $p^a | n$, but no higher power of p divides n. A subgroup of G of order p^a is called a *p-Sylow subgroup*.

It is not a clear that a p-Sylow subgroup must exist. We will prove that for each $p|n$ a p-Sylow subgroup exists.

We first consider and prove a very special case.

Theorem 13.3.3. *Let G be a finite Abelian group, and let p be a prime such that $p||G|$. Then G contains at least one element of order p.*

Proof. Suppose that G is a finite Abelian group of order pn. We use induction on n. If $n = 1$, then G has order p, and hence is cyclic. Therefore, it has an element of order p. Suppose that the theorem is true for all Abelian groups of order pm with $m < n$, and suppose that G has order pn. Suppose that $g \in G$. If the order of g is pt for some integer t, then $g^t \neq 1$, and g^t has order p, proving the theorem in this case. Hence, we may suppose that $g \in G$ has order prime to p, and we show that there must be an element, whose order is a multiple of p, and then use the above argument to get an element of exact order p.

Hence, we have $g \in G$ with order m, where $(m,p) = 1$. Since $m||G| = pn$, we must have $m|n$. Since G is Abelian, $\langle g\rangle$ is normal, and the factor group $G/\langle g\rangle$ is Abelian of order $p(\frac{n}{m}) < pn$. By the inductive hypothesis, $G/\langle g\rangle$ has an element $h\langle g\rangle$ of order p, $h \in G$; hence, $h^p = g^k$ for some k. g^k has order $m_1|m$; therefore, h has order pm_1. Now, as above, h^{m_1} has order p, proving the theorem. □

Therefore, if G is an Abelian group, and if $p|n$, then G contains a subgroup of order p, the cyclic subgroup of order p generated by an element $a \in G$ of order p, whose existence is guaranteed by the above theorem. We now present the first Sylow theorem:

Theorem 13.3.4 (First Sylow theorem). *Let G be a finite group, and let $p||G|$, then G contains a p-Sylow subgroup; that is, a p-Sylow subgroup exists.*

Proof. Let G be a finite group of order pn, and—as above—we do induction on n. If $n = 1$, then G is cyclic, and G is its own maximal p-subgroup; hence, all of G is a p-Sylow subgroup. We assume then that if $|G| = pm$ with $m < n$, then G has a p-Sylow subgroup.

Assume that $|G| = p^t m$ with $(m,p) = 1$. We must show that G contains a subgroup of order p^t. If H is a proper subgroup, whose index is prime to p, then $|H| = p^t m_1$ with $m_1 < m$. Therefore, by the inductive hypothesis, H has a p-Sylow subgroup of order p^t. This will also be a subgroup of G, hence a p-Sylow subgroup of G.

Therefore, we may assume that the index of any proper subgroup H of G must be divisible by p. Now consider the class equation for G,

$$|G| = |Z(G)| + \sum_{g \notin Z(G)} |G : C_G(g)|,$$

where the sum is taken over the distinct centralizers. By assumption, each of the indices are divisible by p and also $p||G|$. Therefore, $p||Z(G)|$. It follows that $Z(G)$ is a finite Abelian group, whose order is divisible by p. From Theorem 13.3.3, there exists an element $g \in Z(G) \subset G$ of order p. Since $g \in Z(G)$, we must have $\langle g\rangle$ normal in G. The factor group $G/\langle g\rangle$ then has order $p^{t-1}m$, and—by the inductive hypothesis—must have a p-Sylow subgroup $\overline{K}$ of order p^{t-1}, hence of index m. By the Correspondence Theorem 10.2.6, there is a subgroup K of G with $\langle g\rangle \subset K$ such that $K/\langle g\rangle \cong \overline{K}$. Therefore, $|K| = p^t$, and K is a p-Sylow subgroup of G. □

On the basis of this theorem, we can now strengthen the result obtained in Theorem 13.3.3.

Theorem 13.3.5 (Cauchy). *If G is a finite group, and if p is a prime such that $p||G|$, then G contains at least one element of order p.*

Proof. Let P be a p-Sylow subgroup of G, and let $|P| = p^t$. If $g \in P, g \neq 1$, then the order of g is p^{t_1}. Then $g^{p^{t_1-1}}$ has order p. □

We have seen that p-Sylow subgroups exist. We now wish to show that any two p-Sylow subgroups are conjugate. This is the content of the second Sylow theorem:

Theorem 13.3.6 (Second Sylow theorem). *Let G be a finite group and p a prime such that $p||G|$. Then any p-subgroup H of G is contained in a p-Sylow subgroup. Furthermore, all p-Sylow subgroups of G are conjugate. That is, if P_1 and P_2 are any two p-Sylow subgroups of G, then there exists an $a \in G$ such that $P_1 = aP_2a^{-1}$.*

Proof. Let Ω be the set of p-Sylow subgroups of G, and let G act on Ω by conjugation. This action will, of course, partition Ω into disjoint orbits. Let P be a fixed p-Sylow subgroup and Ω_P be its orbit under the conjugation action. The size of the orbit is the index of its stabilizer; that is, $|\Omega_P| = |G : \mathrm{Stab}_G(P)|$. Now $P \subset \mathrm{Stab}_G(P)$, and P is a maximal p-subgroup of G. It follows that the index of $\mathrm{Stab}_G(P)$ must be prime to p, and so the number of p-Sylow subgroups conjugate to P is prime to p.

Now let H be a p-subgroup of G, and let H act on Ω_P by conjugation. Ω_P will itself decompose into disjoint orbits under this actions. Furthermore, the size of each orbit is an index of a subgroup of H, hence must be a power of p. On the other hand, the size of the whole orbit is prime to p. Therefore, there must be one orbit that has size exactly 1. This orbit contains a p-Sylow subgroup P', and P' is fixed by H under conjugation; that is, H normalizes P'. It follows that HP' is a subgroup of G, and P' is normal in HP'. From the second isomorphism theorem, we then obtain

$$HP'/P' \cong H/(H \cap P').$$

Since H is a p-group, the size of $H/(H \cap P')$ is a power of p; therefore, so is the size of HP'/P'. But P' is also a p-group, so it follows that HP' also has order a power of p. Now $P' \subset HP'$, but P' is a maximal p-subgroup of G. Hence, $HP' = P'$. This is possible only if $H \subset P'$, proving the first assertion in the theorem. Therefore, any p-subgroup of G is obtained in a p-Sylow subgroup.

Now let H be a p-Sylow subgroup P_1, and let P_1 act on Ω_P. Exactly as in the argument above, $P_1 \subset P'$, where P' is a conjugate of P. Since P_1 and P' are both p-Sylow subgroups, they have the same size; hence, $P_1 = P'$. This implies that P_1 is a conjugate of P. Since P_1 and P are arbitrary p-Sylow subgroups, it follows that all p-Sylow subgroups are conjugate. □

We come now to the last of the three Sylow theorems. This one gives us information concerning the number of p-Sylow subgroups.

Theorem 13.3.7 (Third Sylow theorem). *Let G be a finite group and p a prime such that $p||G|$. Then the number of p-Sylow subgroups of G is of the form $1 + pk$ and divides the order of $|G|$. It follows that if $|G| = p^a m$ with $(p, m) = 1$, then the number of p-Sylow subgroups divides m.*

Proof. Let P be a p-Sylow subgroup, and let P act on Ω, the set of all p-Sylow subgroups, by conjugation. Now P normalizes itself, so there is one orbit, namely, P, having exactly size 1. Every other orbit has size a power of p since the size is the index of a nontrivial subgroup of P, and therefore must be divisible by p. Hence, the size of the Ω is $1+pk$. □

13.4 Some Applications of the Sylow Theorems

We now give some applications of the Sylow theorems. First, we show that the converse of Lagrange's theorem is true for both general p-groups and for finite Abelian groups.

Theorem 13.4.1. *Let G be a group of order p^n, p a prime number. Then G contains at least one normal subgroup of order p^m for each m such that $0 \le m \le n$.*

Proof. We use induction on n. For $n = 1$, the theorem is trivial. By Lemma 10.5.7, any group of order p^2 is Abelian. This, together with Theorem 13.3.3, establishes the claim for $n = 2$.

We now assume the theorem is true for all groups G of order p^k, where $1 \le k < n$, where $n > 2$. Let G be a group of order p^n. From Lemma 10.3.4, G has a nontrivial center of order at least p, hence an element $g \in Z(G)$ of order p. Let $N = \langle g\rangle$. Since $g \in Z(G)$, it follows that N is normal subgroup of order p. Then G/N is of order p^{n-1}, therefore contains (by the induction hypothesis) normal subgroups of orders p^{m-1}, for $0 \le m-1 \le n-1$. These groups are of the form H/N, where the normal subgroup $H \subset G$ contains N and is of order p^m, $1 \le m \le n$, because $|H| = |N|[H : N] = |N| \cdot |H/N|$. □

On the basis of the first Sylow theorem, we see that if G is a finite group, and if $p^k \,||G|$, then G must contain a subgroup of order p^k. One can actually show that, as in the case of Sylow p-groups, the number of such subgroups is of the form $1 + pt$, but we shall not prove this here.

Theorem 13.4.2. *Let G be a finite Abelian group of order n. Suppose that $d|n$. Then G contains a subgroup of order d.*

Proof. Suppose that $n = p_1^{e_1} \cdots p_k^{e_k}$ is the prime factorization of n. Then $d = p_1^{f_1} \cdots p_k^{f_k}$ for some nonnegative $f_1, \dots, f_k$. Now G has p_1-Sylow subgroup H_1 of order $p_1^{e_1}$. Hence, from Theorem 13.4.1, H_1 has a subgroup K_1 of order $p_1^{f_1}$. Similarly, there are subgroups $K_2, \dots, K_k$ of G of respective orders $p_2^{f_2}, \dots, p_k^{f_k}$. Moreover, since the orders are disjoint, $K_i \cap K_j = \{1\}$ if $i \ne j$ and thus $\langle K_1, K_2, \dots, K_k\rangle$ has order $|K_1||K_2| \cdots |K_k| = p_1^{f_1} \cdots p_k^{f_k} = d$. □

In Section 10.5, we examined the classification of finite groups of small orders. Here, we use the Sylow theorems to extend some of this material further.

Theorem 13.4.3. *Let p, q be distinct primes with $p < q$ and q not congruent to 1 modulo p. Then any group of order pq is cyclic. For example, any group of order 15 must be cyclic.*

Proof. Suppose that $|G| = pq$ with $p < q$ and q not congruent to 1 modulo p. The number of q-Sylow subgroups is of the form $1 + qk$ and divides p. Since q is greater than p, this implies that there can be only one; hence, there is a normal q-Sylow subgroup H. Since q is a prime, H is cyclic of order q; therefore, there is an element g of order q.

The number of p-Sylow subgroups is of the form $1 + pk$ and divides q. Since q is not congruent to 1 modulo p, this implies that there also can be only one p-Sylow subgroup; hence, there is a normal p-Sylow subgroup K. Since p is a prime K is cyclic of order p; therefore, there is an element h of order p.

Since p, q are distinct primes $H \cap K = \{1\}$. Consider the element $g^{-1}h^{-1}gh$. Since K is normal, $g^{-1}hg \in K$. Then $g^{-1}h^{-1}gh = (g^{-1}h^{-1}g)h \in K$. But H is also normal, so $h^{-1}gh \in H$. This then implies that $g^{-1}h^{-1}gh = g^{-1}(h^{-1}gh) \in H$; and therefore we have $g^{-1}h^{-1}gh \in K \cap H$. It follows then that $g^{-1}h^{-1}gh = 1$ or $gh = hg$. Since g, h commute, the order of gh is the lcm of the orders of g and h, which is pq. Therefore, G has an element of order pq. Since $|G| = pq$, this implies that G is cyclic. □

In the above theorem, since we assumed that q is not congruent to 1 modulo p, hence $p \neq 2$. In the case where $p = 2$, we get another possibility.

Theorem 13.4.4. *Let p be an odd prime and G a finite group of order $2p$. Then either G is cyclic, or G is isomorphic to the dihedral group of order $2p$; that is, the group of symmetries of a regular p-gon. In this latter case, G is generated by two elements, g and h, which satisfy the relations $g^p = h^2 = (gh)^2 = 1$.*

Proof. As in the proof of Theorem 13.4.3, G must have a normal cyclic subgroup of order p, say $\langle g\rangle$. Since $2 \mid |G|$, the group G must have an element of order 2, say h. Consider the order of gh. By Lagrange's theorem, this element can have order 1, 2, p, $2p$. If the order is 1, then $gh = 1$ or $g = h^{-1} = h$. This is impossible since g has order p, and h has order 2. If the order of gh is p, then from the second Sylow theorem, $gh \in \langle g\rangle$. But this implies that $h \in \langle g\rangle$, which is impossible since every nontrivial element of $\langle g\rangle$ has order p. Therefore, the order of gh is either 2 or $2p$.

If the order of gh is $2p$, then since G has order $2p$, it must be cyclic.

If the order of gh is 2, then within G, we have the relations $g^p = h^2 = (gh)^2 = 1$. Let $H = \langle g, h\rangle$ be the subgroup of G generated by g and h. The relations $g^p = h^2 = (gh)^2 = 1$ imply that H has order $2p$. Since $|G| = 2p$, we get that $H = G$. G is isomorphic to the dihedral group D_p of order $2p$ (see exercises).

In the above description, g represents a rotation of $\frac{2\pi}{p}$ of a regular p-gon about its center, whereas h represents any reflection across a line of symmetry of the regular p-gon. □

Example 13.4.5 (The groups of order 21). Let G be a group of order 21. The number of 7-Sylow subgroups of G is 1, because it is of the form $1 + 7k$ and divides 3. Hence, the 7-Sylow subgroup K is normal and cyclic; that is, $K \lhd G$ and $K = \langle a\rangle$ with a of order 7.

The number of 3-Sylow subgroups is analogously 1 or 7. If it is 1, then we have exactly one element of order 3 in G, and if it is 7, there are 14 elements of order 3 in G.

Let b be an element of order 3. Then $bab^{-1} = a^r$ for some r with $1 \leq r \leq 6$. Now, $a = b^3ab^{-3} = a^{r^3}$; hence, $r^3 = 1$ in $\mathbb{Z}_6$, which implies $r = 1, 2$ or 4. The map $b \mapsto b, a \mapsto a^2$ defines an automorphism of G, because $a^{2^3} = a$. Hence, up to isomorphism, there are exactly two groups of order 21. If $r = 1$, then G is Abelian.

In fact, $G = \langle ab \rangle$ is cyclic of order 21. The group for $r = 2$ can be realized as a subgroup of S_7. Let $a = (1, 2, 3, 4, 5, 6, 7)$ and $b = (2, 3, 5)(4, 7, 6)$. Then $bab^{-1} = a^2$ and $\langle a, b \rangle$ has order 21.

We have looked at the finite fields $\mathbb{Z}_p$. We give an example of a p-Sylow subgroup of a matrix group over $\mathbb{Z}_p$.

Example 13.4.6. Consider $\mathrm{GL}(n, p)$, the group of $n \times n$ invertible matrices over $\mathbb{Z}_p$. If $\{v_1, \ldots, v_n\}$ is a basis for $(\mathbb{Z}_p)^n$ over $\mathbb{Z}_p$, then the size of $\mathrm{GL}(n, p)$ is the number of independent images $\{w_1, \ldots, w_n\}$ of $\{v_1, \ldots, v_n\}$. For w_1 there are $p^n - 1$ choices; for w_2 there are $p^n - p$ choices and so on. It follows that

$$|\mathrm{GL}(n, p)| = (p^n - 1)(p^n - p) \cdots (p^n - p^{n-1}) = p^{1+2+\cdots+(n-1)}m = p^{\frac{n(n-1)}{2}}m$$

with $(p, m) = 1$. Therefore, a p-Sylow subgroup must have size $p^{\frac{n(n-1)}{2}}$.

Let P be the subgroup of upper triangular matrices with 1's on the diagonal. Then P has size $p^{1+2+\cdots+(n-1)} = p^{\frac{n(n-1)}{2}}$, and is therefore a p-Sylow subgroup of $\mathrm{GL}(n, p)$.

The final example is a bit more difficult. We mentioned that a major result on finite groups is the classification of the finite simple groups. This classification showed that any finite simple group is either cyclic of prime order, in one of several classes of groups such as the A_n, $n > 4$, or one of a number of special examples called sporadic groups. One of the major tools in this classification is the following famous result, called the Feit–Thompson theorem, which showed that any finite group G of odd order is solvable and, in addition, if G is not cyclic, then G is nonsimple.

Theorem 13.4.7 (Feit–Thompson theorem). *Any finite group of odd order is solvable.*

The proof of this theorem, one of the major results in algebra in the twentieth century, is way beyond the scope of this book. The proof is actually hundreds of pages in length, when one counts the results used. However, we look at the smallest non-Abelian simple group.

Theorem 13.4.8. *Suppose that G is a simple group of order 60. Then G is isomorphic to A_5. Moreover, A_5 is the smallest non-Abelian finite simple group.*

Proof. Suppose that G is a simple group of order $60 = 2^2 \cdot 3 \cdot 5$. The number of 5-Sylow subgroups is of the form $1 + 5k$ and divides 12. Hence, there is 1 or 6. Since G is assumed simple, and all 5-Sylow subgroups are conjugate, there cannot be only one. Hence, there are 6. Since each of these is cyclic of order 5 they intersect only in the identity. Hence, these 6 subgroups cover 24 distinct elements.

The number of 3-Sylow subgroups is of the form $1+3k$ and divides 20. Hence, there are 1, 4, 10. We claim that there are 10. There cannot be only 1, since G is simple. Suppose there were 4. Let G act on the set of 3-Sylow subgroups by conjugation. Since an action is a permutation, this gives a homomorphism f from G into S_4. By the first isomorphism theorem, $G/\ker(f) \cong \operatorname{im}(f)$.

However, since G is simple, the kernel must be trivial, and this implies that G would imbed into S_4. This is impossible, since $|G| = 60 > 24 = |S_4|$. Therefore, there are 10 3-Sylow subgroups. Since each of these is cyclic of order 3, they intersect only in the identity. Therefore, these 10 subgroups cover 20 distinct elements. Hence, together with the elements in the 5-Sylow subgroups, we have 44 nontrivial elements.

The number of 2-Sylow subgroups is of the form $1+2k$ and divides 15. Hence, there are 1, 3, 5, 15. We claim that there are 5. As before, there cannot be only 1, since G is simple. There cannot be 3, since as for the case of 3-Sylow subgroups, this would imply an imbedding of G into S_3, which is impossible, given $|S_3| = 6$. Suppose that there were 15 2-Sylow subgroups, each of order 4. The intersections would have a maximum of 2 elements. Therefore, each of these would contribute at least 2 distinct elements. This gives a minimum of 30 distinct elements. However, we already have 44 nontrivial elements from the 3-Sylow and 5-Sylow subgroups. Since $|G| = 60$, this is too many. Therefore, G must have 5 2-Sylow subgroups.

Now let G act on the set of 2-Sylow subgroups. This then, as above, implies an imbedding of G into S_5, so we may consider G as a subgroup of S_5. However, the only subgroup of S_5 of order 60 is A_5; therefore, $G \cong A_5$.

The proof that A_5 is the smallest non-Abelian simple group is actually brute force. We show that any group G of order less than 60 either has prime order, or is nonsimple. There are strong tools that we can use. By the Feit–Thompson theorem, we must only consider groups of even order. From Theorem 13.4.4, we do not have to consider orders $2p$. The rest can be done by an analysis using Sylow theory. For example, we show that any group of order 20 is nonsimple. Since $20 = 2^2 \cdot 5$, the number of 5-Sylow subgroups is $1+5k$ and divides 4. Hence, there is only one; therefore, it must be normal, and so G is nonsimple. There is a strong theorem by Burnside, whose proof is usually done with representation theory (see Chapter 22), which says that any group, whose order is divisible by only two primes, is solvable. Therefore, for $|G| = 60$, we only have to show that groups of order $30 = 2\cdot 3\cdot 5$ and $42 = 2\cdot 3\cdot 7$ are nonsimple. This is done in the same manner as the first part of this proof. Suppose $|G| = 30$. The number of 5-Sylow subgroups is of the form $1+5k$ and divides 6. Hence, there are 1 or 6. If G were simple there would have to be 6 covering 24 distinct elements. The number of 3-Sylow subgroups is of the form $1+3k$ and divides 10; hence, there are 1 or 10. If there were 10 these would cover an additional 20 distinct elements, which is impossible, since we already have 24 and G has order 30. Therefore, there is only one, hence a normal 3-Sylow subgroup. It follows that G cannot be simple. The case $|G| = 42$ is even simpler. There must be a normal 7-Sylow subgroup. □

13.5 Exercises

1. Prove Lemma 13.1.3.
2. Let the group G act on itself by conjugation; that is, $g(g_1) = g^{-1}g_1g$. Prove that this is an action on the set G.
3. Show that the dihedral group D_n of order $2n$ has the presentation
$$\langle r, f; r^n = f^2 = (rf)^2 = 1\rangle$$
(see Chapter 14 for group presentations).
4. Show that each group of order ≤ 59 is solvable.
5. Show that there is no simple group of order 84.
6. Let P_1 and P_2 be two different p-Sylow subgroups of a finite group G. Show that P_1P_2 is not a subgroup of G.
7. Let P and Q be two p-Sylow subgroups of the finite group G. If $Z(P)$ is a normal subgroup of Q, then $Z(P) = Z(Q)$.
8. Let G be a finite group. For a prime p the following are equivalent:
   (i) G has exactly one p-Sylow subgroup.
   (ii) The product of any two elements of order p has some order p^k.
9. Let p be a prime and $G = \mathrm{SL}(2,p)$. Let $P = \langle a\rangle$, where $a = \left(\begin{smallmatrix}1 & 1\\ 0 & 1\end{smallmatrix}\right)$.
   (i) Determine the normalizer $N_G(P)$ and the number of p-Sylow subgroups of G.
   (ii) Determine the centralizer $C_G(a)$. How many elements of order p does G have? In how many conjugacy classes can they be decomposed?
   (iii) Show that all subgroups of G of order $p(p-1)$ are conjugate.
   (iv) Show that G has no elements of order $p(p-1)$ for $p \geq 5$.
10. Let G be a finite group and N a normal subgroup such that $|N|$ is a power of p. Show that N is contained in every p-Sylow subgroup of G.
11. Let p be a prime number, and let P and Q be two p-Sylow subgroups of the finite group G such that P is contained in $N_{G(Q)}$. Show that $P = Q$.

14 Free Groups and Group Presentations

14.1 Group Presentations and Combinatorial Group Theory

In discussing the symmetric group on 3 symbols and then the various dihedral groups in Chapters 9, 10, and 11, we came across the concept of a *group presentation*. Roughly, for a group G, a presentation consists of a set of *generators* X for G, so that $G = \langle X \rangle$, and a set of *relations* between the elements of X, from which—in principle—the whole group table can be constructed. In this chapter, we make this concept precise. As we will see, every group G has a presentation, but it is mainly in the case where the group is finite or countably infinite that presentations are most useful. Historically, the idea of group presentations arose out of the attempt to describe the countably infinite fundamental groups that came out of low dimensional topology. The study of groups using group presentations is called *combinatorial group theory*.

Before looking at group presentations in general, we revisit two examples of finite groups and then a class of infinite groups.

Consider the symmetric group on 3 symbols, S_3. We saw that it has the following 6 elements:

$$1 = \begin{pmatrix} 1 & 2 & 3 \\ 1 & 2 & 3 \end{pmatrix}, \quad a = \begin{pmatrix} 1 & 2 & 3 \\ 2 & 3 & 1 \end{pmatrix}, \quad b = \begin{pmatrix} 1 & 2 & 3 \\ 3 & 1 & 2 \end{pmatrix}$$

$$c = \begin{pmatrix} 1 & 2 & 3 \\ 2 & 1 & 3 \end{pmatrix}, \quad d = \begin{pmatrix} 1 & 2 & 3 \\ 3 & 2 & 1 \end{pmatrix}, \quad e = \begin{pmatrix} 1 & 2 & 3 \\ 1 & 3 & 2 \end{pmatrix}.$$

Notice that $a^3 = 1$, $c^2 = 1$, and that $ac = ca^2$. We claim that

$$\langle a, c; a^3 = c^2 = (ac)^2 = 1 \rangle$$

is a presentation for S_3. First, it is easy to show that $S_3 = \langle a, c \rangle$. Indeed,

$$1 = 1, \quad a = a, \quad b = a^2, \quad c = c, \quad d = ac, \quad e = a^2 c,$$

and so a, c generate S_3.

Now from $(ac)^2 = acac = 1$, we get that $ca = a^2 c$. This implies that if we write any sequence (or word in our later language) in a and c, we can also rearrange it so that the only nontrivial powers of a are a and a^2; the only powers of c are c, and all a terms precede c terms. For example,

$$aca^2cac = aca(acac) = a(ca) = a(a^2c) = (a^3)c = c.$$

Therefore, using the three relations from the presentation above, each element of S_3 can be written as $a^\alpha c^\beta$ with $\alpha = 0, 1, 2$ and $\beta = 0, 1$. From this the multiplication of any two elements can be determined.

https://doi.org/10.1515/9783111142524-014

This type of argument exactly applies to all the dihedral groups D_n. We saw that, in general, $|D_n| = 2n$. Since these are the symmetry groups of a regular n-gon, we always have a rotation r of angle $\frac{2\pi}{n}$ about the center of the n-gon. This element r would have order n. Let f be a reflection about any line of symmetry. Then $f^2 = 1$, and rf is a reflection about the rotated line, which is also a line of symmetry. Therefore, $(rf)^2 = 1$. Exactly as for S_3, the relation $(rf)^2 = 1$ implies that $fr = r^{-1}f = r^{n-1}f$. This allows us to always place r terms in front of f terms in any word on r and f. Therefore, the elements of D_n are always of the form

$$r^\alpha f^\beta, \quad \alpha = 0, 1, 2, \ldots, n-1, \ \beta = 0, 1.$$

Moreover, the relations $r^n = f^2 = (rf)^2 = 1$ allow us to rearrange any word in r and f into this form. It follows that $|\langle r, f\rangle| = 2n$; hence, $D_n = \langle r, f\rangle$ together with the relations above. Hence, we obtain the following:

Theorem 14.1.1. *If D_n is the symmetry group of a regular n-gon, then a presentation for D_n is given by*

$$D_n = \langle r, f; r^n = f^2 = (rf)^2 = 1\rangle.$$

(See Section 14.3 *for the concept of group presentations.)*

We now give one class of infinite examples. If G is an infinite cyclic group, so that $G \cong \mathbb{Z}$, then $G = \langle g; \ \rangle$ is a presentation for G. That is, G has a single generator with no relations.

A direct product of n copies of $\mathbb{Z}$ is called a *free Abelian group* of rank n. We will denote this by $\mathbb{Z}^n$. A presentation for $\mathbb{Z}^n$ is then given by

$$\mathbb{Z}^n = \langle x_1, x_2, \ldots, x_n; x_i x_j = x_j x_i \text{ for all } i, j = 1, \ldots, n\rangle.$$

14.2 Free Groups

Crucial to the concept of a group presentation is the idea of a *free group*.

Definition 14.2.1 (Universal mapping property). A group F is *free on a subset X* if every map $f : X \to G$ with G a group can be extended to a unique homomorphism $f : F \to G$. X is called a *free basis* for F. In general, a group F is a *free group* if it is free on some subset X. If X is a free basis for a free group F, we write $F = F(X)$.

We first show that given any set X, there does exist a free group with free basis X. Let $X = \{x_i\}_{i\in I}$ be a set (possibly empty). We will construct a group $F(X)$, which is free with free basis X. First, let X^{-1} be a set disjoint from X, but bijective to X. If $x_i \in X$, then we denote as x_i^{-1} the corresponding element of X^{-1} under the bijection, and say that x_i and x_i^{-1} are *associated.* The set X^{-1} is called the *set of formal inverses* from X, and we call $X \cup X^{-1}$ the *alphabet*. Elements of the alphabet are called *letters*. Hence, a letter has the form $x_i^{\epsilon_1}$, where $\epsilon_i = \pm 1$. A *word* in X is a finite sequence of letters from the alphabet. That is a word has the form

$$w = x_{i_1}^{\epsilon_{i_1}} x_{i_2}^{\epsilon_{i_2}} \cdots x_{i_n}^{\epsilon_{i_n}},$$

where $x_{i_j} \in X$, and $\epsilon_{i_j} = \pm 1$. If $n = 0$, we call it the *empty word*, which we will denote as e. The integer n is called the *length* of the word. Words of the form $x_i x_i^{-1}$ or $x_i^{-1} x_i$ are called *trivial words*. We let $W(X)$ be the set of all words on X.

If $w_1, w_2 \in W(X)$, we say that w_1 is *equivalent* to w_2, denoted as $w_1 \sim w_2$, if w_1 can be converted to w_2 by a finite string of insertions and deletions of trivial words. For example, if $w_1 = x_3 x_4 x_4^{-1} x_2 x_2$ and $w_2 = x_3 x_2 x_2$, then $w_1 \sim w_2$. It is straightforward to verify that this is an equivalence relation on $W(X)$ (see exercises). Let $F(X)$ denote the set of equivalence classes in $W(X)$ under this relation; hence, $F(X)$ is a set of equivalence classes of words from X.

A word $w \in W(X)$ is said to be *freely reduced* or *reduced* if it has no trivial subwords (a subword is a connected sequence within a word). Hence, in the example above, $w_2 = x_3 x_2 x_2$ is reduced, but $w_1 = x_3 x_4 x_4^{-1} x_2 x_2$ is not reduced. There is a unique element of minimal length in each equivalence class in $F(X)$. Furthermore, this element must be reduced or else it would be equivalent to something of smaller length. Two reduced words in $W(X)$ are either equal or not in the same equivalence class in $F(X)$. Hence, $F(X)$ can also be considered as the set of all reduced words from $W(X)$.

Given a word $w = x_{i_1}^{\epsilon_{i_1}} x_{i_2}^{\epsilon_{i_2}} \cdots x_{i_n}^{\epsilon_{i_n}}$, we can find the unique reduced word $\overline{w}$ equivalent to w via the following *free reduction process*. Beginning from the left side of w, we cancel each occurrence of a trivial subword. After all these possible cancellations, we have a word w'. Now we repeat the process again, starting from the left side. Since w has finite length, eventually the resulting word will either be empty or reduced. The final reduced $\overline{w}$ is the *free reduction* of w.

Now we build a multiplication on $F(X)$. If

$$w_1 = x_{i_1}^{\epsilon_{i_1}} x_{i_2}^{\epsilon_{i_2}} \cdots x_{i_n}^{\epsilon_{i_n}}, \quad w_2 = x_{j_1}^{\epsilon_{j_1}} x_{j_2}^{\epsilon_{j_2}} \cdots x_{j_m}^{\epsilon_{j_m}}$$

are two words in $W(X)$, then their *concatenation* $w_1 \star w_2$ is simply placing w_2 after w_1,

$$w_1 \star w_2 = x_{i_1}^{\epsilon_{i_1}} x_{i_2}^{\epsilon_{i_2}} \cdots x_{i_n}^{\epsilon_{i_n}} x_{j_1}^{\epsilon_{j_1}} x_{j_2}^{\epsilon_{j_2}} \cdots x_{j_m}^{\epsilon_{j_m}}.$$

If $w_1, w_2 \in F(X)$, then we define their product as

$$w_1 w_2 = \text{equivalence class of } w_1 \star w_2.$$

That is, we concatenate w_1 and w_2, and the product is the equivalence class of the resulting word. It is easy to show that if $w_1 \sim w_1'$ and $w_2 \sim w_2'$, then $w_1 \star w_2 \sim w_1' \star w_2'$ so that the above multiplication is well defined. Equivalently, we can think of this product in the following way. If w_1, w_2 are reduced words, then to find w_1w_2, first concatenate, and then freely reduce. Notice that if $x_{i_n}^{\epsilon_{i_n}} x_{j_1}^{\epsilon_{j_1}}$ is a trivial word, then it is cancelled when the concatenation is formed. We say then that there is *cancellation* in forming the product w_1w_2. Otherwise, the product is formed without cancellation.

Theorem 14.2.2. *Let X be a nonempty set, and let $F(X)$ be as above. Then $F(X)$ is a free group with free basis X. Furthermore, if $X = \emptyset$, then $F(X) = \{1\}$; if $|X| = 1$, then $F(X) \cong \mathbb{Z}$, and if $|X| \geq 2$, then $F(X)$ is non-Abelian.*

Proof. We first show that $F(X)$ is a group, and then show that it satisfies the universal mapping property on X. We consider $F(X)$ as the set of reduced words in $W(X)$ with the multiplication defined above. Clearly, the empty word acts as the identity element 1. If $w = x_{i_1}^{\epsilon_{i_1}} x_{i_2}^{\epsilon_{i_2}} \cdots x_{i_n}^{\epsilon_{i_n}}$ and $w_1 = x_{i_n}^{-\epsilon_{i_n}} x_{i_{n-1}}^{-\epsilon_{i_{n-1}}} \cdots x_{i_1}^{-\epsilon_{i_1}}$, then both $w \star w_1$ and $w_1 \star w$ freely reduce to the empty word, and so w_1 is the inverse of w. Therefore, each element of $F(X)$ has an inverse. Therefore, to show that $F(X)$ forms a group, we must show that the multiplication is associative. Let

$$w_1 = x_{i_1}^{\epsilon_{i_1}} x_{i_2}^{\epsilon_{i_2}} \cdots x_{i_n}^{\epsilon_{i_n}}, \quad w_2 = x_{j_1}^{\epsilon_{j_1}} x_{j_2}^{\epsilon_{j_2}} \cdots x_{j_m}^{\epsilon_{j_m}}, \quad w_3 = x_{k_1}^{\epsilon_{k_1}} x_{k_2}^{\epsilon_{k_2}} \cdots x_{k_p}^{\epsilon_{k_p}}$$

be three freely reduced words in $F(X)$. We must show that

$$(w_1w_2)w_3 = w_1(w_2w_3).$$

To prove this, we use induction on m, the length of w_2. If $m = 0$, then w_2 is the empty word, hence the identity, and it is certainly true. Now suppose that $m = 1$ so that $w_2 = x_{j_1}^{\epsilon_{j_1}}$. We must consider exactly four cases.

Case (1): There is no cancellation in forming either w_1w_2 or w_2w_3. Put differently, $x_{j_1}^{\epsilon_{j_1}} \neq x_{i_n}^{-\epsilon_{i_n}}$, and $x_{j_1}^{\epsilon_{j_1}} \neq x_{k_1}^{-\epsilon_{k_1}}$. Then the product w_1w_2 is just the concatenation of the words, and so is $(w_1w_2)w_3$. The same is true for $w_1(w_2w_3)$. Therefore, $w_1(w_2w_3) = (w_1w_2)w_3$.

Case (2): There is cancellation in forming w_1w_2, but not in forming w_2w_3. Then if we concatenate all three words, the only cancellation occurs between w_1 and w_2 in either $w_1(w_2w_3)$ or in $(w_1w_2)w_3$; hence, they are equal. Therefore, $w_1(w_2w_3) = (w_1w_2)w_3$.

Case (3): There is cancellation in forming w_2w_3, but not in forming w_1w_2. This is entirely analogous to Case (2). Therefore, $w_1(w_2w_3) = (w_1w_2)w_3$.

Case (4): There is cancellation in forming w_1w_2 and also in forming w_2w_3. Then $x_{j_1}^{\epsilon_{j_1}} = x_{i_n}^{-\epsilon_{i_n}}$ and $x_{j_1}^{\epsilon_{j_1}} = x_{k_1}^{-\epsilon_{k_1}}$. Here,

$$(w_1w_2)w_3 = x_{i_1}^{\epsilon_{i_1}} \cdots x_{i_{n-1}}^{\epsilon_{i_{n-1}}} x_{k_1}^{\epsilon_{k_1}} x_{k_2}^{\epsilon_{k_2}} \cdots x_{k_p}^{\epsilon_{k_p}}.$$

On the other hand,

$$w_1(w_2w_3) = x_{i_1}^{\epsilon_{i_1}} \cdots x_{i_n}^{\epsilon_{i_n}} x_{k_2}^{\epsilon_{k_2}} \cdots x_{k_p}^{\epsilon_{k_p}}.$$

However, these are equal since $x_{i_n}^{\epsilon_{i_n}} = x_{k_1}^{\epsilon_{k_1}}$. Therefore, $w_1(w_2w_3) = (w_1w_2)w_3$.

It follows, inductively, from these four cases, that the associative law holds in $F(X)$; therefore, $F(X)$ forms a group.

Now suppose that $f : X \to G$ is a map from X into a group G. By the construction of $F(X)$ as a set of reduced words this can be extended to a unique homomorphism. If $w \in F$ with $w = x_{i_1}^{\epsilon_{i_1}} \cdots x_{i_n}^{\epsilon_{i_n}}$, then define $f(w) = f(x_{i_1})^{\epsilon_{i_1}} \cdots f(x_{i_n})^{\epsilon_{i_n}}$. Since multiplication in $F(X)$ is concatenation, this defines a homomorphism and again form the construction of $F(X)$, its the only one extending f. This is analogous to constructing a linear transformation from one vector space to another by specifying the images of a basis. Therefore, $F(X)$ satisfies the universal mapping property of Definition 14.2.1. Hence, $F(X)$ is a free group with free basis X.

The final parts of Theorem 14.2.2 are straightforward. If X is empty, the only reduced word is the empty word; hence, the group is just the identity. If X has a single letter, then $F(X)$ has a single generator, and is therefore cyclic. It is easy to see that it must be torsion-free. Therefore, $F(X)$ is infinite cyclic; that is, $F(X) \cong \mathbb{Z}$. Finally, if $|X| \geq 2$, let $x_1, x_2 \in X$. Then $x_1x_2 \neq x_2x_1$, and both are reduced. Therefore, $F(X)$ is non-Abelian. □

The proof of Theorem 14.2.2 provides another way to look at free groups.

Theorem 14.2.3. *F is a free group if and only if there is a generating set X such that every element of F has a unique representation as a freely reduced word on X.*

The structure of a free group is entirely dependent on the cardinality of a free basis. In particular, the cardinality of a free basis X for a free group F is unique, and is called the *rank of F*. If $|X| < \infty$, F is of *finite rank*. If F has rank n and $X = \{x_1, x_2, \ldots, x_n\}$, we say that F is free on $\{x_1, x_2, \ldots, x_n\}$. We denote this by $F(x_1, x_2, \ldots, x_n)$.

Theorem 14.2.4. *If X and Y are sets with the same cardinality, that is, $|X| = |Y|$, then $F(X) \cong F(Y)$, the resulting free groups are isomorphic. Furthermore, if $F(X) \cong F(Y)$, then $|X| = |Y|$.*

Proof. Suppose that $f : X \to Y$ is a bijection from X onto Y. Now $Y \subset F(Y)$, so there is a unique homomorphism $\phi : F(X) \to F(Y)$ extending f. Since f is a bijection, it has an inverse $f^{-1} : Y \to X$, and since $F(Y)$ is free, there is a unique homomorphism ϕ_1 from $F(Y)$ to $F(X)$ extending f^{-1}. Then $\phi\phi_1$ is the identity map on $F(Y)$, and $\phi_1\phi$ is the identity map on $F(X)$. Therefore, ϕ, ϕ_1 are isomorphisms with $\phi = \phi_1^{-1}$.

Conversely, suppose that $F(X) \cong F(Y)$. In $F(X)$, let $N(X)$ be the subgroup generated by all squares in $F(X)$; that is,

$$N(X) = \langle \{g^2 : g \in F(X)\} \rangle.$$

Then $N(X)$ is a normal subgroup, and the factor group $F(X)/N(X)$ is Abelian, where every nontrivial element has order 2 (see exercises). Therefore, $F(X)/N(X)$ can be considered as a vector space over $\mathbb{Z}_2$, the finite field of order 2, with X as a vector space basis. Hence, $|X|$ is the dimension of this vector space. Let $N(Y)$ be the corresponding subgroup of $F(Y)$. Since $F(X) \cong F(Y)$, we would have $F(X)/N(X) \cong F(Y)/N(Y)$; therefore, $|Y|$ is the dimension of the vector space $F(Y)/N(Y)$. Thus, $|X| = |Y|$ from the uniqueness of dimension of vector spaces. □

Expressing elements of $F(X)$ as a reduced word gives a *normal form* for elements in a free group F. As we will see in Section 14.5, this solves what is termed the *word problem* for free groups. Another important concept is the following: a freely reduced word $W = x_{v_1}^{e_1}x_{v_2}^{e_2}\cdots x_{v_n}^{e_n}$ is *cyclically reduced* if $v_1 \neq v_n$, or if $v_1 = v_n$, then $e_1 \neq -e_n$. Clearly then, every element of a free group is conjugate to an element given by a cyclically reduced word. This provides a method to determine conjugacy in free groups.

Theorem 14.2.5. *In a free group F, two elements g_1, g_2 are conjugate if and only if a cyclically reduced word for g_1 is a cyclic permutation of a cyclically reduced word for g_2.*

The theory of free groups has a large and extensive literature. We close this section by stating several important properties. Proofs for these results can be found in [37], [36] or [21].

Theorem 14.2.6. *A free group is torsion-free.*

From Theorem 14.2.4, we can deduce:

Theorem 14.2.7. *An Abelian subgroup of a free group must be cyclic.*

Finally, a celebrated theorem of Nielsen and Schreier states that a subgroup of a free group must be free.

Theorem 14.2.8 (Nielsen–Schreier). *A subgroup of a free group is itself a free group.*

Combinatorially, F is free on X if X is a set of generators for F, and there are no nontrivial relations. In particular, the following hold:

There are several different proofs of this result, see [37], with the most straightforward being topological in nature. We give an outline of a simple topological proof in Section 14.4.

About 1920, Nielsen, using a technique now called Nielsen transformations in his honor, first proved this theorem for finitely generated subgroups. Schreier, shortly after, found a combinatorial method to extend this to arbitrary subgroups. A complete version of the original combinatorial proof appears in [37], and in the notes by Johnson [31].

Schreier's combinatorial proof also allows for a description of the free basis for the subgroup. In particular, let F be free on X, and $H \subset F$ a subgroup. Let $T = \{t_\alpha\}$ be a complete set of right coset representatives for F modulo H with the property that if $t_\alpha = x_{v_1}^{e_1}x_{v_2}^{e_2}\cdots x_{v_n}^{e_n} \in T$, with $\epsilon_i = \pm 1$, then all the initial segments $1, x_{v_1}^{e_1}, x_{v_1}^{e_1}x_{v_2}^{e_2}$, et cetera are also in T. Such a system of coset representatives can always be found, and is called a *Schreier system* or *Schreier transversal* for H. If $g \in F$, let $\overline{g}$ represent its coset representative in T, and further define for $g \in F$ and $t \in T$, $S_{tg} = tg(\overline{tg})^{-1}$. Notice that $S_{tg} \in H$ for all t, g. We then have the following:

Theorem 14.2.9 (Explicit form of Nielsen–Schreier). *Let F be free on X and H a subgroup of F. If T is a Schreier transversal for F modulo H, then H is free on the set*

$$\{S_{tx} : t \in T, x \in X, S_{tx} \neq 1\}.$$

Example 14.2.10. Let F be free on $\{a, b\}$ and $H = F(X^2)$ the normal subgroup of F generated by all squares in F.

Then $F/F(X^2) = \langle a, b; a^2 = b^2 = (ab)^2 = 1\rangle = \mathbb{Z}_2 \times \mathbb{Z}_2$ (see Section 14.3 for the concept of group presentations). It follows that a Schreier system for F modulo H is $\{1, a, b, ab\}$ with $\overline{a} = a, \overline{b} = b$ and $\overline{ba} = ab$. From this it can be shown that H is free on the generating set

$$x_1 = a^2, \quad x_2 = bab^{-1}a^{-1}, \quad x_3 = b^2, \quad x_4 = abab^{-1}, \quad x_5 = ab^2a^{-1}.$$

The theorem also allows for a computation of the rank of H, given the rank of F and the index. Specifically:

Corollary 14.2.11. *Suppose F is free of rank n and $|F : H| = k$. Then H is free of rank $nk - k + 1$.*

From the example, we see that F is free of rank 2, H has index 4, so H is free of rank $2 \cdot 4 - 4 + 1 = 5$.

14.3 Group Presentations

The significance of free groups stems from the following result, which is easily deduced from the definition and will lead us directly to a formal definition of a group presentation. Let G be any group and F the free group on the elements of G considered as a set. The identity map $f : G \to G$ can be extended to a homomorphism of F onto G. Therefore, we have the following:

Theorem 14.3.1. *Every group G is a homomorphic image of a free group. That is, let G be any group. Then $G = F/N$, where F is a free group.*

In the above theorem, instead of taking all the elements of G, we can consider just a set X of generators for G. Then G is a factor group of $F(X)$, $G \cong F(X)/N$. The normal subgroup N is the kernel of the homomorphism from $F(X)$ onto G. We use Theorem 14.3.1 to formally define a group presentation.

If H is a subset of a group G, then the *normal closure* of H denoted by $N(H)$ is the smallest normal subgroup of G containing H. This can be described alternatively in the following manner. The normal closure of H is the subgroup of G generated by all conjugates of elements of H.

Now suppose that G is a group with X, a set of generators for G. We also call X a *generating system* for G. Now let $G = F(X)/N$ as in Theorem 14.3.1 and the comments after it. N is the kernel of the homomorphism $f : F(X) \to G$. It follows that if r is a free group word with $r \in N$, then $r = 1$ in G (under the homomorphism). We then call r a *relator* in G, and the equation $r = 1$ a *relation* in G. Suppose that R is a subset of N such that $N = N(R)$, then R is called a set of *defining relators* for G. The equations $r = 1, r \in R$, are a set of *defining relations* for G. It follows that any relator in G is a product of conjugates of elements of R. Equivalently, $r \in F(X)$ is a relator in G if and only if r can be reduced to the empty word by insertions and deletions of elements of R, and trivial words.

Definition 14.3.2. Let G be a group. Then a *group presentation* for G consists of a set of generators X for G and a set R of defining relators. In this case, we write $G = \langle X; R\rangle$. We could also write the presentation in terms of defining relations as $G = \langle X; r = 1, r \in R\rangle$.

From Theorem 14.3.1, it follows immediately that every group has a presentation. However, in general, there are many presentations for the same group. If $R \subset R_1$, then R_1 is also a set of defining relators.

Lemma 14.3.3. *Let G be a group. Then G has a presentation.*

If $G = \langle X; R\rangle$ and X is finite, then G is said to be *finitely generated*. If R is finite, G is *finitely related*. If both X and R are finite, G is *finitely presented*.

Using group presentations, we get another characterization of free groups.

Theorem 14.3.4. *F is a free group if and only if F has a presentation of the form $F = \langle X; \rangle$.*

Mimicking the construction of a free group from a set X, we can show that to each presentation corresponds a group. Suppose that we are given a supposed presentation $\langle X; R\rangle$, where R is given as a set of words in X. Consider the free group $F(X)$ on X. Define two words w_1, w_2 on X to be equivalent if w_1 can be transformed into w_2 using insertions and deletions of elements of R and trivial words. As in the free group case, this is an equivalence relation. Let G be the set of equivalence classes. If we define multiplication as before, as concatenation followed by the appropriate equivalence class, then G is a group. Furthermore, each $r \in R$ must equal the identity in G so that $G = \langle X; R\rangle$. Notice that here there may be no unique reduced word for an element of G.

Theorem 14.3.5. *Given (X, R), where X is a set and R is a set of words on X. Then there exists a group G with presentation $\langle X; R\rangle$.*

We now give some examples of group presentations:

Example 14.3.6. A free group of rank n has a presentation

$$F_n = \langle x_1, \dots, x_n; \ \rangle.$$

Example 14.3.7. A free Abelian group of rank n has a presentation

$$\mathbb{Z}^n = \langle x_1, \dots, x_n; x_i x_j x_i^{-1} x_j^{-1}, i = 1, \dots, n, j = 1, \dots, n\rangle.$$

Example 14.3.8. A cyclic group of order n has a presentation

$$\mathbb{Z}_n = \langle x; x^n = 1 \rangle.$$

Example 14.3.9. The dihedral groups of order $2n$, representing the symmetry group of a regular n-gon, has a presentation

$$\langle r, f; r^n = 1, f^2 = 1, (rf)^2 = 1 \rangle.$$

14.3.1 The Modular Group

In this section, we give a more complicated example, and then a nice application to number theory.

If R is a commutative ring with identity, then the set of invertible $(n \times n)$-matrices with entries from R forms a group under matrix multiplication called the *n-dimensional general linear group over R*, see [41]. This group is denoted by $\mathrm{GL}(n, R)$. Since $\det(A)\det(B) = \det(AB)$ for square matrices A, B, it follows that the subset of $\mathrm{GL}(n, R)$, consisting of those matrices of determinant 1, forms a subgroup. This subgroup is called the *special linear group over R* and is denoted by $\mathrm{SL}(n, R)$. In this section, we concentrate on $\mathrm{SL}(2, \mathbb{Z})$, or more specifically, a quotient of it, $\mathrm{PSL}(2, \mathbb{Z})$, and find presentations for them. The group $\mathrm{SL}(2, \mathbb{Z})$ then consists of (2×2)-matrices of determinant 1 with integral entries:

$$\mathrm{SL}(2, \mathbb{Z}) = \left\{ \begin{pmatrix} a & b \\ c & d \end{pmatrix} : a, b, c, d \in \mathbb{Z}, ad - bc = 1 \right\}.$$

The group $\mathrm{SL}(2, \mathbb{Z})$ is called the *homogeneous modular group*, and an element of $\mathrm{SL}(2, \mathbb{Z})$ is called a *unimodular matrix*. If G is any group, recall that its *center* $Z(G)$ consists of those elements of G, which commute with all elements of G:

$$Z(G) = \{g \in G : gh = hg, \forall h \in G\}.$$

The group $Z(G)$ is a normal subgroup of G. Hence, we can form the factor group $G/Z(G)$.

For $G = \mathrm{SL}(2, \mathbb{Z})$, the only unimodular matrices that commute with all others are $\pm I = \pm\left(\begin{smallmatrix} 1 & 0 \\ 0 & 1 \end{smallmatrix}\right)$. Therefore, $Z(\mathrm{SL}(2, \mathbb{Z})) = \{I, -I\}$. The quotient

$$\mathrm{SL}(2, \mathbb{Z})/Z(\mathrm{SL}(2, \mathbb{Z})) = \mathrm{SL}(2, \mathbb{Z})/\{I, -I\}$$

is denoted by $\mathrm{PSL}(2, \mathbb{Z})$ and is called the *projective special linear group* or *inhomogeneous modular group*. More commonly, $\mathrm{PSL}(2, \mathbb{Z})$ is just called the *modular group*, and denoted by M.

M arises in many different areas of mathematics, including number theory, complex analysis, and Riemann surface theory and the theory of automorphic forms and functions. M is perhaps the most widely studied single finitely presented group. Complete discussions of M and its structure can be found in the books *Integral Matrices* by M. Newman, see [56], and *Algebraic Theory of the Bianchi Groups* by B. Fine, see [51].

Since $M = \mathrm{PSL}(2,\mathbb{Z}) = \mathrm{SL}(2,\mathbb{Z})/\{I,-I\}$, it follows that each element of M can be considered as $\pm A$, where A is a unimodular matrix. A *projective unimodular matrix* is then

$$\pm\begin{pmatrix} a & b \\ c & d \end{pmatrix}, \quad a,b,c,d \in \mathbb{Z},\ ad-bc=1.$$

The elements of M can also be considered as linear fractional transformations over the complex numbers

$$z' = \frac{az+b}{cz+d}, \quad a,b,c,d \in \mathbb{Z},\ ad-bc=1, \text{ where } z \in \mathbb{C}.$$

Thought of in this way, M forms a *Fuchsian group*, which is a discrete group of isometries of the non-Euclidean hyperbolic plane. The book by Katok, see [33], gives a solid and clear introduction to such groups. This material can also be found in condensed form in [53].

We now determine presentations for both $\mathrm{SL}(2,\mathbb{Z})$ and $M = \mathrm{PSL}(2,\mathbb{Z})$.

Theorem 14.3.10. *The group* $\mathrm{SL}(2,\mathbb{Z})$ *is generated by the elements*

$$X = \begin{pmatrix} 0 & -1 \\ 1 & 0 \end{pmatrix} \quad \textit{and} \quad Y = \begin{pmatrix} 0 & 1 \\ -1 & -1 \end{pmatrix}.$$

Furthermore, a complete set of defining relations for the group in terms of these generators is given by

$$X^4 = Y^3 = YX^2Y^{-1}X^{-2} = I.$$

It follows that $\mathrm{SL}(2,\mathbb{Z})$ *has the presentation*

$$\langle X, Y; X^4 = Y^3 = YX^2Y^{-1}X^{-2} = I\rangle.$$

Proof. We first show that $\mathrm{SL}(2,\mathbb{Z})$ is generated by X and Y; that is, every matrix A in the group can be written as a product of powers of X and Y.

Let

$$U = \begin{pmatrix} 1 & 1 \\ 0 & 1 \end{pmatrix}.$$

Then a direct multiplication shows that $U = XY$, and we show that $\mathrm{SL}(2,\mathbb{Z})$ is generated by X and U, which implies that it is also generated by X and Y. Furthermore,

$$U^n = \begin{pmatrix} 1 & n \\ 0 & 1 \end{pmatrix};$$

therefore, U has infinite order.

Let $A = \left(\begin{smallmatrix} a & b \\ c & d \end{smallmatrix}\right) \in \mathrm{SL}(2, \mathbb{Z})$. Then we have

$$XA = \begin{pmatrix} -c & -d \\ a & b \end{pmatrix}, \quad \text{and} \quad U^k A = \begin{pmatrix} a + kc & b + kd \\ c & d \end{pmatrix}$$

for any $k \in \mathbb{Z}$. We may assume that $|c| \leq |a|$ otherwise start with XA rather than A. If $c = 0$, then $A = \pm U^q$ for some q. If $A = U^q$, then certainly A is in the group generated by X and U. If $A = -U^q$, then $A = X^2 U^q$ since $X^2 = -I$. It follows that here also A is in the group generated by X and U.

Now suppose $c \neq 0$. Apply the Euclidean algorithm to a and c in the following modified way:

$$\begin{aligned} a &= q_0 c + r_1 \\ -c &= q_1 r_1 + r_2 \\ r_1 &= q_2 r_2 + r_3 \\ &\vdots \\ (-1)^n r_{n-1} &= q_n r_n + 0, \end{aligned}$$

where $r_n = \pm 1$ since $(a, c) = 1$. Then

$$XU^{-q_n} \cdots XU^{-q_0} A = \pm U^{q_{n+1}} \quad \text{with } q_{n+1} \in \mathbb{Z}.$$

Therefore,

$$A = X^m U^{q_0} X U^{q_1} \cdots X U^{q_n} X U^{q_{n+1}}$$

with $m = 0, 1, 2, 3$; $q_0, q_1, \ldots, q_{n+1} \in \mathbb{Z}$ and $q_0, \ldots, q_n \neq 0$. Thus, X and U, and hence X and Y generate $\mathrm{SL}(2, \mathbb{Z})$.

We must now show that

$$X^4 = Y^3 = YX^2Y^{-1}X^{-2} = I$$

form a complete set of defining relations for $\mathrm{SL}(2, \mathbb{Z})$, or that every relation on these generators is derivable from these. It is straightforward to see that X and Y do satisfy these relations. Assume then that we have a relation

$$S = X^{\epsilon_1} Y^{\alpha_1} X^{\epsilon_2} Y^{\alpha_2} \cdots Y^{\alpha_n} X^{\epsilon_{n+1}} = I$$

with all $\epsilon_i, \alpha_j \in \mathbb{Z}$. Using the set of relations

$$X^4 = Y^3 = YX^2Y^{-1}X^{-2} = I,$$

we may transform S so that

$$S = X^{\epsilon_1}Y^{\alpha_1}XY^{\alpha_2}\cdots Y^{\alpha_m}X^{\epsilon_{m+1}}$$

with $\epsilon_1, \epsilon_{m+1} = 0, 1, 2$ or 3 and $\alpha_i = 1$ or 2 for $i = 1, \ldots, m$ and $m \geq 0$. Multiplying by a suitable power of X, we obtain

$$Y^{\alpha_1}X\cdots Y^{\alpha_m}X = X^{\alpha} = S_1$$

with $m \geq 0$ and $\alpha = 0, 1, 2$ or 3. Assume that $m \geq 1$, and let

$$S_1 = \begin{pmatrix} a & -b \\ -c & d \end{pmatrix}.$$

We show by induction that

$$a, b, c, d \geq 0, \quad b + c > 0,$$

or

$$a, b, c, d \leq 0, \quad b + c < 0.$$

This claim for the entries of S_1 is true for

$$YX = \begin{pmatrix} 1 & 0 \\ -1 & 1 \end{pmatrix}, \quad \text{and} \quad Y^2X = \begin{pmatrix} -1 & 1 \\ 0 & -1 \end{pmatrix}.$$

Suppose it is correct for $S_2 = \left(\begin{smallmatrix} a_1 & -b_1 \\ -c_1 & d_1 \end{smallmatrix} \right)$. Then

$$YXS_2 = \begin{pmatrix} a_1 & -b_1 \\ -(a_1 + c_1) & b_1 + d_1 \end{pmatrix} \quad \text{and}$$

$$Y^2XS_2 = \begin{pmatrix} -a_1 - c_1 & b_1 + d_1 \\ c_1 & d_1 \end{pmatrix}.$$

Therefore, the claim is correct for all S_1 with $m \geq 1$. This gives a contradiction, for the entries of X^{α} with $\alpha = 0, 1, 2$ or 3 do not satisfy the claim. Hence, $m = 0$, and S can be reduced to a trivial relation by the given set of relations. Therefore, they are a complete set of defining relations, and the theorem is proved. □

Corollary 14.3.11. *The modular group $M = \mathrm{PSL}(2, \mathbb{Z})$ has the presentation*

$$M = \langle x, y; x^2 = y^3 = 1 \rangle.$$

Furthermore, x, y can be taken as the linear fractional transformations

$$x : z' = -\frac{1}{z}, \quad \text{and} \quad y : z' = -\frac{1}{z+1}.$$

Proof. The center of $\mathrm{SL}(2,\mathbb{Z})$ is $\pm I$. Since $X^2 = -I$, setting $X^2 = I$ in the presentation for $\mathrm{SL}_2(\mathbb{Z})$ gives the presentation for M. Writing the projective matrices as linear fractional transformations gives the second statement. □

This corollary says that M is the *free product* of a cyclic group of order 2 and a cyclic group of order 3, a concept we will introduce in Section 14.7.

We note that there is an elementary alternative proof to Corollary 14.3.11 as far as showing that $X^2 = Y^3 = 1$ are a complete set of defining relations. As linear fractional transformations, we have

$$X(z) = -\frac{1}{z}, \quad Y(z) = -\frac{1}{z+1}, \quad Y^2(z) = -\frac{z+1}{z}.$$

Now let

$$\mathbb{R}^+ = \{x \in \mathbb{R} : x > 0\} \quad \text{and} \quad \mathbb{R}^- = \{x \in \mathbb{R} : x < 0\}.$$

Then

$$X(\mathbb{R}^-) \subset \mathbb{R}^+, \quad \text{and} \quad Y^{\alpha}(\mathbb{R}^+) \subset \mathbb{R}^-, \quad \alpha = 1, 2.$$

Let $S \in M$. Using the relations $X^2 = Y^3 = 1$ and a suitable conjugation, we may assume that either $S = 1$ is a consequence of these relations, or that

$$S = Y^{\alpha_1} X Y^{\alpha_2} \cdots X Y^{\alpha_n}$$

with $1 \le \alpha_i \le 2$ and $\alpha_1 = \alpha_n$.

In this second case, if $x \in \mathbb{R}^+$, then $S(x) \in \mathbb{R}^-$; hence, $S \neq 1$.

This type of *ping-pong argument* can be used in many examples, see [36], [21] and [31]. As another example, consider the unimodular matrices

$$A = \begin{pmatrix} 0 & 1 \\ -1 & 2 \end{pmatrix}, \quad B = \begin{pmatrix} 0 & -1 \\ 1 & 2 \end{pmatrix}.$$

Let $\overline{A}, \overline{B}$ denote the corresponding linear fractional transformations in the modular group M. We have

$$A^n = \begin{pmatrix} -n+1 & n \\ -n & n+1 \end{pmatrix}, \quad B^n = \begin{pmatrix} -n+1 & -n \\ n & n+1 \end{pmatrix} \quad \text{for } n \in \mathbb{Z}.$$

In particular, $\overline{A}$ and $\overline{B}$ have infinite order. Now

$$\overline{A}^n(\mathbb{R}^-) \subset \mathbb{R}^+ \quad \text{and} \quad \overline{B}^n(\mathbb{R}^+) \subset \mathbb{R}^-$$

for all $n \neq 0$. The ping-pong argument used for any element of the type

$$S = \overline{A}^{n_1}\overline{B}^{m_1} \cdots \overline{B}^{m_k}\overline{A}^{n_{k+1}}$$

with all $n_i, m_i \neq 0$ and $n_1 + n_{k+1} \neq 0$ shows that $S(x) \in \mathbb{R}^+$ if $x \in \mathbb{R}^-$. It follows that there are no nontrivial relations on $\overline{A}$ and $\overline{B}$; therefore, the subgroup of M generated by $\overline{A}, \overline{B}$ must be a free group of rank 2.

To close this section, we present a significant number of theoretical applications of the modular group. First, we need the following corollary to Corollary 14.3.11:

Corollary 14.3.12. *Let $M = \langle X, Y; X^2 = Y^3 = 1\rangle$ be the modular group. If A is an element of order 2, then A is conjugate to X. If B is an element of order 3, then B is conjugate to either Y or Y^2.*

Definition 14.3.13. Let a, n be relatively prime integers with $a \neq 0, n \geq 1$. Then a is a *quadratic residue* modulo n if there exists an $x \in \mathbb{Z}$ with $x^2 \equiv a \pmod{n}$; that is, $a = x^2 + kn$ for some $k \in \mathbb{Z}$.

The following is called Fermat's two-square theorem.

Theorem 14.3.14 (Fermat's two-square theorem). *Let $n > 0$ be a natural number. Then $n = a^2 + b^2$ with $(a, b) = 1$ if and only if -1 is a quadratic residue modulo n.*

Proof. Suppose -1 is a quadratic residue modulo n, then there exists an x such that $x^2 \equiv -1 \pmod{n}$ or $x^2 = -1 + mn$. This implies that $-x^2 - mn = 1$ so that there must exist a projective unimodular matrix

$$A = \pm\begin{pmatrix} x & n \\ m & -x \end{pmatrix}.$$

It is straightforward that $A^2 = 1$. Therefore, by Corollary 14.3.12, A is conjugate within M to X. Now consider conjugates of X within M. Let $T = \left(\begin{smallmatrix} a & b \\ c & d \end{smallmatrix}\right)$. Then

$$T^{-1} = \begin{pmatrix} d & -b \\ -c & a \end{pmatrix},$$

and

$$TXT^{-1} = \begin{pmatrix} a & b \\ c & d \end{pmatrix}\begin{pmatrix} 0 & 1 \\ -1 & 0 \end{pmatrix}\begin{pmatrix} d & -b \\ -c & a \end{pmatrix} = \pm\begin{pmatrix} -(bd + ac) & a^2 + b^2 \\ -(c^2 + d^2) & bd + ac \end{pmatrix}. \qquad (*)$$

Therefore, any conjugate of X must have the form $(*)$, and thus A also must have the form $(*)$. Therefore, $n = a^2 + b^2$. Furthermore, $(a, b) = 1$ since in finding the form $(*)$, we had $ad - bc = 1$.

Conversely suppose $n = a^2 + b^2$ with $(a,b) = 1$. Then there exist $c, d \in \mathbb{Z}$ with $ad - bc = 1$; hence, there exists a projective unimodular matrix

$$T = \pm \begin{pmatrix} a & b \\ c & d \end{pmatrix}.$$

Then

$$TXT^{-1} = \pm \begin{pmatrix} a & a^2 + b^2 \\ y & -a \end{pmatrix} = \pm \begin{pmatrix} a & n \\ y & -a \end{pmatrix}.$$

This has determinant one, so

$$-a^2 - ny = 1 \implies a^2 = -1 - ny \implies a^2 \equiv -1 \pmod{n}.$$

Therefore, -1 is a quadratic residue modulo n. □

This type of group theoretical proof can be extended in several directions. Kern-Isberner and Rosenberger, see [34], considered groups of matrices of the form

$$U = \begin{pmatrix} a & b\sqrt{N} \\ c\sqrt{N} & d \end{pmatrix}, \quad a, b, c, d, N \in \mathbb{Z}, \ ad - Nbc = 1,$$

or

$$U = \begin{pmatrix} a\sqrt{N} & b \\ c & d\sqrt{N} \end{pmatrix}, \quad a, b, c, d, N \in \mathbb{Z}, \ Nad - bc = 1.$$

They then proved that if

$$N \in \{1, 2, 4, 5, 6, 8, 9, 10, 12, 13, 16, 18, 22, 25, 28, 37, 58\}$$

and $n \in \mathbb{N}$ with $(n, N) = 1$, then the following hold:
(1) If $-N$ is a quadratic residue modulo n and n is a quadratic residue modulo N, then n can be written as $n = x^2 + Ny^2$ with $x, y \in \mathbb{Z}$.
(2) Conversely, if $n = x^2 + Ny^2$ with $x, y \in \mathbb{Z}$ and $(x, y) = 1$, then $-N$ is a quadratic residue modulo n, and n is a quadratic residue modulo N.

The proof of the above results depends on the class number of $\mathbb{Q}(\sqrt{-N})$ (see [34]).

In another direction, Fine [50] and [49] showed that the Fermat two-square property is actually a property satisfied by many rings R. These are called *sum of squares rings*. For example, if $p \equiv 3 \pmod{4}$, then $\mathbb{Z}_{p^n}$ for $n > 1$ is a sum of squares ring.

14.4 Presentations of Subgroups

Given a group presentation $G = \langle X; R\rangle$, it is possible to find a presentation for a subgroup H of G. The procedure to do this is called the *Reidemeister–Schreier process* and is a consequence of the explicit version of the Nielsen–Schreier theorem (Theorem 14.2.9). We give a brief description. A complete description and a verification of its correctness is found in [37], or in [21].

Let G be a group with the presentation $\langle a_1, \dots, a_n; R_1, \dots, R_k\rangle$. Let H be a subgroup of G and T a Schreier system for G modulo H, defined analogously as above.

Reidemeister–Schreier process

Let G, H and T be as above. Then H is generated by the set

$$\{S_{ta_v} : t \in T, a_v \in \{a_1, \dots, a_n\}, S_{ta_v} \neq 1\}$$

with a complete set of defining relations given by conjugates of the original relators rewritten in terms of the subgroup generating set.

To actually rewrite the relators in terms of the new generators, we use a mapping τ on words on the generators of G called the *Reidemeister rewriting process*. This map is defined as follows: If

$$W = a_{v_1}^{e_1} a_{v_2}^{e_2} \cdots a_{v_j}^{e_j} \quad \text{with } e_i = \pm 1 \text{ defines an element of } H$$

then

$$\tau(W) = S_{t_1, a_{v_1}}^{e_1} S_{t_2, a_{v_2}}^{e_2} \cdots S_{t_j, a_{v_j}}^{e_j},$$

where t_i is the coset representative of the initial segment of W preceding a_{v_i}, if $e_i = 1$ and t_i is the representative of the initial segment of W up to and including $a_{v_i}^{-1}$ if $e_i = -1$. The complete set of relators rewritten in terms of the subgroup generators is then given by

$$\{\tau(tR_i t^{-1})\} \quad \text{with } t \in T, \text{ and } R_i \text{ runs over all relators in } G.$$

We present two examples; one with a finite group, and then an important example with a free group, which shows that a countable free group contains free subgroups of arbitrary ranks.

Example 14.4.1. Let $G = A_4$ be the alternating group on 4 symbols. Then a presentation for G is

$$G = A_4 = \langle a, b; a^2 = b^3 = (ab)^3 = 1\rangle.$$

Let $H = A_4'$ be the commutator subgroup. We use the above method to find a presentation for H. Now

$$G/H = A_4/A_4' = \langle a, b; a^2 = b^3 = (ab)^3 = [a, b] = 1\rangle = \langle b; b^3 = 1\rangle.$$

Therefore, $|A_4 : A_4'| = 3$. A Schreier system is then $\{1, b, b^2\}$. The generators for A_4' are then

$$X_1 = S_{1a} = a, \quad X_2 = S_{ba} = bab^{-1}, \quad X_3 = S_{b^2a} = b^2ab,$$

whereas the relations are the following:

1. $\tau(aa) = S_{1a}S_{1a} = X_1^2$
2. $\tau(baab^{-1}) = X_2^2$
3. $\tau(b^2aab^{-2}) = X_3^2$
4. $\tau(bbb) = 1$
5. $\tau(bbbbb^{-1}) = 1$
6. $\tau(b^2bbbb^{-2}) = 1$
7. $\tau(ababab) = S_{1a}S_{ba}S_{b^2a} = X_1X_2X_3$
8. $\tau(babababb^{-1}) = S_{ba}S_{b^2a}S_{1a} = X_2X_3X_1$
9. $\tau(b^2abababb^{-2}) = S_{b^2a}S_{1a}S_{ba} = X_3X_1X_2$

Therefore, after eliminating redundant relations and using $X_3 = X_1X_2$, we get as a presentation for A_4',

$$\langle X_1, X_2; X_1^2 = X_2^2 = (X_1X_2)^2 = 1\rangle.$$

Example 14.4.2. Let $F = \langle x, y; \rangle$ be the free group of rank 2. Let H be the commutator subgroup. Then

$$F/H = \langle x, y; [x, y] = 1\rangle = \mathbb{Z} \times \mathbb{Z}$$

a free Abelian group of rank 2. It follows that H has infinite index in F. As Schreier coset representatives, we can take

$$t_{m,n} = x^m y^n, \quad m = 0, \pm 1, \pm 2, \ldots, \; n = 0, \pm 1, \pm 2, \ldots.$$

The corresponding Schreier generators for H are

$$x_{m,n} = x^m y^n x^{-m} y^{-n}, \quad m = 0, \pm 1, \pm 2, \ldots, \; n = 0, \pm 1, \pm 2, \ldots.$$

The relations are only trivial; therefore, H is free on the countable infinitely many generators above. It follows that a free group of rank 2 contains as a subgroup a free group of countably infinite rank. Since a free group of countable infinite rank contains as subgroups free groups of all finite ranks, it follows that a free group of rank 2 contains as a subgroup a free subgroup of any arbitrary finite rank.

Theorem 14.4.3. *Let F be free of rank 2. Then the commutator subgroup F' is free of countable infinite rank. In particular, a free group of rank 2 contains as a subgroup a free group of any finite rank n.*

Corollary 14.4.4. *Let n, m be any pair of positive integers $n, m \geq 2$ and F_n, F_m free groups of ranks n, m, respectively. Then F_n can be embedded into F_m, and F_m can be embedded into F_n.*

14.5 Geometric Interpretation

Combinatorial group theory has its origins in topology and complex analysis. Especially important in the development is the theory of the fundamental group. This connection is so deep that many people consider combinatorial group theory as the study of the fundamental group—especially the fundamental group of a low-dimensional complex. This connection proceeds in both directions. The fundamental group provides methods and insights to study the topology. In the other direction, the topology can be used to study the groups.

Recall that if X is a topological space, then its *fundamental group* based at a point x_0, denoted by $\pi(X, x_0)$, is the group of all homotopy classes of closed paths at x_0. If X is path-connected, then the fundamental groups at different points are all isomorphic, and we can speak of the fundamental group of X, which we will denote by $\pi(X)$. Historically, group presentations were developed to handle the fundamental groups of spaces, which allowed simplicial or cellular decompositions. In these cases, the presentation of the fundamental group can be read off from the combinatorial decomposition of the space.

An (abstract) *simplicial complex* or *cell complex* K is a topological space consisting of a set of points called the *vertices*, which we will denote by $V(K)$, and collections of subsets of vertices called *simplexes* or *cells*, which have the property that the intersection of any two simplices is again a simplex. If n is the number of vertices in a cell, then $n - 1$ is called its *dimension*. Hence, the set of vertices are the 0-dimensional cells, and a simplex $\{v_1, \ldots, v_n\}$ is an $(n - 1)$-dimensional cell. The 1-dimensional cells are called *edges*. These have the form $\{u, v\}$, where u and v are vertices. One should think of the cells in a geometric manner so that the edges are really edges, the 2-cells are filled triangles (which are equivalent to disks), and so on. The maximum dimension of any cell in a complex K is called the dimension of K. From now on, we will assume that our simplicial complexes are path-connected.

A *graph* Γ is just a 1-dimensional simplicial complex. Hence, Γ consists of just vertices and edges. If K is any complex, then the set of vertices and edges is called the 1-*skeleton* of K. Similarly, all the cells of dimension less than or equal to 2 comprise the 2-skeleton. A connected graph with no closed paths in it is called a *tree*. If K is any complex, then a *maximal tree* in K is a tree that can be contained in no other tree within K.

From the viewpoint of combinatorial group theory what is relevant is that if K is a complex, then a presentation of its fundamental group can be determined from its 2-skeleton and read off directly. In particular the following hold:

Theorem 14.5.1. *Suppose that K is a connected cell complex. Suppose that T is a maximal tree within the 1-skeleton of K. Then a presentation for $\pi(K)$ can be determined in the following manner:*

Generators: *all edges outside of the maximal tree T.*

Relations: (a) $\{u, v\} = 1$ *if* $\{u, v\}$ *is an edge in T.*

(b) $\{u, v\}\{v, w\} = \{u, w\}$ *if u, v, w lie in a simplex of K.*

From this the following is obvious:

Corollary 14.5.2. *The fundamental group of a connected graph is free. Furthermore, its rank is the number of edges outside a maximal tree.*

A connected graph is homotopic to a wedge or bouquet of circles. If there are n circles in a bouquet of circles, then the fundamental group is free of rank n. The converse is also true. A free group can be realized as the fundamental group of a wedge of circles.

An important concept in applying combinatorial group theory is that of a *covering complex.*

Definition 14.5.3. Suppose that K is a complex. Then a complex K_1 is a covering complex for K if there exists a surjection $p : K_1 \to K$ called a *covering map* with the property that for any cell $s \in K$ the inverse image $p^{-1}(s)$ is a union of pairwise disjoint cells in K_1, and p restricted to any of the preimage cells is a homeomorphism.

That is, for each simplex S in K, we have

$$p^{-1}(S) = \bigcup \overline{S_i}$$

and $p : \overline{S_i} \to S$ is a bijection for each i.

The following then becomes clear:

Lemma 14.5.4. *If K_1 is a connected covering complex for K, then K_1 and K have the same dimension.*

What is crucial in using covering complexes to study the fundamental group is that there is a *Galois theory* of covering complexes and maps. The covering map p induces a homomorphism of the fundamental group, which we will also call p. Then we have the following:

Theorem 14.5.5. *Let K_1 be a covering complex of K with covering map p. Then $p(\pi(K_1))$ is a subgroup of $\pi(K)$. Conversely, to each subgroup H of $\pi(K)$, there is a covering complex K_1 with $\pi(K_1) = H$. Hence, there is a one-to-one correspondence between subgroups of the fundamental group of a complex K and covers of K.*

We will see the analog of this theorem in regard to algebraic field extensions in Chapter 15.

A topological space X is *simply connected* if $\pi(X) = \{1\}$. Hence, the covering complex of K corresponding to the identity in $\pi(K)$ is simply connected. This is called the *universal cover* of K since it covers any other cover of K.

Based on Theorem 14.5.1, we get a very simple proof of the Nielsen–Schreier theorem.

Theorem 14.5.6 (Nielsen–Schreier). *Any subgroup of a free group is free.*

Proof. Let F be a free group. Then $F = \pi(K)$, where K is a connected graph. Let H be a subgroup of F. Then H corresponds to a cover K_1 of K. But a cover is also 1-dimensional; hence, $H = \pi(K_1)$, where K_1 is a connected graph. Therefore, H is also free. □

The fact that a presentation of a fundamental group of a simplicial complex is determined by its 2-skeleton going in the other direction also. That is, given an arbitrary presentation, there exists a 2-dimensional complex, whose fundamental group has that presentation. Essentially, given a presentation $\langle X; R\rangle$, we consider a wedge of circles with cardinality $|X|$. We then paste on a 2-cell for each relator W in R bounded by the path corresponding to the word W.

Theorem 14.5.7. *Given an arbitrary presentation $\langle X; R\rangle$, there exists a connected 2-complex K with $\pi(K) = \langle X; R\rangle$.*

We note that the books by Rotman, see [43], and Fine, Moldenhauer, Rosenberger, and Wienke, see [26], have significantly detailed and accessible descriptions of groups and complexes. Cayley, and then Dehn, introduced for each group G a graph, now called Cayley graph, as a tool to apply complexes to the study of G. The Cayley graph is actually tied to a presentation, and not to the group itself. Gromov reversed the procedure and showed that by considering the geometry of the Cayley graph, one could get information about the group. This led to the development of the theory of hyperbolic groups.

In the following, we need a special kind of generating systems for finitely presented groups $G = \langle X; R\rangle$. Let $S \subset G$ be a generating system for G. Then S is called a *valid* generating system if it has the following two properties:

(a) $1 \notin S$ where 1 is the neutral element of G.

(b) the set S is a *symmetric generating system*, that is, if $\gamma \in S$ then also $\gamma^{-1} \in S$.

In the following, the pair (G, S) denotes a finitely presented group G together with a valid generating system S. Given such a pair we define a metric on G with respect to S in the following way. Let (G, S) be a pair as above. Then define $l_S\colon G \to [0, \infty)$ as follows: If $\gamma \in G$, then $l_S(\gamma) = 1$ if $\gamma = 1$, and if $\gamma \neq 1$ then let $l_S(\gamma)$ be the minimal length of a word that is completely constructed of elements from S that represent γ. This length is also called *S-length*.

We now define the desired metric $d_S: G \times G \to [0, \infty)$ via $d_S(\gamma_1, \gamma_2) = l_S(\gamma_1^{-1}\gamma_2)$ and check that d_S is indeed a metric:

1. The equivalence $l_S(\gamma) = 0$ if and only if $\gamma = 1$ implies the equivalence $d_S(\gamma_1, \gamma_2) = 0$ if and only if $\gamma_1 = \gamma_2$.
2. We have $d_S(\gamma_1, \gamma_2) = l_S(\gamma_1^{-1}\gamma_2) = l_S(\gamma_2^{-1}\gamma_1) = d_S(\gamma_2, \gamma_1)$, because S is symmetric.
3. We have $d_S(\gamma_1, \gamma_2) \leq d_S(\gamma_1, \beta) + d_S(\beta, \gamma_2)$ for all $\gamma_1, \gamma_2, \beta \in G$ as $\gamma_1^{-1}\gamma_2 = \gamma_1^{-1}\beta\beta^{-1}\gamma_2$.

We give the following remarks.

1. The metric structure on (G, S) depends on the choice of S. Say $G = \mathbb{Z}$ and $S = \{\pm 1\}$, then $d_S(0, 1) = 1$, and if $S' = \{\pm 2, \pm 3\}$, then $d_{S'}(0, 1) = 2$.
2. The metric structure on (G, S) is induced by the natural metric structure of the *Cayley graph with respect to* (G, S):
   The vertices are elements of G, and two vertices γ_1 and γ_2 are connected by an edge if and only if there exists a $\sigma \in S$ with $\gamma_1\sigma = \gamma_2$. Since $\gamma_1 = \gamma_2\sigma^{-1}$ we get in fact a directed graph called the Cayley graph with respect to (G, S).
   If we parametrize in such a way that any edge of the Cayley graph of (G, S) has length 1, then the metric of (G, S) is induced from that of the Cayley graph of (G, S). Here we extend the metric for the Cayley graph in the usual way for all pairs of points of edges by transforming any edge to an interval of length 1. In this manner the Cayley graph becomes a geodesic metric space. We always consider the Cayley graph in this way which should not lead to misunderstandings. Any closed path represents a relation. If $G = \langle X; R \rangle$ is finitely presented with $1 \notin X$ then we may consider $S = X \cup X^{-1}$ and may call (G, S) the Cayley graph of G without misunderstandings. If we insert a 2-cell for any closed path in the Cayley graph then we obtain a simply connected 2-dimensional complex, the *Cayley complex*.

The construction of the Cayley graph depends on the choice of S as well as on the metric on (G, S). We would like to have an equivalence relation that permits to connect the different metric spaces for G if we alter S.

Definition 14.5.8. Let (X, d) and (X', d') be metric spaces. Then (X, d) are (X', d') are *quasi-isometric*, if there are functions $f: X \to X'$ and $g: X' \to X$ together with constants $\lambda > 0$ and $C \geq 0$, such that

(a) $d'(f(x), f(y)) \leq \lambda d(x, y) + C$ for all $x, y \in X$,
(b) $d(g(x'), g(y')) \leq \lambda d'(x', y') + C$ for all $x', y' \in X'$,
(c) $d(g(f(x)), x) \leq C$ for all $x \in X$, and
(d) $d'(f(g(x')), x') \leq C$ for all $x' \in X'$.

Theorem 14.5.9. *Quasi-isometry is an equivalence relation in the class of metric spaces.*

Proof. Of course quasi-isometry is reflexive and symmetric. We show transitivity. Let (X, d) and (X', d') as well as (X', d') and (X'', d'') be quasi-isometric. Thus we have functions $X \underset{g}{\overset{f}{\rightleftarrows}} X'$, $X' \underset{g'}{\overset{f'}{\rightleftarrows}} X''$, and constants λ, C and λ', C' respectively, such that the conditions (a)–(d) are satisfied. We look for functions $X \underset{g''}{\overset{f''}{\rightleftarrows}} X''$ and constants λ'', C'', such that conditions (a)–(d) are satisfied again. Set $f'' = f' \circ f$ and $g'' = g \circ g'$, $\lambda'' = \lambda\lambda'$ and $C'' = 2C + 2C' + \lambda' C + \lambda C'$. We check the conditions step by step:

(a) Let $x, y \in X$. Then

$$\begin{aligned} d''(f''(x), f''(y)) &= d''(f'(f(x)), f'(f(y))) \\ &\le \lambda' d'(f(x), f(y)) + C' \\ &\le \lambda'(\lambda d(x,y) + C) + C' \\ &= \lambda'\lambda d(x,y) + \lambda' C + C' \\ &\le \lambda'' d(x,y) + C''. \end{aligned}$$

(b) This is analogous.

(c) Let $x \in X$. Then (according to our assumption)

$$d'(g' \circ f'(f(x)), f(x)) \le C',$$

and hence, because of (b),

$$d(g(g' \circ f'(f(x))), g(f(x))) \le \lambda C' + C$$

which gives

$$\begin{aligned} d(g'' \circ f''(x), x) &\le d(g'' \circ f''(x), g(f(x))) + d(g(f(x)), x) \\ &\le (\lambda C' + C) + C = \lambda C' + 2C \\ &\le C''. \end{aligned}$$

(d) This is analogous. □

Theorem 14.5.10. *Let G be a group of finitely presented group with finite valid generating systems S and S′. Then the metric spaces (G, S) and (G, S') are quasi-isometric.*

Proof. We look for suitable f, g, λ, and C. Take $f = \mathrm{id}_{(G,S)}$, $g = \mathrm{id}_{(G,S')}$, $C = 0$, and $\lambda = \max(\{l_{S'}(y) : y \in S\} \cup \{l_S(y') : y' \in S'\})$.

We verify condition (a). Let $x, y \in (G, S)$. Then

$$d_{S'}(f(x), f(y)) = l_{S'}(f(x)^{-1}f(y)) = l_{S'}(x^{-1}y).$$

Our definition of λ permits $l_{S'}(x^{-1}y) \le \lambda l_S(x^{-1}y)$ because, if we write $x^{-1}y$ as a product of elements of S with length k, then we can surely write $x^{-1}y$ as a product (of elements of S') of length $\le \lambda k$. Hence

$$d_{S'}(f(x), f(y)) \le \lambda l_S(x^{-1}y) = \lambda d_S(x, y) + C.$$

The proof of (b) is analogous and that of (c) and (d) is obvious because f and g are inverses for each other. □

We observe: The quasi-isometry class of the metric spaces for (G, S), S finite, is an invariant of the group G and does not depend on the finite generating set S.

We ask: Is this invariant suitable in order to study group theoretical properties of G and to what extent does quasi-isometry preserve group theoretic properties?

We call two finitely presented groups G_1 and G_2 *quasi-isometric*, if the metric spaces for (G_1, S_1), S_1 a valid generating set for G_1, and (G_2, S_2), S_2 a valid generating set for G_2, are quasi-isometric.

Aiming at the motivation of hyperbolic groups we first have to describe a hyperbolic metric space.

Definition 14.5.11. Let (X, d) be a metric space.

1. Let $x_0, x_1 \in X$ with $a = (x_1 - x_0)$. A *geodesic segment* in X starting at x_0 and ending in x_1 is an isometry $g\colon [0, a] \to X$ with $g(0) = x_0$ and $g(a) = x_1$ (recall that an isometry is by definition length preserving). We say that X is a *geodesic space* if for all $x_0, x_1 \in X$ there is a geodesic segment in X starting at x_0 and ending at x_1.
2. A *geodesic triangle* in X with $x, y, z \in X$ as vertices is the union of three geodesic segments with (pairwise) x, y and z as end points.

Note that the definition explicitly allows degenerated triangles, for instance, take $y = z$ and the geodesic segments from x to y and x to z are different.

An example of a geodesic space is the Cayley graph for a finitely presented group. If the Cayley graph is not a tree, then it contains a circle (or embedded loop). Hence, there is more than one geodesic segment allowed between the same pair of points.

We fix the following notation: Let $x_0, x_1 \in X$ for a geodesic space X. Although several geodesic segments in X with start points x_0 and end points x_1 are allowed, we denote by $[x_0, x_1]$ a given geodesic segment with x_0 and x_1 as start and end points.

Definition 14.5.12. Let $\delta \geq 0$. We say that a geodesic space X satisfies the *Rips condition* for the constant δ if for every geodesic triangle $[x, y] \cup [y, z] \cup [z, x]$ in X and for every $u \in [x, y]$ the following holds: $d(u, [y, z] \cup [z, x]) \leq \delta$, see Figure 14.1. We call a geodesic space X *hyperbolic* if it satisfies the Rips condition for a constant $\delta \geq 0$.

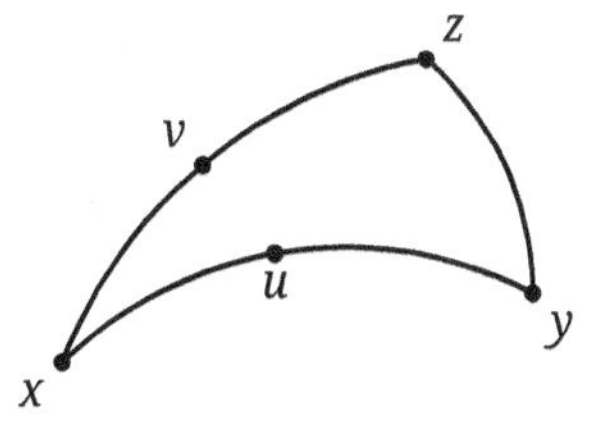

Figure 14.1: Geodesic triangle.

Theorem 14.5.13. *Let X_1 and X_2 be geodesic spaces that are quasi-isometric. If X_1 is hyperbolic then also X_2 is hyperbolic.*

A proof is given in [26]. Hence, quasi-isometries respect hyperbolicity.

Definition 14.5.14. Let Γ be a group of finite type. Γ is called *hyperbolic group* if there is a finite generating system S such that the metric space for (Γ, S)—or the Cayley graph for (Γ, S)—is a hyperbolic space.

According to Theorem 14.5.13 the definition of a hyperbolic group is independent of the choice of a finite generating system X.

Theorem 14.5.15.
1. *Let G_1 be a subgroup of a hyperbolic group G_2 with finite index. Then G_1 is also hyperbolic.*
2. *Let $1 \to \Delta \to G_1 \to G_2 \to 1$ be a short exact sequence with Δ finite and G_2 hyperbolic, that is, $G_2 \cong G_1/\Delta$. Then G_1 is also hyperbolic.*

A proof is given in [26]. Hyperbolic groups have many other important properties (see, for instance, [26]). We end this section with a collection of examples of hyperbolic groups.

Example 14.5.16. The following groups are hyperbolic. For proofs see [26].
1. Finite groups and infinite cyclic groups.
2. Fundamental groups of compact, connected Riemann manifolds. Especially, cocompact Fuchsian and Kleinian groups.
3. One-relator groups with torsion.
4. Free products of finitely many hyperbolic groups, see Section 14.8.
5. A group G of F-type is a group with a presentation

$$G = \langle a_1, \ldots, a_n; a_1^{r_1} = \cdots = a_n^{r_n} = u(a_1, \ldots, a_p)v(a_{p+1}, \ldots, a_n) = 1 \rangle$$

   where $n \geq 2$, $r_i = 0$ or $r_i \geq 2$, $1 \leq p \leq n-1$, $u(a_1, \ldots, a_n)$ a cyclically reduced word in the free product on $a_1, \ldots, a_p$ which is of infinite order, and $v(a_{p+1}, \ldots, a_n)$ is a cyclically reduced word in the free product on $a_{p+1}, \ldots, a_n$ which is of infinite order (see Section 14.8). The group G is hyperbolic unless $u(a_1, \ldots, a_p)$ is a proper power or a product of two elements of order 2 and $v(a_{p+1}, \ldots, a_n)$ also is a proper power or a product of two elements of order 2.
   Especially, free groups of finite rank, oriented surface groups of genus $g \geq 2$ and nonoriented surface groups of genus $g \geq 3$ are of F-type. We remark that a group of F-type is hyperbolic if and only if it has a faithful representation in $\mathrm{PSL}(2, \mathbb{R})$.

14.6 Presentations of Factor Groups

Let G be a group with a presentation $G = \langle X; R \rangle$. Suppose that H is a factor group of G; that is, $H \cong G/N$ for some normal subgroup N of G. We show that a presentation for H is then $H = \langle X; R \cup R_1 \rangle$, where R_1 is a, perhaps additional, system of relators.

Theorem 14.6.1 (Dyck's theorem). *Let $G = \langle X; R \rangle$, and suppose that $H \cong G/N$, where N is a normal subgroup of G. Then a presentation for H is $\langle X; R \cup R_1 \rangle$ for some set of words R_1 on X. Conversely, the presentation $\langle X; R \cup R_1 \rangle$ defines a group, that is, a factor group of G.*

Proof. Since each element of H is a coset of N, they have the form gN for $g \in G$. It is clear then that the images of X generate H. Furthermore, since H is a homomorphic image of G, each relator in R is a relator in H. Let N_1 be a set of elements that generate N, and let R_1 be the corresponding words in the free group on X. Then R_1 is an additional set of relators in H. Hence, $R \cup R_1$ is a set of relators for H. Any relator in H is either a relator in G, hence a consequence of R, or can be realized as an element of G that lies in N, and therefore a consequence of R_1. Therefore, $R \cup R_1$ is a complete set of defining relators for H, and H has the presentation $H = \langle X; R \cup R_1 \rangle$.

Conversely, $G = \langle X; R \rangle$, $G_1 = \langle X; R \cup R_1 \rangle$. Then $G = F(X)/N_1$, where $N_1 = N(R)$, and $G_1 = F(X)/N_2$, where $N_2 = N(R \cup R_1)$. Hence, $N_1 \subset N_2$. The normal subgroup N_2/N_1 of $F(X)/N_1$ corresponds to a normal subgroup of H of G, and therefore by the isomorphism theorem

$$G/H \cong (F(X)/N_1)/(N_2/N_1) \cong F(X)/N_2 \cong G_1. \qquad \square$$

14.7 Decision Problems

We have seen that given any group G, there exists a presentation for it, $G = \langle X; R \rangle$. In the other direction, given any presentation $\langle X; R \rangle$, we have seen that there is a group with that presentation. In principle, every question about a group can be answered via a presentation. However, things are not that simple. Max Dehn in his pioneering work on combinatorial group theory about 1910 introduced the following three fundamental *group decision problems*:

(1) *Word Problem*: Suppose G is a group given by a finite presentation. Is there an algorithm to determine if an arbitrary word w in the generators of G defines the identity element of G?

(2) *Conjugacy Problem*: Suppose G is a group given by a finite presentation. Is there an algorithm to determine if an arbitrary pair of words u, v in the generators of G define conjugate elements of G?

(3) *Isomorphism Problem*: Is there an algorithm to determine, given two arbitrary finite presentations, whether the groups they present are isomorphic or not?

All three of these problems have negative answers in general. That is, for each of these problems one can find a finite presentation, for which these questions cannot be answered algorithmically (see [36]). Attempts for solutions, and for solutions in restricted cases, have been of central importance in combinatorial group theory. For this reason combinatorial group theory has always searched for and studied classes of groups, in which these decision problems are solvable.

For finitely generated free groups, there are simple and elegant solutions to all three problems. If F is a free group on $x_1, \ldots, x_n$ and W is a freely reduced word in $x_1, \ldots, x_n$, then $W \neq 1$ if and only if $L(W) \geq 1$ for $L(W)$ the length of W. Since freely reducing any word to a freely reduced word is algorithmic, this provides a solution to the word problem. Furthermore, a freely reduced word $W = x_{v_1}^{e_1} x_{v_2}^{e_2} \cdots x_{v_n}^{e_n}$ is *cyclically reduced* if $v_1 \neq v_n$, or if $v_1 = v_n$, then $e_1 \neq -e_n$. Clearly then, every element of a free group is conjugate to an element given by a cyclically reduced word called a cyclic reduction. This leads to a solution to the conjugacy problem. Suppose V and W are two words in the generators of F and $\overline{V}, \overline{W}$ are respective cyclic reductions. Then V is conjugate to W if and only if $\overline{V}$ is a cyclic permutation of $\overline{W}$. Finally, two finitely generated free groups are isomorphic if and only if they have the same rank.

14.8 Group Amalgams: Free Products and Direct Products

Closely related to free groups in both form and properties are free products of groups. Let $A = \langle a_1, \ldots ; R_1, \ldots \rangle$ and $B = \langle b_1, \ldots ; S_1, \ldots \rangle$ be two groups. We consider A and B to be disjoint. Then we have the following:

Definition 14.8.1. The *free product* of A and B, denoted by $A * B$, is the group G with the presentation $\langle a_1, \ldots, b_1, \ldots ; R_1, \ldots, S_1, \ldots \rangle$; that is, the generators of G consist of the disjoint union of the generators of A and B with relators taken as the disjoint union of the relators R_i of A and S_j of B. A and B are called the *factors* of G.

In an analogous manner, the concept of a free product can be extended to an arbitrary collection of groups.

Definition 14.8.2. If $A_\alpha = \langle \text{gens}\, A_\alpha ; \text{rels}\, A_\alpha \rangle$, $\alpha \in \mathcal{I}$, is a collection of groups, then their free product $G = *A_\alpha$ is the group, whose generators consist of the disjoint union of the generators of the A_α, and whose relators are the disjoint union of the relators of the A_α.

Free products exist and are nontrivial. In that regard, we have the following:

Theorem 14.8.3. *Let $G = A * B$. Then the maps $A \to G$ and $B \to G$ are injections. The subgroup of G generated by the generators of A has the presentation $\langle$generators of A; relators of $A\rangle$, that is, is isomorphic to A. Similarly for B. Thus, A and B can be considered as subgroups of G. In particular, $A * B$ is nontrivial if A and B are.*

Free products share many properties with free groups. First of all there is a categorical formulation of free products. Specifically we have the following:

Theorem 14.8.4. *A group G is the free product of its subgroups A and B if A and B generate G, and given homomorphisms $f_1 : A \to H, f_2 : B \to H$ into a group H, there exists a unique homomorphism $f : G \to H$, extending f_1 and f_2.*

Secondly, each element of a free product has a *normal form* related to the reduced words of free groups. If $G = A * B$, then a *reduced sequence* or *reduced word* in G is a sequence $g_1 g_2 \dots g_n$, $n \geq 0$, with $g_i \neq 1$, each g_i in either A or B and g_i, g_{i+1} not both in the same factor. Then the following hold:

Theorem 14.8.5. *Each element $g \in G = A * B$ has a unique representation as a reduced sequence. The length n is unique and is called the syllable length. The case $n = 0$ is reserved for the identity.*

A reduced word $g_1 \dots g_n \in G = A * B$ is called *cyclically reduced* if either $n \leq 1$ or $n \geq 2$ and g_1 and g_n are from different factors. Certainly, every element of G is conjugate to a cyclically reduced word.

From this, we obtain several important properties of free products, which are analogous to properties in free groups.

Theorem 14.8.6. *An element of finite order in a free product is conjugate to an element of finite order in a factor. In particular, a finite subgroup of a free product is entirely contained in a conjugate of a factor.*

Theorem 14.8.7. *If two elements of a free product commute, then they are both powers of a single element or are contained in a conjugate of an Abelian subgroup of a factor.*

Finally, a theorem of Kurosh extends the Nielsen–Schreier theorem to free products.

Theorem 14.8.8 (Kurosh). *A subgroup of a free product is also a free product. Explicitly, if $G = A * B$ and $H \subset G$, then*

$$H = F * (*A_\alpha) * (*B_\beta),$$

where F is a free group, $(*A_\alpha)$ is a free product of conjugates of subgroups of A, and $(*B_\beta)$ is a free product of conjugates of subgroups of B.

We note that the rank of F and the number of the other factors can be computed. A complete discussion of these is in [37], [36] and [21].

If A and B are disjoint groups, then we now have two types of products forming new groups out of them: the free product and the direct product. In both these products, the original factors inject. In the free product, there are no relations between elements of A and elements of B, whereas in a direct product, each element of A commutes with each element of B. If $a \in A$ and $b \in B$, a *cross commutator* is $[a, b] = aba^{-1}b^{-1}$. The direct product is a factor group of the free product, and the kernel is precisely the normal subgroup generated by all the cross commutators.

Theorem 14.8.9. *Suppose that A and B are disjoint groups. Then*

$$A \times B = (A \star B)/H,$$

where H is the normal closure in $A \star B$ of all the cross commutators. In particular, a presentation for $A \times B$ is given by

$$A \times B = \langle \text{gens}\, A, \text{gens}\, B; \text{rels}\, A, \text{rels}\, B, [a, b] \textit{ for all } a \in A, b \in B\rangle.$$

This coincides with the concept in Section 10.3.

14.9 Exercises

1. Let X^{-1} be a set disjoint from X, but bijective to X. A *word* in X is a finite sequence of letters from the alphabet. That is, a word has the form

$$w = x_{i_1}^{\epsilon_{i_1}} x_{i_2}^{\epsilon_{i_2}} \cdots x_{i_n}^{\epsilon_{i_n}},$$

   where $x_{i_j} \in X$, and $\epsilon_{i_j} = \pm 1$. Let $W(X)$ be the set of all words on X.
   If $w_1, w_2 \in W(X)$, we say that w_1 is *equivalent* to w_2, denoted by $w_1 \sim w_2$, if w_1 can be converted to w_2 by a finite string of insertions and deletions of trivial words. Verify that this is an equivalence relation on $W(X)$.
2. In $F(X)$, let $N(X)$ be the subgroup generated by all squares in $F(X)$; that is,

$$N(X) = \langle \{g^2 : g \in F(X)\}\rangle.$$

   Show that $N(X)$ is a normal subgroup, and that the factor group $F(X)/N(X)$ is Abelian, where every nontrivial element has order 2.
3. Show that a free group F is torsion-free.
4. Let F be a free group, and $a, b \in F$. Show: If $a^k = b^k$, $k \neq 0$, then $a = b$.
5. Let $F = \langle a, b; \ \rangle$ be a free group with basis $\{a, b\}$. Let $c_i = a^{-i}ba^i$, $i \in \mathbb{Z}$. Show that then $G = \langle c_i, i \in \mathbb{Z}\rangle$ is free with basis $\{c_i \mid i \in \mathbb{Z}\}$.
6. Show that $\langle x, y; x^2y^3, x^3y^4\rangle \cong \langle x; x\rangle = \{1\}$.
7. Let $G = \langle v_1, \ldots, v_n; v_1^2 \cdots v_n^2\rangle$, $n \geq 1$, and $\alpha : G \to \mathbb{Z}_2$ be the epimorphism with $\alpha(v_i) = -1$ for all i. Let U be the kernel of α. Show that then U has a presentation

$$U = \langle x_1, \ldots, x_{n-1}, y_1, \ldots, y_{n-1}; y_1x_1 \cdots y_{n-1}x_{n-1}y_{n-1}^{-1}x_{n-1}^{-1} \cdots y_1^{-1}x_1^{-1}\rangle.$$

8. Let $M = \langle x, y; x^2, y^3\rangle \cong \mathrm{PSL}(2, \mathbb{Z})$ be the modular group. Let M' be the commutator subgroup. Show that M' is a free group of rank 2 with a basis $\{[x, y], [x, y^2]\}$.

15 Finite Galois Extensions

15.1 Galois Theory and the Solvability of Polynomial Equations

As we mentioned in Chapter 1, one of the origins of abstract algebra was the problem of trying to determine a formula for finding the solutions in terms of radicals of a fifth degree polynomial. It was proved first by Ruffini in 1800 and then by Abel that, in general, it is impossible to find a formula in terms of radicals for such a solution. In 1820, Galois extended this and showed that such a formula is impossible for any degree five or greater. In proving this, he laid the groundwork for much of the development of modern abstract algebra, especially field theory and finite group theory. One of the goals of this book has been to present a comprehensive treatment of Galois theory and a proof of the results mentioned above. At this point, we have covered enough general algebra and group theory to discuss Galois extensions and general Galois theory.

In modern terms, *Galois theory* is that branch of mathematics, which deals with the interplay of the algebraic theory of fields, the theory of equations, and finite group theory. This theory was introduced by Evariste Galois about 1830 in his study of the insolvability by radicals of quintic (degree 5) polynomials, a result proved somewhat earlier by Ruffini, and independently by Abel. Galois was the first to see the close connection between field extensions and permutation groups. In doing so, he initiated the study of finite groups. He was the first to use the term group as an abstract concept, although his definition was really just for a closed set of permutations.

The method Galois developed not only facilitated the proof of the insolvability of the quintic and higher powers, but led to other applications, and to a much larger theory.

The main idea of Galois theory is to associate to certain special types of algebraic field extensions called *Galois extensions*, a group called the *Galois group*. The properties of the field extension will be reflected in the properties of the group, which are somewhat easier to examine. Thus, for example, solvability by radicals can be translated into solvability of groups, which was discussed in Chapter 12. Showing that for every polynomial of degree five or greater, there exists a field extension whose Galois group is not solvable proves that there cannot be a general formula for solvability by radicals.

The tie-in to the theory of equations is as follows: If $f(x) = 0$ is a polynomial equation over some field K, we can form the splitting field K. This is usually a Galois extension, and therefore has a Galois group called the *Galois group of the equation*. As before, properties of this group will reflect properties of this equation.

15.2 Automorphism Groups of Field Extensions

To define the Galois group, we must first consider the automorphism group of a field extension. In this section, K, L, M will always be (commutative) fields with additive identity 0 and multiplicative identity 1.

https://doi.org/10.1515/9783111142524-015

Definition 15.2.1. Let $L|K$ be a field extension. Then the set

$$\mathrm{Aut}(L|K) = \{\alpha \in \mathrm{Aut}(L) : \alpha_{|K} = \text{the identity on } K\}$$

is called the set of *automorphisms of* L *over* K. Notice that if $\alpha \in \mathrm{Aut}(L|K)$, then $\alpha(k) = k$ for all $k \in K$.

Lemma 15.2.2. *Let* $L|K$ *be a field extension. Then* $\mathrm{Aut}(L|K)$ *forms a group called the Galois group of* $L|K$.

Proof. $\mathrm{Aut}(L|K) \subset \mathrm{Aut}(L)$. Hence, to show that $\mathrm{Aut}(L|K)$ is a group, we only have to show that its a subgroup of $\mathrm{Aut}(L)$. Now the identity map on L is certainly the identity map on K, so $1 \in \mathrm{Aut}(L|K)$; hence, $\mathrm{Aut}(L|K)$ is nonempty. If $\alpha, \beta \in \mathrm{Aut}(L|K)$, then consider $\alpha^{-1}\beta$. If $k \in K$, then $\beta(k) = k$, and $\alpha(k) = k$, so $\alpha^{-1}(k) = k$.

Therefore, $\alpha^{-1}\beta(k) = k$ for all $k \in K$, and hence $\alpha^{-1}\beta \in \mathrm{Aut}(L|K)$. It follows that $\mathrm{Aut}(L|K)$ is a subgroup of $\mathrm{Aut}(L)$, and therefore a group. □

If $f(x) \in K[x] \setminus K$ and L is the splitting field of $f(x)$ over K, then $\mathrm{Aut}(L|K)$ is also called the *Galois group* of $f(x)$.

Theorem 15.2.3. *If* P *is the prime field of* L, *then* $\mathrm{Aut}(L|P) = \mathrm{Aut}(L)$.

Proof. We must show that any automorphism of a prime field P is the identity. Now if $\alpha \in \mathrm{Aut}(L)$, then $\alpha(1) = 1$, and so $\alpha(n \cdot 1) = n \cdot 1$. Therefore, in P, α fixes all integer multiples of the identity. However, every element of P can be written as a quotient $\frac{m \cdot 1}{n \cdot 1}$ of integer multiples of the identity. Since α is a field homomorphism and α fixes both the top and the bottom, it follows that α will fix every element of this form, and hence fix each element of P. □

For splitting fields, the Galois group is a permutation group on the zeros of the defining polynomial.

Theorem 15.2.4. *Let* $f(x) \in K[x]$ *and* L *the splitting field of* $f(x)$ *over* K. *Suppose that* $f(x)$ *has zeros* $\alpha_1, \dots, \alpha_n \in L$.

(a) *Then each* $\phi \in \mathrm{Aut}(L|K)$ *is a permutation on the zeros. In particular,* $\mathrm{Aut}(L|K)$ *is isomorphic to a subgroup of* S_n *and uniquely determined by the zeros of* $f(x)$.
(b) *If* $f(x)$ *is irreducible, then* $\mathrm{Aut}(L|K)$ *operates transitively on* $\{\alpha_1, \dots, \alpha_n\}$. *Hence, for each* i, j, *there is a* $\phi \in \mathrm{Aut}(L|K)$ *such that* $\phi(\alpha_i) = \alpha_j$.
(c) *If* $f(x) = b(x - \alpha_1) \cdots (x - \alpha_n)$ *with* $\alpha_1, \dots, \alpha_n$ *pairwise distinct and* $\mathrm{Aut}(L|K)$ *operates transitively on* $\alpha_1, \dots, \alpha_n$, *then* $f(x)$ *is irreducible.*

Proof. For the proofs, we use the results of Chapter 8.

For (a), let $\phi \in \mathrm{Aut}(L|K)$. Then, from Theorem 8.1.5, we obtain that ϕ permutes the zeros $\alpha_1, \dots, \alpha_n$. Hence, $\phi_{|\{\alpha_1,\dots,\alpha_n\}} \in S_n$. This map then defines a homomorphism

$$\tau : \mathrm{Aut}(L|K) \to S_n \quad \text{by } \tau(\phi) = \phi_{|\{\alpha_1,\dots,\alpha_n\}}.$$

Furthermore, ϕ is uniquely determined by the images $\phi(a_i)$. It follows that τ is a monomorphism.

We now prove (b). If $f(x)$ is irreducible, then $\operatorname{Aut}(L|K)$ operates transitively on the set $\{a_1, \dots, a_n\}$, again following from Theorem 8.1.5.

Finally, for (c), suppose that $f(x) = b(x - a_1) \cdots (x - a_n)$ with $a_1, \dots, a_n$ distinct and $f \in \operatorname{Aut}(L|K)$ operates transitively on $a_1, \dots, a_n$. Now, assume that $f(x) = g(x)h(x)$ with $g(x), h(x) \in K[x] \setminus K$. Without loss of generality, let a_1 be a zero of $g(x)$ and a_n be a zero of $h(x)$.

Let $\alpha \in \operatorname{Aut}(L|K)$ with $\alpha(a_1) = a_n$. However, $\alpha(g(x)) = g(x)$; that is, $\alpha(a_1)$ is a zero of $\alpha(g(x)) = g(x)$, which gives a contradiction since a_n is not a zero of $g(x)$. Therefore, $f(x)$ must be irreducible. □

Example 15.2.5. Let $f(x) = (x^2 - 2)(x^2 - 3) \in \mathbb{Q}[x]$. The field $L = \mathbb{Q}(\sqrt{2}, \sqrt{3})$ is the spitting field of $f(x)$.

Over L, we have

$$f(x) = (x + \sqrt{2})(x - \sqrt{2})(x + \sqrt{3})(x - \sqrt{3}).$$

We want to determine the Galois group $\operatorname{Aut}(L|\mathbb{Q}) = \operatorname{Aut}(L) = G$.

Lemma 15.2.6. *The Galois group G above is the Klein 4-group.*

Proof. First, we show that $|\operatorname{Aut}(L)| \le 4$. Let $\alpha \in \operatorname{Aut}(L)$. Then α is uniquely determined by $\alpha(\sqrt{2})$ and $\alpha(\sqrt{3})$, and

$$\alpha(2) = 2 = (\sqrt{2})^2 = \alpha(\sqrt{2}^2) = (\alpha(\sqrt{2}))^2.$$

Hence, $\alpha(\sqrt{2}) = \pm\sqrt{2}$. Analogously, $\alpha(\sqrt{3}) = \pm\sqrt{3}$. From this it follows that $|\operatorname{Aut}(L)| \le 4$. Furthermore, $\alpha^2 = 1$ for any $\alpha \in G$.

Next we show that the polynomial $f(x) = x^2 - 3$ is irreducible over $K = \mathbb{Q}(\sqrt{2})$. Assume that $x^2 - 3$ were reducible over K. Then $\sqrt{3} \in K$. This implies that $\sqrt{3} = \frac{a}{b} + \frac{c}{d}\sqrt{2}$ with $a, b, c, d \in \mathbb{Z}$ and $b \ne 0 \ne d$, and $\gcd(c, d) = 1$. Then $bd\sqrt{3} = ad + bc\sqrt{2}$, hence $3b^2d^2 = a^2b^2 + 2b^2c^2 + 2\sqrt{2}adbc$. Since $bd \ne 0$, this implies that we must have $ac = 0$. If $c = 0$, then $\sqrt{3} = \frac{a}{b} \in \mathbb{Q}$, a contradiction. If $a = 0$, then $\sqrt{3} = \frac{c}{d}\sqrt{2}$, which implies $3d^2 = 2c^2$. It follows from this that $3 \mid \gcd(c, d) = 1$, again a contradiction.

Hence $f(x) = x^2 - 3$ is irreducible over $K = \mathbb{Q}(\sqrt{2})$.

Since L is the splitting field of $f(x)$ and $f(x)$ is irreducible over K, then there exists an automorphism $\alpha \in \operatorname{Aut}(L)$ with $\alpha(\sqrt{3}) = -\sqrt{3}$ and $\alpha_{|K} = I_K$; that is, $\alpha(\sqrt{2}) = \sqrt{2}$. Analogously, there is a $\beta \in \operatorname{Aut}(L)$ with $\beta(\sqrt{2}) = -\sqrt{2}$ and $\beta(\sqrt{3}) = \sqrt{3}$.

Clearly, $\alpha \ne \beta$, $\alpha\beta = \beta\alpha$ and $\alpha \ne \alpha\beta \ne \beta$. It follows that $\operatorname{Aut}(L) = \{1, \alpha, \beta, \alpha\beta\}$, completing the proof. □

15.3 Finite Galois Extensions

We now define (finite) Galois extensions. First, we introduce the concept of a *fix field*. Let K be a field and G a subgroup of $\mathrm{Aut}(K)$. Define the set

$$\mathrm{Fix}(K,G) = \{k \in K : g(k) = k \text{ for all } g \in G\}.$$

Theorem 15.3.1. *For a $G \subset \mathrm{Aut}(K)$, the set $\mathrm{Fix}(K,G)$ is a subfield of K called the fix field of G over K.*

Proof. $1 \in K$ is in $\mathrm{Fix}(K,G)$, so $\mathrm{Fix}(K,G)$ is not empty. Let $k_1, k_2 \in \mathrm{Fix}(K,G)$, and let $g \in G$. Then $g(k_1 \pm k_2) = g(k_1) \pm g(k_2)$ since g is an automorphism.

Then $g(k_1) \pm g(k_2) = k_1 \pm k_2$, and it follows that $k_1 \pm k_2 \in \mathrm{Fix}(K,G)$. In an analogous manner, $k_1 k_2^{-1} \in \mathrm{Fix}(K,G)$ if $k_2 \neq 0$; therefore, $\mathrm{Fix}(K,G)$ is a subfield of K. □

Using the concept of a fix field, we define a finite Galois extension.

Definition 15.3.2. The extension $L|K$ is a *(finite) Galois extension* if there exists a finite subgroup $G \subset \mathrm{Aut}(L)$ such that $K = \mathrm{Fix}(L,G)$.

We now give some examples of finite Galois extensions:

Lemma 15.3.3. *Let $L = \mathbb{Q}(\sqrt{2}, \sqrt{3})$ and $K = \mathbb{Q}$. Then $L|K$ is a Galois extension.*

Proof. Let $G = \mathrm{Aut}(L|K)$. From the example in the previous section, there are automorphisms $\alpha, \beta \in G$ with

$$\alpha(\sqrt{3}) = -\sqrt{3}, \quad \alpha(\sqrt{2}) = \sqrt{2} \quad \text{and} \quad \beta(\sqrt{2}) = -\sqrt{2}, \quad \beta(\sqrt{3}) = \sqrt{3}.$$

We have

$$\mathbb{Q}(\sqrt{2}, \sqrt{3}) = \{c + d\sqrt{3} : c, d \in \mathbb{Q}(\sqrt{2})\}.$$

Let $t = a_1 + b_1\sqrt{2} + (a_2 + b_2\sqrt{2})\sqrt{3} \in \mathrm{Fix}(L,G)$.

Then applying β, we have

$$t = \beta(t) = a_1 - b_1\sqrt{2} + (a_2 - b_2\sqrt{2})\sqrt{3}.$$

It follows that $b_1 + b_2\sqrt{3} = 0$; that is, $b_1 = b_2 = 0$ since $\sqrt{3} \notin \mathbb{Q}$. Therefore, $t = a_1 + a_2\sqrt{3}$. Applying α, we have $\alpha(t) = a_1 - a_2\sqrt{3}$, and hence $a_2 = 0$. Therefore, $t = a_1 \in \mathbb{Q}$. Hence $\mathbb{Q} = \mathrm{Fix}(L,G)$, and $L|K$ is a Galois extension. □

Lemma 15.3.4. *Let $L = \mathbb{Q}(2^{\frac{1}{4}})$ and $K = \mathbb{Q}$. Then $L|K$ is not a Galois extension.*

Proof. Suppose that $\alpha \in \mathrm{Aut}(L)$ and $a = 2^{\frac{1}{4}}$. Then a is a zero of $x^4 - 2$, and hence $\alpha(a) = 2^{\frac{1}{4}}$ or $\alpha(a) = i2^{\frac{1}{4}} \notin L$ since $i \notin L$ or $\alpha(a) = -2^{\frac{1}{4}}$ or $\alpha(a) = -i2^{\frac{1}{4}} \notin L$ since $i \notin L$. In particular, $\alpha(\sqrt{2}) = \sqrt{2}$; therefore,

$$\mathrm{Fix}(L, \mathrm{Aut}(L)) = \mathbb{Q}(\sqrt{2}) \neq \mathbb{Q}.$$

□

15.4 The Fundamental Theorem of Galois Theory

We now state the fundamental theorem of Galois theory. This theorem describes the interplay between the Galois group and Galois extensions. In particular, the result ties together subgroups of the Galois group and intermediate fields between L and K.

Theorem 15.4.1 (Fundamental theorem of Galois theory). *Let $L|K$ be a Galois extension with Galois group $G = \operatorname{Aut}(L|K)$. For each intermediate field E, let $\tau(E)$ be the subgroup of G fixing E. Then the following hold:*

(1) *τ is a bijection between intermediate fields containing K and subgroups of G.*
(2) *$L|K$ is a finite extension, and if M is an intermediate field, then $|L : M| = |\operatorname{Aut}(L|M)|$ and $|M : K| = |\operatorname{Aut}(L|K) : \operatorname{Aut}(L|M)|$.*
(3) *If M is an intermediate field, then the following hold:*
    (a) *$L|M$ is always a Galois extension.*
    (b) *$M|K$ is a Galois extension if and only if $\operatorname{Aut}(L|M)$ is a normal subgroup of $\operatorname{Aut}(L|K)$.*
(4) *If M is an intermediate field and $M|K$ is a Galois extension we have the following:*
    (a) *$\alpha(M) = M$ for all $\alpha \in \operatorname{Aut}(L|K)$.*
    (b) *The map $\phi : \operatorname{Aut}(L|K) \to \operatorname{Aut}(M|K)$ with $\phi(\alpha) = \alpha_{|M} = \beta$ is an epimorphism.*
    (c) *$\operatorname{Aut}(M|K) = \operatorname{Aut}(L|K)/\operatorname{Aut}(L|M)$.*
(5) *The lattice of subfields of L containing K is the inverted lattice of subgroups of $\operatorname{Aut}(L|K)$.*

We will prove this main result via a series of theorems, and then combine them all.

Theorem 15.4.2. *Let G be a group, K a field, and $\alpha_1, \dots, \alpha_n$ pairwise distinct group homomorphisms from G to $K^\star$, the multiplicative group of K. Then $\alpha_1, \dots, \alpha_n$ are linearly independent elements of the K-vector space of all homomorphisms from G to K.*

Proof. We use induction on n. If $n = 1$ and $k\alpha_1 = 0$ with $k \in K$, then $0 = k\alpha_1(1) = k \cdot 1$, and hence $k = 0$. Now suppose that $n \geq 2$, and suppose that each $n - 1$ of the $\alpha_1, \dots, \alpha_n$ are linearly independent over K. If

$$\sum_{i=1}^{n} k_i \alpha_i = 0, \quad k_i \in K, \tag{$*$}$$

then we must show that all $k_i = 0$. Since $\alpha_1 \neq \alpha_n$, there exists an $a \in G$ such that $\alpha_1(a) \neq \alpha_n(a)$. Let $g \in G$ and apply the sum above to ag. We get

$$\sum_{i=1}^{n} k_i(\alpha_i(a))(\alpha_i(g)) = 0. \tag{$**$}$$

Now multiply equation $(*)$ by $\alpha_n(a) \in K$ to get

$$\sum_{i=1}^{n} k_i(\alpha_n(a))(\alpha_i(g)) = 0. \tag{$***$}$$

If we subtract equation $(***)$ from equation $(**)$, then the last term vanishes and we have an equation in the $n-1$ homomorphism $\alpha_1, \dots, \alpha_{n-1}$. Since these are linearly independent, we obtain

$$k_1(\alpha_1(a)) - k_1(\alpha_n(a)) = 0$$

for the coefficient for α_1. Since $\alpha_1(a) \neq \alpha_n(a)$, we must have $k_1 = 0$. Now $\alpha_2, \dots, \alpha_{n-1}$ are by assumption linearly independent, so $k_2 = \cdots = k_n = 0$ also. Hence, all the coefficients must be zero, and therefore the mappings are independent. □

Theorem 15.4.3. *Let $\alpha_1, \dots, \alpha_n$ be pairwise distinct monomorphisms from the field K into the field K'. Let*

$$L = \{k \in K : \alpha_1(k) = \alpha_2(k) = \cdots = \alpha_n(k)\}.$$

Then L is a subfield of K with $|L : K| \geq n$.

Proof. Certainly L is a field. Assume that $r = |K : L| < n$, and let $\{a_1, \dots, a_r\}$ be a basis of the L-vector space K. We consider the following system of linear equations with r equations and n unknowns:

$$(\alpha_1(a_1))x_1 + \cdots + (\alpha_n(a_1))x_n = 0$$

$$\vdots$$

$$(\alpha_1(a_r))x_1 + \cdots + (\alpha_n(a_r))x_n = 0.$$

Since $r < n$, there exists a nontrivial solution $(x_1, \dots, x_n) \in (K')^n$.

Let $a \in K$. Then

$$a = \sum_{j=1}^{r} l_j a_j \quad \text{with } l_j \in L.$$

From the definition of L, we have

$$\alpha_1(l_j) = \alpha_i(l_j) \quad \text{for } i = 2, \dots, n.$$

Then with our nontrivial solution $(x_1, \dots, x_n)$, we have

$$\sum_{i=1}^{n} x_i(\alpha_i(a)) = \sum_{i=1}^{n} x_i\left(\sum_{j=1}^{r} \alpha_i(l_j)\alpha_i(a_j)\right) = \sum_{j=1}^{r} (\alpha_1(l_j)) \sum_{i=1}^{n} x_i(\alpha_i(a_j)) = 0$$

since $\alpha_1(l_j) = \alpha_i(l_j)$ for $i = 2, \dots, n$. This holds for all $a \in K$, and hence $\sum_{i=1}^{n} x_i \alpha_i = 0$, contradicting Theorem 15.4.2. Therefore, our assumption that $|K : L| < n$ must be false, and hence $|K : L| \geq n$. □

Definition 15.4.4. Let K be a field and G a finite subgroup of $\mathrm{Aut}(K)$. The map $\mathrm{tr}_G : K \to K$, given by

$$\mathrm{tr}_G(k) = \sum_{\alpha \in G} \alpha(k),$$

is called the *G-trace of K*.

Theorem 15.4.5. *Let K be a field and G a finite subgroup of* $\mathrm{Aut}(K)$. *Then*

$$\{0\} \neq \mathrm{tr}_G(K) \subset \mathrm{Fix}(K, G).$$

Proof. Let $\beta \in G$. Then

$$\beta(\mathrm{tr}_G(k)) = \sum_{\alpha \in G} \beta\alpha(k) = \sum_{\alpha \in G} \alpha(k) = \mathrm{tr}_G(k).$$

Therefore, $\mathrm{tr}_G(K) \subset \mathrm{Fix}(K, G)$.

Now assume that $\mathrm{tr}_G(k) = 0$ for all $k \in K$. Then $\sum_{\alpha \in G} \alpha(k) = 0$ for all $k \in K$. It follows that $\sum_{\alpha \in G} \alpha$ is the zero map; hence, the set of all $\alpha \in G$ are linearly dependent as elements of the K-vector space of all maps from K to K. This contradicts Theorem 15.4.2, and hence the trace cannot be the zero map. □

Theorem 15.4.6. *Let K be a field and G a finite subgroup of* $\mathrm{Aut}(K)$. *Then*

$$|K : \mathrm{Fix}(K, G)| = |G|.$$

Proof. Let $L = \mathrm{Fix}(K, G)$, and suppose that $|G| = n$. From Theorem 15.4.3, we know that $|K : L| \geq n$. We must show that $|K : L| \leq n$.

Suppose that $G = \{\alpha_1, \dots, \alpha_n\}$. To prove the result, we show that if $m > n$ and $a_1, \dots, a_m \in K$, then $a_1, \dots, a_m$ are linearly dependent.

We consider the system of equations

$$(\alpha_1^{-1}(a_1))x_1 + \cdots + (\alpha_1^{-1}(a_m))x_m = 0$$

$$\vdots$$

$$(\alpha_n^{-1}(a_1))x_1 + \cdots + (\alpha_n^{-1}(a_m))x_m = 0.$$

Since $m > n$, there exists a nontrivial solution $(y_1, \dots, y_m) \in K^m$. Suppose that $y_l \neq 0$. Using Theorem 15.4.5, we can choose $k \in K$ with $\mathrm{tr}_G(k) \neq 0$. Define

$$(x_1, \dots, x_m) = k y_l^{-1}(y_1, \dots, y_m).$$

This m-tuple $(x_1, \dots, x_m)$ is then also a nontrivial solution of the system of equations considered above.

Then we have

$$\mathrm{tr}_G(x_l) = \mathrm{tr}_G(k) \quad \text{since } x_l = k.$$

Now we apply α_i to the i-th equation to obtain

$$\begin{aligned} a_1(\alpha_1(x_1)) + \cdots + a_m(\alpha_1(x_m)) &= 0 \\ &\vdots \\ a_1(\alpha_n(x_1)) + \cdots + a_m(\alpha_n(x_m)) &= 0. \end{aligned}$$

Summation leads to

$$0 = \sum_{j=1}^{m} a_j \sum_{i=1}^{n} (\alpha_i(x_j)) = \sum_{j=1}^{m} (\mathrm{tr}_G(x_j)) a_j$$

by definition of the G-trace. Hence, $a_1, \ldots, a_m$ are linearly dependent over L since $\mathrm{tr}_G(x_l) \neq 0$. Therefore, $|K : L| \leq n$. Combining this with Theorem 15.4.3, we get that $|K : L| = n = |G|$. □

Theorem 15.4.7. *Let K be a field and G a finite subgroup of* $\mathrm{Aut}(K)$. *Then*

$$\mathrm{Aut}(K|\mathrm{Fix}(K, G)) = G.$$

Proof. We have $G \subset \mathrm{Aut}(K|\mathrm{Fix}(K, G))$. Since if $g \in G$, then $g \in \mathrm{Aut}(K)$, and g fixes $\mathrm{Fix}(K, G)$ by definition. Therefore, we must show that $\mathrm{Aut}(K|\mathrm{Fix}(K, G)) \subset G$.

Assume then that there exists an $\alpha \in \mathrm{Aut}(K|\,\mathrm{Fix}(K, G))$ with $\alpha \notin G$. Suppose, as in the previous proof, $|G| = n$ and $G = \{\alpha_1, \ldots, \alpha_n\}$ with $\alpha_1 = 1$. Now

$$\begin{aligned} \mathrm{Fix}(K, G) &= \{a \in K : a = \alpha_2(a) = \cdots = \alpha_n(a)\} \\ &= \{a \in K : \alpha(a) = a = \alpha_2(a) = \cdots = \alpha_n(a)\}. \end{aligned}$$

From Theorem 15.4.3, we have that $|K : \mathrm{Fix}(K, G)| \geq n+1$. However, from Theorem 15.4.6, $|K : \mathrm{Fix}(K, G)| = n$, getting a contradiction. □

Suppose that $L|K$ is a Galois extension. We now establish that the map τ between intermediate fields $K \subset E \subset L$ and subgroups of $\mathrm{Aut}(L|K)$ is a bijection.

Theorem 15.4.8. *Let $L|K$ be a Galois extension. Then we have the following:*

(1) $\mathrm{Aut}(L|K)$ *is finite and*

$$\mathrm{Fix}(L, \mathrm{Aut}(L|K)) = K.$$

(2) *If $H \subset \mathrm{Aut}(L|K)$, then*

$$\mathrm{Aut}(L|\mathrm{Fix}(L, H)) = H.$$

Proof. If $(L|K)$ is a Galois extension, there exists a finite subgroup G of $\mathrm{Aut}(L)$ with $K = \mathrm{Fix}(K, G)$. From Theorem 15.4.7, we have $G = \mathrm{Aut}(L|K)$. In particular, $\mathrm{Aut}(L|K)$ is finite, and $K = \mathrm{Fix}(L, \mathrm{Aut}(L|K))$.

Now, let $H \subset \mathrm{Aut}(L|K)$. From the first part, H is finite, and then $\mathrm{Aut}(L|\mathrm{Fix}(L, H)) = H$ from Theorem 15.4.7. □

Theorem 15.4.9. *Let $L|K$ be a field extension. Then the following are equivalent:*
(1) *$L|K$ is a Galois extension.*
(2) $|L : K| = |\mathrm{Aut}(L|K)| < \infty$.
(3) $|\mathrm{Aut}(L|K)| < \infty$, *and* $K = \mathrm{Fix}(L, \mathrm{Aut}(L|K))$.

Proof. (1) ⇒ (2): Now, from Theorem 15.4.8, $|\mathrm{Aut}(L|K)| < \infty$, and $\mathrm{Fix}(L, \mathrm{Aut}(L|K)) = K$. Therefore, from Theorem 15.4.6, $|L : K| = |\mathrm{Aut}(L|K)|$.

(2) ⇒ (3): Let $G = \mathrm{Aut}(L|K)$. Then $K \subset \mathrm{Fix}(L, G) \subset L$. From Theorem 15.4.6, we have

$$\big|L : \mathrm{Fix}(L, G)\big| = |G| = |L : K|.$$

(3) ⇒ (1) follows directly from the definition completing the proof. □

We now show that if $L|K$ is a Galois extension, then $L|M$ is also a Galois extension for any intermediate field M.

Theorem 15.4.10. *Let $L|K$ be a Galois extension and $K \subset M \subset L$ be an intermediate field. Then $L|M$ is always a Galois extension, and*

$$|M : K| = \big|\mathrm{Aut}(L|K) : \mathrm{Aut}(L|M)\big|.$$

Proof. Let $G = \mathrm{Aut}(L|K)$. Then, from Theorem 15.4.9, $|G| < \infty$, and $K = \mathrm{Fix}(L, G)$. Define $H = \mathrm{Aut}(L|M)$ and $M' = \mathrm{Fix}(L, H)$. We must show that $M' = M$ for then $L|M$ is a Galois extension.

Since the elements of H fix M, we have $M \subset M'$. Let $G = \bigcup_{i=1}^{r} \alpha_i H$, a disjoint union of the cosets of H. Let $\alpha_1 = 1$, and define $\beta_i = \alpha_{i|_M}$. The $\beta_1, \ldots, \beta_r$ are pairwise distinct for if $\beta_i = \beta_j$; that is $\alpha_{i|_M} = \alpha_{j|_M}$. Then $\alpha_j^{-1}\alpha_i \in H$, so α_i and α_j are in the same coset.

We claim that

$$\{a \in M : \beta_1(a) = \cdots = \beta_r(a)\} = M \cap \mathrm{Fix}(L, G).$$

Moreover, from Theorem 15.4.9, we know that

$$M \cap \mathrm{Fix}(L, G) = M \cap K = K.$$

To establish the claim, it is clear that

$$M \cap \mathrm{Fix}(L, G) \subset \{a \in M : \beta_1(a) = \cdots = \beta_r(a)\},$$

since

$$a = \beta_i(a) = \alpha_i(a) \quad \text{for } \alpha_i \in G,\ a \in K.$$

Hence, we must show that

$$\{a \in M : \beta_1(a) = \cdots = \beta_r(a)\} \subset M \cap \mathrm{Fix}(L, G).$$

To do this, we must show that $\alpha(b) = b$ for all $\alpha \in G$, $b \in M$. We have $\alpha \in \alpha_i H$ for some i, and hence $\alpha = \alpha_i \gamma$ for $\gamma \in H$. We obtain then

$$\alpha(b) = \alpha_i(\gamma(b)) = \alpha_i(b) = \beta_i(b) = b,$$

proving the inclusion and establishing the claim.

Now, from Theorem 15.4.3, $|M : K| \geq r$. From the degree formula, we get

$$|L : M'||M' : M||M : K| = |L : K| = |G| = |G : H||H| = r|L : M'|,$$

since, from Theorem 15.4.9, $|L : K| = |G|$ and $|H| = |L : M'|$. Therefore, $|M : M'| = 1$. Hence, $M = M'$, since $|M : K| \geq r$. Now

$$|M : K| = |G : H| = |\mathrm{Aut}(L|K) : \mathrm{Aut}(L|M)|,$$

completing the proof. □

Lemma 15.4.11. *Let $L|K$ be a field extension and $K \subset M \subset L$ be an intermediate field. If $\alpha \in \mathrm{Aut}(L|K)$, then*

$$\mathrm{Aut}(L|\alpha(M)) = \alpha\,\mathrm{Aut}(L|M)\alpha^{-1}.$$

Proof. Now, $\beta \in \mathrm{Aut}(L|\alpha(M))$ if and only if $\beta(\alpha(a)) = \alpha(a)$ for all $a \in M$. This occurs if and only if $\alpha^{-1}\beta\alpha(a) = a$ for all $a \in M$, which is true if and only if $\beta \in \alpha\,\mathrm{Aut}(L|M)\alpha^{-1}$. □

Lemma 15.4.12. *Let $L|K$ be a Galois extension and $K \subset M \subset L$ be an intermediate field. Suppose that $\alpha(M) = M$ for all $\alpha \in \mathrm{Aut}(L|K)$. Then*

$$\phi : \mathrm{Aut}(L|K) \to \mathrm{Aut}(M|K) \quad \text{with } \phi(\alpha) = \alpha_{|M}$$

is an epimorphism with kernel $\ker(\phi) = \mathrm{Aut}(L|M)$.

Proof. It is clear that ϕ is a homomorphism with $\ker(\phi) = \mathrm{Aut}(L|M)$ (see exercises). We must show that it is an epimorphism.

Let $G = \mathrm{im}(\phi)$. Since $L|K$ is a Galois extension, we get that

$$\mathrm{Fix}(M, G) = \mathrm{Fix}(L, \mathrm{Aut}(L|K)) \cap M = K \cap M = K.$$

Then, from Theorem 15.4.8, we have

$$\mathrm{Aut}(M|K) = \mathrm{Aut}(M|\mathrm{Fix}(M, G)) = G,$$

and therefore ϕ is an epimorphism. □

Theorem 15.4.13. *Let $L|K$ be a Galois extension and $K \subset M \subset L$ be an intermediate field. Then the following are equivalent:*
(1) *$M|K$ is a Galois extension.*
(2) *If $\alpha \in \operatorname{Aut}(L|K)$, then $\alpha(M) = M$.*
(3) *$\operatorname{Aut}(L|M)$ is a normal subgroup of $\operatorname{Aut}(L|K)$.*

Proof. (1) ⇒ (2): Suppose that $M|K$ is a Galois extension. Let $\operatorname{Aut}(M|K) = \{\alpha_1, \dots, \alpha_r\}$. Consider the α_i as monomorphisms from M into L. Let $\alpha_{r+1} : M \to L$ be a monomorphism with $\alpha_{r+1_{|K}} = 1$. Then

$$\{a \in M : \alpha_1(a) = \alpha_2(a) = \cdots = \alpha_r(a) = \alpha_{r+1}(a)\} = K,$$

since $M|K$ is a Galois extension. Therefore, from Theorem 15.4.3, we have that if the $\alpha_1, \dots, \alpha_r, \alpha_{r+1}$ are distinct, then

$$|M : K| \geq r + 1 > r = |\operatorname{Aut}(M|K)| = |M : K|,$$

giving a contradiction. Hence, if $\alpha_{r+1} \in \operatorname{Aut}(L|K)$ is arbitrary, then $\alpha_{r+1_{|M}} \in \{\alpha_1, \dots, \alpha_r\}$; that is, α_{r+1} fixes M.

(2) ⇒ (1): Suppose that if $\alpha \in \operatorname{Aut}(L|K)$, then $\alpha(M) = M$. The map $\phi : \operatorname{Aut}(L|K) \to \operatorname{Aut}(M|K)$ with $\phi(\alpha) = \alpha_{|M}$ is surjective. Since $L|K$ is a Galois extension, then $\operatorname{Aut}(L|K)$ is finite. Therefore, also $H = \operatorname{Aut}(M|K)$ is finite. To prove (1) then, it is sufficient to show that $K = \operatorname{Fix}(M, H)$.

The field $K \subset \operatorname{Fix}(M, H)$ from the definition of the fix field. Hence, we must show that $\operatorname{Fix}(M, H) \subset K$. Assume that there exists an $\alpha \in \operatorname{Aut}(L|K)$ with $\alpha(a) \neq a$ for some $a \in \operatorname{Fix}(M, H)$. Recall that $L|K$ is a Galois extension, and therefore $\operatorname{Fix}(L, \operatorname{Aut}(L|K)) = K$. Define $\beta = \alpha_{|M}$. Then $\beta \in H$, since $\alpha(M) = M$ and our original assumption. Then $\beta(a) \neq a$, contradicting $a \in \operatorname{Fix}(M, H)$. Therefore, $K = \operatorname{Fix}(M, H)$, and $M|K$ is a Galois extension.

(2) ⇒ (3): Suppose that if $\alpha \in \operatorname{Aut}(L|K)$, then $\alpha(M) = M$. Then $\operatorname{Aut}(L|M)$ is a normal subgroup of $\operatorname{Aut}(L|K)$ follows from Lemma 15.4.12, since $\operatorname{Aut}(L|M)$ is the kernel of ϕ.

(3) ⇒ (2): Suppose that $\operatorname{Aut}(L|M)$ is a normal subgroup of $\operatorname{Aut}(L|K)$. Let $\alpha \in \operatorname{Aut}(L|K)$, then from our assumption and Lemma 15.4.11, we get that

$$\operatorname{Aut}(L|\alpha(M)) = \operatorname{Aut}(L|M).$$

Now $L|M$ and $L|\alpha(M)$ are Galois extensions by Theorem 15.4.10. Therefore,

$$\alpha(M) = \operatorname{Fix}(L, \operatorname{Aut}(L|\alpha(M)) = \operatorname{Fix}(L, \operatorname{Aut}(L|M)) = M,$$

completing the proof. □

We now combine all of these results to give the proof of Theorem 15.4.1, the fundamental theorem of Galois theory.

Proof of Theorem 15.4.1. Let $L|K$ be a Galois extension.

For (1), let $G \subset \mathrm{Aut}(L|K)$. Both G and $\mathrm{Aut}(L|K)$ are finite from Theorem 15.4.8. Furthermore, $G = \mathrm{Aut}(L|\mathrm{Fix}(L, G))$ from Theorem 15.4.7. Now let M be an intermediate field of $L|K$. Then $L|M$ is a Galois extension from Theorem 15.4.10, and then $\mathrm{Fix}(L, \mathrm{Aut}(L|M)) = M$ from Theorem 15.4.8.

For (2), let M be an intermediate field of $L|K$. From Theorem 15.4.10, $L|M$ is a Galois extension. From Theorem 15.4.9, we have $|L : M| = |\mathrm{Aut}(L|M)|$. Applying Theorem 15.4.10, we get the result on indices

$$|M : K| = \big|\mathrm{Aut}(L|K) : \mathrm{Aut}(L|M)\big|.$$

For (3), let M be an intermediate field of $L|K$. From Theorem 15.4.10, we have that $L|M$ is a Galois extension, hence (a) holds. From Theorem 15.4.13, $M|K$ is a Galois extension if and only if $\mathrm{Aut}(L|M)$ is a normal subgroup of $\mathrm{Aut}(L|K)$, that is, (b) holds.

For (4), let $M|K$ be a Galois extension. Assertion (a) holds because $\alpha(M) = M$ for all $\alpha \in \mathrm{Aut}(L|K)$ by Theorem 15.4.13. The map $\phi : \mathrm{Aut}(L|K) \to \mathrm{Aut}(M|K)$ with $\phi(\alpha) = \alpha_{|M} = \beta$ is an epimorphism by Lemma 15.4.12 and Theorem 15.4.13, hence (b) holds. Assertion (c), that is, $\mathrm{Aut}(M|K) = \mathrm{Aut}(L|K)/\mathrm{Aut}(L|M)$, follows directly from the group isomorphism theorem.

That the lattice of subfields of L containing K is the inverted lattice of subgroups of $\mathrm{Aut}(L|K)$ follows directly from the previous results, this shows (5) and finishes the proof. □

In Chapter 8, we looked at Example 8.1.7. Here, we analyze it further using the Galois theory.

Example 15.4.14. Let $f(x) = x^3 - 7 \in \mathbb{Q}[x]$. This has no zeros in $\mathbb{Q}$, and since it is of degree 3, it follows that it must be irreducible in $\mathbb{Q}[x]$.

Let $\omega = -\frac{1}{2} + \frac{\sqrt{3}}{2}i \in \mathbb{C}$. Then it is easy to show by computation that

$$\omega^2 = -\frac{1}{2} - \frac{\sqrt{3}}{2}i \quad \text{and} \quad \omega^3 = 1.$$

Therefore, the three zeros of $f(x)$ in $\mathbb{C}$ are

$$a_1 = 7^{1/3}, \quad a_2 = \omega(7^{1/3}), \quad a_3 = \omega^2(7^{1/3}).$$

Hence, $L = \mathbb{Q}(a_1, a_2, a_3)$ is the splitting field of $f(x)$. Since the minimal polynomial of all three zeros over $\mathbb{Q}$ is the same $f(x)$, it follows that

$$\mathbb{Q}(a_1) \cong \mathbb{Q}(a_2) \cong \mathbb{Q}(a_3).$$

Since $\mathbb{Q}(a_1) \subset \mathbb{R}$ and a_2, a_3 are nonreal, it is clear that $a_2, a_3 \notin \mathbb{Q}(a_1)$.

Suppose that $\mathbb{Q}(a_2) = \mathbb{Q}(a_3)$. Then $\omega = a_3 a_2^{-1} \in \mathbb{Q}(a_2)$, and so $7^{1/3} = \omega^{-1} a_2 \in \mathbb{Q}(a_2)$. Hence, $Q(a_1) \subset \mathbb{Q}(a_2)$; therefore, $\mathbb{Q}(a_1) = \mathbb{Q}(a_2)$ since they are of the same degree over $\mathbb{Q}$. This contradiction shows that $\mathbb{Q}(a_2)$ and $\mathbb{Q}(a_3)$ are distinct.

By computation, we have $a_3 = a_1^{-1}a_2^2$, and hence

$$L = \mathbb{Q}(a_1, a_2, a_3) = \mathbb{Q}(a_1, a_2) = \mathbb{Q}(7^{1/3}, \omega).$$

Now the degree of L over $\mathbb{Q}$ is

$$|L : \mathbb{Q}| = |Q(7^{1/3}, \omega) : \mathbb{Q}(\omega)||\mathbb{Q}(\omega) : \mathbb{Q}|.$$

Now $|\mathbb{Q}(\omega) : \mathbb{Q}| = 2$, since the minimal polynomial of ω over $\mathbb{Q}$ is $x^2 + x + 1$. Since no zero of $f(x)$ lies in $\mathbb{Q}(\omega)$, and the degree of $f(x)$ is 3, it follows that $f(x)$ is irreducible over $\mathbb{Q}(\omega)$. Therefore, we have that the degree of L over $\mathbb{Q}(\omega)$ is 3. Hence, $|L : \mathbb{Q}| = (2)(3) = 6$.

Clearly then, we have the following lattice of intermediate fields:

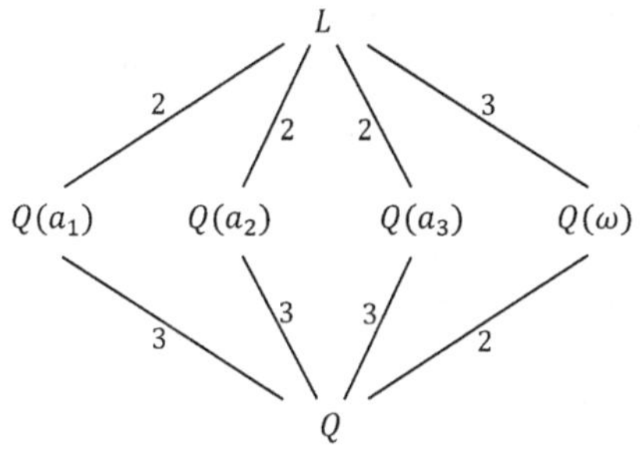

The question then arises as to whether these are all the intermediate fields. The answer is yes, which we now prove.

Let $G = \mathrm{Aut}(L|\mathbb{Q}) = \mathrm{Aut}(L)$. ($\mathrm{Aut}(L|\mathbb{Q}) = \mathrm{Aut}(L)$, since $\mathbb{Q}$ is a prime field.) Now $G \cong S_3$. G acts transitively on $\{a_1, a_2, a_3\}$, since f is irreducible. Let $\delta : \mathbb{C} \to \mathbb{C}$ be the automorphism of $\mathbb{C}$ taking each element to its complex conjugate; that is, $\delta(z) = \overline{z}$. Then $\delta(f) = f$ and $\delta_{|L} \in G$ (Theorem 8.2.2). Since $a_1 \in \mathbb{R}$, we get that $\delta_{|\{a_1,a_2,a_3\}} = (a_2, a_3)$, the 2-cycle that maps a_2 to a_3 and a_3 to a_2. Since G is transitive on $\{a_1, a_2, a_3\}$, there is a $\tau \in G$ with $\tau(a_1) = a_2$.

Case 1: $\tau(a_3) = a_3$. Then $\tau = (a_1, a_2)$, and $(a_1, a_2)(a_2, a_3) = (a_1, a_2, a_3) \in G$.

Case 2: $\tau(a_3) \neq a_3$. Then τ is a 3-cycle. In either case, G is generated by a transposition and a 3-cycle. Hence, G is all of S_3. Then $L|\mathbb{Q}$ is a Galois extension from Theorem 15.4.9, since $|G| = |L : \mathbb{Q}|$.

The subgroups of S_3 are as follows:

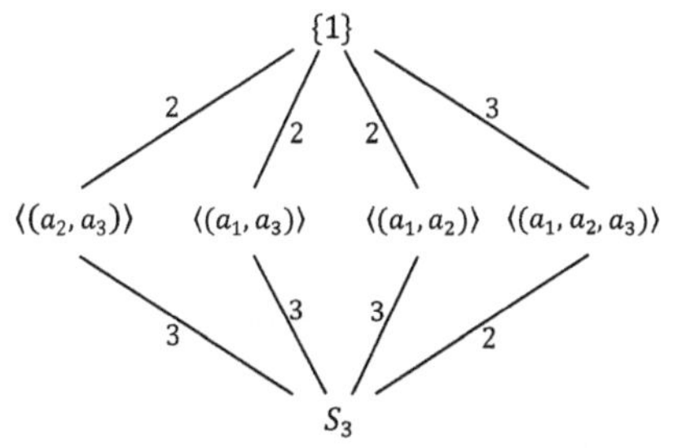

Hence, the above lattice of fields is complete. $L|\mathbb{Q}$, $\mathbb{Q}|\mathbb{Q}$, $\mathbb{Q}(\omega)|\mathbb{Q}$ and $L|\mathbb{Q}(a_i)$ are Galois extensions, whereas $\mathbb{Q}(a_i)|\mathbb{Q}$ with $i = 1, 2, 3$ are not Galois extensions.

15.5 Exercises

1. Let $K \subset M \subset L$ be a chain of fields, and let $\phi : \mathrm{Aut}(L|K) \to \mathrm{Aut}(M|K)$ be defined by $\phi(\alpha) = \alpha_{|M}$. Show that ϕ is an epimorphism with kernel $\ker(\phi) = \mathrm{Aut}(L|M)$.
2. Show that $\mathbb{Q}(5^{\frac{1}{4}})|\mathbb{Q}(\sqrt{5})$ and $\mathbb{Q}(\sqrt{5})|\mathbb{Q}$ are Galois extensions, and $\mathbb{Q}(5^{\frac{1}{4}})|\mathbb{Q}$ is not a Galois extension.
3. Let $L|K$ be a field extension and $u, v \in L$ algebraic over K with $|K(u) : K| = m$ and $|K(v) : K| = n$. If m and n are coprime, then $|K(u, v) : K| = n \cdot m$.
4. Let p, q be prime numbers with $p \neq q$. Let $L = \mathbb{Q}(\sqrt{p}, q^{\frac{1}{3}})$. Show that $L = \mathbb{Q}(\sqrt{p} \cdot q^{\frac{1}{3}})$. Determine a basis of L over $\mathbb{Q}$ and the minimal polynomial of $\sqrt{p} \cdot q^{\frac{1}{3}}$.
5. Let $K = \mathbb{Q}(2^{\frac{1}{n}})$ with $n \geq 2$.
   (i) Determine the number of $\mathbb{Q}$-embeddings $\sigma : K \to \mathbb{R}$. Show that for each such embedding, we have $\sigma(K) = K$.
   (ii) Determine $\mathrm{Aut}(K|\mathbb{Q})$.
6. Let $\alpha = \sqrt{5 + 2\sqrt{5}}$.
   (i) Determine the minimal polynomial of α over $\mathbb{Q}$.
   (ii) Show that $\mathbb{Q}(\alpha)|\mathbb{Q}$ is a Galois extension.
   (iii) Determine $\mathrm{Aut}(\mathbb{Q}(\alpha)|\mathbb{Q})$.
7. Let K be a field of prime characteristic p, and let $f(x) = x^p - x + a \in K$ be an irreducible polynomial. Let $L = K(v)$, where v is a zero of $f(x)$.
   (i) If α is a zero of $f(x)$, then also $\alpha + 1$ is.
   (ii) $L|K$ is a Galois extension.
   (iii) There is exactly one K-automorphism σ of L with $\sigma(v) = v + 1$.
   (iv) The Galois group $\mathrm{Aut}(L|K)$ is cyclic with generating element σ.

16 Separable Field Extensions

16.1 Separability of Fields and Polynomials

In the previous chapter, we introduced and examined Galois extensions. Recall that $L|K$ is a *Galois extension* if there exists a finite subgroup $G \subset \operatorname{Aut}(L)$ with $K = \operatorname{Fix}(L, G)$. The following questions logically arise:

(1) Under what conditions is a field extension $L|K$ a Galois extension?
(2) If $L|K$ is a Galois extension when L is the splitting field of a polynomial $f(x) \in K[x]$?

In this chapter, we consider these questions and completely characterize Galois extensions. To do this, we must introduce separable extensions.

Definition 16.1.1. Let K be a field. Then a nonconstant polynomial $f(x) \in K[x]$ is called *separable over* K if each irreducible factor of $f(x)$ has only simple zeros in its splitting field.

We now extend this definition to field extensions.

Definition 16.1.2. Let $L|K$ be a field extension and $a \in L$. Then a is *separable over* K if a is a zero of a separable polynomial. The field extension $L|K$ is a *separable field extension*, or just *separable* if all $a \in L$ are separable over K. In particular, a separable extension is an algebraic extension.

Finally, we consider fields, where every nonconstant polynomial is separable.

Definition 16.1.3. A field K is *perfect* if each nonconstant polynomial in $K[x]$ is separable over K.

The following is straightforward from the definitions: An element a is separable over K if and only if its minimal polynomial $m_a(x)$ is separable.

If $f(x) \in K[x]$, then $f(x) = \sum_{i=0}^{n} k_i x^i$ with $k_i \in K$. The *formal derivative* of $f(x)$ is then $f'(x) = \sum_{i=1}^{n} i k_i x^{i-1}$. As in ordinary Calculus, we have the usual differentiation rules

$$(f(x) + g(x))' = f'(x) + g'(x)$$

and

$$(f(x)g(x))' = f'(x)g(x) + f(x)g'(x)$$

for $f(x), g(x) \in K[x]$.

Lemma 16.1.4. *Let K be a field and $f(x)$ an irreducible nonconstant polynomial in $K[x]$. Then $f(x)$ is separable if and only if its formal derivative is nonzero.*

https://doi.org/10.1515/9783111142524-016

Proof. Let L be the splitting field of $f(x)$ over K. Let $f(x) = (x-a)^r g(x)$, where $(x-a)$ does not divide $g(x)$. Then

$$f'(x) = (x-a)^{r-1}(rg(x) + (x-a)g'(x)).$$

If $f'(x) \neq 0$, then a is a zero of $f(x)$ in L over K of multiplicity $m \geq 2$ if and only if $(x-a)|f(x)$, and also $(x-a)|f'(x)$.

Let $f(x)$ be a separable polynomial over $K[x]$, and let a be a zero of $f(x)$ in L. Then if $f(x) = (x-a)^r g(x)$ with $(x-a)$ not dividing $g(x)$, we must have $r = 1$. Then

$$f'(x) = g(x) + (x-a)g'(x).$$

If $g'(x) = 0$, then $f'(x) = g(x) \neq 0$. Now suppose that $g'(x) \neq 0$. Assume that $f'(x) = 0$; then, necessarily, $(x-a)|g(x)$ giving a contradiction. Therefore, $f'(x) \neq 0$.

Conversely, suppose that $f'(x) \neq 0$. Assume that $f(x)$ is not separable. Then both $f(x)$ and $f'(x)$ have a common zero $a \in L$. Let $m_a(x)$ be the minimal polynomial of a in $K[x]$. Then $m_a(x)|f(x)$, and $m_a(x)|f'(x)$. Since $f(x)$ is irreducible, then the degree of $m_a(x)$ must equal the degree of $f(x)$. But $m_a(x)$ must also have the same degree as $f'(x)$, which is less than that of $f(x)$, giving a contradiction. Therefore, $f(x)$ must be separable. □

We now consider the following example of a nonseparable polynomial over the finite field $\mathbb{Z}_p$ of p elements. We will denote this field now as $\mathrm{GF}(p)$, the Galois field of p elements.

Example 16.1.5. Let $K = \mathrm{GF}(p)$ and $L = K(t)$, the field of rational functions in t over K. Consider the polynomial $f(x) = x^p - t \in L[x]$.

Now $K[t]/tK[t] \cong K$. Since K is a field, this implies that $tK[t]$ is a maximal ideal, and hence a prime ideal in $K[t]$ with prime element $t \in K[t]$ (see Theorem 3.2.7). By the Eisenstein criteria, $f(x)$ is an irreducible polynomial in $L[x]$ (see Theorem 4.4.8). However, $f'(x) = px^{p-1} = 0$, since $\mathrm{char}(K) = p$. Therefore, $f(x)$ is not separable.

16.2 Perfect Fields

We now consider when a field K is perfect. First, we show that, in general, any field of characteristic 0 is perfect. In particular, the rationals $\mathbb{Q}$ are perfect, and hence any extension of the rationals is separable.

Theorem 16.2.1. *Each field K of characteristic zero is perfect.*

Proof. Suppose that K is a field with $\mathrm{char}(K) = 0$. Suppose that $f(x)$ is a nonconstant polynomial in $K[x]$. Then $f'(x) \neq 0$. If $f(x)$ is irreducible, then $f(x)$ is separable from Lemma 16.1.4. Therefore, by definition, each nonconstant polynomial $f(x) \in K[x]$ is separable. □

We remark that in the original motivation for Galois theory, the ground field was the rationals $\mathbb{Q}$. Since this has characteristic zero, it is perfect and all extensions are separable. Hence, the question of separability did not arise until the question of extensions of fields of prime characteristic arose.

Corollary 16.2.2. *Any finite extension of the rationals $\mathbb{Q}$ is separable.*

We now consider the case of prime characteristic.

Theorem 16.2.3. *Let K be a field with $\operatorname{char}(K) = p \neq 0$. If $f(x)$ is a nonconstant polynomial in $K[x]$, then the following are equivalent:*
(1) $f'(x) = 0$.
(2) *$f(x)$ is a polynomial in x^p; that is, there is a $g(x) \in K[x]$ with $f(x) = g(x^p)$.*

If in (1) and (2) $f(x)$ is irreducible, then $f(x)$ is not separable over K if and only if $f(x)$ is a polynomial in x^p.

Proof. Let $f(x) = \sum_{i=1}^{n} a_i x^i$. Then $f'(x) = 0$ if and only if $p|i$ for all i with $a_i \neq 0$. But this is equivalent to

$$f(x) = a_0 + a_p x^p + \cdots + a_m x^{mp}.$$

If $f(x)$ is irreducible, then $f(x)$ is not separable if and only if $f'(x) = 0$ from Lemma 16.1.4. □

Theorem 16.2.4. *Let K be a field with $\operatorname{char}(K) = p \neq 0$. Then the following are equivalent:*
(1) *K is perfect.*
(2) *Each element in K has a p-th root in K.*
(3) *The Frobenius homomorphism $\tau : x \mapsto x^p$ is an automorphism of K.*

Proof. First we show that (1) implies (2). Suppose that K is perfect, and $a \in K$. Then $x^p - a$ is separable over K. Let $g(x) \in K[x]$ be an irreducible factor of $x^p - a$. Let L be the splitting field of $g(x)$ over K, and b a zero of $g(x)$ in L. Then $b^p = a$. Furthermore, $x^p - b^p = (x - b)^p \in L[x]$, since the characteristic of K is p. Hence, $g(x) = (x - b)^s$, and then s must equal 1 since $g(x)$ is irreducible. Therefore, $b \in K$, and b is a p-th root of a.

Now we show that (2) implies (3). Recall that the Frobenius homomorphism τ is injective (see Theorem 1.8.8). We must show that it is also surjective. Let $a \in K$, and let b be a p-th root of a so that $a = b^p$. Then $\tau(b) = b^p = a$, and τ is surjective.

Finally, we show that (3) implies (1). Let $\tau : x \mapsto x^p$ be surjective. It follows that each $a \in K$ has a p-th root in K. Now let $f(x) \in K[x]$ be irreducible. Assume that $f(x)$ is not separable. From Theorem 16.2.3, there is a $g(x) \in K[x]$ with $f(x) = g(x^p)$; that is,

$$f(x) = a_0 + a_1 x^p + \cdots + a_m x^{mp}.$$

Let $b_i \in K$ with $a_i = b_i^p$. Then

$$f(x) = b_o^p + b_1^p x^p + \cdots + b_m^p x^{mp} = (b_0 + b_1 x + \cdots + b_m x^m)^p.$$

However, this is a contradiction since $f(x)$ is irreducible. Therefore, $f(x)$ is separable, completing the proof. □

Theorem 16.2.5. *Let K be a field with $\operatorname{char}(K) = p \neq 0$. Then each element of K has at most one p-th power in K.*

Proof. Suppose that $b_1, b_2 \in K$ with $b_1^p = b_2^p = a$. Then

$$0 = b_1^p - b_2^p = (b_1 - b_2)^p.$$

Since K has no zero divisors, it follows that $b_1 = b_2$. □

16.3 Finite Fields

In this section, we consider finite fields. In particular, we show that if K is a finite field, then $|K| = p^m$ for some prime p and natural number $m > 0$. Moreover, we show that if K_1, K_2 are finite fields with $|K_1| = |K_2|$, then $K_1 \cong K_2$. Hence, there is a unique finite field for each possible order.

Notice that if K is a finite field, then by necessity $\operatorname{char} K = p \neq 0$. We first show that, in this case, K is always perfect.

Theorem 16.3.1. *A finite field is perfect.*

Proof. Let K be a finite field of characteristic $p > 0$. Then the Frobenius map τ is surjective since it is injective and K is finite. Therefore, K is perfect from Theorem 16.2.4. □

Next we show that each finite field has order p^m for some prime p and natural number $m > 0$.

Lemma 16.3.2. *Let K be a finite field. Then $|K| = p^m$ for some prime p and natural number $m > 0$.*

Proof. Let K be a finite field with characteristic $p > 0$. Then K can be considered as a vector space over $K = \mathrm{GF}(p)$, and hence of finite dimension since $|K| < \infty$. If $\alpha_1, \ldots, \alpha_m$ is a basis, then each $f \in K$ can be written as $f = c_1\alpha_1 + \cdots + c_n\alpha_m$ with each $c_i \in \mathrm{GF}(p)$. Hence, there are p choices for each c_i, and therefore p^m choices for each f. □

In Theorem 9.5.16, we proved that any finite subgroup of the multiplicative group of a field is cyclic. If K is a finite field, then its multiplicative subgroup $K^\star$ is finite, and hence cyclic.

Lemma 16.3.3. *Let K be a finite field. Then its multiplicative subgroup $K^\star$ is cyclic.*

If K is a finite field with order p^m, then its multiplicative subgroup $K^\star$ has order $p^m - 1$. Then, from Lagrange's theorem, each nonzero element to the power p^m is the identity. Therefore, we have the result.

Lemma 16.3.4. *Let K be a field of order p^m. Then each $\alpha \in K$ is a zero of the polynomial $x^{p^m} - x$. In particular, if $\alpha \neq 0$, then α is a zero of $x^{p^m-1} - 1$.*

If K is a finite field of order p^m, it is a finite extension of GF(p). Since the multiplicative group is cyclic, we must have $K = \mathrm{GF}(p)(\alpha)$ for some $\alpha \in K$. From this, we obtain that for a given possible finite order, there is only one finite field up to isomorphism.

Theorem 16.3.5. *Let K_1, K_2 be finite fields with $|K_1| = |K_2|$. Then $K_1 \cong K_2$.*

Proof. Let $|K_1| = |K_2| = p^m$. From the remarks above, $K_1 = \mathrm{GF}(p)(\alpha)$, where α has order $p^m - 1$ in $K_1^\star$. Similarly, $K_2 = \mathrm{GF}(p)(\beta)$, where β also has order $p^m - 1$ in $K_2^\star$. Hence, $\mathrm{GF}(p)(\alpha) \cong \mathrm{GF}(p)(\beta)$, and therefore $K_1 \cong K_2$. □

In Lemma 16.3.2, we saw that if K is a finite field, then $|K| = p^n$ for some prime p and positive integer n. We now show that given a prime power p^n, there does exist a finite field of that order.

Theorem 16.3.6. *Let p be a prime and $n > 0$ a natural number. Then there exists a field K of order p^n.*

Proof. Given a prime p, consider the polynomial $g(x) = x^{p^n} - x \in \mathrm{GF}(p)[x]$. Let K be the splitting field of this polynomial over GF(p). Since a finite field is perfect, K is a separable extension, and hence all the zeros of $g(x)$ are distinct in K.

Let F be the set of p^n distinct zeros of $g(x)$ within K. Let $a, b \in F$. Since

$$(a \pm b)^{p^n} = a^{p^n} \pm b^{p^n} \quad \text{and} \quad (ab)^{p^n} = a^{p^n} b^{p^n},$$

it follows that F forms a subfield of K. However, F contains all the zeros of $g(x)$, and since K is the smallest extension of GF(p) containing all the zeros of $g(x)$, we must have $K = F$. Since F has p^n elements, it follows that the order of K is p^n. □

Combining Theorems 16.3.5 and 16.3.6, we get the following summary result, indicating that up to isomorphism there exists one and only one finite field of order p^n.

Theorem 16.3.7. *Let p be a prime and $n > 0$ a natural number. Then up to isomorphism, there exists a unique finite field of order p^n.*

16.4 Separable Extensions

In this section, we consider some properties of separable extensions.

Theorem 16.4.1. *Let K be a field with $K \subset L$ and L algebraically closed. Let $\alpha : K \to L$ be a monomorphism. Then the number of monomorphisms $\beta : K(a) \to L$ with $\beta_{|K} = \alpha$ is equal to the number of pairwise distinct zeros in L of the minimal polynomial m_a of a over K.*

Proof. Let β be as in the statement of the theorem. Then β is uniquely determined by $\beta(a)$, and $\beta(a)$ is a zero of the polynomial $\beta(m_a(x)) = \alpha(m_a(x))$. Now let a' be a zero of $\alpha(m_a(x))$ in L. Then there exists a $\beta : K(a) \to L$ with $\beta(a) = a'$ from Theorem 7.1.4. Therefore, α has exactly as many extensions β as $\alpha(m_a(x))$ has pairwise distinct zeros in L. The number of pairwise distinct zeros of $\alpha(m_a(x))$ is equal to the number of pairwise distinct zeros of $m_a(x)$. This can be seen as follows: Let L_0 be a splitting field of $m_a(x)$ and $L_1 \subset L$ a splitting field of $\alpha(m_a(x))$. From Theorems 8.1.5 and 8.1.6, there is an isomorphism $\psi : L_0 \to L_1$, which maps the zeros of $m_a(x)$ onto the zeros of $\alpha(m_a(x))$. □

Lemma 16.4.2. *Let $L|K$ be a finite extension with $L \subset \overline{L}$, and $\overline{L}$ algebraically closed. In particular, $L = K(a_1, \ldots, a_n)$, where the a_i are algebraic over K. Let p_i be the number of pairwise distinct zeros of the minimal polynomial m_{a_i} of a_i over $K(a_1, \ldots, a_{n-1})$ in $\overline{L}$. Then there are exactly $p_1, \ldots, p_n$ monomorphisms $\beta : L \to \overline{L}$ with $\beta_{|K} = 1_K$.*

Proof. From Theorem 16.4.1, there are exactly p_1 monomorphisms $\alpha : K(a_1) \to \overline{L}$ with $\alpha_{|K}$ equal to the identity on K. Each such α has exactly p_2 extensions of the identity on K to $K(a_1, a_2)$. We now continue in this manner. □

Theorem 16.4.3. *Let $L|K$ be a field extension with M an intermediate field. If $a \in L$ is separable over K, then it is also separable over M.*

Proof. This follows directly from the fact that the minimal polynomial of a over M divides the minimal polynomial of a over K. □

Theorem 16.4.4. *Let $L|K$ be a field extension. Then the following are equivalent:*
(1) *$L|K$ is finite and separable.*
(2) *There are finitely many separable elements $a_1, \ldots, a_n$ over K with $K = K(a_1, \ldots, a_n)$.*
(3) *$L|K$ is finite, and if $L \subset \overline{L}$ with $\overline{L}$ algebraically closed, then there are exactly $[L : K]$ monomorphisms $\alpha : L \to \overline{L}$ with $\alpha_{|K} = 1_K$.*

Proof. That (1) implies (2) follows directly from the definitions. We show then that (2) implies (3). Let $L = K(a_1, \ldots, a_n)$, where $a_1, \ldots, a_n$ are separable elements over K. The extension $L|K$ is finite (see Theorem 5.3.4).

Let p_i be the number of pairwise distinct zeros in $\overline{L}$ of the minimal polynomial $m_{a_i}(x) = f_i(x)$ of a_i over $K(a_1, \ldots, a_{i-1})$. Then

$$p_i \le \deg(f_i) = |K(a_1, \ldots, a_i) : K(a_1, \ldots, a_{i-1})|.$$

Hence, $p_i = \deg(f_i(x))$ since a_i is separable over $K(a_1, \ldots, a_{i-1})$ from Theorem 16.4.3. Therefore, $[L : K] = p_1 \cdots p_n$ is equal to the number of monomorphisms $\alpha : L \to \overline{L}$ with $\alpha_{|K}$, the identity on K.

Finally, we show that (3) implies (1). Suppose then the conditions of (3). Since $L|K$ is finite, there are finitely many $a_1, \dots, a_n \in L$ with $L = K(a_1, \dots, a_n)$. Let p_i and $f_i(x)$ be as in the proof above, and hence $p_i \le \deg(f_i(x))$. By assumption we have

$$[L : K] = p_1 \cdots p_n$$

equal to the number of monomorphisms $\alpha : L \to \overline{L}$ with $\alpha_{|K}$, the identity on K. Also

$$[L : K] = p_1 \cdots p_n \le \deg(f_1(x)) \cdots \deg(f_n(x)) = [L : K].$$

Hence, $p_i = \deg(f_i(x))$. Therefore, by definition, each a_i is separable over K.

To complete the proof, we must show that $L|K$ is separable. Inductively, it suffices to prove that $K(a_1)|K$ is separable over K whenever a_1 is separable over K, and not in K. This is clear if $\operatorname{char}(K) = 0$, because K is perfect.

Suppose then that $\operatorname{char}(K) = p > 0$. First, we show that $K(a_1^p) = K(a_1)$. Certainly, $K(a_1^p) \subset K(a_1)$. Assume that $a_1 \notin K(a_1^p)$. Then $g(x) = x^p - a_1^p$ is the minimal polynomial of a_1 over K. This follows from the fact that $x^p - a_1^p = (x - a_1)^p$, and hence there can be no irreducible factor of $x^p - a_1^p$ of the form $(x - a_1)^m$ with $m < p$ and $m|p$.

However, it follows then, in this case, that $g'(x) = 0$, contradicting the separability of a_1 over K. Therefore, $K(a_1) = K(a_1^p)$.

Let $E = K(a_1)$, then also $E = K(E^p)$, where E^p is the field generated by the p-th powers of E. Now let $b \in E = K(a_1)$. We must show that the minimal polynomial of b, say $m_b(x)$, is separable over K. Assume that $m_b(x)$ is not separable over K. Then

$$m_b(x) = \sum_{i=0}^{k} b_i x^{pi}, \quad b_i \in K,\ b_k = 1$$

from Theorem 16.2.3. We have

$$b_0 + b_1 b^p + \cdots + b_k b^{pk} = 0.$$

Therefore, the elements $1, b^p, \dots, b^{pk}$ are linearly dependent over K.

Since $K(a_1) = E = K(E^p)$, we find that $1, b, \dots, b^k$ are linearly dependent also, since if they were independent the p-th powers would also be independent. However, this is not possible, since $k < \deg(m_b(x))$. Therefore, $m_b(x)$ is separable over K, and hence $K(a_1)|K$ is separable. Altogether $L|K$ is then finite and separable, completing the proof. □

Theorem 16.4.5. *Let $L|K$ be a field extension, and let M be an intermediate field. Then the following are equivalent:*

(1) *$L|K$ is separable.*

(2) *$L|M$ and $M|K$ are separable.*

Proof. We first show that (1) implies (2): If $L|K$ is separable then $L|M$ is separable by Theorem 16.4.3, and $M|K$ is separable.

Now suppose (2), and let $M|K$ and $L|M$ be separable. Let $a \in L$, and let

$$m_a(x) = f(x) = b_0 + \cdots + b_{n-1}x^{n-1} + x^n$$

be the minimal polynomial of a over M. Then $f(x)$ is separable. Let

$$M' = K(b_1, \ldots, b_{n-1}).$$

We have $K \subset M' \subset M$, and hence $M'|K$ is separable, since $M|K$ is separable. Furthermore, a is separable over M', since $f(x)$ is separable, and $f(x) \in M'[x]$. From Theorem 16.4.1, there are $m = \deg(f(x)) = [M'(a) : M']$ extensions of $\alpha : M' \to \overline{M}$ with $\overline{M}$ the algebraic closure of M'. Since $M'|K$ is separable and finite, there are $[M' : K]$ monomorphisms $\alpha : M' \to \overline{M}$ from Theorem 16.4.4. Altogether, there are $[M'(a) : K]$ monomorphisms $\alpha : M' \to \overline{M}$ with $\alpha_{|K}$, the identity on K. Therefore, $M'(a)|K$ is separable from Theorem 16.4.4. Hence, a is separable over K, and then $L|K$ is separable. Therefore, (2) implies (1). □

Theorem 16.4.6. *Let $L|K$ be a field extension, and let $S \subset L$ such that all elements of S are separable over K. Then $K(S)|K$ is separable, and $K[S] = K(S)$.*

Proof. Let W be the set of finite subsets of S. Let $T \in W$. From Theorem 16.4.4, we obtain that $K(T)|K$ is separable. Since each element of $K(S)$ is contained in some $K(T)$, we have that $K(S)|K$ is separable. Since all elements of S are algebraic, we have that $K[S] = K(S)$. □

Theorem 16.4.7. *Let $L|K$ be a field extension. Then there exists in L a uniquely determined maximal field M with the property that $M|K$ is separable. If $a \in L$ is separable over M, then $a \in M$. M is called the separable hull of K in L.*

Proof. Let S be the set of all elements in L, which are separable over K. We now define $M = K(S)$. Then $M|K$ is separable from Theorem 16.4.6. Now, let $a \in L$ be separable over M. Then $M(a)|M$ is separable from Theorem 16.4.4. Furthermore, $M(a)|K$ is separable from Theorem 16.4.5. It follows that $a \in M$. □

16.5 Separability and Galois Extensions

We now completely characterize Galois extensions $L|K$ as finite, normal, separable extensions.

Theorem 16.5.1. *Let $L|K$ be a field extension. Then the following are equivalent:*
(1) *$L|K$ is a Galois extension.*
(2) *L is the splitting field of a separable polynomial in $K[x]$.*
(3) *$L|K$ is finite, normal, and separable.*

Therefore, we may characterize Galois extensions of a field K as finite, normal, and separable extensions of K.

Proof. Recall from Theorem 8.2.2 that an extension $L|K$ is *normal* if the following hold:
(1) $L|k$ is algebraic, and
(2) each irreducible polynomial $f(x) \in K[x]$ that has a zero in L splits into linear factors in $L[x]$.

Now suppose that $L|K$ is a Galois extension. Then $L|K$ is finite from Theorem 15.4.1. Let $L = K(b_1, \dots, b_m)$ and $m_{b_i}(x) = f_i(x)$ be the minimal polynomial of b_i over K. Let $a_{i_1}, \dots, a_{i_n}$ be the pairwise distinct elements from

$$H_i = \{\alpha(b_i) : \alpha \in \operatorname{Aut}(L|K)\}.$$

Define

$$g_i(x) = (x - a_{i_1}) \cdots (x - a_{i_n}) \in L[x].$$

If $\alpha \in \operatorname{Aut}(L|K)$, then $\alpha(g_i) = g_i$, since α permutes the elements of H_i. This means that the coefficients of $g_i(x)$ are in $\operatorname{Fix}(L, \operatorname{Aut}(L|K)) = K$. Furthermore, $g_i(x) \in K[x]$, because b_i is one of the a_{i_j}, and $f_i(x)|g_i(x)$. The group $\operatorname{Aut}(L|K)$ acts transitively on $\{a_{i_1}, \dots, a_{i_n}\}$ by the choice of $a_{i_1}, \dots, a_{i_n}$. Therefore, each $g_i(x)$ is irreducible (see Theorem 15.2.4). It follows that $f_i(x) = g_i(x)$. Now, $f_i(x)$ has only simple zeros in L; that is, no zero has multiplicity ≥ 2, and hence $f_i(x)$ splits over L. Thus, L is a splitting field of $f(x) = f_1(x) \cdots f_m(x)$, and $f(x)$ is separable by definition. Hence, (1) implies (2).

Now suppose that L is a splitting field of the separable polynomial $f(x) \in K[x]$, and $L|K$ is finite. From Theorem 16.4.4, we get that $L|K$ is separable, since $L = K(a_1, \dots, a_n)$ with each a_i separable over K. Therefore, $L|K$ is normal from Definition 8.2.1. Hence, (2) implies (3).

Finally, suppose that $L|K$ is finite, normal, and separable. Since $L|K$ is finite and separable from Theorem 16.4.4, there exist exactly $[L : K]$ monomorphisms $\alpha : L \to \overline{L}$, $\overline{L}$, the algebraic closure of L, with $\alpha_{|K}$ the identity on K. Since $L|K$ is normal, these monomorphisms are already automorphisms of L from Theorem 8.2.2.

Hence, $[L : K] \leq |\operatorname{Aut}(L|K)|$. Furthermore, $|L : K| \geq |\operatorname{Aut}(L|K)|$ from Theorem 15.4.3. Combining these, we have $[L : K] = \operatorname{Aut}(L|K)$, and hence $L|K$ is a Galois extension from Theorem 15.4.9. Therefore, (3) implies (1), completing the proof. □

Recall that any field of characteristic 0 is perfect, and therefore any finite extension is separable. Applying this to $\mathbb{Q}$ implies that the Galois extensions of the rationals are precisely the splitting fields of polynomials.

Corollary 16.5.2. *The Galois extensions of the rationals are precisely the splitting fields of polynomials in* $\mathbb{Q}[x]$.

Theorem 16.5.3. *Let $L|K$ be a finite, separable field extension. Then there exists an extension field M of L such that $M|K$ is a Galois extension.*

Proof. Let $L = K(a_1,\ldots,a_n)$ with all a_i separable over K. Let $f_i(x)$ be the minimal polynomial of a_i over K. Then each $f_i(x)$, and hence also $f(x) = f_1(x)\cdots f_n(x)$, is separable over K. Let M be the splitting field of $f(x)$ over K. Then $M|K$ is a Galois extension from Theorem 16.5.1. □

Example 16.5.4. Let $K = \mathbb{Q}$ be the rationals, and let $f(x) = x^4 - 2 \in \mathbb{Q}[x]$. From Chapter 8, we know that $L = \mathbb{Q}(\sqrt[4]{2}, i)$ is a splitting field of $f(x)$. By the Eisenstein criteria, $f(x)$ is irreducible, and $[L : \mathbb{Q}] = 8$. Moreover,

$$\sqrt[4]{2},\ i\sqrt[4]{2},\ -\sqrt[4]{2},\ -i\sqrt[4]{2}$$

are the zeros of $f(x)$. Since the rationals are perfect, $f(x)$ is separable. $L|K$ is a Galois extension by Theorem 16.5.1. From the calculations in Chapter 15, we have

$$|\mathrm{Aut}(L|K)| = |\mathrm{Aut}(L)| = [L : K] = 8.$$

Let

$$G = \mathrm{Aut}(L|K) = \mathrm{Aut}(L|\mathbb{Q}) = \mathrm{Aut}(L).$$

We want to determine the subgroup lattice of the Galois group G. We show $G \cong D_4$, the dihedral group of order 8. Since there are 4 zeros of $f(x)$, and G permutes these, G must be a subgroup of S_4, and since the order is 8, G is a 2-Sylow subgroup of S_4. From this, we have that

$$G = \langle (2,4), (1,2,3,4) \rangle.$$

If we let $\tau = (2,4)$ and $\sigma = (1,2,3,4)$, we get the isomorphism between G and D_4. From Theorem 14.1.1, we know that $D_4 = \langle r, f ; r^4 = f^2 = (rf)^2 = 1 \rangle$.

This can also be seen in the following manner. Let

$$a_1 = \sqrt[4]{2}, \quad a_2 = i\sqrt[4]{2}, \quad a_3 = -\sqrt[4]{2}, \quad a_4 = -i\sqrt[4]{2}.$$

Let $\alpha \in G$. α is determined if we know $\alpha(\sqrt[4]{2})$ and $\alpha(i)$. The possibilities for $\alpha(i)$ are i or $-i$; that is, the zeros of $x^2 + 1$.

The possibilities for $\sqrt[4]{2}$ are the 4 zeros of $f(x) = x^4 - 2$. Hence, we have 8 possibilities for α. These are exactly the elements of the group G. We have $\delta, \tau \in G$ with

$$\delta(\sqrt[4]{2}) = i\sqrt[4]{2}, \quad \delta(i) = i$$

and

$$\tau(\sqrt[4]{2}) = \sqrt[4]{2}, \quad \tau(i) = -i.$$

It is straightforward to show that δ has order 4, τ has order 2, and $\delta\tau$ has order 2. These define a group of order 8 isomorphic to D_4, and since G has 8 elements, this must be all of G.

We now look at the subgroup lattice of G, and then the corresponding field lattice. Let δ and τ be as above. Then G has 5 subgroups of order 2

$$\{1,\delta^2\},\ \{1,\tau\},\ \{1,\delta\tau\},\ \{1,\delta^2\tau\},\ \{1,\delta^3\tau\}.$$

Of these only $\{1,\delta^2\}$ is normal in G.

G has 3 subgroups of order 4

$$\{1,\delta,\delta^2,\delta^3\},\ \{1,\delta^2,\tau,\tau\delta^2\},\ \{1,\delta^2,\delta\tau,\delta^3\tau\},$$

and all are normal since they all have index 2.

Hence, we have the following subgroup lattice:

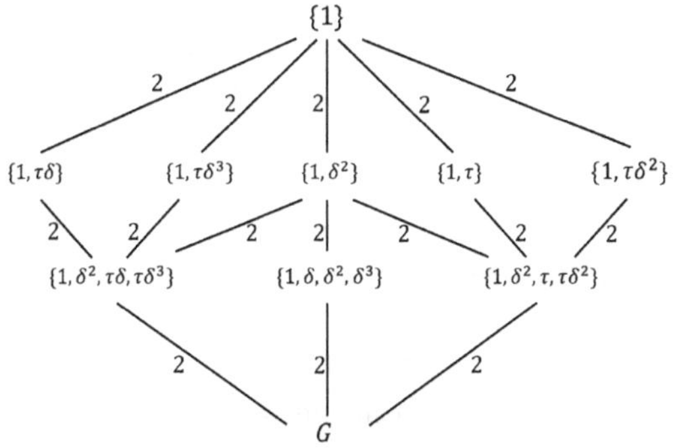

From this we construct the lattice of fields and intermediate fields. Since there are 10 proper subgroups of G from the fundamental theorem of Galois theory, there are 10 intermediate fields in $L|\mathbb{Q}$, namely, the fix fields $\operatorname{Fix}(L,H)$, where H is a proper subgroup of G. In the identification, the extension field corresponding to the whole group G is the ground field $\mathbb{Q}$ (recall that the lattice of fields is the inverted lattice of the subgroups), whereas the extension field corresponding to the identity is the whole field L. We now consider the other proper subgroups. Let δ,τ be as before.

(1) Let $M_1 = \operatorname{Fix}(L,\{1,\tau\})$. Now, $\{1,\tau\}$ fixes $\mathbb{Q}(\sqrt[4]{2})$ elementwise such that $\mathbb{Q}(\sqrt[4]{2}) \subset M_1$. Furthermore, $[L:M_1] = |\{1,\tau\}| = 2$, and hence $[L:\mathbb{Q}(\sqrt[4]{2})] = 2$. Hence, $M_1 = \mathbb{Q}(\sqrt[4]{2})$.

(2) Consider $M_2 = \operatorname{Fix}(L,\{1,\tau\delta\})$. We have the following:

$$\begin{aligned}
\tau\delta(\sqrt[4]{2}) &= \tau(i\sqrt[4]{2}) = -i\sqrt[4]{2}\\
\tau\delta(i\sqrt[4]{2}) &= \tau(-\sqrt[4]{2}) = -\sqrt[4]{2}\\
\tau\delta(-\sqrt[4]{2}) &= \tau(-i\sqrt[4]{2}) = i\sqrt[4]{2}\\
\tau\delta(-i\sqrt[4]{2}) &= \tau(\sqrt[4]{2}) = \sqrt[4]{2}.
\end{aligned}$$

It follows that $\tau\delta$ fixes $(1-i)\sqrt[4]{2}$, and hence $M_2 = \mathbb{Q}((1-i)\sqrt[4]{2})$.

(3) Consider $M_3 = \operatorname{Fix}(L, \{1, \tau\delta^2\})$. The map $\tau\delta^2$ interchanges a_1 and a_3 and fixes a_2 and a_4. Therefore, $M_3 = \mathbb{Q}(i\sqrt[4]{2})$.

In an analogous manner, we can then consider the other 5 proper subgroups and corresponding intermediate fields. We get the following lattice of fields and subfields:

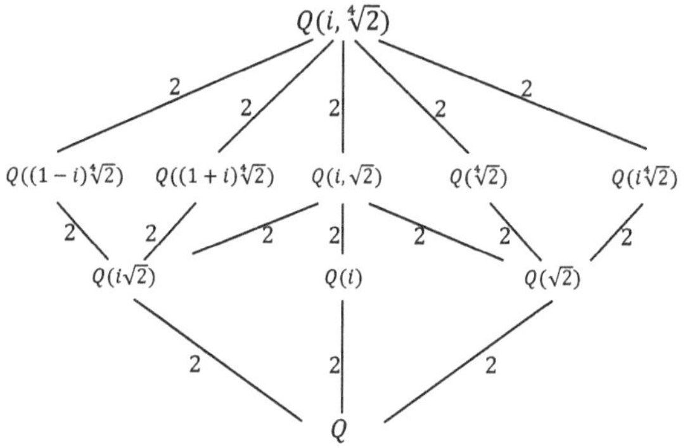

16.6 The Primitive Element Theorem

In this section, we describe finite separable field extensions as simple extensions. It follows that a Galois extension is always a simple extension.

Theorem 16.6.1 (Primitive element theorem). *Let $L = K(\gamma_1, \dots, \gamma_n)$, and suppose that each γ_i is separable over K. Then there exists a $\gamma_0 \in L$ such that $L = K(\gamma_0)$. The element γ_0 is called a primitive element.*

Proof. Suppose first that K is a finite field. Then L is also a finite field, and therefore $L^\star = \langle \gamma_0 \rangle$ is cyclic. Therefore, $L = K(\gamma_0)$, and the theorem is proved if K is a finite field.

Now suppose that K is infinite. Inductively, it suffices to prove the theorem for $n = 2$. Hence, let $\alpha, \beta \in L$ be separable over K. We must show that there exists a $\gamma \in L$ with $K(\alpha, \beta) = K(\gamma)$.

Let $\overline{L}$ be the splitting field of the polynomial $m_\alpha(x)m_\beta(x)$ over L, where $m_\alpha(x), m_\beta(x)$ are, respectively, the minimal polynomials of α, β over K. In $\overline{L}[x]$, we have the following:

$$m_\alpha(x) = (x - \alpha_1)(x - \alpha_2) \cdots (x - \alpha_s) \quad \text{with } \alpha = \alpha_1$$
$$m_\beta(x) = (x - \beta_1)(x - \beta_2) \cdots (x - \beta_t) \quad \text{with } \beta = \beta_1.$$

By assumption the α_i and the β_j are, respectively, pairwise distinct.

For each pair (i, j) with $1 \le i \le s$, $2 \le j \le t$, the equation

$$\alpha_1 + z\beta_1 = \alpha_i + z\beta_j$$

has exactly one solution $z \in \overline{L}$, since $\beta_j - \beta_1 \neq 0$ if $j \geq 2$. Since K is infinite, there exists a $c \in K$ with

$$\alpha_1 + c\beta_1 \neq \alpha_i + c\beta_j$$

for all i, j with $1 \leq i \leq s$, $2 \leq j \leq t$. With such a value $c \in K$, we define

$$\gamma = \alpha + c\beta = \alpha_1 + c\beta_1.$$

We claim that $K(\alpha, \beta) = K(\gamma)$ holds. It suffices to show that $\beta \in K(\gamma)$, for then $\alpha = \gamma - c\beta \in K(\gamma)$. This implies that $K(\alpha, \beta) \subset K(\gamma)$, and since $\gamma \in K(\alpha, \beta)$, it follows that $K(\alpha, \beta) = K(\gamma)$. To show that $\beta \in K(\gamma)$, we first define $f(x) = m_\alpha(\gamma - cx)$, and let $d(x) = \gcd(f(x), m_\beta(x))$. We may assume that $d(x)$ is monic. We show that $d(x) = x - \beta$. Then $\beta \in K(\gamma)$, since $d(x) \in K(\gamma)[x]$.

Assume first that $d(x) = 1$. Then $\gcd(f(x), m_\beta(x)) = 1$, and $f(x)$ and $m_\beta(x)$ are also relatively prime in $\overline{L}[x]$. This is a contradiction, since $f(x)$ and $m_\beta(x)$ have the common zero $\beta \in \overline{L}$, and hence the common divisor $x - \beta$.

Therefore, $d(x) \neq 1$, so $\deg(d(x)) \geq 1$.

The polynomial $d(x)$ is a divisor of $m_\beta(x)$, and hence $d(x)$ splits into linear factors of the form $x - \beta_j$, $1 \leq j \leq t$ in $\overline{L}[x]$. The proof is completed if we can show that no linear factor of the form $x - \beta_j$ with $2 \leq j \leq t$ is a divisor of $f(x)$. That is, we must show that $f(\beta_j) \neq 0$ in $\overline{L}$ if $j \geq 2$.

Now $f(\beta_j) = m_\alpha(\gamma - c\beta_j) = m_\alpha(\alpha_1 + c\beta_1 - c\beta_j)$. Suppose that $f(\beta_j) = 0$ for some $j \geq 2$. This would imply that $\alpha_i = \alpha_1 + c\beta_1 - c\beta_j$; that is, $\alpha_1 + c\beta_1 = \alpha_j + c\beta_j$ for $j \geq 2$. This contradicts the choice of the value c. Therefore, $f(\beta_j) \neq 0$ if $j \geq 2$, completing the proof. □

In the above theorem, it is sufficient to assume that $n - 1$ of $\gamma_1, \ldots, \gamma_n$ are separable over K. The proof is similar. We only need that the $\beta_1, \ldots, \beta_t$ are pairwise distinct if β is separable over K to show that $K(\alpha, \beta) = K(\gamma)$ for some $\gamma \in L$.

If K is a perfect field, then every finite extension is separable. Therefore, we get the following corollary:

Corollary 16.6.2. *Let $L|K$ be a finite extension with K a perfect field. Then $L = K(\gamma)$ for some $\gamma \in L$.*

Corollary 16.6.3. *Let $L|K$ be a finite extension with K a perfect field. Then there exist only finitely many intermediate fields E with $K \subset E \subset L$.*

Proof. Since K is a perfect field, we have $L = K(\gamma)$ for some $\gamma \in L$. Let $m_\gamma(x) \in K[x]$ be the minimal polynomial of γ over K, and let $\overline{L}$ be the splitting field of $m_\gamma(x)$ over K. Then $\overline{L}|K$ is a Galois extension; hence, there are only finitely many intermediate fields between K and $\overline{L}$. Therefore, also only finitely many fields between K and L. □

Suppose that $L|K$ is algebraic. Then, in general, $L = K(\gamma)$ for some $\gamma \in L$ if and only if there exist only finitely many intermediate fields E with $K \subset E \subset L$.

This condition on intermediate fields implies that $L|K$ is finite if $L|K$ is algebraic. Hence, we have proved this result, in the case that K is perfect. The general case is discussed in the book of S. Lang [13].

16.7 Exercises

1. Let $f(x) = x^4-8x^3+24x^2-32x+14 \in \mathbb{Q}[x]$, and let $v \in \mathbb{C}$ be a zero of f. Let $\alpha := v(4-v)$, and K a splitting field of f over $\mathbb{Q}$. Show the following:
   (i) f is irreducible over $\mathbb{Q}$, and $f(x) = f(4-x)$.
   (ii) There is exactly one automorphism σ of $\mathbb{Q}(v)$ with $\sigma(v) = 4-v$.
   (iii) $L := \mathbb{Q}(\alpha)$ is the Fix field of σ and $|L : \mathbb{Q}| = 2$.
   (iv) Determine the minimal polynomial of α over $\mathbb{Q}$ and determine α.
   (v) $|\mathbb{Q}(v) : L| = 2$, and determine the minimal polynomial of v over L; also determine v and all other zeros of $f(x)$.
   (vi) Determine the degree of $|K : \mathbb{Q}|$.
   (vii) Determine the structure of $\mathrm{Aut}(K|\mathbb{Q})$.
2. Let $L|K$ be a field extension and $f \in K[x]$ a separable polynomial. Let Z be a splitting field of f over L and Z_0 a splitting field of f over K. Show that $\mathrm{Aut}(Z|L)$ is isomorphic to a subgroup of $\mathrm{Aut}(Z_0|K)$.
3. Let $L|K$ be a field extension and $v \in L$. For each element $c \in K$ it is $K(v+c) = K(v)$. For $c \neq 0$, it is $K(cv) = K(v)$.
4. Let $v = \sqrt{2}+\sqrt{3}$ and let $K = \mathbb{Q}(v)$. Show that $\sqrt{2}$ and $\sqrt{3}$ are presentable as a $\mathbb{Q}$-linear combination of $1, v, v^2, v^3$. Conclude that $K = \mathbb{Q}(\sqrt{2}, \sqrt{3})$.
5. Let L be the splitting field of x^3-5 over $\mathbb{Q}$ in $\mathbb{C}$. Determine a primitive element t of L over $\mathbb{Q}$.

17 Applications of Galois Theory

As we mentioned in Chapter 1, Galois theory was originally developed as part of the proof that polynomial equations of degree 5 or higher over the rationals cannot be solved by formulas in terms of radicals. In this chapter, we do this first and prove the insolvability of the quintic polynomials by radicals. To do this, we must examine in detail what we call radical extensions.

We then return to some geometric material we started in Chapter 6. There, using general field extensions, we proved the impossibility of certain geometric compass and straightedge constructions. Here, we use Galois theory to consider constructible n-gons.

Finally, we will use Galois theory to present a proof of the fundamental theorem of algebra, which says, essentially, that the complex number field $\mathbb{C}$ is algebraically closed.

In Chapter 17, we always assume that K is a field of characteristic 0; in particular, K is perfect. We remark that some parts of Sections 17.1–17.4 go through for finite fields of characteristic $p > 3$.

17.1 Field Extensions by Radicals

We would like to use Galois theory to prove the insolvability by radicals of polynomial equations of degree 5 or higher. To do this we must introduce *extensions by radicals* and *solvability by radicals*.

Definition 17.1.1. Let $L|K$ be a field extension.

(1) Each zero of a polynomial $x^n - a \in K[x]$ in L is called a *radical* (over K). We denote it by $\sqrt[n]{a}$ (if a more detailed identification is not necessary).

(2) L is called a *simple extension of K by a radical* if $L = K(\sqrt[n]{a})$ for some $a \in K$.

(3) L is called an *extension of K by radicals* if there is a chain of fields

$$K = L_0 \subset L_1 \subset \cdots \subset L_m = L$$

such that each L_i is a simple extension of L_{i-1} by a radical for each $i = 1, \ldots, m$.

(4) Let $f(x) \in K[x]$. Then the equation $f(x) = 0$ is *solvable by radicals*, or just *solvable*, if the splitting field of $f(x)$ over K is contained in an extension of K by radicals.

In proving the insolvability of the quintic polynomial, we will look for necessary and sufficient conditions for the solvability of polynomial equations. Our main result will be that if $f(x) \in K[x]$, then $f(x) = 0$ is solvable over K if the Galois group of the splitting field of $f(x)$ over K is a solvable group (see Chapter 11).

In the remainder of this section, we assume that *all fields have characteristic zero*. The next theorem gives a characterization of simple extensions by radicals:

https://doi.org/10.1515/9783111142524-017

Theorem 17.1.2. *Let $L|K$ be a field extension and $n \in \mathbb{N}$. Assume that the polynomial $x^n - 1$ splits into linear factors in $K[x]$ so that K contains all the n-th roots of unity.*

Then $L = K(\sqrt[n]{a})$ for some $a \in K$ if and only if L is a Galois extension over K, and if $\mathrm{Aut}(L|K) = \mathbb{Z}/m\mathbb{Z}$ *for some $m \in \mathbb{N}$ with $m|n$.*

Proof. The n-th roots of unity, that is, the zeros of the polynomial $x^n - 1 \in K[x]$, form a cyclic multiplicative group $\mathcal{F} \subset K^\star$ of order n, since each finite subgroup of the multiplicative group $K^\star$ of K is cyclic, and $|\mathcal{F}| = n$. We call an n-th root of unity ω *primitive* if $\mathcal{F} = \langle\omega\rangle$.

Now let $L = K(\sqrt[n]{a})$ with $a \in K$; that is, $L = K(\beta)$ with $\beta^n = a \in K$. Let ω be a primitive n-th root of unity. With this β, the elements $\omega\beta, \omega^2\beta, \dots, \omega^n\beta = \beta$ are zeros of $x^n - a$. Hence, the polynomial $x^n - a$ splits into linear factors over L; hence, $L = K(\beta)$ is a splitting field of $x^n - a$ over K. It follows that $L|K$ is a Galois extension.

Let $\sigma \in \mathrm{Aut}(L|K)$. Then $\sigma(\beta) = \omega^\nu\beta$ for some $0 < \nu \le n$. The element ω^ν is uniquely determined by σ, and we may write $\omega^\nu = \omega_\sigma$.

Consider the map $\phi : \mathrm{Aut}(L|K) \to \mathcal{F}$ given by $\sigma \to \omega_\sigma$, where ω_σ is defined as above by $\sigma(\beta) = \omega_\sigma\beta$. If $\tau, \sigma \in \mathrm{Aut}(L|K)$, then

$$\sigma\tau(\beta) = \sigma(\omega_\tau)\sigma(\beta) = \omega_\tau\omega_\sigma\beta,$$

because $\omega_\tau \in K$.

Therefore, $\phi(\sigma\tau) = \phi(\sigma)\phi(\tau)$; hence, ϕ is a homomorphism. The kernel $\ker(\phi)$ contains all the K-automorphisms of L, for which $\sigma(\beta) = \beta$. However, since $K = K(\beta)$, it follows that $\ker(\phi)$ contains only the identity. The Galois group $\mathrm{Aut}(L|K)$ is, therefore, isomorphic to a subgroup of $\mathcal{F}$. Since $\mathcal{F}$ is cyclic of order n, we have that $\mathrm{Aut}(L|K)$ is cyclic of order m for some $m|n$, completing one way in the theorem.

Conversely, first suppose that $L|K$ is a Galois extension with $\mathrm{Aut}(L|K) = \mathbb{Z}_n$, a cyclic group of order n. Let σ be a generator of $\mathrm{Aut}(L|K)$. This is equivalent to

$$\mathrm{Aut}(L|K) = \{\sigma, \sigma^2, \dots, \sigma^n = 1\}.$$

Let ω be a primitive n-th root of unity. Then, by assumption, $\omega \in K$, $\sigma(\omega) = \omega$, and $\mathcal{F} = \{\omega, \omega^2, \dots, \omega^n = 1\}$. Furthermore, the pairwise distinct automorphism σ^ν, $\nu = 1, 2, \dots, n$, of L are linearly independent; that is, there exists an $\eta \in L$ such that

$$\omega \star \eta = \sum_{\nu=1}^{n} \omega^\nu\sigma^\nu(\eta) \neq 0.$$

The element $\omega \star \eta$ is called the *Lagrange resolvent* of ω by η. We fix such an element $\eta \in L$. Then we get, since $\sigma(\omega) = \omega$,

$$\sigma(\omega \star \eta) = \sum_{\nu=1}^{n} \omega^\nu\sigma^{\nu+1}(\eta) = \omega^{-1}\sum_{\nu=1}^{n} \omega^{\nu+1}\sigma^{\nu+1}(\eta) = \omega^{-1}\sum_{\nu=2}^{n+1} \omega^\nu\sigma^\nu(\eta)$$

$$= \omega^{-1} \sum_{\nu=1}^{n} \omega^{\nu} \sigma^{\nu}(\eta) = \omega^{-1}(\omega \star \eta).$$

Moreover, $\sigma^{\mu}(\omega \star \eta) = \omega^{-\mu}(\omega \star \eta)$, $\mu = 1, 2, \ldots, n$. Hence, the only K-automorphism of L, which fixes $\omega \star \eta$ is the identity. Therefore, $\mathrm{Aut}(L|K(\omega \star \eta)) = \{1\}$; hence, $L = K(\omega \star \eta)$ by the fundamental theorem of Galois theory.

Furthermore,

$$\sigma((\omega \star \eta)^n) = (\sigma(\omega \star \eta))^n = (\omega^{-1}(\omega \star \eta))^n = \omega^{-n}(\omega \star \eta)^n = (\omega \star \eta)^n.$$

Therefore, $(\omega \star \eta)^n \in \mathrm{Fix}(L, \mathrm{Aut}(L|K)) = K$, again from the fundamental theorem of Galois theory. If $a = (\omega \star \eta)^n \in K$, then first $a \in K$, and second $L = K(\sqrt[n]{a}) = K(\omega \star \eta)$. This proves the result in the case where $m = n$. We now use this to prove it in general.

Finally, suppose that $L|K$ is a Galois extension with $\mathrm{Aut}(L|K) = \mathbb{Z}_m$, a cyclic group of order m, where $n = qm$ for some $q \geq 1$. If $n = qm$, then $L = K(\sqrt[m]{b})$ for some $b \in K$ by the above argument. Hence, $L = K(\beta)$ with $\beta^m \in K$. Then certainly, $a = \beta^n = (\beta^m)^q \in K$; therefore, $L = K(\beta) = K(\sqrt[n]{a})$ for some $a \in K$, completing the general case. □

We next show that every extension by radicals is contained in a Galois extension by radicals.

Theorem 17.1.3. *Each extension L of K by radicals is contained in a Galois extension $\tilde{L}$ of K by radicals. This means that there is an extension $\tilde{L}$ of K by radicals with $L \subset \tilde{L}$, and $\tilde{L}|K$ is a Galois extension.*

Proof. We use induction on the degree $m = [L : K]$. Suppose that $m = 1$. If $L = K(\sqrt[n]{a})$, then if ω is a primitive n-th root of unity, define $\tilde{K} = K(\omega)$ and $\tilde{L} = \tilde{K}(\sqrt[n]{a})$. We then get the chain $K \subset \tilde{K} \subset \tilde{L}$ with $L \subset \tilde{L}$, and $\tilde{L}|K$ is a Galois extension. This last statement is due to the fact that $\tilde{L}$ is the splitting field of the polynomial $x^n - a \in K[x]$ over K. Hence, the theorem is true if $m = 1$.

Now suppose that $m \geq 2$, and suppose that the theorem is true for all extensions F of K by radicals with $[F : K] < m$.

Since $m \geq 2$ by the definition of extension by radicals, there exists a simple extension $L|E$ by a radical. That is, there exists a field E with

$$K \subset E \subset L, \quad [L : E] \geq 2$$

and $L = E(\sqrt[n]{a})$ for some $a \in E$, $n \in \mathbb{N}$. Now $[E : K] < m$. Therefore, by the inductive hypothesis, there exists a Galois extension by radicals $\tilde{E}$ of K with $E \subset \tilde{E}$.

Let $G = \mathrm{Aut}(\tilde{E}|K)$ and $\tilde{L}$ be the splitting field of the polynomial $f(x) = m_a(x^n) \in K[x]$ over $\tilde{E}$, where $m_a(x)$ is the minimal polynomial of a over K. We show that $\tilde{L}$ has the desired properties.

Now $\sqrt[n]{a} \in L$ is a zero of the polynomial $f(x)$, and $E \subset \tilde{E} \subset \tilde{L}$. Therefore, $\tilde{L}$ contains an E-isomorphic image of $L = K(\sqrt[n]{a})$; hence, we may consider $\tilde{L}$ as an extension of L.

Since $\tilde{E}$ is a Galois extension of K, the polynomial $f(x)$ may be factored as

$$f(x) = (x^n - \alpha_1) \cdots (x^n - \alpha_s)$$

with $\alpha_i \in \tilde{E}$ for $i = 1, \dots, s$. All zeros of $f(x)$ in $\tilde{L}$ are radicals over $\tilde{E}$. Therefore, $\tilde{L}$ is an extension by radicals of $\tilde{E}$. Since $\tilde{E}$ is also an extension by radicals of K, we obtain that $\tilde{L}$ is an extension by radicals of K.

Since $\tilde{E}$ is a Galois extension of K, we have that $\tilde{E}$ is a splitting field of a polynomial $g(x) \in K[x]$. Furthermore, $\tilde{L}$ is a splitting field of $f(x) \in K[x]$ over $\tilde{E}$. Altogether then, we have that $\tilde{L}$ is a splitting field of $f(x)g(x) \in K[x]$ over K. Therefore, $\tilde{L}$ is a Galois extension of K, completing the proof. □

We will eventually show that a polynomial equation is solvable by radicals if and only if the corresponding Galois group is a solvable group. We now begin to find conditions, where the Galois group is solvable.

Lemma 17.1.4. *Let $K = L_0 \subset L_1 \subset \cdots \subset L_r = L$ be a chain of fields such that the following hold:*
(i) *L is a Galois extension of K.*
(ii) *L_j is a Galois extension of L_{j-1} for $j = 1, \dots, r$.*
(iii) *$G_j = \operatorname{Aut}(L_j|L_{j-1})$ is Abelian for $j = 1, \dots, r$.*

Then $G = \operatorname{Aut}(L|K)$ is solvable.

Proof. We prove the lemma by induction on r. If $r = 0$, then $G = \{1\}$, and there is nothing to prove. Suppose then that $r \geq 1$, and assume that the lemma holds for all such chains of fields with a length $r' < r$. Since $L_1|K$ is a Galois extension, then $\operatorname{Aut}(L_1|K)$ is a normal subgroup of G by the fundamental theorem of Galois theory. Moreover,

$$G_1 = \operatorname{Aut}(L_1|K) = G/\operatorname{Aut}(L|L_1).$$

Since G_1 is an Abelian group, it is solvable, and by assumption $\operatorname{Aut}(L|L_1)$ is solvable. Therefore, G is solvable (see Theorem 12.1.4). □

Lemma 17.1.5. *Let $L|K$ be a field extension. Let $\tilde{K}$ and $\tilde{L}$ be the splitting fields of the polynomial $x^n - 1 \in K[x]$ over K and L, respectively. Since $K \subset L$, we have $\tilde{K} \subset \tilde{L}$. Then the following hold:*
(1) *If $\sigma \in \operatorname{Aut}(\tilde{L}|L)$, then $\sigma_{|\tilde{K}} \in \operatorname{Aut}(\tilde{K}|K)$, and the map*

$$\operatorname{Aut}(\tilde{L}|L) \to \operatorname{Aut}(\tilde{K}|K), \quad \text{given by } \sigma \mapsto \sigma_{|\tilde{K}},$$

is an injective homomorphism.
(2) *Suppose that in addition $L|K$ is a Galois extension. Then $\tilde{L}|K$ is also a Galois extension. If furthermore, $\sigma \in \operatorname{Aut}(\tilde{L}|\tilde{K})$, then $\sigma_{|L} \in \operatorname{Aut}(L|K)$, and*

$$\mathrm{Aut}(\tilde{L}|\tilde{K}) \to \mathrm{Aut}(L|K), \quad \textit{given by } \sigma \mapsto \sigma_{|L},$$

is an injective homomorphism.

Proof. (1) Let ω be a primitive nth root of unity. Then $\tilde{K} = K(\omega)$, and $\tilde{L} = L(\omega)$. Each $\sigma \in \mathrm{Aut}(\tilde{L}|L)$ maps ω onto a primitive nth root of unity, and fixes $K \subset L$ elementwise. Hence, from $\sigma \in \mathrm{Aut}(\tilde{L}|L)$, we get that $\sigma_{|\tilde{K}} \in \mathrm{Aut}(\tilde{K}|K)$. Certainly, the map $\sigma \mapsto \sigma_{|\tilde{K}}$ defines a homomorphism $\mathrm{Aut}(\tilde{L}|L) \to \mathrm{Aut}(\tilde{K}|K)$. Let $\sigma_{|\tilde{K}} = 1$ with $\sigma \in \mathrm{Aut}(\tilde{L}|L)$. Then $\sigma(\omega) = \omega$; therefore, we have already that $\sigma = 1$, since $\tilde{L} = L(\omega)$.

(2) If L is the splitting field of a polynomial $g(x)$ over K, then $\tilde{L}$ is the splitting field of $g(x)(x^n - 1)$ over K. Hence, $\tilde{L}|K$ is a Galois extension. Therefore, $K \subset L \subset \tilde{L}$, and $L|K, \tilde{L}|L$ and $\tilde{L}|K$ are all Galois extensions. Therefore, from the fundamental theorem of Galois theory

$$\mathrm{Aut}(L|K) = \{\sigma_{|L} : \sigma \in \mathrm{Aut}(\tilde{L}|K)\}.$$

In particular, $\sigma_{|L} \in \mathrm{Aut}(L|K)$ if $\sigma \in \mathrm{Aut}(\tilde{L}|\tilde{K})$. Certainly, the map $\mathrm{Aut}(\tilde{L}|\tilde{K}) \to \mathrm{Aut}(L|K)$, given by $\sigma \mapsto \sigma_{|L}$, is a homomorphism. From $\sigma \in \mathrm{Aut}(\tilde{L}|\tilde{K})$, we get that $\sigma(\omega) = \omega$, where—as above—ω is a primitive nth root of unity. Therefore, if $\sigma_{|L} = 1$, then already, $\sigma = 1$, since $\tilde{L} = L(\omega)$. Hence, the map is injective. □

17.2 Cyclotomic Extensions

Very important in the solvability by radicals problem are the splitting fields of the polynomials $x^n - 1$ over $\mathbb{Q}$. These are called *cyclotomic fields.*

Definition 17.2.1. The splitting field of the polynomial $x^n - 1 \in \mathbb{Q}[x]$ with $n \geq 2$ is called the nth *cyclotomic field* denoted by k_n.

We have $k_n = \mathbb{Q}(\omega)$, where ω is a primitive nth root of unity. For example, consider $\omega = e^{\frac{2\pi i}{n}}$ over $\mathbb{Q}$. $k_n|\mathbb{Q}$ is a Galois extension, and the Galois group $\mathrm{Aut}(k_n|\mathbb{Q})$ is the set of automorphisms $\sigma_m : \omega \to \omega^m$ with $1 \leq m \leq n$ and $\gcd(m, n) = 1$.

To understand this group G, we need the following concept: A *prime residue class modulo n* is a residue class $a + n\mathbb{Z}$ with $\gcd(a, n) = 1$. The set of the prime residue classes modulo n is just the set of invertible elements with respect to multiplication of the $\mathbb{Z}/n\mathbb{Z}$. This forms a multiplicative group that we denote by $(\mathbb{Z}/n\mathbb{Z})^\star = P_n$. We have $|P_n| = \phi(n)$, where $\phi(n)$ is the Euler phi-function. If $G = \mathrm{Aut}(k_n|\mathbb{Q})$, then $G \cong P_n$ under the map $\sigma_m \mapsto m + n\mathbb{Z}$. If $n = p$ is a prime number, then $G = \mathrm{Aut}(k_p|\mathbb{Q})$ is cyclic with $|G| = p - 1$.

If $n = p^2$, then $|G| = |\mathrm{Aut}(k_{p^2}|\mathbb{Q})| = p(p - 1)$, since

$$\frac{x^{p^2-1}}{x-1}\frac{x-1}{x^p-1} = x^{p(p-1)} + x^{p(p-1)-1} + \cdots + 1.$$

Lemma 17.2.2. *Let K be a field and $\tilde{K}$ be the splitting field of $x^n - 1$ over K. Then* $\mathrm{Aut}(\tilde{K}|K)$ *is Abelian.*

Proof. We apply Lemma 17.1.5 for the field extension $K|\mathbb{Q}$. This can be done since the characteristic of K is zero, and $\mathbb{Q}$ is the prime field of K. It follows that $\mathrm{Aut}(\tilde{K}|K)$ is isomorphic to a subgroup of $\mathrm{Aut}(\tilde{\mathbb{Q}}|\mathbb{Q})$ from part (1) of Lemma 17.1.5. But $\tilde{\mathbb{Q}} = k_n$, and hence $\mathrm{Aut}(\tilde{\mathbb{Q}}|\mathbb{Q})$ is Abelian. Therefore, $\mathrm{Aut}(\tilde{K}|K)$ is Abelian. □

17.3 Solvability and Galois Extensions

In this section, we prove that solvability by radicals is equivalent to the solvability of the Galois group.

Theorem 17.3.1. *Let $L|K$ be a Galois extension of K by radicals. Then $G =$* $\mathrm{Aut}(L|K)$ *is a solvable group.*

Proof. Suppose that $L|K$ is a Galois extension. Then we have a chain of fields

$$K = L_0 \subset L_1 \subset \cdots \subset L_r = L$$

such that $L_j = L_{j-1}(\sqrt[n_j]{a_j})$ for some $a_j \in L_j$. Let $n = n_1 \cdots n_r$, and let $\tilde{L}_j$ be the splitting field of the polynomial $x^n - 1 \in K[x]$ over L_j for each $j = 0, 1, \ldots, r$. Then $\tilde{L}_j = \tilde{L}_{j-1}(\sqrt[n_j]{a_j})$, and we get the chain

$$K \subset \tilde{K} = \tilde{L}_0 \subset \tilde{L}_1 \subset \cdots \subset \tilde{L}_r = \tilde{L}.$$

From part (2) of Lemma 17.1.5, we get that $\tilde{L}|K$ is a Galois extension. Furthermore, $\tilde{L}_j|\tilde{L}_{j-1}$ is a Galois extension with $\mathrm{Aut}(\tilde{L}_j|\tilde{L}_{j-1})$ cyclic from Theorem 17.1.2. In particular, $\mathrm{Aut}(\tilde{L}_j|\tilde{L}_{j-1})$ is Abelian. The group $\mathrm{Aut}(\tilde{K}|K)$ is Abelian from Lemma 17.2.2. Therefore, we may apply Lemma 17.1.4 to the chain

$$K \subset \tilde{K} = \tilde{L}_0 \subset \cdots \subset \tilde{L}_r = \tilde{L}.$$

Therefore, $\tilde{G} = \mathrm{Aut}(\tilde{L}|K)$ is solvable. The group $G = \mathrm{Aut}(L|K)$ is a homomorphic image of $\tilde{G}$ from the fundamental theorem of Galois theory. Since homomorphic images of solvable groups are still solvable (see Theorem 12.1.3), it follows that G is solvable. □

Lemma 17.3.2. *Let $L|K$ be a Galois extension, and suppose that $G =$* $\mathrm{Aut}(L|K)$ *is solvable. Assume further that K contains all q-th roots of unity for each prime divisor q of $m = [L : K]$. Then L is an extension of K by radicals.*

Proof. Let $L|K$ be a Galois extension, and suppose that $G = \mathrm{Aut}(L|K)$ is solvable; also assume that K contains all the q-th roots of unity for each prime divisor q of $m = [L : K]$. We prove the result by induction on m.

If $m = 1$, then $L = K$, and the result is clear. Now suppose that $m \geq 2$, and assume that the result holds for all Galois extensions $L'|K'$ with $[L' : K'] < m$. Now $G = \mathrm{Aut}(L|K)$ is solvable, and G is nontrivial since $m \geq 2$. Let q be a prime divisor of m. From Lemma 12.1.2 and Theorem 13.3.5, it follows that there is a normal subgroup H of G with G/H cyclic of order q. Let $E = \mathrm{Fix}(L, H)$. From the fundamental theorem of Galois theory, $E|K$ is a Galois extension with $\mathrm{Aut}(E|K) \cong G/H$, and hence $\mathrm{Aut}(E|K)$ is cyclic of order q. From Theorem 17.1.2, $E|K$ is a simple extension of K by a radical. The proof is completed if we can show that L is an extension of E by radicals.

The extension $L|E$ is a Galois extension, and the group $\mathrm{Aut}(L|E)$ is solvable, since it is a subgroup of $G = \mathrm{Aut}(L|K)$. Each prime divisor p of $[L : E]$ is also a prime divisor of $m = [L : K]$ by the degree formula. Hence, as an extension of K, the field E contains all the p-th roots of unity. Finally,

$$[L : E] = \frac{[L : K]}{[E : K]} = \frac{m}{q} < m.$$

Therefore, $L|E$ is an extension of E by radicals from the inductive assumption, completing the proof. □

17.4 The Insolvability of the Quintic Polynomial

We are now able to prove the insolvability of the quintic polynomial. This is one of the most important applications of Galois theory. As aforementioned, we do this by equating the solvability of a polynomial equation by radicals to the solvability of the Galois group of the splitting field of this polynomial.

Theorem 17.4.1. *Let K be a field of characteristic 0, and let $f(x) \in K[x]$. Suppose that L is the splitting field of $f(x)$ over K. Then the polynomial equation $f(x) = 0$ is solvable by radicals if and only if $\mathrm{Aut}(L|K)$ is solvable.*

Proof. Suppose first that $f(x) = 0$ is solvable by radicals. Then L is contained in an extension L' of K by radicals. Hence, L is contained in a Galois extension $\tilde{L}$ of K by radicals from Theorem 17.1.3. The group $\tilde{G} = \mathrm{Aut}(\tilde{L}|K)$ is solvable from Theorem 17.3.1. Furthermore, $L|K$ is a Galois extension. Therefore, the Galois group $\mathrm{Aut}(L|K)$ is solvable as a subgroup of $\tilde{G}$.

Conversely, suppose that the group $\mathrm{Aut}(L|K)$ is solvable. Let $q_1, \ldots, q_r$ be the prime divisors of $m = [K : K]$, and let $n = q_1 \cdots q_r$. Let $\tilde{K}$ and $\tilde{L}$ be the splitting fields of the polynomial $x^n - 1 \in K[x]$ over K and L, respectively. We have $\tilde{K} \subset \tilde{L}$. From part (2) of Lemma 17.1.5, we have that $\tilde{L}|K$ is a Galois extension, and $\mathrm{Aut}(\tilde{L}|\tilde{K})$ is isomorphic to a subgroup of $\mathrm{Aut}(L|K)$. From this, we first obtain that $[\tilde{L} : \tilde{K}] = |\mathrm{Aut}(\tilde{L}|\tilde{K})|$ is a divisor of $[L : K] = |\mathrm{Aut}(L|K)|$. Hence, each prime divisor q of $[\tilde{L} : \tilde{K}]$ is also a prime divisor of $[L : K]$. Therefore, $\tilde{L}$ is an extension by radicals of $\tilde{K}$ by Lemma 17.3.2. Since $\tilde{K} = K(\omega)$, where ω is a primitive n-th root of unity, we obtain that $\tilde{L}$ is also an extension of K by radicals. Therefore, L is contained in an extension $\tilde{L}$ of K by radicals; therefore, $f(x) = 0$ is solvable by radicals. □

Corollary 17.4.2. *Let K be a field of characteristic 0, and let $f(x) \in K[x]$ be a polynomial of degree m with $1 \leq m \leq 4$. Then the equation $f(x) = 0$ is solvable by radicals.*

Proof. Let L be the splitting field of $f(x)$ over K. The Galois group $\mathrm{Aut}(L|K)$ is isomorphic to the subgroup of the symmetric group S_m. Now the group S_4 is solvable via the chain

$$\{1\} \subset \mathbb{Z}_2 \subset D_2 \subset A_4 \subset S_4,$$

where $\mathbb{Z}_2$ is the cyclic group of order 2, and D_2 is the Klein 4-group, which is isomorphic to $\mathbb{Z}_2 \times \mathbb{Z}_2$. Because $S_m \subset S_4$ for $1 \leq m \leq 4$, it follows that $\mathrm{Aut}(L|K)$ is solvable. From Theorem 17.4.1, the equation $f(x) = 0$ is solvable by radicals. □

Corollary 17.4.2 uses the general theory to show that any polynomial equation of degree less than or equal to 4 is solvable by radicals. This, however, does not provide explicit formulas for the solutions. We present these below:

Let K be a field of characteristic 0, and let $f(x) \in K[x]$ be a polynomial of degree m with $1 \leq m \leq 4$. As mentioned above, we assume that K is the splitting field of the respective polynomial.

Case (1): If $\deg(f(x)) = 1$, then $f(x) = ax + b$ with $a, b \in K$ and $a \neq 0$. A zero is then given by $k = -\frac{b}{a}$.

Case (2): If $\deg(f(x)) = 2$, then $f(x) = ax^2 + bx + c$ with $a, b, c \in K$ and $a \neq 0$. The zeros are then given by the quadratic formula

$$k = \frac{-b \pm \sqrt{b^2 - 4ac}}{2a}.$$

We note that the quadratic formula holds over any field of characteristic not equal to 2. Whether there is a solution within the field K then depends on whether $b^2 - 4ac$ has a square root within K.

For the cases of degrees 3 and 4, we have the general forms of what are known as *Cardano's formulas*.

Case (3): If $\deg(f(x)) = 3$, then $f(x) = ax^3 + bx^2 + cx + d$ with $a, b, c, d \in K$ and $a \neq 0$. Dividing through by a, we may assume, without loss of generality, that $a = 1$.

By a substitution $x = y - \frac{b}{3}$, the polynomial is transformed into

$$g(y) = y^3 + py + q \in K[y].$$

Let L be the splitting field of $g(y)$ over K, and let $\alpha \in L$ be a zero of $g(y)$ so that

$$\alpha^3 + p\alpha + q = 0.$$

If $p = 0$, then $\alpha = \sqrt[3]{-q}$ so that $g(y)$ has the three zeros

$$\sqrt[3]{-q}, \quad \omega\sqrt[3]{-q}, \quad \omega^2\sqrt[3]{-q},$$

where ω is a primitive third root of unity, $\omega^3 = 1$ with $\omega \neq \omega^2$.

Now let $p \neq 0$, and let β be a zero of $x^2 - \alpha x - \frac{p}{3}$ in a suitable extension L' of L. We have $\beta \neq 0$, since $p \neq 0$. Hence, $\alpha = \beta - \frac{p}{3\beta}$. Putting this into the transformed cubic equation

$$\alpha^3 + p\alpha + q = 0,$$

we get

$$\beta^3 - \frac{p^3}{27\beta^3} + q = 0.$$

Define $\gamma = \beta^3$ and $\delta = (\frac{-p}{3\beta})^3$ so that

$$\gamma + \delta + q = 0.$$

Then

$$\gamma^2 + q\gamma - \left(\frac{p}{3}\right)^3 = 0 \quad \text{and} \quad -\frac{p^3}{27\delta} + \delta + q = 0 \quad \text{and} \quad \delta^2 + q\delta - \left(\frac{p}{3}\right)^3 = 0.$$

Hence, the zeros of the polynomial

$$x^2 + qx - \left(\frac{p}{3}\right)^3$$

are

$$\gamma, \delta = -\frac{q}{2} \pm \sqrt{\left(\frac{q}{2}\right)^2 + \left(\frac{p}{3}\right)^3}.$$

If we have $\gamma = \delta$, then both are equal to $-\frac{q}{2}$, and

$$\sqrt{\left(\frac{q}{2}\right)^2 + \left(\frac{p}{3}\right)^3} = 0.$$

Then from the definitions of γ, δ, we have $\gamma = \beta^3$, and $\delta = (\frac{-p}{3\beta})^3$. From above, $\alpha = \beta - \frac{p}{3\beta}$. Therefore, we get α by finding the cube roots of γ and δ.

There are certain possibilities and combinations with these cube roots, but because of the conditions, the cube roots of γ and δ are not independent. We must satisfy the condition

$$\sqrt[3]{\gamma}\sqrt[3]{\delta} = \beta\frac{-p}{3\beta} = -\frac{p}{3}.$$

Therefore, we get the final result:

The zeros of $g(y) = y^3 + py + q$ with $p \neq 0$ are

$$u + v, \quad \omega u + \omega^2 v, \quad \omega^2 u + \omega v,$$

where ω is a primitive third root of unity, and

$$u = \sqrt[3]{-\frac{q}{2} + \sqrt{\left(\frac{q}{2}\right)^2 + \left(\frac{p}{3}\right)^3}} \quad \text{and} \quad v = \sqrt[3]{-\frac{q}{2} - \sqrt{\left(\frac{q}{2}\right)^2 + \left(\frac{p}{3}\right)^3}}.$$

The above is known as the *cubic formula*, or *Cardano's formula*.

Case (4): If $\deg(f(x)) = 4$, then $f(x) = ax^4 + bx^3 + cx^2 + dx + e$ with $a, b, c, d, e \in K$ and $a \neq 0$. Dividing through by a, we may assume without loss of generality that $a = 1$.

By a substitution $x = y - \frac{b}{4}$, the polynomial $f(x)$ is transformed into

$$g(y) = y^4 + py^2 + qy + r.$$

We have to find the zeros of $g(y)$. Let x_1, x_2, x_3, x_4 be the solutions in the splitting field of the polynomial

$$y^4 + py^2 + qy + r = 0.$$

Then

$$0 = y^4 + py^2 + qy + r = (y - x_1)(y - x_2)(y - x_3)(y - x_4).$$

If we compare the coefficients, we get the following:

$$\begin{aligned} 0 &= x_1 + x_2 + x_3 + x_4, \\ p &= x_1x_2 + x_1x_3 + x_1x_4 + x_2x_3 + x_2x_4 + x_3x_4, \\ -q &= x_1x_2x_3 + x_1x_2x_4 + x_1x_3x_4 + x_2x_3x_4, \\ r &= x_1x_2x_3x_4. \end{aligned}$$

We define

$$\begin{aligned} y_1 &= (x_1 + x_2)(x_3 + x_4), \\ y_2 &= (x_1 + x_3)(x_2 + x_4), \\ y_3 &= (x_1 + x_4)(x_2 + x_3). \end{aligned}$$

From $x_1 + x_2 + x_3 + x_4 = 0$, we get

$$y_1 = -(x_1 + x_2)^2 = -(x_3 + x_4)^2, \quad \text{because } x_1 + x_2 = -(x_3 + x_4),$$
$$y_2 = -(x_1 + x_3)^2 = -(x_2 + x_4)^2, \quad \text{because } x_1 + x_3 = -(x_2 + x_4),$$
$$y_3 = -(x_1 + x_4)^2 = -(x_2 + x_3)^2, \quad \text{because } x_1 + x_4 = -(x_2 + x_3).$$

Let $y^3 + fy^2 + gy + h = 0$ be the cubic equation with the solutions y_1, y_2, and y_3. This polynomial $y^3 + fy^2 + gy + h$ is called the *cubic resolvent* of the equation of degree four.

If we compare the coefficients, we get the following:

$$f = -y_1 - y_2 - y_3,$$
$$g = y_1y_2 + y_1y_3 + y_2y_3,$$
$$h = -y_1y_2y_3.$$

Direct calculations leads to

$$f = -2p,$$
$$g = p^2 - 4r,$$
$$h = q^2.$$

Hence, the equation

$$y^3 - 2py^2 + (p^2 - 4r)y + q^2 = 0$$

is the resolvent of $y^4 + py^2 + qy + r = 0$. We now calculate the solutions y_1, y_2, y_3 of $y^3 - 2py^2 + (p^2 - 4r)y + q^2 = 0$ using Cardano's formula.

Then we substitute backwards, and get the following:

$$x_1 + x_2 = -(x_3 + x_4) = \pm\sqrt{-y_1},$$
$$x_1 + x_3 = -(x_2 + x_4) = \pm\sqrt{-y_2},$$
$$x_1 + x_4 = -(x_2 + x_3) = \pm\sqrt{-y_3}.$$

We add these equations, and get

$$3x_1 + x_2 + x_3 + x_4 = 2x_1 = \pm\sqrt{-y_1} \pm \sqrt{-y_2} \pm \sqrt{-y_3} \quad \Rightarrow \quad x_1 = \frac{\pm\sqrt{-y_1} \pm \sqrt{-y_2} \pm \sqrt{-y_3}}{2}.$$

The formulas for x_2, x_3, and x_4 follow analogously, and are of the same type as that for x_1.

By variation of the signs we get eight numbers $\pm x_1, \pm x_2, \pm x_3$ and $\pm x_4$. Four of them are the solutions of the equation

$$y^4 + py^3 + qy + r = 0.$$

The correct ones we get by putting into the equation. They are as follows:

$$x_1 = \frac{1}{2}(\sqrt{-y_1} + \sqrt{-y_2} + \sqrt{-y_3}),$$
$$x_2 = \frac{1}{2}(\sqrt{-y_1} - \sqrt{-y_2} - \sqrt{-y_3}),$$
$$x_3 = \frac{1}{2}(-\sqrt{-y_1} + \sqrt{-y_2} - \sqrt{-y_3}),$$
$$x_4 = \frac{1}{2}(-\sqrt{-y_1} - \sqrt{-y_2} + \sqrt{-y_3}).$$

The following theorem is due to Abel; it shows the insolvability of the general degree 5 polynomial over the rationals $\mathbb{Q}$.

Theorem 17.4.3. *Let L be the splitting field of the polynomial $f(x) = x^5 - 2x^4 + 2 \in \mathbb{Q}[x]$ over $\mathbb{Q}$. Then $\operatorname{Aut}(L|K) = S_5$, the symmetric group on 5 letters. Since S_5 is not solvable, the equation $f(x) = 0$ is not solvable by radicals.*

Proof. The polynomial $f(x)$ is irreducible over $\mathbb{Q}$ by the Eisenstein criterion. Furthermore, $f(x)$ has five zeros in the complex numbers $\mathbb{C}$ by the fundamental theorem of algebra (see Section 17.6). We claim that $f(x)$ has exactly 3 real zeros and 2 nonreal zeros, which then necessarily are complex conjugates. In particular, the 5 zeros are pairwise distinct.

To see the claim, notice first that $f(x)$ has at least 3 real zeros from the intermediate value theorem. As a real function, $f(x)$ is continuous, $f(-1) = -1 < 0$, $f(0) = 2 > 0$, so it must have a real zero between -1 and 0. Furthermore, we have $f(\frac{3}{2}) = -\frac{81}{3} < 0$ and $f(2) = 2 > 0$. Hence, there must be distinct real zeros between 0 and $\frac{3}{2}$, and between $\frac{3}{2}$ and 2. Suppose that $f(x)$ has more than 3 real zeros. Then $f'(x) = x^3(5x - 8)$ has at least 3 pairwise distinct real zeros from Rolle's theorem. But $f'(x)$ clearly has only 2 real zeros, so this is not the case. Therefore, $f(x)$ has exactly 3 real zeros, and hence 2 nonreal zeros that are complex conjugates.

Let L be the splitting field of $f(x)$. The field L lies in $\mathbb{C}$, and the restriction of the map $\delta : z \mapsto \bar{z}$ of $\mathbb{C}$ to L maps the set of zeros of $f(x)$ onto themselves. Therefore, δ is an automorphism of L. The map δ fixes the 3 real zeros and transposes the 2 nonreal zeros. From this, we now show that $\operatorname{Aut}(L|\mathbb{Q}) = \operatorname{Aut} L = G = S_5$, the full symmetric group on 5 symbols. Clearly, $G \subset S_5$, since G acts as a permutation group on the 5 zeros of $f(x)$.

Since δ transposes the 2 nonreal zeros, G (as a permutation group) contains at least one transposition. Since $f(x)$ is irreducible, G acts transitively on the zeros of $f(x)$. Let x_0 be one of the zeros of $f(x)$, and let G_{x_0} be the stabilizer of x_0.

Since G acts transitively, x_0 has five images under G; therefore, the index of the stabilizer must be 5 (see Chapter 10):

$$5 = [G : G_{x_0}],$$

which—by Lagrange's theorem—must divide the order of G. Therefore, from the Sylow theorems, G contains an element of order 5. Hence, G contains a 5-cycle and a transposition; therefore, by Theorem 11.4.3, it follows that $G = S_5$. Since S_5 is not solvable, it follows that $f(x)$ cannot be solved by radicals. □

Since Abel's theorem shows that there exists a degree 5 polynomial that cannot be solved by radicals, it follows that there can be no formula like Cardano's formula in terms of radicals for degree 5.

Corollary 17.4.4. *There is no general formula for solving by radicals a fifth degree polynomial over the rationals.*

We now show that this result can be further extended to any degree greater than 5.

Theorem 17.4.5. *For each $n \geq 5$, there exist polynomials $f(x) \in \mathbb{Q}[x]$ of degree n, for which the equation $f(x) = 0$ is not solvable by radicals.*

Proof. Let $f(x) = x^{n-5}(x^5 - 2x^4 + 2)$, and let L be the splitting field of $f(x)$ over $\mathbb{Q}$. Then $\mathrm{Aut}(L|\mathbb{Q}) = \mathrm{Aut}(L)$ contains a subgroup that is isomorphic to S_5. It follows that $\mathrm{Aut}(L)$ is not solvable; therefore, the equation $f(x) = 0$ is not solvable by radicals. □

This immediately implies the following:

Corollary 17.4.6. *There is no general formula for solving by radicals polynomial equations over the rationals of degree* 5 *or greater.*

17.5 Constructibility of Regular *n*-Gons

In Chapter 6, we considered certain geometric material related to field extensions. There, using general field extensions, we proved the impossibility of certain geometric compass and straightedge constructions. In particular, there were four famous insolvable (to the Greeks) construction problems. The first is the *squaring of the circle*. This problem is, given a circle, to construct using straightedge and compass a square having an area equal to that of the given circle. The second is the *doubling of the cube*. This problem is, given a cube of given side length, to construct, using a straightedge and compass, a side of a cube having double the volume of the original cube. The third problem is the *trisection of an angle*. This problem is to trisect a given angle using only a straightedge and compass. The final problem is the *construction of a regular n-gon*. This problems asks which regular n-gons could be constructed using only straightedge and compass. In Chapter 6, we proved the impossibility of the first 3 problems. Here, we use Galois theory to consider constructible n-gons.

Recall that a *Fermat number* is a positive integer of the form

$$F_n = 2^{2^n} + 1, \quad n = 0, 1, 2, 3, \ldots.$$

If a particular F_m is prime, it is called a *Fermat prime*.

Fermat believed that all the numbers in this sequence were primes. In fact, F_0, F_1, F_2, F_3, F_4 are all prime, but F_5 is composite and divisible by 641 (see exercises). It is still an open question whether or not there are infinitely many Fermat primes. It has been conjectured that there are only finitely many. On the other hand, if a number of the form $2^n + 1$ is a prime for some integer n, then it must be a Fermat prime; that is, n must be a power of 2.

We first need the following:

Theorem 17.5.1. *Let $p = 2^n + 1$, $n = 2^s$ with $s \geq 0$ be a Fermat prime. Then there exists a chain of fields*

$$\mathbb{Q} = L_0 \subset L_1 \subset \cdots \subset L_n = k_p,$$

where k_p is the p-th cyclotomic field such that

$$[L_j : L_{j-1}] = 2$$

for $j = 1, \ldots, n$.

Proof. The extension $k_p|\mathbb{Q}$ is a Galois extension, and $[k_p : \mathbb{Q}] = p - 1$. Furthermore, $\mathrm{Aut}(k_p)$ is cyclic of order $p - 1 = 2^n$. Hence, there is a chain of subgroups

$$\{1\} = U_n \subset U_{n-1} \subset \cdots \subset U_0 = \mathrm{Aut}(k_p)$$

with $[U_{j-1} : U_j] = 2$ for $j = 1, \ldots, n$. From the fundamental theorem of Galois theory, the fields $L_j = \mathrm{Fix}(k_p, U_j)$ with $j = 0, \ldots, n$ have the desired properties. □

The following corollaries describe completely the constructible n-gons, tying them to Fermat primes.

Corollary 17.5.2. *Consider the numbers $0, 1$, that is, a unit line segment or a unit circle. A regular p-gon with $p \geq 3$ prime is constructible from $\{0, 1\}$ using a straightedge and compass if and only if $p = 2^{2^s} + 1$, $s \geq 0$ is a Fermat prime.*

Proof. From Theorem 6.3.13, we have that if a regular p-gon is constructible with a straightedge and compass, then p must be a Fermat prime. The sufficiency follows from Theorem 17.5.1. □

We now extend this to general n-gons. Let $m, n \in \mathbb{N}$. Assume that we may construct from $\{0, 1\}$ a regular n-gon and a regular m-gon. In particular, this means that we may construct the real numbers $\cos(\frac{2\pi}{n})$, $\sin(\frac{2\pi}{n})$, $\cos(\frac{2\pi}{m})$, and $\sin(\frac{2\pi}{m})$. If the $\gcd(m, n) = 1$, then we may construct from $\{0, 1\}$ a regular mn-gon.

To see this, notice that

$$\cos\left(\frac{2\pi}{n} + \frac{2\pi}{m}\right) = \cos\left(\frac{2(n+m)\pi}{nm}\right) = \cos\left(\frac{2\pi}{n}\right)\cos\left(\frac{2\pi}{m}\right) - \sin\left(\frac{2\pi}{n}\right)\sin\left(\frac{2\pi}{m}\right),$$

and

$$\sin\left(\frac{2\pi}{n}+\frac{2\pi}{m}\right)=\sin\left(\frac{2(n+m)\pi}{nm}\right)=\sin\left(\frac{2\pi}{n}\right)\cos\left(\frac{2\pi}{m}\right)+\cos\left(\frac{2\pi}{n}\right)\sin\left(\frac{2\pi}{m}\right).$$

Therefore, we may construct from $\{0,1\}$ the numbers $\cos(\frac{2\pi}{mn})$ and $\sin(\frac{2\pi}{mn})$, because $\gcd(n+m,mn)=1$. Therefore, we may construct from $\{0,1\}$ a regular mn-gon.

Now let $p\geq 3$ be a prime. Then $[k_{p^2}:\mathbb{Q}]=p(p-1)$, which is not a power of 2. Therefore, from $\{0,1\}$ it is not possible to construct a regular p^2-gon. Hence, altogether we have the following:

Corollary 17.5.3. *Consider the numbers* $0,1$, *that is, a unit line segment or a unit circle. A regular n-gon with* $n\in\mathbb{N}$ *is constructible from* $\{0,1\}$ *using a straightedge and compass if and only if*
(i) $n=2^m$, $m\geq 0$ *or*
(ii) $n=2^m p_1p_2\cdots p_r$, $m\geq 0$, *and the* p_i *are pairwise distinct Fermat primes.*

Proof. Certainly we may construct a 2^m-gon. Furthermore, if $r,s\in\mathbb{N}$ with $\gcd(r,s)=1$, and if we can construct a regular rs-gon, then clearly, we may construct a regular r-gon and a regular s-gon. □

17.6 The Fundamental Theorem of Algebra

In this section we present a Galois theoretic proof of the fundamental theorem of Algebra that we have first studied in Section 7.3.

Theorem 17.6.1. *Each nonconstant polynomial* $f(x)\in\mathbb{C}[x]$, *where* $\mathbb{C}$ *is the field of complex numbers, has a zero in* $\mathbb{C}$. *Therefore,* $\mathbb{C}$ *is an algebraically closed field.*

Proof. Let $f(x)\in\mathbb{C}[x]$ be a nonconstant polynomial, and let K be the splitting field of $f(x)$ over $\mathbb{C}$. Since the characteristic of the complex numbers $\mathbb{C}$ is zero, this will be a Galois extension of $\mathbb{C}$. Since $\mathbb{C}$ is a finite extension of $\mathbb{R}$, this field K would also be a Galois extension of $\mathbb{R}$. The fundamental theorem of algebra asserts that K must be $\mathbb{C}$ itself, and hence the fundamental theorem of algebra is equivalent to the fact that any nontrivial Galois extension of $\mathbb{C}$ must be $\mathbb{C}$.

Let K be any finite extension of $\mathbb{R}$ with $|K:\mathbb{R}|=2^m q$, $(2,q)=1$. If $m=0$, then K is an odd-degree extension of $\mathbb{R}$. Since K is separable over $\mathbb{R}$, from the primitive element theorem, it is a simple extension, and hence $K=\mathbb{R}(\alpha)$, where the minimal polynomial $m_\alpha(x)$ over $\mathbb{R}$ has odd degree. However, odd-degree real polynomials always have a real zero, and therefore $m_\alpha(x)$ is irreducible only if its degree is one. But then, $\alpha\in\mathbb{R}$, and $K=\mathbb{R}$. Therefore, if K is a nontrivial finite extension of $\mathbb{R}$ of degree $2^m q$, we must have $m>0$. This shows more generally that there are no odd-degree finite extensions of $\mathbb{R}$.

Suppose that K is a degree 2 extension of $\mathbb{C}$. Then $K = \mathbb{C}(a)$ with $\deg m_a(x) = 2$, where $m_a(x)$ is the minimal polynomial of a over $\mathbb{C}$. But from the quadratic formula complex, quadratic polynomials always have zeros in $\mathbb{C}$, so a contradiction. Therefore, $\mathbb{C}$ has no degree 2 extensions.

Now, let K be a Galois extension of $\mathbb{C}$. Then K is also Galois over $\mathbb{R}$. Suppose that $|K : \mathbb{R}| = 2^m q$, $(2, q) = 1$. From the argument above, we must have $m > 0$. Consider the Galois group $G = \mathrm{Gal}(K/\mathbb{R})$. Then $|G| = 2^m q$, $m > 0$, $(2, q) = 1$. Thus, G has a 2-Sylow subgroup of order 2^m and index q (see Theorem 13.3.4). This would correspond to an intermediate field E with $|K : E| = 2^m$ and $|E : \mathbb{R}| = q$. However, then E is an odd-degree finite extension of $\mathbb{R}$. It follows that $q = 1$ and $E = \mathbb{R}$. Therefore, $|K : \mathbb{R}| = 2^m$, and $|G| = 2^m$.

Now, $|K : \mathbb{C}| = 2^{m-1}$ and suppose $G_1 = \mathrm{Gal}(K/\mathbb{C})$. This is a 2-group. If it were not trivial, then from Theorem 13.4.1 there would exist a subgroup of order 2^{m-2} and index 2. This would correspond to an intermediate field E of degree 2 over $\mathbb{C}$. However, from the argument above, $\mathbb{C}$ has no degree 2 extensions. It follows then that G_1 is trivial; that is, $|G_1| = 1$, so $|K : \mathbb{C}| = 1$, and $K = \mathbb{C}$, completing the proof. □

The fact that $\mathbb{C}$ is algebraically closed limits the possible algebraic extensions of the reals.

Corollary 17.6.2. *Let K be a finite field extension of the real numbers $\mathbb{R}$. Then $K = \mathbb{R}$ or $K = \mathbb{C}$.*

Proof. Since $|K : \mathbb{R}| < \infty$ by the primitive element theorem, $K = \mathbb{R}(a)$ for some $a \in K$. Then the minimal polynomial $m_a(x)$ of a over $\mathbb{R}$ is in $\mathbb{R}[x]$, and hence in $\mathbb{C}[x]$. Therefore, from the fundamental theorem of algebra it has a zero in $\mathbb{C}$. Hence, $a \in \mathbb{C}$. If $a \in \mathbb{R}$, then $K = \mathbb{R}$, if not, then $K = \mathbb{C}$. □

17.7 Exercises

1. For $f(x) \in \mathbb{Q}[x]$ with

$$f(x) = x^6 - 12x^4 + 36x^2 - 50$$
$$(f(x) = 4x^4 - 12x^2 + 20x - 3)$$

   determine for each complex zero a of $f(x)$ a finite number of radicals $\gamma_i = \beta_i^{\frac{1}{m_i}}$, $i = 1, \ldots, r$, and a presentation of a as a rational function in $\gamma_1, \ldots, \gamma_r$ over $\mathbb{Q}$ such that γ_{i+1} is irreducible over $\mathbb{Q}(\gamma_1, \ldots, \gamma_i)$, and $\beta_{i+1} \in \mathbb{Q}(\gamma_1, \ldots, \gamma_i)$ for $i = 0, \ldots, r-1$.
2. Let K be a field of prime characteristic p. Let $n \in \mathbb{N}$ and K_n be the splitting field of $x^n - 1$ over K. Show that $\mathrm{Aut}(K_n|K)$ is cyclic.
3. Let $f(x) = x^4 - x + 1 \in \mathbb{Z}[x]$. Show the following:
   (i) f has a real zero.
   (ii) f is irreducible over $\mathbb{Q}$.

(iii) If $u+iv$ $(u,v \in \mathbb{R})$ is a zero of f in $\mathbb{C}$, then $g = x^3-4x-1$ is the minimal polynomial of $4u^2$ over $\mathbb{Q}$.
(iv) The Galois group of f over $\mathbb{Q}$ has an element of order 3.
(v) No zero $a \in \mathbb{C}$ of f is constructible from the points 0 and 1 with straightedge and compass.
4. Show that each polynomial $f(x)$ over $\mathbb{R}$ decomposes in linear factors and quadratic factors $(f(x) = d(x-a_1)\cdot(x-a_2)\cdots(x^2+b_1x+c_1)\cdot(x^2+b_2x+c_2)\cdots, d \in \mathbb{R})$.
5. Let E be a finite (commutative) field extension of $\mathbb{R}$. Then $E \cong \mathbb{R}$, or $E \cong \mathbb{C}$.
6. (Vieta) Show that $y^3-py = q$ reduces to the form $4z^3-3z = c$ by a suitable substitution $y = mz$.
7. Suppose that $|a+id| = |c+id|$ and $|a+ib|^3 = c+id$. Show that the relation between a and c is $4a^3-3a = c$.
8. Show the identity of Bombelli:

$$\sqrt[3]{(2 \pm \sqrt{-121})} = 2 \pm \sqrt{-1},$$

and apply it on the equation $x^4 = 15x + 4$.
9. Solve the following equations:
(a) $x^3 - 2x + 3 = 0$.
(b) $x^4 + 2x^3 + 3x^2 - x - 2 = 0$.
10. Let $n \geq 1$ be a natural number and x an indeterminate over $\mathbb{C}$. Consider the polynomial $x^n - 1 \in \mathbb{Z}[x]$. In $\mathbb{C}[x]$ it decomposes in linear factors:

$$x^n - 1 = (x-\xi_1)(x-\xi_2)\cdots(x-\xi_n),$$

where the complex numbers

$$\xi_\nu = e^{2\pi i \frac{\nu}{n}} = \cos\frac{2\pi\nu}{n} + i\cdot\sin\frac{2\pi\nu}{n}, \quad 1 \leq \nu \leq n,$$

are all (different) n-th roots of unity, that is, especially $\xi_n = 1$. These ξ_ν form a multiplicative cyclic group $G = \{\xi_1, \xi_2, \ldots, \xi_n\}$ generated by ξ_1. It is $\xi_\nu = \xi_1^\nu$.
An n-th root of unity ξ_ν is called a *primitive n-th root of unity*, if ξ_ν is not an m-th root of unity for any $m < n$.
Show that the following are equivalent:
(i) ξ_ν is a primitive n-th root of unity.
(ii) ξ_ν is a generating element of G.
(iii) $\gcd(\nu, n) = 1$.
11. The polynomial $\phi_n(x) \in \mathbb{C}[x]$, whose zeros are exactly the primitive n-th roots of unity, is called the *n-th cyclotomic polynomial*. With Exercise 6 it is

$$\phi_n(x) = \prod_{\substack{1\leq\nu\leq n\\ \gcd(\nu,n)=1}} (x-\xi_\nu) = \prod_{\substack{1\leq\nu\leq n\\ \gcd(\nu,n)=1}} (x - e^{2\pi i\frac{\nu}{n}}).$$

The degree of $\phi_n(x)$ is the number of the integers $\{1, \dots, n\}$, which are coprime to n. Show the following:

(i) $x^n - 1 = \prod_{\substack{d \geq 1 \\ d|n}} \phi_d(x)$.

(ii) $\phi_n(x) \in \mathbb{Z}[x]$ for all $n \geq 1$.

(iii) $\phi_n(x)$ is irreducible over $\mathbb{Q}$ (and therefore also over $\mathbb{Z}$) for all $n \geq 1$.

12. Show that the Fermat numbers F_0, F_1, F_2, F_3, F_4 are all prime but F_5 is composite and divisible by 641.

18 The Theory of Modules

18.1 Modules over Rings

Recall that a vector space V over a field K is an Abelian group V with a scalar multiplication $\cdot : K \times V \to V$, satisfying the following:

(1) $f(v_1 + v_2) = fv_1 + fv_2$ for $f \in K$ and $v_1, v_2 \in V$.
(2) $(f_1 + f_2)v = f_1 v + f_2 v$ for $f_1, f_2 \in K$ and $v \in V$.
(3) $(f_1 f_2)v = f_1(f_2 v)$ for $f_1, f_2 \in K$ and $v \in V$.
(4) $1v = v$ for $v \in V$.

Vector spaces are the fundamental algebraic structures in linear algebra, and the study of linear equations. Vector spaces have been crucial in our study of fields and Galois theory, since any field extension is a vector space over any subfield. In this context, the degree of a field extension is just the dimension of the extension field as a vector space over the base field. If we modify the definition of a vector space to allow scalar multiplication from an arbitrary ring, we obtain a more general structure called a *module*. We will formally define this below. Modules generalize vector spaces, but the fact that the scalars do not necessarily have inverses makes the study of modules much more complicated. Modules will play an important role in both the study of rings and the study of Abelian groups. In fact, any Abelian group is a module over the integers $\mathbb{Z}$ so that modules, besides being generalizations of vector spaces, can also be considered as generalizations of Abelian groups.

In this chapter, we will introduce the theory of modules. In particular, we will extend to modules the basic algebraic properties such as the isomorphism theorems, which have been introduced earlier in presenting groups, rings, and fields. We restrict ourselves to commutative rings, so that throughout R is always a commutative ring. If R has an identity 1, then we always consider only the case that $1 \neq 0$. Throughout this chapter, we use letters $\mathfrak{a}, \mathfrak{b}, \mathfrak{c}, \mathfrak{m}, \ldots$ for ideals in R. For principal ideals, we write $\langle a \rangle$ or aR for the ideal generated by $a \in R$. We note, however, that the definition can be extended to include modules over noncommutative rings (see Chapter 22). In this case, we would speak of left modules and right modules.

Definition 18.1.1. Let $R = (R, +, \cdot)$ a commutative ring and $M = (M, +)$ an Abelian group. M together with a scalar multiplication $\cdot : R \times M \to M$, $(\alpha, x) \mapsto \alpha x$, is called a *R-module* or module over R if the following axioms hold:

(M1) $(\alpha + \beta)x = \alpha x + \beta x$,
(M2) $\alpha(x + y) = \alpha x + \alpha y$, and
(M3) $(\alpha\beta)x = \alpha(\beta x)$ for all $\alpha, \beta \in R$ and $x, y \in M$.

If R has an identity 1, then M is called an *unitary R-module*, if in addition

(M4) $1 \cdot x = x$ for all $x \in M$ holds.

https://doi.org/10.1515/9783111142524-018

In the following, R always is a commutative ring. If R contains an identity 1, then M always is an unitary R-module. If R has an identity 1, then we always assume $1 \neq 0$.

As usual, we have the rules:

$$0 \cdot x = 0, \quad a \cdot 0 = 0, \quad -(ax) = (-a)x = a(-x),$$

for all $a \in R$ and for all $x \in M$.

We next present a series of examples of modules.

Example 18.1.2. (1) If $R = K$ is a field, then a K-module is a K-vector space.
(2) Let $G = (G, +)$ be an Abelian group. If $n \in \mathbb{Z}$ and $x \in G$, then nx is defined as usual:

$$\begin{aligned} 0 \cdot x &= 0, \\ nx &= \underbrace{x + \cdots + x}_{n\text{-times}} \quad \text{if } n > 0, \quad \text{and} \\ nx &= (-n)(-x) \quad \text{if } n < 0. \end{aligned}$$

Then G is an unitary $\mathbb{Z}$-module via the scalar multiplication

$$\cdot : \mathbb{Z} \times G \to G, \quad (n, x) \mapsto nx.$$

(3) Let S be a subring of R. Then, via $(s, r) \mapsto sr$, the ring R itself becomes an S-module.
(4) Let K be a field, V a K-vector space, and $f : V \to V$ a linear map of V.
Let $p = \sum_i a_i t^i \in K[t]$. Then $p(f) := \sum_i a_i f^i$ defines a linear map of V, and V is an unitary $K[t]$-module via the scalar multiplication

$$K[t] \times V \to V, \quad (p, v) \mapsto pv := p(f)(v).$$

(5) If R is a commutative ring and $\mathfrak{a}$ is an ideal in R, then $\mathfrak{a}$ is a module over R.

Basic to all algebraic theory is the concept of substructures. Next we define *submodules*.

Definition 18.1.3. Let M be an R-module. $\emptyset \neq U \subset M$ is called a *submodule* of M if
(UMI) $(U, +) < (M, +)$ and
(UMII) $a \in R, u \in U \Rightarrow au \in U$; that is, $RU \subset U$.

Example 18.1.4. (1) In an Abelian group G, considered as a $\mathbb{Z}$-module, the subgroups are precisely the submodules.
(2) The submodules of R, considered as a R-module, are precisely the ideals.
(3) $Rx := \{ax : a \in R\}$ is a submodule of M for each $x \in M$.
(4) Let K be a field, V a K-vector space, and $f : V \to V$ a linear map of V. Let U be a submodule of V, considered as a $K[t]$-module as above.
Then the following holds:
(a) $U < V$.

(b) $pU = p(f)U \subset U$ for all $p \in K[t]$. In particular, $aU \subset U$ for $p = a \in K$ and $tU = f(U) \subset U$ for $p = t$; that is, U is an f-invariant subspace.
Also, on the other hand, $p(f)U \subset U$ for all $p \in K[t]$ if U is an f-invariant subspace.

We next extend to modules the concept of a generating system. For a single generator, as with groups, this is called *cyclic*.

Definition 18.1.5. A submodule U of the R-module M is called *cyclic* if there exists an $x \in M$ with $U = Rx$.

Example 18.1.4.(3) (above) is an example for a cyclic submodule.

As in vector spaces, groups, and rings, the following constructions are standard leading us to generating systems.

(1) Let M be a R-module and $\{U_i : i \in I\}$ a family of submodules. Then $\bigcap_{i \in I} U_i$ is a submodule of M.
(2) Let M be a R-module. If $A \subset M$, then we define

$$\langle A \rangle := \bigcap \{U : U \text{ submodule of M with } A \subset U\}.$$

$\langle A \rangle$ is the smallest submodule of M, which contains A. If R has an identity 1, then $\langle A \rangle$ is the set of all linear combinations $\sum_i \alpha_i a_i$ with all $\alpha_i \in R$, all $a_i \in A$. This holds because M is unitary, and $na = n(1 \cdot a) = (n \cdot 1)a$ for $n \in \mathbb{Z}$ and $a \in A$; that is, we may consider the pseudoproduct na as a real product in the module. Especially, if R has an identity 1, then $aR = \langle \{a\} \rangle =: \langle a \rangle$.

Definition 18.1.6. Let R have an identity 1. If $M = \langle A \rangle$, then A is called a generating system of M. M is called finitely generated if there are $a_1, \ldots, a_n \in M$ with $M = \langle \{a_1, \ldots, a_n\} \rangle =: \langle a_1, \ldots, a_n \rangle$.

The following is clear:

Lemma 18.1.7. *Let U_i be submodules of M, $i \in I$, I an index set. Then*

$$\left\langle \bigcup_{i \in I} U_i \right\rangle = \left\{ \sum_{i \in L} a_i : a_i \in U_i, L \subset I \text{ finite} \right\}.$$

We write $\langle \bigcup_{i \in I} U_i \rangle =: \sum_{i \in I} U_i$ and call this submodule the sum of the U_i. A sum $\sum_{i \in I} U_i$ is called a direct sum if for each representation of 0, as $0 = \sum a_i$, $a_i \in U_i$, it follows that all $a_i = 0$. This is equivalent to $U_i \cap \sum_{i \neq j} U_j = 0$ for all $i \in I$.

Notation: $\bigoplus_{i \in I} U_i$; and if $I = \{1, \ldots, n\}$, then we also write $U_1 \oplus \cdots \oplus U_n$.

In analogy with our previously defined algebraic structure, we extend to modules the concepts of quotient modules and module homomorphisms.

Definition 18.1.8. Let U be a submodule of the R-module M. Let M/U be the factor group. We define a (well defined) scalar multiplication:

$$R \times M/U \to M/U, \quad \alpha(x + U) := \alpha x + U.$$

With this M/U is a R-module, the *factor module* or *quotient module* of M by U. In M/U, we have the operations

$$(x + U) + (y + U) = (x + y) + U,$$

and

$$\alpha(x + U) = \alpha x + U.$$

A module M over a ring R can also be considered as a module over a quotient ring of R. The following is straightforward to verify (see exercises):

Lemma 18.1.9. *Let $\mathfrak{a} \lhd R$ an ideal in R and M a R-module. The set of all finite sums of the form $\sum \alpha_i x_i$, $\alpha_i \in \mathfrak{a}$, $x_i \in M$, is a submodule of M, which we denote by $\mathfrak{a}M$. The factor group $M/\mathfrak{a}M$ becomes a $R/\mathfrak{a}$-module via the well defined scalar multiplication*

$$(\alpha + \mathfrak{a})(m + \mathfrak{a}M) = \alpha m + \mathfrak{a}M.$$

If here R has an identity 1 and $\mathfrak{a}$ is a maximal ideal, then $M/\mathfrak{a}M$ becomes a vector space over the field $K = R/\mathfrak{a}$.

We next define module homomorphisms:

Definition 18.1.10. Let R be a ring and M, N be R-modules. A map $f : M \to N$ is called a *R-module homomorphism* (or R-linear) if

$$f(x + y) = f(x) + f(y) \quad \text{and} \quad f(\alpha x) = \alpha f(x)$$

for all $\alpha \in R$ and all $x, y \in M$. Endo-, epi-, mono-, iso- and automorphisms are defined analogously via the corresponding properties of the maps. If $f : M \to N$ and $g : N \to P$ are module homomorphisms, then $g \circ f : M \to P$ is also a module homomorphism. If $f : M \to N$ is an isomorphism, then also $f^{-1} : N \to M$.

We define kernel and image in the usual way:

$$\ker(f) := \{x \in M : f(x) = 0\},$$

and

$$\operatorname{im}(f) := f(M) = \{f(x) : x \in M\}.$$

The set $\ker(f)$ is a submodule of M, and $\operatorname{im}(f)$ is a submodule of N. As usual, f is injective if and only if $\ker(f) = \{0\}$.

If U is a submodule of M, then the map $x \mapsto x + U$ defines a module epimorphism (the canonical epimorphism) from M onto M/U with kernel U.

There are module isomorphism theorems. The proofs are straightforward extensions of the corresponding proofs for groups and rings.

Theorem 18.1.11 (Module isomorphism theorems). *Let M, N be R-modules.*

(1) *If $f : M \to N$ is a module homomorphism, then*

$$f(M) \cong M/\ker(f).$$

(2) *If U, V are submodules of the R-module M, then*

$$U/(U \cap V) \cong (U + V)/V.$$

(3) *If U and V are submodules of the R-module M with $U \subset V \subset M$, then*

$$(M/U)/(V/U) \cong M/V.$$

For the proofs, as for groups, just consider the map $f : U + V \to U/(U \cap V)$, $u + v \mapsto u + (U \cap V)$, which is well defined because $U \cap V$ is a submodule of U; then we have $\ker(f) = V$.

Note that $\alpha \mapsto \alpha\rho$, $\rho \in R$ fixed, defines a module homomorphism $R \to R$ if we consider R itself as a R-module.

18.2 Annihilators and Torsion

In this section, we define torsion for an R-module and a very important subring of R called the annihilator.

Definition 18.2.1. Let M be an R-module. For a fixed $a \in M$, consider the module homomorphism $\lambda_a : R \to M$, $\lambda_a(\alpha) := \alpha a$ where we consider R as an R-module. We call $\ker(\lambda_a)$ the *annihilator of a* denoted by $\operatorname{Ann}(a)$; that is,

$$\operatorname{Ann}(a) = \{\alpha \in R : \alpha a = 0\}.$$

Lemma 18.2.2. *The annihilator* $\operatorname{Ann}(a)$ *is a submodule of R and the module isomorphism theorem* (1) *gives* $R/\operatorname{Ann}(a) \cong Ra$.

We next extend the annihilator to whole submodules of M:

Definition 18.2.3. Let U be a submodule of the R-module M. The *annihilator* $\operatorname{Ann}(U)$ is defined to be

$$\operatorname{Ann}(U) := \{\alpha \in R : \alpha u = 0 \text{ for all } u \in U\}.$$

As for single elements, since $\operatorname{Ann}(U) = \bigcap_{u \in U} \operatorname{Ann}(u)$, then $\operatorname{Ann}(U)$ is a submodule of R. If $\rho \in R$, $u \in U$, then $\rho u \in U$; that means, if $u \in \operatorname{Ann}(U)$, then also $\rho u \in \operatorname{Ann}(U)$, because $(\alpha\rho)u = \alpha(\rho u) = 0$. Hence, $\operatorname{Ann}(U)$ is an ideal in R.

Suppose that G is an Abelian group. Then as aforementioned, G is a $\mathbb{Z}$-module. An element $g \in G$ is a *torsion element*, or has *finite order* if $ng = 0$ for some $n \in \mathbb{N}$. The set $\operatorname{Tor}(G)$ consists of all the torsion elements in G. An Abelian group is torsion-free if $\operatorname{Tor}(G) = \{0\}$.

Lemma 18.2.4. *Let G be an Abelian group. Then $\operatorname{Tor}(G)$ is a subgroup of G, and the factor group $G/\operatorname{Tor}(G)$ is torsion-free.*

We extend this concept now to general modules:

Definition 18.2.5. The R-module M is called *faithful* if $\operatorname{Ann}(M) = \{0\}$. We call an element $a \in M$ a *torsion element*, or element of finite order, if $\operatorname{Ann}(a) \neq \{0\}$. A module without torsion elements $\neq 0$ is called *torsion-free*. If the R-module M is torsion-free, then R has no zero divisors $\neq 0$.

Theorem 18.2.6. *Let R be an integral domain and M an R-module (by our agreement M is unitary). Let $\operatorname{Tor}(M) = T(M)$ be the set of torsion elements of M. Then $\operatorname{Tor}(M)$ is a submodule of M, and $M/\operatorname{Tor}(M)$ is torsion-free.*

Proof. If $m \in \operatorname{Tor}(M)$, $\alpha \in \operatorname{Ann}(m)$, $\alpha \neq 0$, and $\beta \in R$, then we get

$$\alpha(\beta m) = (\alpha\beta)m = (\beta\alpha)m = \beta(\alpha m) = 0;$$

that is, $\beta m \in \operatorname{Tor}(M)$, because $\alpha\beta \neq 0$ if $\beta \neq 0$ (R is an integral domain). Let m' another element of $\operatorname{Tor}(M)$ and $0 \neq \alpha' \in \operatorname{Ann}(m')$. Then $\alpha\alpha' \neq 0$, and

$$\alpha\alpha'(m + m') = \alpha\alpha' m + \alpha\alpha' m' = \alpha'(\alpha m) + \alpha(\alpha' m') = 0;$$

that is, $m + m' \in \operatorname{Tor}(M)$. Therefore, $\operatorname{Tor}(M)$ is a submodule.

Now, let $m + \operatorname{Tor}(M)$ be a torsion element in $M/\operatorname{Tor}(M)$. Let $\alpha \in R$, $\alpha \neq 0$ with $\alpha(m + \operatorname{Tor}(M)) = \alpha m + \operatorname{Tor}(M) = \operatorname{Tor}(M)$. Then $\alpha m \in \operatorname{Tor}(M)$. Hence, there exists a $\beta \in R$, $\beta \neq 0$, with $0 = \beta(\alpha m) = (\beta\alpha)m$. Since $\beta\alpha \neq 0$, we get that $m \in \operatorname{Tor}(M)$, and the torsion element $m + \operatorname{Tor}(M)$ is trivial. □

18.3 Direct Products and Direct Sums of Modules

Let M_i, $i \in I \neq \emptyset$, be a family of R-modules. On the direct product

$$P = \prod_{i \in I} M_i = \left\{ f : I \to \bigcup_{i \in I} M_i : f(i) \in M_i \text{ for all } i \in I \right\},$$

we define the module operations

$$+ : P \times P \to P \quad \text{and} \quad \cdot : R \times P \to P$$

via

$$(f + g)(i) := f(i) + g(i) \quad \text{and} \quad (af)(i) := af(i).$$

Together with these operations, $P = \prod_{i\in I} M_i$ is an R-module, *the direct product of the* M_i. If we identify f with the I-tuple of the images $f = (f_i)_{i\in I}$, then the sum and the scalar multiplication are componentwise. If $I = \{1, \ldots, n\}$ and $M_i = M$ for all $i \in I$, then we write, as usual, $M^n = \prod_{i\in I} M_i$.

We make the agreement that $\prod_{i\in I=\emptyset} M_i := \{0\}$.

$\bigoplus_{i\in I} M_i := \{f \in \prod_{i\in I} M_i : f(i) = 0 \text{ for almost all } i\}$ ("for almost all i" means that there are at most finitely many i with $f(i) \neq 0$) is a submodule of the direct product, called the *direct sum of the* M_i. If $I = \{1, \ldots, n\}$, then we write $\bigoplus_{i\in I} M_i = M_1 \oplus \cdots \oplus M_n$. Here, $\prod_{i=1}^n M_i = \bigoplus_{i=1}^n M_i$ for finite I.

Theorem 18.3.1. (1) *If $\pi \in S_I$ is a permutation of I, then*

$$\prod_{i\in I} M_i \cong \prod_{i\in I} M_{\pi(i)},$$

and

$$\bigoplus_{i\in I} M_i \cong \bigoplus_{i\in I} M_{\pi(i)}.$$

(2) *If $I = \dot\bigcup_{j\in J} I_j$, the disjoint union, then*

$$\prod_{i\in I} M_i \cong \prod_{j\in J}\Big(\prod_{i\in I_j} M_i\Big),$$

and

$$\bigoplus_{i\in I} M_i \cong \bigoplus_{j\in J}\Big(\bigoplus_{i\in I_j} M_i\Big).$$

Proof. For (1), consider the map $f \mapsto f \circ \pi$.

For (2), consider the map $f \mapsto \bigcup_{j\in J} f_j$, where $f_j \in \prod_{i\in I_j} M_i$ is the restriction of f onto I_j, and $\bigcup_{j\in J} f_j$ is on J, defined by $(\bigcup_{j\in J} f_j)(k) := f_k = f(k)$. □

Let $I \neq \emptyset$. If $M = \prod_{i\in I} M_i$, then we get in a natural manner module homomorphisms $\pi_i : M \to M_i$ via $f \mapsto f(i)$; π_i is called the *projection onto the ith component.* In duality, we define module homomorphisms $\delta_i : M_i \to \bigoplus_{i\in I} M_i \subset \prod_{i\in I} M_i$ via $\delta_i(m_i) = (n_j)_{j\in I}$, where $n_j = 0$ if $i \neq j$ and $n_i = m_i$. δ_i is called *the ith canonical injection.* If $I = \{1, \ldots, n\}$, then $\pi_i(a_1, \ldots, a_i, \ldots, a_n) = a_i$, and $\delta_i(m_i) = (0, \ldots, 0, m_i, 0, \ldots, 0)$.

We now consider universal properties.

Theorem 18.3.2 (Universal properties). *Let $A, M_i, i \in I \neq \emptyset$, be R-modules.*

(1) *If $\phi_i : A \to M_i, i \in I$, are module homomorphisms, then there exists exactly one module homomorphism $\phi : A \to \prod_{i \in I} M_i$ such that, for each i, the following diagram commutes:*

$$
\begin{array}{ccc}
\prod_{i \in I} M_i & \xrightarrow{\quad \pi_j \quad} & M_j \\
\phi \nwarrow & & \nearrow \phi_j \\
& A &
\end{array}
$$

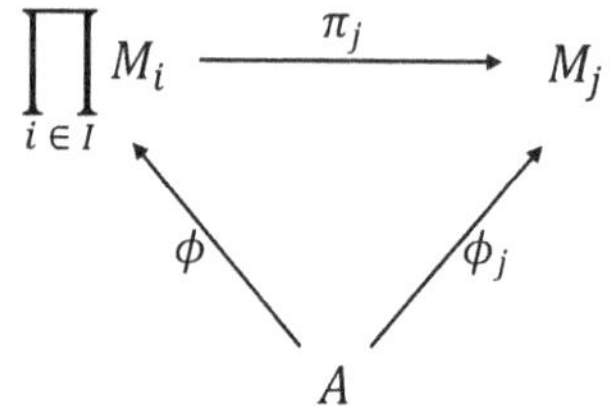

that is, $\phi_j = \pi_j \circ \phi$ where π_j is the jth projection.

(2) *If $\Psi_i : M_i \to A, i \in I$, are module homomorphisms then there exists exactly one module homomorphism $\Psi : \bigoplus_{i \in I} M_i \to A$ such that for each $j \in J$ the following diagram commutes:*

$$
\begin{array}{ccc}
\bigoplus_{i \in I} M_i & \xleftarrow{\quad \delta_j \quad} & M_j \\
\psi \searrow & & \swarrow \psi_j \\
& A &
\end{array}
$$

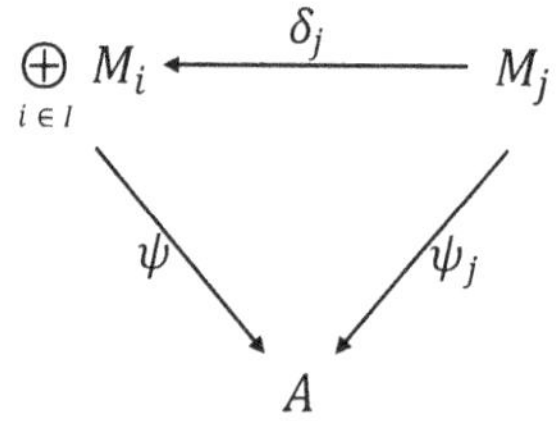

that is, $\Psi_j = \Psi \circ \delta_j$ where δ_j is the jth canonical injection.

Proof. We first consider (1). If there is such ϕ, then the jth component of $\phi(a)$ is equal $\phi_j(a)$, because $\pi_j \circ \phi = \phi_j$. Hence, define $\phi(a) \in \prod_{i \in I} M_i$ via $\phi(a)(i) := \phi_i(a)$, and ϕ is the desired map.

We now prove (2). If there is such a Ψ with $\Psi \circ a_j = \Psi_j$, then

$$
\Psi(x) = \Psi((x_i)) = \Psi\left(\sum_{i \in I} \delta_i(x_i)\right) = \sum_{i \in I} \Psi \circ \delta_i(x_i) = \sum_{i \in I} \Psi_i(x_i).
$$

Hence, define $\Psi((x_i)) = \sum_{i \in I} \Psi_i(x_i)$, and Ψ is the desired map (recall that the sum is well defined). $\qquad \square$

18.4 Free Modules

If V is a vector space over a field K, then V always has a basis over K, which may be infinite. Despite the similarity to vector spaces, because the scalars may not have inverses, this is not necessarily true for modules.

We now define a basis for a module. Those modules that actually have a basis are called *free modules*.

Let R be a ring with identity 1, M be a unitary R-module, and $S \subset M$. Each finite sum $\sum a_i s_i$, the $a_i \in R$, and the $s_i \in S$, is called a *linear combination in* S. Since M is unitary, and $S \neq \emptyset$, then $\langle S\rangle$ is exactly the set of all linear combinations in S. In the following, we assume that $S \neq \emptyset$. If $S = \emptyset$, then $\langle S\rangle = \langle\emptyset\rangle = \{0\}$, and this case is not interesting. For convention, in the following, we always assume $m_i \neq m_j$ if $i \neq j$ in a finite sum $\sum a_i m_i$ with all $a_i \in R$ and all $m_i \in M$.

Definition 18.4.1. A finite set $\{m_1, \ldots, m_n\} \subset M$ is called *linear independent* or *free (over R)* if a representation $0 = \sum_{i=1}^{n} a_i m_i$ implies always $a_i = 0$ for all $i \in \{1, \ldots, n\}$; that is, 0 can be represented only trivially on $\{m_1, \ldots, m_n\}$. A nonempty subset $S \subset M$ is called *free (over R)* if each finite subset of S is free.

Definition 18.4.2. Let M be an R-module (as above).

(1) $S \subset M$ is called a *basis* of M if
   (a) $M = \langle S\rangle$, and
   (b) S is free (over R).

(2) If M has a basis, then M is called a *free R-module*. If S is a basis of M, then M is called *free on S*, or *free with basis S*.

In this sense, we can consider $\{0\}$ as a free module with basis $\emptyset$.

Example 18.4.3. 1. $R \times R = R^2$, as an R-module, is free with basis $\{(1,0), (0,1)\}$.

2. More generally, let $I \neq \emptyset$. Then $\bigoplus_{i\in I} R_i$ with $R_i = R$ for all $i \in I$ is free with basis $\{\epsilon_i : I \to R : \epsilon_i(j) = \delta_{ij},\ i,j \in I\}$, where

$$\delta_{ij} = \begin{cases} 0 & \text{if } i \neq j, \\ 1 & \text{if } i = j. \end{cases}$$

   In particular, if $I = \{1, \ldots, n\}$, then $R^n = \{(a_1, \ldots, a_n) : a_i \in R\}$ is free with basis $\{\epsilon_i = (\underbrace{0, \ldots, 0}_{i-1}, 1, 0, \ldots, 0); 1 \le i \le n\}$.

3. Let G be an Abelian group. If G, as a $\mathbb{Z}$-module, is free on $S \subset G$, then G is called a *free Abelian group with basis S*. If $|S| = n < \infty$, then $G \cong \mathbb{Z}^n$.

Theorem 18.4.4. *The R-module M is free on S if and only if each $m \in M$ can be written uniquely in the form $\sum a_i s_i$ with $a_i \in R$, $s_i \in S$. This is exactly the case, where $M = \bigoplus_{s\in S} Rs$ is the direct sum of the cyclic submodules Rs, and each Rs is module isomorphic to R.*

Proof. If S is a basis then each $m \in M$ can be written as $m = \sum a_i s_i$, because $M = \langle S\rangle$. This representation is unique, because if $\sum a_i s_i = \sum \beta_i s_i$, then $\sum (a_i - \beta_i)s_i = 0$; that is, $a_i - \beta_i = 0$ for all i. If, on the other side, we assume that the representation is unique, then we get from $\sum a_i s_i = 0 = \sum 0 \cdot s_i$ that all $a_i = 0$, and therefore M is free on S.

The rest of the theorem, essentially, is a rewriting of the definition. If each $m \in M$ can be written as $m = \sum a_i s_i$, then $M = \sum_{s \in S} Rs$. If $x \in Rs' \cap \sum_{s \in S, s \neq s'} Rs$ with $s' \in S$, then $x = \alpha' s' = \sum_{s_i \neq s', s_i \in S} \alpha_i s_i$, and $0 = \alpha' s' - \sum_{s_i \neq s', s_i \in S} \alpha_i s_i$. Therefore, $\alpha' = 0$, and $\alpha_i = 0$ for all i. This gives $M = \bigoplus_{s \in S} Rs$. The cyclic modules Rs are isomorphic to $R/\operatorname{Ann}(s)$, and $\operatorname{Ann}(s) = \{0\}$ in the free modules. On the other side such modules are free on S. □

Corollary 18.4.5. (1) *M is free on $S \Leftrightarrow M \cong \bigoplus_{s \in S} R_s$, $R_s = R$ for all $s \in S$.*
(2) *If M is finitely generated and free, then there exists an $n \in \mathbb{N}_0$ such that*

$$M \cong R^n = \underbrace{R \oplus \cdots \oplus R}_{n\text{-times}}.$$

Proof. Part (1) is clear. For (2), let $M = \langle x_1, \ldots, x_r \rangle$ and S a basis of M. Each x_i is uniquely representable on S, as $x_i = \sum_{s_i \in S} \alpha_i s_i$. Since the x_i generates M, $m = \sum \beta_i x_i = \sum \beta_i \alpha_j s_j$ for arbitrary $m \in M$, and we need only finitely many s_j to generate M. Hence, S is finite. □

Theorem 18.4.6. *Let R be a commutative ring with identity 1, and M a free R-module. Then any two bases of M have the same cardinality.*

Proof. The ring R contains a maximal ideal $\mathfrak{m}$, and $R/\mathfrak{m}$ is a field (see Theorems 2.3.2 and 2.4.2). Then $M/\mathfrak{m}M$ is a vector space over $R/\mathfrak{m}$. From $M \cong \bigoplus_{s \in S} Rs$ with basis S, we get $\mathfrak{m}M \cong \bigoplus_{s \in S} \mathfrak{m}s$; hence,

$$M/\mathfrak{m}M \cong \left(\bigoplus_{s \in S} Rs\right)/\mathfrak{m}M \cong \bigoplus_{s \in S}(Rs/\mathfrak{m}M) \cong \bigoplus_{s \in S} R/\mathfrak{m}.$$

Therefore, the $R/\mathfrak{m}$-vector space $M/\mathfrak{m}M$ has a basis of the cardinality of S. This gives the result. □

Let R be a commutative ring with identity 1, and M a free R-module. The cardinality of a basis is an invariant of M, called the *rank of M* or *dimension of M*.

If $\operatorname{rank}(M) = n < \infty$, then this means $M \cong R^n$.

Theorem 18.4.7. *Each R-module is a (module-)homomorphic image of a free R-module.*

Proof. Let M be a R-module. We consider $F := \bigoplus_{m \in M} R_m$ with $R_m = R$ for all $m \in M$. F is a free R-module. The map $f : F \to M, f((\alpha_m)_{m \in M}) = \sum \alpha_m m$, defines a surjective module homomorphism. □

Theorem 18.4.8. *Let F, M be R-modules, and let F be free. Let $f : M \to F$ be a module epimorphism. Then there exists a module homomorphism $g : F \to M$ with $f \circ g = \mathrm{id}_F$, and we have $M = \ker(f) \oplus g(F)$.*

Proof. Let S be a basis of F. By the axiom of choice, there exists for each $s \in S$ an element $m_s \in M$ with $f(m_s) = s$ (f is surjective). We define the map $g : F \to M$ via $s \mapsto m_s$ linearly; that is, $g(\sum_{s_i \in S} \alpha_i s_i) = \sum_{s_i \in S} \alpha_i m_{s_i}$. Since F is free, the map g is well defined. Obviously, $f \circ g(s) = f(m_s) = s$ for $s \in S$; that means $f \circ g = \mathrm{id}_F$, because F is free on S. For each $m \in M$, we have also $m = g \circ f(m) + (m - g \circ f(m))$, where $g \circ f(m) = g(f(m)) \in g(F)$. Since $f \circ g = \mathrm{id}_F$, the elements of the form $m - g \circ f(m)$ are in the kernel of f. Therefore, $M = g(F) + \ker(f)$. Now let $x \in g(F) \cap \ker(f)$. Then $x = g(y)$ for some $y \in F$ and $0 = f(x) = f \circ g(y) = y$, and hence $x = 0$. Therefore, the sum is direct: $M = g(F) \oplus \ker(f)$. □

Corollary 18.4.9. *Let M be an R-module and N a submodule such that M/N is free. Then there is a submodule N' of M with $M = N \oplus N'$.*

Proof. Apply the above theorem for the canonical map $\pi : M \to M/N$ with $\ker(\pi) = N$. □

18.5 Modules over Principal Ideal Domains

We now specialize to the case of modules over principal ideal domains. For the remainder of this section, R is always a principal ideal domain $\neq \{0\}$. We now use the notation $(a) := aR$, $a \in R$, for the principal ideal aR.

Theorem 18.5.1. *Let M be a free R-module of finite rank over the principal ideal domain R. Then each submodule U is free of finite rank, and* $\mathrm{rank}(U) \leq \mathrm{rank}(M)$.

Proof. We prove the theorem by induction on $n = \mathrm{rank}(M)$. The theorem certainly holds if $n = 0$. Now let $n \geq 1$, and assume that the theorem holds for all free R-modules of rank $< n$. Let M be a free R-module of rank n with basis $\{x_1, \ldots, x_n\}$. Let U be a submodule of M. We represent the elements of U as linear combination of the basis elements $x_1, \ldots, x_n$, and we consider the set of coefficients of x_1 for the elements of U:

$$\mathfrak{a} = \left\{\beta \in R : \beta x_1 + \sum_{i=2}^{n} \beta_i x_i \in U\right\}.$$

Certainly $\mathfrak{a}$ is an ideal in R. Since R is a principal ideal domain, we have $\mathfrak{a} = (\alpha_1)$ for some $\alpha_1 \in R$. Let $u \in U$ be an element in U, which has α_1 as its first coefficient; that is

$$u = \alpha_1 x_1 + \sum_{i=2}^{n} \alpha_i x_i \in U.$$

Let $v \in U$ be arbitrary. Then

$$v = \rho(\alpha_1 x_1) + \sum_{i=2}^{n} \rho_i x_i.$$

Hence, $v - \rho u \in U' := U \cap M'$, where M' is the free R-module with basis $\{x_2, \ldots, x_n\}$. By induction, U' is a free submodule of M' with a basis $\{y_1, \ldots, y_t\}$, $t \leq n - 1$. If $\alpha_1 = 0$, then $\mathfrak{a} = (0)$, and $U = U'$, and there is nothing to prove. Now let $\alpha_1 \neq 0$. We show that $\{u, y_1, \ldots, y_t\}$ is a basis of U. $v - \rho u$ is a linear combination of the basis elements of U'; that is, $v - \rho u = \sum_{i=1}^{t} \eta_i y_i$ uniquely. Hence, $v = \rho u + \sum_{i=1}^{t} \eta_i y_i$, and $U = \langle u, y_1, \ldots, y_t \rangle$.

Now let be $0 = \gamma u + \sum_{i=1}^{t} \mu_i y_i$. We write u and the y_i as linear combinations in the basis elements $x_1, \ldots, x_n$ of M. There is only an x_1-portion in γu. Hence,

$$0 = \gamma a_1 x_1 + \sum_{i=2}^{n} \mu_i' x_i.$$

Therefore, first $\gamma a_1 x_1 = 0$; that is, $\gamma = 0$, because R has no zero divisor $\neq 0$, and furthermore, $\mu_2' = \cdots = \mu_n' = 0$. That means, $\mu_1 = \cdots = \mu_t = 0$. □

Let R be a principal ideal domain. Then the annihilator $\operatorname{Ann}(x)$ in R-modules M has certain further properties. Let $x \in M$. By definition

$$\operatorname{Ann}(x) = \{\alpha \in R : \alpha x = 0\} \lhd R, \quad \text{an ideal in } R,$$

hence $\operatorname{Ann}(x) = (\delta_x)$. If $x = 0$, then $(\delta_x) = R$. δ_x is called the *order of* x and (δ_x) the *order ideal of* x. δ_x is uniquely determined up to units in R (that is, up to elements η with $\eta\eta' = 1$ for some $\eta' \in R$). For a submodule U of M, we call $\operatorname{Ann}(U) = \bigcap_{u \in U} (\delta_u) = (\mu)$, the *order ideal* of U.

In an Abelian group G, considered as a $\mathbb{Z}$-module, this order for elements corresponds exactly to the order as group elements if we choose $\delta_x \geq 0$ for $x \in G$.

Theorem 18.5.2. *Let R be a principal ideal domain and M be a finitely generated torsion-free R-module. Then M is free.*

Proof. Let $M = \langle x_1, \ldots, x_n \rangle$ torsion-free and R a principal ideal domain. Each submodule $\langle x_i \rangle = R x_i$ is free, because M is torsion-free. We call a subset $S \subset \langle x_1, \ldots, x_n \rangle$ free if the submodule $\langle S \rangle$ is free. Since $\langle x_i \rangle$ is free, there exist such nonempty subsets. Under all free subsets $S \subset \langle x_1, \ldots, x_n \rangle$, we choose one with a maximal number of elements. We may assume that $\{x_1, \ldots, x_s\}$, $1 \leq s \leq n$, is such a maximal set—after possible renaming. If $s = n$, then the theorem holds. Now, let $s < n$. By the choice of s, the sets $\{x_1, \ldots, x_s, x_j\}$ with $s < j \leq n$ are not free. Hence, there are $\alpha_j \in R$, and $\alpha_i \in R$, not all 0, with

$$\alpha_j x_j = \sum_{i=1}^{s} \alpha_i x_i, \quad \alpha_j \neq 0, \; s < j \leq n.$$

For the product $\alpha := \alpha_{s+1} \cdots \alpha_n \neq 0$, we get $\alpha x_j \in R x_1 \oplus \cdots \oplus R x_s =: F$, $s < j \leq n$, because $\alpha x_i \in F$ for $1 \leq i \leq s$. Altogether, we get $\alpha M \subset F$. αM is a submodule of the free R-module F of rank s. By Theorem 18.5.1, we have that αM is free. Since $\alpha \neq 0$, and M is torsion-free, the map $M \to \alpha M$, $x \mapsto \alpha x$, defines an (module) isomorphism; that is, $M \cong \alpha M$. Therefore, also M is free. □

We remind that for an integral domain R, the set

$$\operatorname{Tor}(M) = T(M) = \{x \in M : \exists \alpha \in R, \alpha \neq 0, \text{ with } \alpha x = 0\}$$

of the torsion elements of an R-module M, is a submodule with torsion-free factor module $M/T(M)$.

Corollary 18.5.3. *Let R be a principal ideal domain and M be a finitely generated R-module. Then $M = T(M) \oplus F$ with a free submodule $F \cong M/T(M)$.*

Proof. $M/T(M)$ is a finitely generated, torsion-free R-module, and hence free. By Corollary 18.4.9, we have $M = T(M) \oplus F, F \cong M/T(M)$. □

From now on, we are interested in the case where $M \neq \{0\}$ is a torsion R-module; that is, $M = T(M)$. Let R be a principal ideal domain and $M = T(M)$ an R-module. Let $M \neq \{0\}$ and finitely generated. As above, let δ_x be the order of $x \in M$, unique up to units in R, and let $(\delta_x) = \{\alpha \in R : \alpha x = 0\}$ be the order ideal of x.

Let $(\mu) = \bigcap_{x \in M}(\delta_x)$ be the *order ideal of* M. Since $(\mu) \subset (\delta_x)$, we have $\delta_x | \mu$ for all $x \in M$. Since principal ideal domains are unique factorization domains, if $\mu \neq 0$, then there can not be many essentially different orders (that means, different up to units). Since $M \neq \{0\}$ and finitely generated, we have in any case $\mu \neq 0$, because if $M = \langle x_1, \ldots, x_n \rangle$, $\alpha_i x_i = 0$ with $\alpha_i \neq 0$, then $\alpha M = \{0\}$ if $\alpha := \alpha_1 \cdots \alpha_n \neq 0$.

Lemma 18.5.4. *Let R be a principal ideal domain and $M \neq \{0\}$ be an R-module with $M = T(M)$.*
(1) *If the orders δ_x and δ_y of $x, y \in M$ are relatively prime; that is,* $\gcd(\delta_x, \delta_y) = 1$, *then* $(\delta_{x+y}) = (\delta_x \delta_y)$.
(2) *Let δ_z be the order of $z \in M$, $z \neq 0$. If $\delta_z = \alpha\beta$ with* $\gcd(\alpha, \beta) = 1$, *then there exist $x, y \in M$ with $z = x + y$ and $(\delta_x) = (\alpha)$, $(\delta_y) = (\beta)$.*

Proof. (1) Since $\delta_x\delta_y(x + y) = \delta_x\delta_y x + \delta_x\delta_y y = \delta_y\delta_x x + \delta_x\delta_y y = 0$, we get $(\delta_x\delta_y) \subset (\delta_{x+y})$. On the other hand, from $\delta_x x = 0$ and $\delta_{x+y}(x + y) = 0$, we get $0 = \delta_x\delta_{x+y}(x + y) = \delta_x\delta_{x+y}y$; that means, $\delta_x\delta_{x+y} \in (\delta_y)$, and hence $\delta_y | \delta_x\delta_{x+y}$. Since $\gcd(\delta_x, \delta_y) = 1$, we have $\delta_y | \delta_{x+y}$. Analogously $\delta_x | \delta_{x+y}$. Hence, $\delta_x\delta_y | \delta_{x+y}$, and $(\delta_{x+y}) \subset (\delta_x\delta_y)$.

(2) Let $\delta_z = \alpha\beta$ with $\gcd(\alpha, \beta) = 1$. Then there are $\rho, \sigma \in R$ with $1 = \rho\alpha + \sigma\beta$. Therefore, we get

$$z = 1 \cdot z = \underbrace{\rho\alpha z}_{=:y} + \underbrace{\sigma\beta z}_{=:x} = y + x = x + y.$$

Since $\alpha x = \alpha\sigma\beta z = \sigma\delta_z z = 0$, we get $\alpha \in (\delta_z)$; that means, $\delta_x | \alpha$. On the other hand, from $0 = \delta_x x = \sigma\beta\delta_x z$, we get $\delta_z | \sigma\beta\delta_x$, and hence $\alpha\beta | \sigma\beta\delta_x$, because $\delta_z = \alpha\beta$. Therefore, $\alpha | \sigma\delta_x$. From $\gcd(\alpha, \sigma) = 1$, we get $\alpha | \delta_x$. Therefore, α is associated to δ_x; that is $\alpha = \delta_x \epsilon$ with ϵ a unit in R, and furthermore, $(\alpha) = (\delta_x)$. Analogously, $(\beta) = (\delta_y)$. □

In Lemma 18.5.4, we do not need $M = T(M)$. We only need $x, y, z \in M$ with $\delta_x \neq 0$, $\delta_y \neq 0$ and $\delta_z \neq 0$, respectively.

Corollary 18.5.5. *Let R be a principal ideal domain and $M \neq \{0\}$ be an R-module with $M = T(M)$.*
1. *Let $x_1, \ldots, x_n \in M$ be pairwise different and pairwise relatively prime orders $\delta_{x_i} = \alpha_i$. Then $y = x_1 + \cdots + x_n$ has order $\alpha := \alpha_1 \cdots \alpha_n$.*
2. *Let $0 \neq x \in M$ and $\delta_x = \epsilon\pi_1^{k_1} \cdots \pi_n^{k_n}$ be a prime decomposition of the order δ_x of x (ϵ a unit in R and the π_i pairwise nonassociate prime elements in R), where $n > 0$, $k_i > 0$. Then there exist x_i, $i = 1, \ldots, n$, with δ_{x_i} associated with $\pi_i^{k_i}$ and $x = x_1 + \cdots + x_n$.*

This is exercise 7.

18.6 The Fundamental Theorem for Finitely Generated Modules

In Section 10.4, we described the following result called the basis theorem for finite Abelian groups. In the following, we give a complete proof in detail; an elementary proof is given in Chapter 19:

Theorem 18.6.1 (Theorem 10.4.1, basis theorem for finite Abelian groups). *Let G be a finite Abelian group. Then G is a direct product of cyclic groups of prime power order.*

This allowed us, for a given finite order n, to present a complete classification of Abelian groups of order n. In this section, we extend this result to general modules over principal ideal domains. As a consequence, we obtain the fundamental decomposition theorem for finitely generated (not necessarily finite) Abelian groups, which finally proves Theorem 10.4.1. In the next chapter, we present a separate proof of this in a slightly different format.

Definition 18.6.2. Let R be a principal ideal domain and M be an R-module. Let $\pi \in R$ be a prime element. $M_\pi := \{x \in M : \exists k \geq 0 \text{ with } \pi^k x = 0\}$ is called the *π-primary component* of M. If $M = M_\pi$ for some prime element $\pi \in R$, then M is called *π-primary*.

We have the following:
1. M_π is a submodule of M.
2. The primary components correspond to the p-subgroup in Abelian groups.

Theorem 18.6.3. *Let R be a principal ideal domain and $M \neq \{0\}$ be an R-module with $M = T(M)$. Then M is the direct sum of its π-primary components.*

Proof. $x \in M$ has finite order δ_x. Let $\delta_x = \epsilon\pi_1^{k_1} \cdots \pi_n^{k_n}$ be a prime decomposition of δ_x. By Corollary 18.5.5, we have that $x = \sum x_i$ with $x_i \in M_{\pi_i}$. That means, $M = \sum_{\pi \in P} M_\pi$, where P is the set of the prime elements of R. Let $y \in M_\pi \cap \sum_{\sigma \in P, \sigma \neq \pi} M_\sigma$; that is, $\delta_y = \pi^k$ for some $k \geq 0$ and $y = \sum x_i$ with $x_i \in M_{\sigma_i}$. That means, $\delta_{x_i} = \sigma^{l_i}$ for some $l_i \geq 0$. By Corollary 18.5.5, we get that y has the order $\prod_{\sigma_i \neq \pi} \sigma_i^{l_i}$; that means, π^k is associated to $\prod_{\sigma_i \neq \pi} \sigma_i^{l_i}$. Therefore, $k = l_i = 0$ for all i, and the sum is direct. □

If R is a principal ideal domain and $\{0\} \neq M = T(M)$ a finitely generated torsion R-module, then there are only finitely many π-primary components. That is to say, for the prime elements, π with $\pi|\mu$, where (μ) is the order ideal of M.

Corollary 18.6.4. *Let R be a principal ideal domain and $\{0\} \neq M$ be a finitely generated torsion R-module. Then M has only finitely many nontrivial primary components $M_{\pi_1}, \dots, M_{\pi_n}$, and we have*

$$M = \bigoplus_{i=1}^{n} M_{\pi_i}.$$

Hence, we have a reduction of the decomposition problem to the primary components.

Theorem 18.6.5. *Let R be a principal ideal domain, $\pi \in R$ a prime element, and $M \neq \{0\}$ a R-module with $\pi^k M = \{0\}$; furthermore, let $m \in M$ with $(\delta_m) = (\pi^k)$. Then there exists a submodule $N \subset M$ with $M = Rm \oplus N$.*

Proof. By Zorn's lemma, the set $\{U : U$ submodule of M and $U \cap Rm = \{0\}\}$ has a maximal element N. This set is nonempty, because it contains $\{0\}$. We consider $M' := N \oplus Rm \subset M$, and have to show that $M' = M$. Assume that $M' \neq M$. Then there exists a $x \in M$ with $x \notin M'$, especially $x \notin N$. Then N is properly contained in the submodule $Rx + N = \langle x, N\rangle$. By our choice of N, we get $A := (Rx + N) \cap Rm \neq \{0\}$. If $z \in A$, $z \neq 0$, then $z = \rho m = \alpha x + n$ with $\rho, \alpha \in R$ and $n \in N$. Since $z \neq 0$, we have $\rho m \neq 0$; also $x \neq 0$, because otherwise $z \in Rm \cap N = \{0\}$; α is not a unit in R, because otherwise $x = \alpha^{-1}(\rho m - n) \in M'$. Hence we have: If $x \in M$, $x \notin M'$, then there exist $\alpha \in R$, $\alpha \neq 0$, α not a unit in R, $\rho \in R$ with $\rho m \neq 0$, and $n \in N$ such that

$$\alpha x = \rho m + n. \tag{$\star$}$$

In particular, $\alpha x \in M'$.

Now let $\alpha = \epsilon \pi_1 \cdots \pi_r$ be a prime decomposition. We consider one after the other the elements $x, \pi_r x, \pi_{r-1}\pi_r x, \dots, \epsilon\pi_1 \cdots \pi_r x = \alpha x$. We have $x \notin M'$, but $\alpha x \in M'$; hence, there exists an $y \notin M'$ with $\pi_i y \in N + Rm$.

1. $\pi_i \neq \pi$, π the prime element in the statement of the theorem. Then we have $\gcd(\pi_i, \pi^k) = 1$; hence, there are $\sigma, \sigma' \in R$ with $\sigma\pi_i + \sigma'\pi^k = 1$, and we get
$$Rm = (R\pi_i + R\pi^k)m = \pi_i Rm,$$
because $\pi^k m = 0$. Therefore, $\pi_i y \in M' = N \oplus Rm = N + \pi_i Rm$.
2. $\pi_i = \pi$. Then we write πy as $\pi y = n + \lambda m$ with $n \in N$ and $\lambda \in R$. This is possible, because $\pi y \in M'$. Since $\pi^k M = \{0\}$, we get $0 = \pi^{k-1} \cdot \pi y = \pi^{k-1} n + \pi^{k-1}\lambda m$. Therefore, $\pi^{k-1} n = \pi^{k-1}\lambda m = 0$, because $N \cap Rm = \{0\}$. In particular, we get $\pi^{k-1}\lambda \in (\delta_m)$; that is, $\pi^k | \pi^{k-1}\lambda$, and hence $\pi|\lambda$. Therefore, $\pi y = n + \lambda m = n + \pi\lambda' m \in N + \pi Rm$, $\lambda' \in R$.

Hence, in any case, we have $\pi_i y \in N + \pi_i Rm$; that is, $\pi_i y = n + \pi_i z$ with $n \in N$ and $z \in Rm$. It follows that $\pi_i(y - z) = n \in N$.

$y - z$ is not an element of M', because $y \notin M'$. By ($\star$), we have, therefore, $\alpha, \beta \in R$, $\beta \neq 0$ not a unit in R with $\beta(y - z) = n' + \alpha m$, $\alpha m \neq 0$, $n' \in N$. We write $z' = \alpha m$, then $z' \in Rm$, $z' \neq 0$, and $\beta(y-z) = n' + z'$. So, we have the equations $\beta(y-z) = n' + z'$, $z' \neq 0$, and

$$\pi_i(y - z) = n. \qquad (\star\star)$$

We have $\gcd(\beta, \pi_i) = 1$, because otherwise $\pi_i | \beta$ and, hence, $\beta(y - z) \in N$ and $z' = 0$, because $N \cap Rm = \{0\}$. Then there exist γ, γ' with $\gamma\pi_i + \gamma'\beta = 1$. In ($\star\star$), we multiply the first equation with γ' and the second with γ.

Addition gives $y - z \in N \oplus Rm = M'$, and hence $y \in M'$, which contradicts $y \notin M'$. Therefore, $M = M'$. □

Theorem 18.6.6. *Let R be a principal ideal domain, $\pi \in R$ a prime element, and $M \neq \{0\}$ a finitely generated π-primary R-module. Then there exist finitely many $m_1, \dots, m_s \in M$ with $M = \bigoplus_{i=1}^{s} Rm_i$.*

Proof. Let $M = \langle x_1, \dots, x_n \rangle$. Each x_i has an order π^{k_i}. We may assume that $k_1 = \max\{k_1, k_2, \dots, k_n\}$, possibly after renaming. We have $\pi^{k_i} x_i = 0$ for all i. Since $x_i^{k_1} = (x_i^{k_i})^{k_1 - k_i}$, we have also $\pi^{k_1} M = 0$, and also $(\delta_{x_1}) = (\pi^{k_1})$. Then $M = Rx_1 \oplus N$ for some submodule $N \subset M$ by Theorem 18.6.5. Now $N \cong M/Rx_1$, and M/Rx_1 is generated by the elements $x_2 + Rx_1, \dots, x_n + Rx_1$. Hence, N is finitely generated by $n - 1$ elements, and certainly N is π-primary. This proves the result by induction. □

Since $R_{m_i} \cong R/\operatorname{Ann}(m_i)$, and $\operatorname{Ann}(m_i) = (\delta_{m_i}) = (\pi^{k_i})$, we get the following extension of Theorem 18.6.6:

Theorem 18.6.7. *Let R be a principal ideal domain, $\pi \in R$ a prime element, and $M \neq \{0\}$ a finitely generated π-primary R-module. Then there exist finitely many $k_1, \dots, k_s \in \mathbb{N}$ with*

$$M \cong \bigoplus_{i=0}^{s} R/(\pi^{k_i}),$$

and M is, up to isomorphism, uniquely determined by $(k_1, \dots, k_s)$.

Proof. The first part, that is, a description as $M \cong \bigoplus_{i=0}^{s} R/(\pi^{k_i})$, follows directly from Theorem 18.6.6. Now, let

$$M \cong \bigoplus_{i=0}^{n} R/(\pi^{k_i}) \cong \bigoplus_{i=0}^{m} R/(\pi^{l_i}).$$

We may assume that $k_1 \geq k_2 \geq \cdots \geq k_n > 0$, and $l_1 \geq l_2 \geq \cdots \geq l_m > 0$. We consider first the submodule $N := \{x \in M : \pi x = 0\}$. Let $M = \bigoplus_{i=1}^{n} R/(\pi^{k_i})$.

If we then write $x = \sum(r_i + (\pi^{k_i}))$, we have $\pi x = 0$ if and only if $r_i \in (\pi^{k_i - 1})$; that is, $N \cong \bigoplus_{i=1}^{n} (\pi^{k_i-1})/(\pi^{k_i}) \cong \bigoplus_{i=1}^{n} R/(\pi)$, because $\pi^{k-1}R/\pi^k R \cong R/\pi R$.

Since $(\alpha + (\pi))x = \alpha x$ if $\pi x = 0$, we get that N is an $R/(\pi)$-module, and hence a vector space over the field $R/(\pi)$. From the decompositions

$$N \cong \bigoplus_{i=1}^{n} R/(\pi) \quad \text{and, analogously,} \quad N \cong \bigoplus_{i=1}^{m} R/(\pi),$$

we get

$$n = \dim_{R/(\pi)} N = m. \tag{$\star\star\star$}$$

Assume that there is an i with $k_i < l_i$ or $l_i < k_i$. Without loss of generality, assume that there is an i with $k_i < l_i$.

Let j be the smallest index, for which $k_j < l_j$. Then (because of the ordering of the k_i)

$$M' := \pi^{k_j} M \cong \bigoplus_{i=1}^{n} \pi^{k_j} R/\pi^{k_i} R \cong \bigoplus_{i=1}^{j-1} \pi^{k_j} R/\pi^{k_i} R,$$

because if $i > j$, then $\pi^{k_j} R/\pi^{k_i} R = \{0\}$.

We now consider $M' = \pi^{k_j} M$ with respect to the second decomposition; that is, $M' \cong \bigoplus_{i=1}^{m} \pi^{k_j} R/\pi^{l_i} R$. By our choice of j, we have $k_j < l_j \leq l_i$ for $1 \leq i \leq j$.

Therefore, in this second decomposition, the first j summands $\pi^{k_j} R/\pi^{l_i} R$ are unequal $\{0\}$; that is, $\pi^{k_j} R/\pi^{l_i} R \neq \{0\}$ if $1 \leq i \leq j$. The remaining summands are $\{0\}$, or of the form $R/\pi^s R$. Hence, altogether, on the one hand, M' is a direct sum of $j - 1$ cyclic submodules, and, on the other hand, a direct sum of $t \geq j$ nontrivial submodules. But this contradicts the above result $(\star\star\star)$ about the number of direct sums for finitely generated π-primary modules, because, certainly, M' is also finitely generated and π-primary. Therefore, $k_i = l_i$ for $i = 1, \ldots, n$. This proves the theorem. □

Theorem 18.6.8 (Fundamental theorem for finitely generated modules over principal ideal domains). *Let R be a principal ideal domain and $M \neq \{0\}$ be a finitely generated (unitary) R-module. Then there exist prime elements $\pi_1, \ldots, \pi_r \in R$, $0 \leq r < \infty$ and numbers $k_1, \ldots, k_r \in t\mathbb{N}$, $t \in \mathbb{N}_0$ such that*

$$M \cong R/(\pi_1^{k_1}) \oplus R/(\pi_2^{k_2}) \oplus \cdots \oplus R/(\pi_r^{k_r}) \oplus \underbrace{R \oplus \cdots \oplus R}_{t\text{-times}},$$

and M is, up to isomorphism, uniquely determined by $(\pi_1^{k_1}, \ldots, \pi_r^{k_r}, t)$.

The prime elements π_i are not necessarily pairwise different (up to units in R); that means, it can be $\pi_i = \epsilon \pi_j$ for $i \neq j$, where ϵ is a unit in R.

Proof. The proof is a combination of the preceding results. The free part of M is isomorphic to $M/T(M)$, and the rank of $M/T(M)$, which we call here t, is uniquely determined, because two bases of $M/T(M)$ have the same cardinality. Therefore, we may restrict ourselves on torsion modules. Here, we have a reduction to π-primary modules, because in a decomposition $M = \bigoplus_i R/(\pi_i^{k_i})$ is $M_\pi = \bigoplus_{\pi_i=\pi} R/(\pi_i^{k_i})$, the π-primary component of M (an isomorphism certainly maps a π-primary component onto a π-primary component). Therefore, it is only necessary, now, to consider π-primary modules M. The uniqueness statement now follows from Theorem 18.6.8: □

Since Abelian groups can be considered as $\mathbb{Z}$-modules, and $\mathbb{Z}$ is a principal ideal domain, we get the following corollary. We will restate this result in the next chapter and prove a different version of it.

Theorem 18.6.9 (Fundamental theorem for finitely generated Abelian groups). *Let $\{0\} \neq G = (G,+)$ be a finitely generated Abelian group. Then there exist prime numbers $p_1,\dots,p_r$, $0 \le r < \infty$, and numbers $k_1,\dots,k_r \in \mathbb{N}$, $t \in \mathbb{N}_0$ such that*

$$G \cong \mathbb{Z}/(p_1^{k_1}\mathbb{Z}) \oplus \cdots \oplus \mathbb{Z}/(p_r^{k_r}\mathbb{Z}) \oplus \underbrace{\mathbb{Z} \oplus \cdots \oplus \mathbb{Z}}_{t\text{-times}},$$

and G is, up to isomorphism, uniquely determined by $(p_1^{k_1},\dots,p_r^{k_r},t)$.

18.7 Exercises

1. Let M and N be isomorphic modules over a commutative ring R. Then $\mathrm{End}_R(M)$ and $\mathrm{End}_R(N)$ are isomorphic rings. ($\mathrm{End}_R(M)$ is the set of all R-modules endomorphisms of M.)
2. Let R be an integral domain and M an R-module with $M = \mathrm{Tor}(M)$ (torsion module). Show that $\mathrm{Hom}_R(M,R) = 0$. ($\mathrm{Hom}_R(M,R)$ is the set of all R-module homomorphisms from M to R.)
3. Prove the isomorphism theorems for modules (1), (2), and (3) in Theorem 18.1.11.
4. Let M, M', N be R-modules, R a commutative ring. Show the following:
   (i) $\mathrm{Hom}_R(M \oplus M', N) \cong \mathrm{Hom}_R(M,N) \times \mathrm{Hom}_R(M',N)$.
   (ii) $\mathrm{Hom}_R(N, M \times M') \cong \mathrm{Hom}_R(N,M) \oplus \mathrm{Hom}_R(N,M')$.
5. Show that two free R-modules having bases, whose cardinalities are equal are isomorphic.
6. Let M be an unitary R-module (R a commutative ring), and let $\{m_1,\dots,m_s\}$ be a finite subset of M. Show that the following are equivalent:
   (i) $\{m_1,\dots,m_s\}$ generates M freely.
   (ii) $\{m_1,\dots,m_s\}$ is linearly independent and generates M.
   (iii) Every element $m \in M$ is uniquely expressible in the form $m = \sum_{i=1}^{s} r_i m_i$ with $r_i \in R$.
   (iv) Each Rm_i is torsion-free, and $M = Rm_1 \oplus \cdots \oplus Rm_s$.
7. Let R be a principal domain and $M \neq \{0\}$ be an R-module with $M = T(M)$.
   (i) Let $x_1,\dots,x_n \in M$ be pairwise different and pairwise relatively prime orders $\delta_{x_i} = \alpha_i$. Then $y = x_1 + \cdots + x_n$ has order $\alpha := \alpha_1 \dots \alpha_n$.

(ii) Let $0 \neq x \in M$ and $\delta_x = \epsilon\pi_1^{k_1} \cdots \pi_n^{k_n}$ be a prime decomposition of the order δ_x of x (ϵ a unit in R and the π_i pairwise nonassociate prime elements in R), where $n > 0$, $k_i > 0$. Then there exist x_i, $i = 1, \dots, n$, with δ_{x_i} associated with $\pi_i^{k_i}$ and $x = x_1 + \cdots + x_n$.

19 Finitely Generated Abelian Groups

19.1 Finite Abelian Groups

In Chapter 10, we described the theorem below that completely provides the structure of finite Abelian groups. As we saw in Chapter 18, this result is a special case of a general result on modules over principal ideal domains.

Theorem 19.1.1 (Theorem 10.4.1, basis theorem for finite Abelian groups). *Let G be a finite Abelian group. Then G is a direct product of cyclic groups of prime power order.*

We review two examples that show how this theorem leads to the classification of finite Abelian groups. In particular, this theorem allows us, for a given finite order n, to present a complete classification of Abelian groups of order n.

Since all cyclic groups of order n are isomorphic to $(\mathbb{Z}_n, +)$, $\mathbb{Z}_n = \mathbb{Z}/n\mathbb{Z}$, we will denote a cyclic group of order n by $\mathbb{Z}_n$.

Example 19.1.2. Classify all Abelian groups of order 60. Let G be an Abelian group of order 60. From Theorem 10.4.1, G must be a direct product of cyclic groups of prime power order. Now $60 = 2^2 \cdot 3 \cdot 5$, so the only primes involved are 2, 3, and 5. Hence, the cyclic groups involved in the direct product decomposition of G have order either 2, 4, 3, or 5 (by Lagrange's theorem they must be divisors of 60). Therefore, G must be of the form

$$G \cong \mathbb{Z}_4 \times \mathbb{Z}_3 \times \mathbb{Z}_5,$$

or

$$G \cong \mathbb{Z}_2 \times \mathbb{Z}_2 \times \mathbb{Z}_3 \times \mathbb{Z}_5.$$

Hence, up to isomorphism, there are only two Abelian groups of order 60.

Example 19.1.3. Classify all Abelian groups of order 180. Let G be an Abelian group of order 180. Now $180 = 2^2 \cdot 3^2 \cdot 5$, so the only primes involved are 2, 3, and 5. Hence, the cyclic groups involved in the direct product decomposition of G have order either 2, 4, 3, 9, or 5 (by Lagrange's theorem they must be divisors of 180). Therefore, G must be of the form

$$G \cong \mathbb{Z}_4 \times \mathbb{Z}_9 \times \mathbb{Z}_5$$
$$G \cong \mathbb{Z}_2 \times \mathbb{Z}_2 \times \mathbb{Z}_9 \times \mathbb{Z}_5$$
$$G \cong \mathbb{Z}_4 \times \mathbb{Z}_3 \times \mathbb{Z}_3 \times \mathbb{Z}_5$$
$$G \cong \mathbb{Z}_2 \times \mathbb{Z}_2 \times \mathbb{Z}_3 \times \mathbb{Z}_3 \times \mathbb{Z}_5.$$

Therefore, up to isomorphism, there are four Abelian groups of order 180.

https://doi.org/10.1515/9783111142524-019

The proof of Theorem 19.1.1 involves the lemmas that follow. We refer back to Chapter 10 or Chapter 18 for the proofs. Notice how these lemmas mirror the results for finitely generated modules over principal ideal domains considered in the last chapter.

Lemma 19.1.4. *Let G be a finite Abelian group, and let $p||G|$, where p is a prime. Then all the elements of G, whose orders are a power of p form a normal subgroup of G. This subgroup is called the p-primary component of G, which we will denote by G_p.*

Lemma 19.1.5. *Let G be a finite Abelian group of order n. Suppose that $n = p_1^{e_1} \cdots p_k^{e_k}$ with $p_1, \ldots, p_k$ distinct primes.*

Then

$$G \cong G_{p_1} \times \cdots \times G_{p_k},$$

where G_{p_i} is the p_i-primary component of G.

Theorem 19.1.6 (Basis theorem for finite Abelian groups). *Let G be a finite Abelian group. Then G is a direct product of cyclic groups of prime power order.*

19.2 The Fundamental Theorem: p-Primary Components

In this section, we use the fundamental theorem for finitely generated modules over principal ideal domains to extend the basis theorem for finite Abelian groups to the more general case of finitely generated Abelian groups. We also consider the decomposition into p-primary components, mirroring our result in the finite case. In the next section, we present a different form of the basis theorem with a more elementary proof.

In Chapter 18, we proved the following:

Theorem 19.2.1 (Fundamental theorem for finitely generated modules over principal ideal domains). *Let R be a principal ideal domain and $M \neq \{0\}$ be a finitely generated (unitary) R-module. Then there exist prime elements $\pi_1, \ldots, \pi_r \in R$, $0 \leq r < \infty$ and numbers $k_1, \ldots, k_r \in \mathbb{N}$, $t \in \mathbb{N}_0$, such that*

$$M \cong R/(\pi_1^{k_1}) \oplus R/(\pi_2^{k_2}) \oplus \cdots \oplus R/(\pi_r^{k_r}) \oplus \underbrace{R \oplus \cdots \oplus R}_{t\text{-times}},$$

and M is, up to isomorphism, uniquely determined by $(\pi_1^{k_1}, \ldots, \pi_r^{k_r}, t)$.

The prime elements π_i are not necessarily pairwise different (up to units in R); that means, it can be $\pi_i = \epsilon\pi_j$ for $i \neq j$, where ϵ is a unit in R.

Since Abelian groups can be considered as $\mathbb{Z}$-modules, and $\mathbb{Z}$ is a principal ideal domain, we get the following corollary, which is extremely important in its own right.

Theorem 19.2.2 (Fundamental theorem for finitely generated Abelian groups). *Suppose $\{0\} \neq G = (G, +)$ is a finitely generated Abelian group. Then there exist prime numbers $p_1, \ldots, p_r$, $0 \leq r < \infty$, and numbers $k_1, \ldots, k_r \in \mathbb{N}$, $t \in \mathbb{N}_0$, such that*

$$G \cong \mathbb{Z}/(p_1^{k_1}\mathbb{Z}) \oplus \cdots \oplus \mathbb{Z}/(p_r^{k_r}\mathbb{Z}) \oplus \underbrace{\mathbb{Z} \oplus \cdots \oplus \mathbb{Z}}_{t\text{-times}},$$

and G is, up to isomorphism, uniquely determined by $(p_1^{k_1}, \ldots, p_r^{k_r}, t)$.

Notice that the number t of infinite components is unique. This is called the *rank* or *Betti number* of the Abelian group G. This number plays an important role in the study of homology and cohomology groups in topology.

If $G = \mathbb{Z} \times \mathbb{Z} \times \cdots \times \mathbb{Z} = \mathbb{Z}^r$ for some r, we call G a *free Abelian* group of rank r. Notice that if an Abelian group G is torsion-free, then the p-primary components are just the identity. It follows that, in this case, G is a free Abelian group of finite rank. Again, using module theory, it follows that subgroups of this must also be free Abelian and of smaller or equal rank. Notice the distinction between free Abelian groups and absolutely free groups (see Chapter 14). In the free group case, a non-Abelian free group of finite rank contains free subgroups of all possible countable ranks. In the free Abelian case, however, the subgroups have smaller or equal rank. We summarize these comments as follows:

Theorem 19.2.3. *Let $G \neq \{0\}$ be a finitely generated torsion-free Abelian group. Then G is a free Abelian group of finite rank r; that is, $G \cong \mathbb{Z}^r$. Furthermore, if H is a subgroup of G, then H is also free Abelian and the rank of H is smaller than or equal to the rank of G.*

19.3 The Fundamental Theorem: Elementary Divisors

In this section, we present the fundamental theorem of finitely generated Abelian groups in a slightly different form, and present an elementary proof of it.

In the following, G is always a finitely generated Abelian group. We use the addition "+" for the binary operation; that is,

$$+ : G \times G \to G, \quad (x, y) \mapsto x + y.$$

We also write ng instead of g^n, and use 0 as the symbol for the identity element in G; that is, $0 + g = g$ for all $g \in G$. $G = \langle g_1, \ldots, g_t \rangle$, $0 \leq t < \infty$. That is, G is (finitely) generated by $g_1, \ldots, g_t$, is equivalent to the fact that each $g \in G$ can be written in the form $g = n_1 g_1 + n_2 g_2 + \cdots + n_t g_t$, $n_i \in \mathbb{Z}$. A *relation* between the g_i with coefficients $n_1, \ldots, n_t$ is then each an equation of the form $n_1 g_1 + \cdots + n_t g_t = 0$. A relation is called *nontrivial* if $n_i \neq 0$ for at least one i. A system R of relations in G is called a *system of defining relations*, if each relation in G is a consequence of R. The elements $g_1, \ldots, g_t$ are called *integrally linear independent* if there are no nontrivial relations between them.

A finite generating system $\{g_1, \dots, g_t\}$ of G is called a *minimal generating system* if there is no generating system with $t-1$ elements.

Certainly, each finitely generated group has a minimal generating system. In what follow, we always assume that our finitely generated Abelian group G is unequal $\{0\}$; that is, G is nontrivial.

As above, we may consider G as a finitely generated $\mathbb{Z}$-module, and in this sense, the subgroups of G are precisely the submodules. Hence, it is clear what we mean if we call G a direct product $G = U_1 \times \cdots \times U_s$ of its subgroups $U_1, \dots, U_s$; namely, each $g \in G$ can be written as $g = u_1 + u_2 + \cdots + u_s$ with $u_i \in U_i$ and

$$U_i \cap \left(\prod_{j=1, j \neq i}^{s} U_j \right) = \{0\}.$$

To emphasize the little difference between Abelian groups and $\mathbb{Z}$-modules, here we use the notation "direct product" instead of "direct sum". Considered as $\mathbb{Z}$-modules, for finite index sets $I = \{1, \dots, s\}$, we have anyway

$$\prod_{i=1}^{s} U_i = \bigoplus_{i=1}^{s} U_i.$$

Finally, we use the notation $\mathbb{Z}_n$ instead of $\mathbb{Z}/n\mathbb{Z}$, $n \in \mathbb{N}$. In general, we use Z_n to be a cyclic group of order n.

The aim in this section is to prove the following:

Theorem 19.3.1 (Basis theorem for finitely generated Abelian groups). *Let $G \neq \{0\}$ be a finitely generated Abelian group. Then G is a direct product*

$$G \cong Z_{k_1} \times \cdots \times Z_{k_r} \times U_1 \times \cdots \times U_s,$$

$r \geq 0$, $s \geq 0$, of cyclic subgroups with $|Z_{k_i}| = k_i$ for $i = 1, \dots, r$, $k_i | k_{i+1}$ for $i = 1, \dots, r-1$ and $U_j \cong \mathbb{Z}$ for $j = 1, \dots, s$. Here, the numbers $k_1, \dots, k_r$, r, and s are uniquely determined by G; that means, if $k_1', \dots, k_r', r'$ and s' are the respective numbers for a second analogous decomposition of G, then $r = r'$, $k_1 = k_1', \dots, k_r = k_r'$, and $s = s'$.

The numbers k_i are called the *elementary divisors* of G.

We can have $r = 0$, or $s = 0$ (but not both, because $G \neq \{0\}$). If $s > 0$, $r = 0$, then G is a *free Abelian group of rank* s (exactly the same rank if you consider G as a free $\mathbb{Z}$-module of rank s). If $s = 0$, then G is finite. In fact, $s = 0$ if and only if G is finite.

We first prove some preliminary results:

Lemma 19.3.2. *Let $G = \langle g_1, \dots, g_t \rangle$, $t \geq 2$, an Abelian group.*
Then also $G = \langle g_1 + \sum_{i=2}^{t} m_i g_i, g_2, \dots, g_t \rangle$ for arbitrary $m_2, \dots, m_t \in \mathbb{Z}$.

Lemma 19.3.3. *Let G be a finitely generated Abelian group. Among all nontrivial relations between elements of minimal generating systems of G, we choose one relation,*

$$m_1 g_1 + \cdots + m_t g_t = 0 \qquad (\star)$$

with smallest possible positive coefficient, and let this smallest coefficient be m_1. *Let*

$$n_1 g_1 + \cdots + n_t g_t = 0 \qquad (\star\star)$$

be another relation between the same generators $g_1, \ldots, g_t$. *Then*

(1) $m_1 | n_1$, *and*

(2) $m_1 | m_i$ *for* $i = 1, 2, \ldots, t$.

Proof. For (1), assume $m_1 \nmid n_1$. Then $n_1 = q m_1 + m_1'$ with $0 < m_1' < m_1$. If we multiply the relation $(\star)$ with q and subtract the resulting relation from the relation $(\star\star)$, then we get a relation with a coefficient $m_1' < m_1$, contradicting the choice of m_1. Hence, $m_1 | n_1$.

For (2), assume $m_1 \nmid m_2$. Then $m_2 = q m_1 + m_2'$ with $0 < m_2' < m_2$. $\{g_1 + q g_2, g_2, \ldots, g_t\}$ is a minimal generating system, which satisfies the relation

$$m_1(g_1 + q g_2) + m_2' g_2 + m_3 g_3 + \cdots + m_t g_t = 0,$$

and this relation has a coefficient $m_2' < m_1$. This again contradicts the choice of m_1. Hence, $m_1 | m_2$, and furthermore, $m_1 | m_i$ for $i = 1, \ldots, t$. □

Lemma 19.3.4 (Invariant characterization of k_r for finite Abelian groups G). *Consider the group* $G = Z_{k_1} \times \cdots \times Z_{k_r}$ *with* Z_{k_i} *finite cyclic of order* $k_i \geq 2$, $i = 1, \ldots, r$ *and* $k_i | k_{i+1}$ *for* $i = 1, \ldots, r-1$. *Then* k_r *is the smallest natural number* n *such that* $ng = 0$ *for all* $g \in G$. k_r *is called the exponent or the maximal order of* G.

Proof. Let $g \in G$ arbitrary; that is, $g = n_1 g_1 + \cdots + n_r g_r$ with $g_i \in Z_{k_i}$. Then $k_i g_i = 0$ for $i = 1, \ldots, r$ by the theorem of Fermat. Since $k_i | k_r$, we get $k_r g = n_1 k_1 g_1 + \cdots + n_r k_r g_r = 0$. Let $a \in G$ with $Z_{k_r} = \langle a \rangle$. Then the order of a is k_r and, hence, $na \neq 0$ for all $0 < n < k_r$. □

Lemma 19.3.5 (Invariant characterization of s). *Let* $G = Z_{k_1} \times \cdots \times Z_{k_r} \times U_1 \times \cdots \times U_s$, $s > 0$, *where the* Z_{k_i} *are finite cyclic groups of order* k_i, *and the* U_j *are infinite cyclic groups. Then,* s *is the maximal number of integrally linear independent elements of* G; s *is called the rank of* G.

Proof. Let $g_i \in U_i$, $g_i \neq 0$, for $i = 1, \ldots, s$. Then the $g_1, \ldots, g_s$ are integrally linear independent, because from $n_1 g_1 + \cdots + n_s g_s = 0$, the $n_i \in \mathbb{Z}$, we get

$$n_1 g_1 \in U_1 \cap (U_2 \times \cdots \times U_s) = \{0\}.$$

Hence, $n_1 g_1 = 0$; that is, $n_1 = 0$, because g_1 has infinite order. Analogously, we get $n_2 = \cdots = n_s = 0$.

Let $g_1, \ldots, g_{s+1} \in G$. We look for integers $x_1, \ldots, x_{s+1}$, not all 0, such that a relation $\sum_{i=1}^{s+1} x_i g_i = 0$ holds. Let $Z_{k_i} \in \langle a_i \rangle$, $U_j = \langle b_j \rangle$. Then we may write each g_i as

$$g_i = m_{i1}a_1 + \cdots + m_{ir}a_r + n_{i1}b_1 + \cdots + n_{is}b_s$$

for $i = 1, \dots, s+1$, where $m_{ij}a_j \in Z_{k_j}$, and $n_{il}b_l \in U_l$.

Case 1: all $m_{ij}a_j = 0$. Then $\sum_{i=1}^{s+1} x_i g_i = 0$ is equivalent to

$$\sum_{i=1}^{s+1} x_i \left(\sum_{j=1}^{s} n_{ij} b_j \right) = \sum_{j=1}^{s} \left(\sum_{i=1}^{s+1} n_{ij} x_i \right) b_j = 0.$$

The system $\sum_{i=1}^{s+1} n_{ij} x_i = 0, j = 1, \dots, s$, of linear equations has at least one nontrivial rational solution $(x_1, \dots, x_{s+1})$, because we have more unknowns than equations. Multiplication with the common denominator gives a nontrivial integral solution $(x_1, \dots, x_{s+1}) \in \mathbb{Z}^{s+1}$. For this solution, we get

$$\sum_{i=1}^{s+1} x_i g_i = 0.$$

Case 2: $m_{ij}a_j$ arbitrary. Let $k \neq 0$ be a common multiple of the orders k_j of the cyclic groups $Z_{k_j}, j = 1, \dots, r$. Then

$$kg_i = \underbrace{m_{i1}ka_1}_{=0} + \cdots + \underbrace{m_{ir}ka_r}_{=0} + n_{i1}kb_1 + \cdots + n_{is}kb_s$$

for $i = 1, \dots, s+1$. By case 1, the $kg_1, \dots, kg_{s+1}$ are integrally linear dependent; that is, we have integers $x_1, \dots, x_{s+1}$, not all 0, with $\sum_{i=1}^{s+1} x_i(kg_i) = 0 = \sum_{i=1}^{s+1}(x_i k) g_i$, and the $x_i k$ are not all 0. Hence, also $g_1, \dots, g_{s+1}$ are integrally linear dependent. □

Lemma 19.3.6. *Let $G := Z_{k_1} \times \cdots \times Z_{k_r} \cong Z_{k'_1} \times \cdots \times Z_{k'_{r'}} =: G'$, the $Z_{k_i}, Z_{k'_j}$ cyclic groups of orders $k_i \neq 1$ and $k'_j \neq 1$, respectively, and $k_i | k_{i+1}$ for $i = 1, \dots, r-1$ and $k'_j | k'_{j+1}$ for $j = 1, \dots, r'-1$. Then $r = r'$, and $k_1 = k'_1, k_2 = k'_2, \dots, k_r = k'_r$.*

Proof. We prove this lemma by induction on the group order $|G| = |G'|$. Certainly, Lemma 19.3.6 holds if $|G| \leq 2$, because then, either $G = \{0\}$, and here $r = r' = 0$, or $G \cong \mathbb{Z}_2$, and here $r = r' = 1$. Now let $|G| > 2$. Then, in particular, $r \geq 1$. Inductively we assume that Lemma 19.3.6 holds for all finite Abelian groups of order less than $|G|$. By Lemma 19.3.4 the number k_r is invariantly characterized, that is, from $G \cong G'$ follows $k_r = k'_{r'}$, that is especially, $Z_{k_r} \cong Z_{k'_{r'}}$. Then $G/Z_{k_r} \cong G/Z_{k'_{r'}}$, that is,

$$Z_{k_1} \times \cdots \times Z_{k_{r-1}} \cong Z_{k'_1} \times \cdots \times Z_{k'_{r'-1}}.$$

Inductively, $r - 1 = r' - 1$; that is, $r = r'$, and $k_1 = k'_1, \dots, k_{r-1} = k'_{r'-1}$. □

We can now present the main result, which we state again, and its proof.

Theorem 19.3.7 (Basis theorem for finitely generated Abelian groups). *Let $G \neq \{0\}$ be a finitely generated Abelian group. Then G is a direct product*

$$G \cong Z_{k_1} \times \cdots \times Z_{k_r} \times U_1 \times \cdots \times U_s, \quad r \geq 0,\ s \geq 0,$$

of cyclic subgroups with $|Z_{k_i}| = k_i$ *for* $i = 1, \dots, r$, $k_i | k_{i+1}$ *for* $i = 1, \dots, r-1$, *and* $U_j \cong \mathbb{Z}$ *for* $j = 1, \dots, s$. *Here, the numbers* $k_1, \dots, k_r, r$, *and* s *are uniquely determined by* G; *that means, are* $k_1', \dots, k_r', r'$, *and* s', *the respective numbers for a second analogous decomposition of* G. *Then* $r = r'$, $k_1 = k_1', \dots, k_r = k_r'$, *and* $s = s'$.

Proof. We first prove the existence of the given decomposition. Let $G \neq \{0\}$ be a finitely generated Abelian group. Let t, $0 < t < \infty$, be the number of elements in a minimal generating system of G. We have to show that G is decomposable as a direct product of t cyclic groups with the given description. We prove this by induction on t. If $t = 1$, then the basis theorem is correct. Now let $t \geq 2$, and assume that the assertion holds for all Abelian groups with less then t generators.

Case 1: There does not exist a minimal generating system of G, which satisfies a nontrivial relation. Let $\{g_1, \dots, g_t\}$ be an arbitrary minimal generating system for G. Let $U_i = \langle g_i \rangle$. Then all U_i are infinite cyclic, and we have $G = U_1 \times \cdots \times U_t$, because if, for instance, $U_1 \cap (U_2 + \cdots + U_t) \neq \{0\}$, then we must have a nontrivial relation between the $g_1, \dots, g_t$.

Case 2: There exist minimal generating systems of G, which satisfy nontrivial relations. Among all nontrivial relations between elements of minimal generating systems of G, we choose one relation,

$$m_1 g_1 + \cdots + m_t g_t = 0 \tag{$\star$}$$

with smallest possible positive coefficient. Without loss of generality, let m_1 be this coefficient. By Lemma 19.3.3, we get $m_2 = q_2 m_1, \dots, m_t = q_t m_1$. Now,

$$\left\{ g_1 + \sum_{i=2}^{t} q_i g_i, g_2, \dots, g_t \right\}$$

is a minimal generating system of G by Lemma 19.3.2. Define $h_1 = g_1 + \sum_{i=2}^{t} q_i g_i$, then $m_1 h_1 = 0$. If $n_1 h_1 + n_2 g_2 + \cdots + n_t g_t = 0$ is an arbitrary relation between $h_1, g_2, \dots, g_t$, then $m_1 | n_1$ by Lemma 19.3.3; hence, $n_1 h_1 = 0$. Define $H_1 := \langle h_1 \rangle$, and $G' = \langle g_2, \dots, g_t \rangle$. Then $G = H_1 \times G'$. This we can see as follows: First, each $g \in G$ can be written as $g = m_1 h_1 + m_2 g_2 + \cdots + m_t g_t = m_1 h_1 + g'$ with $g' \in G'$. Also $H_1 \cap G' = \{0\}$, because $m_1 h_1 = g' \in G'$ implies a relation $n_1 h_1 + n_2 g_2 + \cdots + n_t g_t = 0$, and from this we get, as above, $n_1 h_1 = g' = 0$. Now, inductively, $G' = Z_{k_2} \times \cdots \times Z_{k_r} \times U_1 \times \cdots \times U_s$ with Z_{k_i} a cyclic group of order k_i, $i = 2, \dots, r$, $k_i | k_{i+1}$ for $i = 2, \dots, r-2$, $U_j \cong \mathbb{Z}$ for $j = 1, \dots, s$, and $(r-1) + s = t-1$; that is, $r + s = t$. Furthermore, $G = H_1 \times G'$, where H_1 is cyclic of order m_1. If $r \geq 2$ and $Z_{k_2} = \langle h_2 \rangle$, then we get a nontrivial relation

$$\underbrace{m_1 h_1}_{=0} + \underbrace{k_2 h_2}_{=0} = 0,$$

since $k_2 \neq 0$. Again $m_1 | k_2$ by Lemma 19.3.3. This gives the desired decomposition.

We now prove the uniqueness statement.

Case 1: G is finite Abelian. Then the claim follows from Lemma 19.3.6.

Case 2: G is arbitrary finitely generated and Abelian. Let $T := \{x \in G : |x| < \infty\}$; that is, the set of elements of G of finite order. Since G is Abelian, T is a subgroup of G, the so called *torsion subgroup of G*. If, as above,

$$G = Z_{k_1} \times \cdots \times Z_{k_r} \times U_1 \times \cdots \times U_s,$$

then $T = Z_{k_1} \times \cdots \times Z_{k_r}$, because an element $b_1 + \cdots + b_r + c_1 + \cdots + c_s$, $b_i \in Z_{k_i}$, $c_j \in U_j$ has finite order if and only if all $c_j = 0$. That means: $Z_{k_1} \times \cdots \times Z_{k_r}$ is independent of the special decomposition, uniquely determined by G; hence, also the numbers $r, k_1, \dots, k_r$ by Lemma 19.3.6. Finally, the number s, the rank of G, is uniquely determined by Lemma 19.3.5. □

As a corollary, we get the fundamental theorem for finitely generated Abelian groups as given in Theorem 19.2.1.

Theorem 19.3.8. *Let $\{0\} \neq G = (G, +)$ be a finitely generated Abelian group. Then there exist prime numbers $p_1, \dots, p_r$, $0 \leq r < \infty$, and numbers $k_1, \dots, k_r \in \mathbb{N}$, $t \in \mathbb{N}_0$ such that*

$$G \cong \mathbb{Z}_{p_1^{k_1}} \times \cdots \times \mathbb{Z}_{p_r^{k_r}} \times \underbrace{\mathbb{Z} \times \cdots \times \mathbb{Z}}_{t\text{-times}},$$

and G is, up to isomorphism, uniquely determined by $(p_1^{k_1}, \dots, p_r^{k_r}, t)$.

Proof. For the existence, we only have to show that $\mathbb{Z}_{mn} \cong \mathbb{Z}_m \times \mathbb{Z}_n$ if $\gcd(m, n) = 1$. For this, we write $U_n = \langle m + mn\mathbb{Z} \rangle < \mathbb{Z}_{mn}$, $U_m = \langle n + nm\mathbb{Z} \rangle < \mathbb{Z}_{mn}$, and $U_n \cap U_m = \{mn\mathbb{Z}\}$, because $\gcd(m, n) = 1$. Furthermore, there are $h, k \in \mathbb{Z}$ with $1 = hm + kn$. Hence, $l + mn\mathbb{Z} = hlm + mn\mathbb{Z} + kln + mn\mathbb{Z}$, and therefore $\mathbb{Z}_{mn} = U_n \times U_m \cong \mathbb{Z}_n \times \mathbb{Z}_m$.

For the uniqueness statement, we may reduce the problem to the case $|G| = p^k$ for a prime number p and $k \in \mathbb{N}$. But here the result follows directly from Lemma 19.3.6. □

From this proof, we automatically get the Chinese remainder theorem for the case $\mathbb{Z}_n = \mathbb{Z}/n\mathbb{Z}$.

Theorem 19.3.9 (Chinese remainder theorem). *Let $m_1, \dots, m_r \in \mathbb{N}$ with $r \geq 2$ and $\gcd(m_i, m_j) = 1$, for $i \neq j$. Define $m := m_1 \cdots m_r$.*

(1) *$\pi : \mathbb{Z}_m \to \mathbb{Z}_{m_1} \times \cdots \times \mathbb{Z}_{m_r}$, $a + m\mathbb{Z} \mapsto (a + m_1\mathbb{Z}, \dots, a + m_r\mathbb{Z})$, defines a ring isomorphism.*

(2) *The restriction of π on the multiplicative group of the prime residue classes defines a group isomorphism $\mathbb{Z}_m^\star \to \mathbb{Z}_{m_1}^\star \times \cdots \times \mathbb{Z}_{m_r}^\star$.*

(3) *For $a_1, \dots, a_r \in \mathbb{Z}$, there exists modulo m exactly one $x \in \mathbb{Z}$ with $x \equiv a_i \pmod{m_i}$ for $i = 1, \dots, r$.*

Recall that for $k \in \mathbb{N}$, a prime residue class is defined by $a + k\mathbb{Z}$ with $\gcd(a, k) = 1$. The set of prime residue classes modulo k is certainly a multiplicative group.

Proof. By Theorem 19.3.1, we get that π is an additive group isomorphism, which can be extended directly to a ring isomorphism via

$$(a + m\mathbb{Z})(b + m\mathbb{Z}) \mapsto (ab + m_1\mathbb{Z}, \ldots, ab + m_r\mathbb{Z}).$$

The remaining statements are now obvious. □

Let $A(n)$ be the number of nonisomorphic finite Abelian groups that have order $n = p_1^{k_1} \cdots p_r^{k_r}$, $r \geq 1$, with pairwise different primes $p_1, \ldots, p_r$ and $k_1, \ldots, k_r \in \mathbb{N}$. By Theorem 19.2.2, we have $A(n) = A(p_1^{k_1}) \cdots A(p_r^{k_r})$. Hence, to calculate $A(n)$, we have to calculate $A(p^m)$ for a prime number p and a natural number $m \in \mathbb{N}$. Again, by Theorem 19.2.2, we get $G \cong \mathbb{Z}_{p^{m_1}} \times \cdots \times \mathbb{Z}_{p^{m_k}}$, all $m_i \geq 1$, if G is Abelian of order p^m. If we compare the orders, we get $m = m_1 + \cdots + m_k$. We may order the m_i by size. A k-tuple $(m_1, \ldots, m_k)$ with $0 < m_1 \leq m_2 \leq \cdots \leq m_k$ and $m_1 + m_2 + \cdots + m_k = m$ is called a *partition of* m. From above, each Abelian group of order p^m gives a partition $(m_1, \ldots, m_k)$ of m for some k with $1 \leq k \leq m$. On the other hand, each partition $(m_1, \ldots, m_k)$ of m gives an Abelian group of order p^m, namely $\mathbb{Z}_{p^{m_1}} \times \cdots \times \mathbb{Z}_{p^{m_k}}$. Theorem 19.2.2 shows that different partitions give nonisomorphic groups. If we define $p(m)$ to be the number of partitions of m, then we get the following: $A(p^m) = p(m)$, and $A(p_1^{k_1} \cdots p_r^{k_r}) = p(k_1) \cdots p(k_r)$.

19.4 Exercises

1. Let H be a finite generated Abelian group, which is the homomorphic image of a torsion-free Abelian group of finite rank n. Show that H is the direct sum of $\leq n$ cyclic groups.
2. Determine (up to isomorphism) all groups of order p^2 (p prime) and all Abelian groups of order ≤ 15.
3. Let G be an Abelian group with generating elements $a_1, \ldots, a_4$ and defining relations

$$\begin{aligned} 5a_1 + 4a_2 + a_3 + 5a_4 &= 0 \\ 7a_1 + 6a_2 + 5a_3 + 11a_4 &= 0 \\ 2a_1 + 2a_2 + 10a_3 + 12a_4 &= 0 \\ 10a_1 + 8a_2 - 4a_3 + 4a_4 &= 0. \end{aligned}$$

   Express G as a direct product of cyclic groups.
4. Let G be a finite Abelian group and $u = \prod_{g \in G} g$, the product of all elements of G. Show: If G has exactly one element a of order 2, then $u = a$, otherwise $u = e$. Conclude from this the theorem of Wilson:

$$(p-1)! \equiv -1((\operatorname{mod} p)) \quad \text{for each prime } p.$$

5. Let p be a prime and G a finite Abelian p-group; that is, the order of all elements of G is finite and a power of p. Show that G is cyclic, if G has exactly one subgroup of order p. Is the statement still correct if G is not Abelian?

20 Integral and Transcendental Extensions

20.1 The Ring of Algebraic Integers

Recall that a complex number α is an *algebraic number* if it is algebraic over the rational numbers $\mathbb{Q}$. That is, α is a zero of a polynomial $p(x) \in \mathbb{Q}[x]$. If $\alpha \in \mathbb{C}$ is not algebraic, then it is a *transcendental number.*

We will let $\mathcal{A}$ denote the totality of algebraic numbers within the complex numbers $\mathbb{C}$, and $\mathcal{T}$ the set of transcendentals, so that $\mathbb{C} = \mathcal{A} \cup \mathcal{T}$. The set $\mathcal{A}$ is the algebraic closure of $\mathbb{Q}$ within $\mathbb{C}$.

The set $\mathcal{A}$ of algebraic numbers forms a subfield of $\mathbb{C}$ (see Chapter 5), and the subset $\mathcal{A}' = \mathcal{A} \cap \mathbb{R}$ of real algebraic numbers forms a subfield of $\mathbb{R}$. The field $\mathcal{A}$ is an algebraic extension of the rationals $\mathbb{Q}$. However, the degree is infinite.

Since each rational is algebraic, it is clear that there are algebraic numbers. Furthermore, there are irrational algebraic numbers, $\sqrt{2}$ for example, since it is a zero of the irreducible polynomial $x^2 - 2$ over $\mathbb{Q}$. In Chapter 5, we proved that there are uncountably infinitely many transcendental numbers (Theorem 5.5.3). However, it is very difficult to prove that any particular real or complex number is actually transcendental. In Theorem 5.5.4, we showed that the real number

$$c = \sum_{j=1}^{\infty} \frac{1}{10^{j!}}$$

is transcendental.

In this section, we examine a special type of algebraic number called an *algebraic integer*. These are the algebraic numbers that are zeros of monic integral polynomials. The set of all such algebraic integers forms a subring of $\mathbb{C}$. The proofs in this section can be found in [53].

After we do this, we extend the concept of an algebraic integer to a general context and define integral ring extensions. We then consider field extensions that are nonalgebraic—transcendental field extensions. Finally, we will prove that the familiar numbers e and π are transcendental.

Definition 20.1.1. An *algebraic integer* is a complex number α, that is, a zero of a *monic* integral polynomial. That is, $\alpha \in \mathbb{C}$ is an algebraic integer if there exists $f(x) \in \mathbb{Z}[x]$ with $f(x) = x^n + b_{n-1}x^{n-1} + \cdots + b_0$, $b_i \in \mathbb{Z}$, $n \geq 1$, and $f(\alpha) = 0$.

An algebraic integer is clearly an algebraic number. The following are clear:

Lemma 20.1.2. *If $\alpha \in \mathbb{C}$ is an algebraic integer, then all its conjugates, $\alpha_1, \ldots, \alpha_n$, over $\mathbb{Q}$ are also algebraic integers.*

Lemma 20.1.3. *$\alpha \in \mathbb{C}$ is an algebraic integer if and only if $m_\alpha \in \mathbb{Z}[x]$.*

https://doi.org/10.1515/9783111142524-020

To prove the converse of this lemma, we need the concept of a *primitive integral polynomial*. This is a polynomial $p(x) \in \mathbb{Z}[x]$ such that the GCD of all its coefficients is 1. The following can be proved (see exercises or Chapter 4):

(1) If $f(x)$ and $g(x)$ are primitive, then so is $f(x)g(x)$.
(2) If $f(x) \in \mathbb{Z}[x]$ is monic, then it is primitive.
(3) If $f(x) \in \mathbb{Q}[x]$, then there exists a rational number c such that $f(x) = cf_1(x)$ with $f_1(x)$ primitive.

Now suppose $f(x) \in \mathbb{Z}[x]$ is a monic polynomial with $f(\alpha) = 0$. Let $p(x) = m_\alpha(x)$. Then $p(x)$ divides $f(x)$ so $f(x) = p(x)q(x)$.

Let $p(x) = c_1p_1(x)$ with $p_1(x)$ primitive, and let $q(x) = c_2q_1(x)$ with $q_1(x)$ primitive. Then

$$f(x) = cp_1(x)q_1(x).$$

Since $f(x)$ is monic, it is primitive; hence $c = 1$, so $f(x) = p_1(x)q_1(x)$.

Since $p_1(x)$, and $q_1(x)$ are integral and their product is monic, they both must be monic. Since $p(x) = c_1p_1(x)$, and they are both monic, it follows that $c_1 = 1$. Hence, $p(x) = p_1(x)$. Therefore, $p(x) = m_\alpha(x)$ is integral.

When we speak of algebraic integers, we will refer to the ordinary integers as *rational integers*. The next lemma shows the close ties between algebraic integers and rational integers.

Lemma 20.1.4. *If α is an algebraic integer and also rational, then it is a rational integer.*

The following ties algebraic numbers in general to corresponding algebraic integers. Notice that if $q \in \mathbb{Q}$, then there exists a rational integer n such that $nq \in \mathbb{Z}$. This result generalizes this simple idea.

Theorem 20.1.5. *If θ is an algebraic number, then there exists a rational integer $r \neq 0$ such that $r\theta$ is an algebraic integer.*

We saw that the set $\mathcal{A}$ of all algebraic numbers is a subfield of $\mathbb{C}$. In the same manner, the set $\mathcal{I}$ of all algebraic integers forms a subring of $\mathcal{A}$. First, an extension of the following result on algebraic numbers.

Lemma 20.1.6. *Suppose $\alpha_1, \ldots, \alpha_n$ form the set of conjugates over $\mathbb{Q}$ of an algebraic integer α. Then any integral symmetric function of $\alpha_1, \ldots, \alpha_n$ is a rational integer.*

Theorem 20.1.7. *The set $\mathcal{I}$ of all algebraic integers forms a subring of $\mathcal{A}$.*

We note that $\mathcal{A}$, the field of algebraic numbers, is precisely the quotient field of the ring of algebraic integers.

An *algebraic number field* is a finite extension of $\mathbb{Q}$ within $\mathbb{C}$. Since any finite extension of $\mathbb{Q}$ is a simple extension, each algebraic number field has the form $K = \mathbb{Q}(\theta)$ for some algebraic number θ.

Let $K = \mathbb{Q}(\theta)$ be an algebraic number field, and let $R_K = K \cap \mathcal{I}$. Then R_K forms a subring of K called the algebraic integers, or integers of K. An analysis of the proof of Theorem 20.1.5 shows that each $\beta \in K$ can be written as

$$\beta = \frac{\alpha}{r}$$

with $\alpha \in R_K$ and $r \in \mathbb{Z}$.

These rings of algebraic integers share many properties with the rational integers. Whereas there may not be unique factorization into primes, there is always prime factorization.

Theorem 20.1.8. *Let K be an algebraic number field and R_K its ring of integers. Then each $\alpha \in R_K$ is either 0, a unit, or can be factored into a product of primes.*

We stress again that the prime factorization need not be unique. However, from the existence of a prime factorization, we can extend Euclid's original proof of the infinitude of primes (see [53]) to obtain the following:

Corollary 20.1.9. *There exist infinitely many primes in R_K for any algebraic number ring R_K.*

Just as any algebraic number field is finite-dimensional over $\mathbb{Q}$, we will see that each R_K is of finite degree over $\mathbb{Q}$. That is, if K has degree n over $\mathbb{Q}$, we show that there exists $\omega_1, \dots, \omega_n$ in R_K such that each $\alpha \in R_K$ is expressible as

$$\alpha = m_1\omega_1 + \dots + m_n\omega_n,$$

where $m_1, \dots, m_n \in \mathbb{Z}$.

Definition 20.1.10. An *integral basis* for R_K is a set of integers $\omega_1, \dots, \omega_t \in R_K$ such that each $\alpha \in R_K$ can be expressed uniquely as

$$\alpha = m_1\omega_1 + \dots + m_t\omega_t,$$

where $m_1, \dots, m_t \in \mathbb{Z}$.

The finite degree comes from the following result that shows there does exist an integral basis (see [53]):

Theorem 20.1.11. *Let R_K be the ring of integers in the algebraic number field K of degree n over $\mathbb{Q}$. Then there exists at least one integral basis for R_K.*

20.2 Integral Ring Extensions

We now extend the concept of an algebraic integer to general ring extensions. We first need the idea of an R-algebra, where R is a commutative ring with identity $1 \neq 0$.

Definition 20.2.1. Let R be a commutative ring with an identity $1 \neq 0$. An *R-algebra* or *algebra over R* is a unitary R-module A, in which there is an additional multiplication such that the following hold

(1) A is a ring with respect to the addition and this multiplication.
(2) $(rx)y = x(ry) = r(xy)$ for all $r \in R$ and $x, y \in A$.

As examples of R-algebras, first consider $R = K$, where K is a field, set $A = M(n, K)$, the set of all $(n \times n)$-matrices over K. Then $M(n, K)$ is a K-algebra. Furthermore, the set of polynomials $K[x]$ is also a K-algebra.

We now define *ring extensions*. Let A be a ring, not necessarily commutative, with an identity $1 \neq 0$, and R be a commutative subring of A, which contains 1. Assume that R is contained in the center of A; that is, $rx = xr$ for all $r \in R$ and $x \in A$. We then call A a *ring extension of R* and write $A|R$. If $A|R$ is a ring extension, then A is an R-algebra in a natural manner.

Let A be an R-algebra with an identity $1 \neq 0$. Then we have the canonical ring homomorphism $\phi : R \to A, r \mapsto r \cdot 1$. The image $R' := \phi(R)$ is a subring of the center of A, and R' contains the identity element of A. Then $A|R'$ is a ring extension (in the above sense). Hence, if A is a R-algebra with an identity $1 \neq 0$, then we may consider R as a subring of A and $A|R$ as a ring extension.

We now will extend to the general context of ring extensions the ideas of integral elements and integral extensions. As above, let R be a commutative ring with an identity $1 \neq 0$, and let A be an R-algebra.

Definition 20.2.2. An element $a \in A$ is said to be *integral* over R, or integrally dependent over R, if there is a monic polynomial $f(x) = x^n + a_{n-1}x^{n-1} + \cdots + a_0 \in R[x]$ of degree $n \geq 1$ over R with $f(a) = a^n + a_{n-1}a^{n-1} + \cdots + a_0 = 0$. That is, a is integral over R if it is a zero of a monic polynomial of degree ≥ 1 over R.

An equation that an integral element satisfies is called *integral equation of a over R*. If A has an identity $1 \neq 0$, then we may write $a^0 = 1$ and $\sum_{i=0}^n a_i a^i$ with $a_n = 1$.

Example 20.2.3. 1. Let $E|K$ be a field extension. $a \in E$ is integral over K if and only if a is algebraic over K. If K is the quotient field of an integral domain R, and $a \in E$ is algebraic over K. Then there exists an $\alpha \in R$ with αa integral over R, because if $0 = a_n a^n + \cdots + a_0$, thus, $0 = (a_n a)^n + \cdots + a_n^{n-1} a_0$.
2. The elements of $\mathbb{C}$, which are integral over $\mathbb{Z}$ are precisely the algebraic integers over $\mathbb{Z}$, that is, the zeros of monic polynomials over $\mathbb{Z}$.

Theorem 20.2.4. *Let R be as above and A an R-algebra with an identity $1 \neq 0$. If A is, as an R-module, finitely generated, then each element of A is integral over R.*

Proof. Let $\{b_1, \dots, b_n\}$ be a finite generating system of A, as an R-module. We may assume that $b_1 = 1$, otherwise add 1 to the system. As explained in the preliminaries, without loss of generality, we may assume that $R \subset A$. Let $a \in A$. For each $1 \le j \le n$, we have an equation $ab_j = \sum_{k=1}^{n} a_{kj} b_k$ for some $a_{kj} \in R$. In other words,

$$\sum_{k=1}^{n} (a_{kj} - \delta_{jk} a) b_k = 0 \qquad (\star\star)$$

for $j = 1, \dots, n$, where

$$\delta_{jk} = \begin{cases} 0 & \text{if } j \ne k, \\ 1 & \text{if } j = k. \end{cases}$$

Define $\gamma_{jk} := a_{kj} - \delta_{jk} a$ and $C = (\gamma_{jk})_{j,k}$. C is an $(n \times n)$-matrix over the commutative ring $R[a]$. Recall that $R[a]$ has an identity element. Let $\tilde{C} = (\tilde{\gamma}_{jk})_{j,k}$ be the complementary matrix of C (see for instance [9]). Then $\tilde{C}C = (\det C)E_n$. From $(\star\star)$, we get

$$0 = \sum_{j=1}^{n} \tilde{\gamma}_{ij} \left(\sum_{k=1}^{n} \gamma_{jk} b_k \right) = \sum_{k=1}^{n} \sum_{j=1}^{n} \tilde{\gamma}_{ij} \gamma_{jk} b_k = \sum_{k=1}^{n} (\det C) \delta_{ik} b_k = (\det C) b_i$$

for all $1 \le i \le n$. Since $b_1 = 1$, we have necessarily that $\det C = \det(a_{jk} - \delta_{jk} a)_{j,k} = 0$ (recall that $\delta_{jk} = \delta_{kj}$). Hence, a is a zero of the monic polynomial $f(x) = \det(\delta_{jk} x - a_{jk})$ in $R[x]$ of degree $n \ge 1$. Therefore, a is integral over R. □

Definition 20.2.5. A ring extension $A|R$ is called an *integral extension* if each element of A is integral over R. A ring extension $A|R$ is called *finite* if A, as a R-module, is finitely generated.

Recall that finite field extensions are algebraic extensions. As an immediate consequence of Theorem 20.2.4, we get the corresponding result for ring extensions.

Theorem 20.2.6. *Each finite ring extension $A|R$ is an integral extension.*

Theorem 20.2.7. *Let A be an R-algebra with an identity $1 \ne 0$. If $a \in A$, then the following are equivalent:*
(1) *a is integral over R.*
(2) *The subalgebra $R[a]$ is, as an R-module, finitely generated.*
(3) *There exists a subalgebra A' of A, which contains a, and which is, as an R-module, finitely generated.*

A subalgebra of an algebra over R is a submodule, which is also a subring.

Proof. (1) implies (2): We have $R[a] = \{g(a) : g \in R[x]\}$. Let $f(a) = 0$ be an integral equation of a over R. Since f is monic, by the division algorithm, for each $g \in R[x]$, there are $h, r \in R[x]$ with $g = h \cdot f + r$ and $r = 0$, or $r \neq 0$ and $\deg(r) < \deg(f) =: n$. Let $r \neq 0$. Since $g(a) = r(a)$, we get that $\{1, a, \dots, a^{n-1}\}$ is a generating system for the R-module $R[a]$.

(2) implies (3): Take $A' = R[a]$.

(3) implies (1): Use Theorem 20.2.4 for A'. □

For the remainder of this chapter, all rings are commutative with an identity $1 \neq 0$.

Theorem 20.2.8. *Let $A|R$ and $B|A$ be finite ring extensions. Then also $B|R$ is finite.*

Proof. From $A = Re_1 + \cdots + Re_m$, and $B = Af_1 + \cdots + Af_n$, we get $B = Re_1f_1 + \cdots + Re_mf_n$. □

Theorem 20.2.9. *Let $A|R$ be a ring extension. Then the following are equivalent:*
(1) *There are finitely many, over R integral elements $a_1, \dots, a_m$ in A such that*

$$A = R[a_1, \dots, a_m].$$

(2) *$A|R$ is finite.*

Proof. (2) ⇒ (1): We only need to take for $a_1, \dots, a_m$ a generating system of A as an R-module, and the result holds, because $A = Ra_1 + \cdots + Ra_m$, and each a_i is integral over R by Theorem 20.2.4.

(1) ⇒ (2): We use induction for m. If $m = 0$, then there is nothing to prove. Now let $m \geq 1$, and assume that (1) holds. Define $A' = R[a_1, \dots, a_{m-1}]$. Then $A = A'[a_m]$, and a_m is integral over A'. $A|A'$ is finite by Theorem 20.2.7. By the induction assumption, $A'|R$ is finite. Then $A|R$ is finite by Theorem 20.2.8. □

Definition 20.2.10. Let $A|R$ be a ring extension. Then the subset

$$C = \{a \in A : a \text{ is integral over } R\} \subset A$$

is called the *integral closure of R in A*.

Theorem 20.2.11. *Let $A|R$ be a ring extension. Then the integral closure of R in A is a subring of A with $R \subset A$.*

Proof. $R \subset C$, because $a \in R$ is a zero of the polynomial $x - a$. Let $a, b \in C$. We consider the subalgebra $R[a, b]$ of the R-algebra A. $R[a, b]|R$ is finite by Theorem 20.2.9. Hence, by Theorem 20.2.4, all elements from $R[a, b]$ are integral over R; that is, $R[a, b] \subset C$. In particular, $a + b$, $a - b$, and ab are in C. □

We extend to ring extensions the idea of a closure:

Definition 20.2.12. Let $A|R$ a ring extension. R is called *integrally closed in A*, if R itself is its integral closure in R; that is, $R = C$, the integral closure of R in A.

Theorem 20.2.13. *For each ring extension $A|R$, the integral closure C of R in A, is integrally closed in A.*

Proof. Let $a \in A$ be integral over C. Then $a^n + \alpha_{n-1}a^{n-1} + \cdots + \alpha_0 = 0$ for some $\alpha_i \in C$, $n \geq 1$. Then a is also integral over the R-subalgebra $A' = R[\alpha_0, \ldots, \alpha_{n-1}]$ of C, and $A'|R$ is finite. Furthermore, $A'[a]|A$ is finite. Hence, $A'[a]|R$ is finite. By Theorem 20.2.4, then $a \in A'[a]$ is already integral over R, that is, $a \in C$. □

Theorem 20.2.14. *Let $A|R$ and $B|A$ be ring extensions. If $A|R$ and $B|A$ are integral extensions, then also $B|R$ is an integral extension (and certainly vice versa).*

Proof. Let C be the integral closure of R in B. We have $A \subset C$, since $A|R$ is integral. Together with $B|A$, we also have that $B|C$ is integral. By Theorem 20.2.13, we get that C is integrally closed in B. Hence, $B = C$. □

We now consider integrally closed integral domains.

Definition 20.2.15. An integral domain R is called *integrally closed* if R is integrally closed in its quotient field K.

Theorem 20.2.16. *Each unique factorization domain R is integrally closed.*

Proof. Let $\alpha \in K$ and $\alpha = \frac{a}{b}$ with $a, b \in R$, $a \neq 0$. Since R is a unique factorization domain, we may assume that a and b are relatively prime. Let α be integral over R. Then we have over R an integral equation $\alpha^n + a_{n-1}\alpha^{n-1} + \cdots + a_0 = 0$ for α. Multiplication with b^n gives $a^n + ba_{n-1} + \cdots + b^n a_0 = 0$. Hence, b is a divisor of a^n. Since a and b are relatively prime in R, we have that b is a unit in R. Hence, $\alpha = \frac{a}{b} \in R$. □

Theorem 20.2.17. *Let R be an integral domain and K its quotient field. Let $E|K$ be a finite field extension. Let R be integrally closed and $\alpha \in E$ integral over R. Then the minimal polynomial $g \in K[x]$ of α over K has only coefficients of R.*

Proof. Let $g \in K[x]$ be the minimal polynomial of α over K (recall that g is monic by definition). Let $\bar{E}$ be an algebraic closure of E. Then $g(x) = (x - \alpha_1)\cdots(x - \alpha_n)$ with $\alpha_1 = \alpha$ over $\bar{E}$. There are K-isomorphisms $\sigma_i : K(\alpha) \to \bar{E}$ with $\sigma_i(\alpha) = \alpha_i$. Hence, all α_i are also integral over R. Since all coefficients of g are polynomial expressions $C_j(\alpha_1, \ldots, \alpha_n)$ in the α_i, we get that all coefficients of g are integral over R (see Theorem 20.2.11). Now $g \in R[x]$, because $g \in K[x]$, and R is integrally closed. □

Theorem 20.2.18. *Let R be an integrally closed integral domain and K its quotient field. Let $f, g, h \in K[x]$ be monic polynomials over K with $f = gh$.*

If $f \in R[x]$, then also $g, h \in R[x]$.

Proof. Let E be the splitting field of f over K. Over E, we have $f(x) = (x - \alpha_1)\cdots(x - \alpha_n)$. Since f is monic, all α_k are integral over R (see the proof of Theorem 20.2.17). Since $f = gh$, there are $I, J \subset \{1, \ldots, n\}$ with $g(x) = \prod_{i \in I}(x - \alpha_i)$ and $h(x) = \prod_{j \in J}(x - \alpha_j)$. As polynomial expressions in the α_i, $i \in I$, and α_j, $j \in J$, respectively, the coefficients of g and h, respectively, are integral over R. On the other hand, all these coefficients are in K, and R is integrally closed. Hence, $g, h \in R[x]$. □

Theorem 20.2.19. *Let $E|R$ be an integral ring extension. If E is a field, then also R is a field.*

Proof. Let $a \in R \setminus \{0\}$. The element $\frac{1}{a} \in E$ satisfies an integral equation

$$\left(\frac{1}{a}\right)^n + a_{n-1}\left(\frac{1}{a}\right)^{n-1} + \cdots + a_0 = 0$$

over R. Multiplication with a^{n-1} gives

$$\frac{1}{a} = -a_{n-1} - a_{n-2}a - \cdots - a_0 a^{n-1} \in R.$$

Hence, R is a field. □

20.3 Transcendental Field Extensions

Recall that a transcendental number is an element of $\mathbb{C}$ that is not algebraic over $\mathbb{Q}$. More generally, if $E|K$ is a field extension, then an element $\alpha \in E$ is *transcendental* over K if it is not algebraic; that is, it is not a zero of any polynomial $f(x) \in K[x]$. Since finite extensions are algebraic, clearly $E|K$ will contain transcendental elements only if $[E : K] = \infty$. However, this is not sufficient. The field $\mathcal{A}$ of algebraic numbers is algebraic over $\mathbb{Q}$, but infinite dimensional over $\mathbb{Q}$. We now extend the idea of a transcendental number to that of a transcendental extension.

Let $K \subset E$ be fields; that is, $E|K$ is a field extension. Let M be a subset of E. The *algebraic cover* of M in E is defined to be the algebraic closure $H(M)$ of $K(M)$ in E; that is, $H_{K,E}(M) = H(M) = \{\alpha \in E : \alpha \text{ algebraic over } K(M)\}$.

$H(M)$ is a field with $K \subset K(M) \subset H(M) \subset E$. $\alpha \in E$ is called *algebraically dependent on M (over K)* if $\alpha \in H(M)$; that is, if α is algebraic over $K(M)$.

The following are clear:

1. $M \subset H(M)$,
2. $M \subset M'$ implies $H(M) \subset H(M')$, and
3. $H(H(M)) = H(M)$.

Definition 20.3.1. (a) M is said to be *algebraically independent (over K)* if $\alpha \notin H(M \setminus \{\alpha\})$ for all $\alpha \in M$; that is, if each $\alpha \in M$ is transcendental over $K(M \setminus \{\alpha\})$.

(b) M is said to be *algebraically dependent (over K)* if M is not algebraically independent.

The proofs of the statements in the following lemma are straightforward:

Lemma 20.3.2. (1) *M is algebraically dependent if and only if there exists an $\alpha \in M$, which is algebraic over $K(M \setminus \{\alpha\})$.*

(2) *Let $\alpha \in M$. Then $\alpha \in H(M \setminus \{\alpha\}) \Leftrightarrow H(M) = H(M \setminus \{\alpha\})$.*

(3) *If $\alpha \notin M$ and α is algebraic over $K(M)$, then $M \cup \{\alpha\}$ is algebraically dependent.*

(4) *M is algebraically dependent if and only if there is a finite subset in M, which is algebraically dependent.*
(5) *M is algebraically independent if and only if each finite subset of M is algebraically independent.*
(6) *M is algebraically independent if and only if the following holds: If $\alpha_1, \dots, \alpha_n$ are finitely many, pairwise different elements of M, then the canonical homomorphism $\phi : K[x_1, \dots, x_n] \to E, f(x_1, \dots, x_n) \mapsto f(\alpha_1, \dots, \alpha_n)$ is injective; or in other words, for all $f \in K[x_1, \dots, x_n]$, we have that $f = 0$ iff $f(\alpha_1, \dots, \alpha_n) = 0$. That is, there is no nontrivial algebraic relation between the $\alpha_1, \dots, \alpha_n$ over K.*
(7) *Let $M \subset E$, $\alpha \in E$. If M is algebraically independent and $M \cup \{\alpha\}$ algebraically dependent, then $\alpha \in H(M)$; that is, α is algebraically dependent on M.*
(8) *Let $M \subset E$, $B \subset M$. If B is maximal algebraically independent, that is, if $\alpha \in M \setminus B$, then $B \cup \{\alpha\}$ is algebraically dependent, thus $M \subset H(B)$. That is, each element of M is algebraic over $K(B)$.*

We will show that any field extension can be decomposed into a transcendental extension over an algebraic extension. We need the idea of a transcendence basis.

Definition 20.3.3. $B \subset E$ is called a *transcendence basis* of the field extension $E|K$ if the following two conditions are satisfied:
1. $E = H(B)$, that is, the extension $E|K(B)$ is algebraic.
2. B is algebraically independent over K.

Theorem 20.3.4. *If $B \subset E$, then the following are equivalent:*
(1) *B is a transcendence basis of $E|K$.*
(2) *If $B \subset M \subset E$ with $H(M) = E$, then B is a maximal algebraically independent subset of M.*
(3) *There exists a subset $M \subset E$ with $H(M) = E$, which contains B as a maximal algebraically independent subset.*

Proof. (1) implies (2): Let $\alpha \in M \setminus B$. We have to show that $B \cup \{\alpha\}$ is algebraically dependent. But this is clear, because $\alpha \in H(B) = E$.

(2) implies (3): We just take $M = E$.

(3) implies (1): We have to show that $H(B) = E$. Certainly, $M \subset H(B)$.

Hence, $E = H(M) \subset H(H(B)) = H(B) \subset E$. □

We next show that any field extension does have a transcendence basis:

Theorem 20.3.5. *Each field extension $E|K$ has a transcendence basis. More concretely, if there is a subset $M \subset E$ such that $E|K(M)$ is algebraic and if there is a subset $C \subset M$, which is algebraically independent, then there exists a transcendence basis B of $E|K$ with $C \subset B \subset M$.*

Proof. We have to extend C to a maximal algebraically independent subset B of M. By Theorem 20.3.4, such a B is a transcendence basis of $E|K$. If M is finite, then such a B certainly exists. Now let M be not finite. We argue analogously as for the existence of a basis of a vector space, for instance, with Zorn's lemma: If a partially ordered, nonempty set S is inductive, then there exist maximal elements in S. Here, a partially ordered, nonempty set S is said to be inductive if every totally ordered subset of S has an upper bound in S. The set N of all algebraically independent subsets of M, which contain C is partially ordered with respect to "$\subset$", and $N \neq \emptyset$, because $C \in N$. Let $K \neq \emptyset$ be an ascending chain in N; that is, given an ascending chain $\emptyset \neq Y_1 \subset Y_2 \subset \cdots$ in N. The union $U = \bigcup_{Y \in K} Y$ is also algebraically independent. Hence, there exists a maximal algebraically independent subset $B \subset M$ with $C \subset B$. □

Theorem 20.3.6. *Let $E|K$ be a field extension and M a subset of E, for which $E|K(M)$ is algebraic. Let C be an arbitrary subset of E, which is algebraically independent on K. Then there exists a subset $M' \subset M$ with $C \cap M' = \emptyset$ such that $C \cup M'$ is a transcendence basis of $E|K$.*

Proof. Take $M \cup C$, and define $M' := B \setminus C$ in Theorem 20.3.5. □

Theorem 20.3.7. *Let B, B' be two transcendence bases of the field extension $E|K$. Then there is a bijection $\phi : B \to B'$. In other words, any two transcendence bases of $E|K$ have the same cardinal number.*

Proof. (a) If B is a transcendental basis of $E|K$ and M is a subset of E such that $E|K(M)$ is algebraic, then we may write $B = \bigcup_{\alpha \in M} B_\alpha$ with finite sets B_α. In particular, if B is infinite, then the cardinal number of B is not bigger than the cardinal number of M.

(b) Let B and B' be two transcendence bases of $E|K$. If B and B' are both infinite, then B and B' have the same cardinal number by (a) and the theorem by Schroeder–Bernstein [10]. We now prove Theorem 20.3.7 for the case that $E|K$ has a finite transcendence basis. Let B be finite with n elements. Let C be an arbitrary algebraically independent subset in E over K with m elements. We show that $m \leq n$. Let $C = \{\alpha_1, \ldots, \alpha_m\}$ with $m \geq n$. We show, by induction, that for each integer k, $0 \leq k \leq n$, there are subsets $B \supsetneq B_1 \supsetneq \cdots \supsetneq B_k$ of B such that $\{\alpha_1, \ldots, \alpha_k\} \cup B_k$ is a transcendence basis of $E|K$, and $\{\alpha_1, \ldots, \alpha_k\} \cap B_k = \emptyset$. For $k = 0$, we take $B_0 = B$, and the statement holds. Assume now that the statement is correct for $0 \leq k < n$. By Theorems 20.3.4 and 20.3.5, there is a subset B_{k+1} of $\{\alpha_1, \ldots, \alpha_k\} \cup B_k$ such that $\{\alpha_1, \ldots, \alpha_{k+1}\} \cup B_{k+1}$ is a transcendence basis of $E|K$, and $\{\alpha_1, \ldots, \alpha_{k+1}\} \cap B_{k+1} = \emptyset$. Then necessarily, $B_{k+1} \subset B_k$. Assume $B_k = B_{k+1}$. Then on the one hand, $B_k \cup \{\alpha_1, \ldots, \alpha_{k+1}\}$ is algebraic independent because $B_k = B_{k+1}$. On the other hand, also $B_k \cup \{\alpha_1, \ldots, \alpha_k\} \cup \{\alpha_{k+1}\}$ is algebraically dependent, which gives a contradiction. Hence, $B_{k+1} \subsetneq B_k$. Now B_k has at most $n - k$ elements. Therefore, $B_n = \emptyset$; that is, $\{\alpha_1, \ldots, \alpha_n\} = \{\alpha_1, \ldots, \alpha_n\} \cup B_n$ is a transcendence basis of $E|K$. Because $C = \{\alpha_1, \ldots, \alpha_m\}$ is algebraically independent, we cannot have $m > n$. Thus, $m \leq n$, and B and B' have the same number of elements, because B' must also be finite. □

Since the cardinality of any transcendence basis for a field extension $E|K$ is the same, we can define the transcendence degree.

Definition 20.3.8. The *transcendence degree* $\operatorname{trgd}(E|K)$ of a field extension is the cardinal number of one (and hence of each) transcendence basis of $E|K$. A field extension $E|K$ is called *purely transcendental*, if $E|K$ has a transcendence basis B with $E = K(B)$.

We note the following facts:

(1) If $E|K$ is purely transcendental and $B = \{a_1, \dots, a_n\}$ is a transcendence basis of $E|K$, then E is K-isomorphic to the quotient field of the polynomial ring $K[x_1, \dots, x_n]$ of the independence indeterminates $x_1, \dots, x_n$.

(2) K is algebraically closed in E if $E|K$ is purely transcendental.

(3) By Theorem 20.3.4, the field extension $E|K$ has an intermediate field F, $K \subset F \subset E$, such that $F|K$ is purely transcendental, and $E|F$ is algebraic. Certainly F is not uniquely determined.
For example, take $\mathbb{Q} \subset F \subset \mathbb{Q}(i, \pi)$, and for F, we may take $F = \mathbb{Q}(\pi)$, and also $F = \mathbb{Q}(i\pi)$, for instance.

(4) $\operatorname{trgd}(\mathbb{R}|\mathbb{Q}) = \operatorname{trgd}(\mathbb{C}|\mathbb{Q}) = \operatorname{card} \mathbb{R}$, the cardinal number of $\mathbb{R}$. This holds, because the set of the algebraic numbers (over $\mathbb{Q}$) is countable.

Theorem 20.3.9. *Let $E|K$ be a field extension and F an arbitrary intermediate field, that is, $K \subset F \subset E$. Let B be a transcendence basis of $F|K$ and B' a transcendence base of $E|F$. Then $B \cap B' = \emptyset$, and $B \cup B'$ is a transcendence basis of $E|K$.*

In particular, $\operatorname{trgd}(E|K) = \operatorname{trgd}(E|F) + \operatorname{trgd}(F|K)$.

Proof. (1) Assume $a \in B \cap B'$. As an element of F, then a is algebraic over $F(B') \setminus \{a\}$. But this gives a contradiction, because $a \in B'$, and B' is algebraically independent over F.

(2) $F|K(B)$ is an algebraic extension, and also $F(B')|K(B \cup B') = K(B)(B')$. Since the relation “algebraic extension” is transitive, we have that $E|K(B \cup B')$ is algebraic.

(3) Finally, we have to show that $B \cup B'$ is algebraically independent over K. By Theorems 20.3.5 and 20.3.6, there is a subset B'' of $B \cup B'$ with $B \cap B'' = \emptyset$ such that $B \cup B''$ is a transcendence basis of $E|K$. We have $B'' \subset B'$, and have to show that $B' \subset B''$. Assume that there is an $a \in B'$ with $a \notin B''$. Then a is algebraic over $K(B \cup B'') = K(B)(B'')$, and hence algebraic over $F(B'')$. Since $B'' \subset B'$, we have that a is algebraically independent over F, which gives a contradiction. Hence, $B'' = B'$. □

Theorem 20.3.10 (Noether's normalization theorem). *Let K be a field and $A = K[a_1, \dots, a_n]$. Then there exist elements $u_1, \dots, u_m$, $0 \le m \le n$, in A with the following properties:*

(1) *$K[u_1, \dots, u_m]$ is K-isomorphic to the polynomial ring $K[x_1, \dots, x_m]$ of the independent indeterminates $x_1, \dots, x_m$.*

(2) *The ring extension $A|K[u_1, \dots, u_m]$ is an integral extension, that is, for each*

$$a \in A \setminus K[u_1, \dots, u_m]$$

there exists a monic polynomial

$$f(x) = x^n + \alpha_{n-1}x^{n-1} + \cdots + \alpha_0 \in K[u_1, \ldots, u_m][x]$$

of degree $n \geq 1$ with

$$f(a) = a^n + \alpha_{n-1}a^{n-1} + \cdots + \alpha_0 = 0.$$

In particular, $A|K[u_1, \ldots, u_m]$ is finite.

Proof. Without loss of generality, let the $a_1, \ldots, a_n$ be pairwise different. We prove the theorem by induction on n. If $n = 1$, then there is nothing to show. Now, let $n \geq 2$, and assume that the statement holds for $n-1$. If there is no nontrivial algebraic relation $f(a_1, \ldots, a_n) = 0$ over K between the $a_1, \ldots, a_n$, then there is nothing to show. Hence, let there exist a polynomial $f \in K[x_1, \ldots, x_n]$ with $f \neq 0$ and $f(a_1, \ldots, a_n) = 0$. Let $f = \sum_{\nu=(\nu_1,\ldots,\nu_n)} c_\nu x_1^{\nu_1} \cdots x_n^{\nu_n}$. Let $\mu_2, \mu_3, \ldots, \mu_n$ be natural numbers, which we specify later. Define $b_2 = a_2 - a_1^{\mu_2}$, $b_3 = a_3 - a_1^{\mu_3}, \ldots, b_n = a_n - a_1^{\mu_n}$. Then $a_i = b_i + a_1^{\mu_i}$ for $2 \leq i \leq n$, hence, $f(a_1, b_2 + a_1^{\mu_2}, \ldots, b_n + a_1^{\mu_n}) = 0$. We write $R := K[x_1, \ldots, x_n]$ and consider the polynomial ring $R[y_2, \ldots, y_n]$ of the $n-1$ independent indeterminates $y_2, \ldots, y_n$ over R. In $R[y_2, \ldots, y_n]$, we consider the polynomial $f(x_1, y_2 + x_1^{\mu_2}, \ldots, y_n + x_1^{\mu_n})$. We may rewrite this polynomial as

$$\sum_{\nu=(\nu_1,\ldots,\nu_n)} c_\nu x_1^{\nu_1+\mu_2\nu_2+\cdots+\mu_n\nu_n} + g(x_1, y_2, \ldots, y_n)$$

with a polynomial $g(x_1, y_2, \ldots, y_n)$, for which, as a polynomial in x_1 over $K[y_2, \ldots, y_n]$, the degree in x_1 is smaller than the degree of $\sum_{\nu=(\nu_1,\ldots,\nu_n)} c_\nu x_1^{\nu_1+\mu_2\nu_2+\cdots+\mu_n\nu_n}$, provided that we may choose the $\mu_2, \ldots, \mu_n$ in such a way that this really holds. We now specify the $\mu_2, \ldots, \mu_n$. We write $\mu := (1, \mu_2, \ldots, \mu_n)$, and define the scalar product $\mu\nu = 1 \cdot \nu_1 + \mu_2\nu_2 + \cdots + \mu_n\nu_n$. Choose $p \in \mathbb{N}$ with $p > \deg(f) = \max\{\nu_1 + \cdots + \nu_n : c_\nu \neq 0\}$. We now take $\mu = (1, p, p^2, \ldots, p^{n-1})$. If $\nu = (\nu_1, \ldots, \nu_n)$ with $c_\nu \neq 0$ and $\nu' = (\nu_1', \ldots, \nu_n')$ with $c_{\nu'}' \neq 0$ are different n-tuples then indeed $\mu\nu \neq \mu\nu'$ because $\nu_i, \nu_i' < p$ for all i, $1 \leq i \leq n$. This follows from the uniqueness of the p-adic expression of a natural number. Hence, we may choose $\mu_2, \ldots, \mu_n$ such that $f(x_1, y_2 + x_1^{\mu_2}, \ldots, y_n + x_1^{\mu_n}) = cx_1^N + h(x_1, y_2, \ldots, y_n)$ with $c \in K, c \neq 0$, and $h \in K[y_2, \ldots, y_n][x_1]$ has in x_1 a degree $< N$. If we divide by c and take $a_1, b_2, \ldots, b_n$ for $x_1, y_2, \ldots, y_n$, then we get an integral equation of a_1 over $K[b_2, \ldots, b_n]$. Therefore, the ring extension $A = K[a_1, \ldots, a_n]|K[b_2, \ldots, b_n]$ is integral (see Theorem 20.2.9), $a_i = b_i + a_1^{\mu_i}$ for $2 \leq i \leq n$. By induction, there exist elements $u_1, \ldots, u_m$ in $K[b_2, \ldots, b_n]$ with the following properties:

1. $K[u_1, \ldots, u_m]$ is a polynomial ring of the m independent indeterminates $u_1, \ldots, u_m$, and
2. $K[b_2, \ldots, b_n]|K[u_1, \ldots, u_m]$ is integral.

Hence, also $A|K[u_1, \ldots, u_m]$ is integral by Theorem 20.2.14. □

Corollary 20.3.11. *Let $E|K$ be a field extension. If $E = K[a_1, \dots, a_n]$ for $a_1, \dots, a_n \in E$, then $E|K$ is algebraic.*

Proof. By Theorem 20.3.10, we have that E contains a polynomial ring $K[u_1, \dots, u_m]$, $0 \le m \le n$, of the m independent indeterminates $u_1, \dots, u_m$ as a subring, for which $E|K[u_1, \dots, u_m]$ is integral. We claim that then already $K[u_1, \dots, u_m]$ is a field. To prove that, let $a \in K[u_1, \dots, u_m]$, $a \ne 0$. The element $a^{-1} \in E$ satisfies an integral equation $(a^{-1})^n + a_{n-1}(a^{-1})^{n-1} + \dots + a_0 = 0$ over $K[u_1, \dots, u_m] =: R$. Hence,

$$a^{-1} = -a_{n-1} - a_{n-2}a - \dots - a_0 a^{n-1} \in R.$$

Therefore, R is a field, which proves the claim. This is possible only for $m = 0$, and then $E|K$ is integral; here, that is algebraic. □

20.4 The Transcendence of e and π

Although we have shown that within $\mathbb{C}$, there are continuously many transcendental numbers, we have only shown that one particular number is transcendental. In this section, we prove that the numbers e and π are transcendental. We start with e.

Theorem 20.4.1. *e is a transcendental number, that is, transcendental over $\mathbb{Q}$.*

Proof. Let $f(x) \in \mathbb{R}[x]$ with the degree of $f(x) = m \ge 1$.

Let $z_1 \in \mathbb{C}$, $z_1 \ne 0$, and $\gamma : [0,1] \to \mathbb{C}$, $\gamma(t) = tz_1$. Let

$$I(z_1) = \int_\gamma e^{z_1 - z} f(z)dz = \Big(\int_0^{z_1}\Big)_\gamma e^{z_1 - z} f(z)dz.$$

By $(\int_0^{z_1})_\gamma$, we mean the integral from 0 to z_1 along γ. Recall that

$$\Big(\int_0^{z_1}\Big)_\gamma e^{z_1 - z} f(z)dz = -f(z_1) + e^{z_1} f(0) + \Big(\int_0^{z_1}\Big)_\gamma e^{z_1 - z} f'(z)dz.$$

It follows then by repeated partial integration that

(1) $I(z_1) = e^{z_1} \sum_{j=0}^m f^{(j)}(0) - \sum_{j=0}^m f^{(j)}(z_1)$.

Let $|f|(x)$ be the polynomial we get if we replace the coefficients of $f(x)$ by their absolute values. Since $|e^{z_1 - z}| \le e^{|z_1 - z|} \le e^{|z_1|}$, we get

(2) $|I(z_1)| \le |z_1| e^{|z_1|} |f|(|z_1|)$.

Now assume that e is an algebraic number; that is,

(3) $q_0 + q_1 e + \dots + q_n e^n = 0$ for $n \ge 1$ and integers $q_0 \ne 0, q_1, \dots, q_n$, and the greatest common divisor of $q_0, q_1, \dots, q_n$, is equal to 1.

For a detailed proof of these facts see for instance [52]. We consider now the polynomial $f(x) = x^{p-1}(x-1)^p \cdots (x-n)^p$ with p a sufficiently large prime number, and we consider $I(z_1)$ with respect to this polynomial. Let $J = q_0 I(0) + q_1 I(1) + \cdots + q_n I(n)$.

From (1) and (3), we get that

$$J = -\sum_{j=0}^{m}\sum_{k=0}^{n} q_k f^{(j)}(k),$$

where $m = (n+1)p - 1$, since $(q_0 + q_1 e + \cdots + q_n e^n)(\sum_{j=0}^{m} f^{(j)}(0)) = 0$.

Now, $f^{(j)}(k) = 0$ if $j < p$, $k > 0$, and if $j < p-1$, then $k = 0$. Hence, $f^{(j)}(k)$ is an integer that is divisible by $p!$ for all j, k, except for $j = p-1$, $k = 0$. Furthermore, $f^{(p-1)}(0) = (p-1)!(-1)^{np}(n!)^p$. Hence, if $p > n$, then $f^{(p-1)}(0)$ is an integer divisible by $(p-1)!$, but not by $p!$. It follows that J is a nonzero integer that is divisible by $(p-1)!$ if $p > |q_0|$ and $p > n$. So let $p > n$, $p > |q_0|$, so that $|J| \geq (p-1)!$. Now, $|f|(k) \leq (2n)^m$. Together with (2), we then get that $|J| \leq |q_1| e |f|(1) + \cdots + |q_n| n e^n |f|(n) \leq c^p$ for a number c independent of p. It follows that

$$(p-1)! \leq |J| \leq c^p;$$

that is,

$$1 \leq \frac{|J|}{(p-1)!} \leq c \frac{c^{p-1}}{(p-1)!}.$$

This gives a contradiction, since $\frac{c^{p-1}}{(p-1)!} \to 0$ as $p \to \infty$. □

We now move on to the transcendence of π. We first need the following lemma:

Lemma 20.4.2. *Suppose $\alpha \in \mathbb{C}$ is an algebraic number and $f(x) = a_n x^n + \cdots + a_0$, $n \geq 1$, $a_n \neq 0$, and all $a_i \in \mathbb{Z}$ ($f(x) \in \mathbb{Z}[x]$) with $f(\alpha) = 0$. Then $a_n \alpha$ is an algebraic integer.*

Proof.

$$\begin{aligned} a_n^{n-1} f(x) &= a_n^n x^n + a_n^{n-1} a_{n-1} x^{n-1} + \cdots + a_n^{n-1} a_0 \\ &= (a_n x)^n + a_{n-1}(a_n x)^{n-1} + \cdots + a_n^{n-1} a_0 \\ &= g(a_n x) = g(y) \in \mathbb{Z}[y], \end{aligned}$$

where $y = a_n x$, and $g(y)$ is monic. Then $g(a_n \alpha) = 0$; hence, $a_n \alpha$ is an algebraic integer. □

Theorem 20.4.3. *π is a transcendental number, that is, transcendental over $\mathbb{Q}$.*

Proof. Assume that π is an algebraic number. Then $\theta = i\pi$ is also algebraic. Consider the conjugates $\theta_1 = \theta, \theta_2, \ldots, \theta_d$ of θ. Suppose

$$p(x) = q_0 + q_1 x + \cdots + q_d x^d \in \mathbb{Z}[x], \quad q_d > 0, \quad \text{and} \quad \gcd(q_0, \ldots, q_d) = 1$$

is the entire minimal polynomial of θ over $\mathbb{Q}$. Then $\theta_1 = \theta, \theta_2, \ldots, \theta_d$ are the zeros of this polynomial. Let $t = q_d$. Then from Lemma 20.4.2, $t\theta_i$ is an algebraic integer for all i. From $e^{i\pi} + 1 = 0$, and from $\theta_1 = i\pi$, we get that

$$(1 + e^{\theta_1})(1 + e^{\theta_2}) \cdots (1 + e^{\theta_d}) = 0.$$

The product on the left side can be written as a sum of 2^d terms e^{ϕ}, where

$$\phi = \epsilon_1\theta_1 + \cdots + \epsilon_d\theta_d,$$

$\epsilon_j = 0$ or 1. Let n be the number of terms $\epsilon_1\theta_1 + \cdots + \epsilon_d\theta_d$ that are nonzero. Call these $\alpha_1, \ldots, \alpha_n$. We then have an equation

$$q + e^{\alpha_1} + \cdots + e^{\alpha_n} = 0$$

with $q = 2^d - n > 0$. Recall that all $t\alpha_i$ are algebraic integers, and we consider the polynomial

$$f(x) = t^{np}x^{p-1}(x - \alpha_1)^p \cdots (x - \alpha_n)^p$$

with p a sufficiently large prime integer. We have $f(x) \in \mathbb{R}[x]$, since the α_i are algebraic numbers, and the elementary symmetric polynomials in $\alpha_1, \ldots, \alpha_n$ are rational numbers.

Let $I(z_1)$ be defined as in the proof of Theorem 20.4.1, and now let

$$J = I(\alpha_1) + \cdots + I(\alpha_n).$$

From (1) in the proof of Theorem 20.4.1 and (4), we get

$$J = -q\sum_{j=0}^{m} f^{(j)}(0) - \sum_{j=0}^{m}\sum_{k=1}^{n} f^{(j)}(\alpha_k),$$

with $m = (n + 1)p - 1$.

Now, $\sum_{k=1}^{n} f^{(j)}(\alpha_k)$ is a symmetric polynomial in $t\alpha_1, \ldots, t\alpha_n$ with integer coefficients, since the $t\alpha_i$ are algebraic integers. It follows from the main theorem on symmetric polynomials that $\sum_{j=0}^{m}\sum_{k=1}^{n} f^{(j)}(\alpha_k)$ is an integer. Furthermore, $f^{(j)}(\alpha_k) = 0$ for $j < p$. Hence, $\sum_{j=0}^{m}\sum_{k=1}^{n} f^{(j)}(\alpha_k)$ is an integer divisible by $p!$. Now, $f^{(j)}(0)$ is an integer divisible by $p!$ if $j \neq p - 1$, and $f^{(p-1)}(0) = (p - 1)!(-t)^{np} \times (\alpha_1 \cdots \alpha_n)^p$ is an integer divisible by $(p - 1)!$, but not divisible by $p!$ if p is sufficiently large. In particular, this is true if $p > |t^n(\alpha_1 \cdots \alpha_n)|$ and also $p > q$.

From (2) in the proof of Theorem 20.4.1, we get that

$$|J| \le |\alpha_1|e^{|\alpha_1|}|f|(|\alpha_1|) + \cdots + |\alpha_n|e^{|\alpha_n|}|f|(|\alpha_n|) \le c^p$$

for some number c independent of p.

As in the proof of Theorem 20.4.1, this gives us

$$(p-1)! \le |J| \le c^p;$$

that is,

$$1 \le \frac{|J|}{(p-1)!} \le c\frac{c^{p-1}}{(p-1)!}.$$

This, as before, gives a contradiction, since $\frac{c^{p-1}}{(p-1)!} \to 0$ as $p \to \infty$. Therefore, π is transcendental. □

20.5 Exercises

1. A polynomial $p(x) \in \mathbb{Z}[x]$ is primitive if the GCD of all its coefficients is 1. Prove the following:
   (i) If $f(x)$ and $g(x)$ are primitive, then so is $f(x)g(x)$.
   (ii) If $f(x) \in \mathbb{Z}[x]$ is monic, then it is primitive.
   (iii) If $f(x) \in \mathbb{Q}[x]$, then there exists a rational number c such that $f(x) = cf_1(x)$ with $f_1(x)$ primitive.
2. Let d be a square-free integer and $K = \mathbb{Q}(\sqrt{d})$ be a quadratic field. Let R_K be the subring of K of the algebraic integers of K. Show the following:
   (i) $R_K = \{m + n\sqrt{d} : m, n \in \mathbb{Z}\}$ if $d \equiv 2 \pmod 4$ or $d \equiv 3 \pmod 4$. $\{1, \sqrt{d}\}$ is an integral basis for R_K.
   (ii) $R_K = \{m + n\frac{1+\sqrt{d}}{2} : m, n \in \mathbb{Z}\}$ if $d \equiv 1 \pmod 4$. $\{1, \frac{1+\sqrt{d}}{2}\}$ is an integral basis for R_K.
   (iii) If $d < 0$, then there are only finitely many units in R_K.
   (iv) If $d > 0$, then there are infinitely many units in R_K.
3. Let $K = \mathbb{Q}(\alpha)$ with $\alpha^3 + \alpha + 1 = 0$ and R_K the subring of the algebraic integers in K. Show that:
   (i) $\{1, \alpha, \alpha^2\}$ is an integral basis for R_K.
   (ii) $R_K = \mathbb{Z}[\alpha]$.
4. Let $A|R$ be an integral ring extension. If A is an integral domain and R a field, then A is also a field.
5. Let $A|R$ be an integral extension. Let $\mathcal{P}$ be a prime ideal of A and $\mathfrak{p}$ be a prime ideal of R such that $\mathcal{P} \cap R = \mathfrak{p}$. Show that:
   (i) If $\mathfrak{p}$ is maximal in R, then $\mathcal{P}$ is maximal in A.
   (*Hint*: consider $A/\mathcal{P}$.)
   (ii) If $\mathcal{P}_0$ is another prime ideal of A with $\mathcal{P}_0 \cap R = \mathfrak{p}$ and $\mathcal{P}_0 \subset \mathcal{P}$, then $\mathcal{P} = \mathcal{P}_0$.
   (*Hint*: we may assume that A is an integral domain, and $\mathcal{P} \cap R = \{0\}$, otherwise go to $A/\mathcal{P}$.)
6. Show that for a field extension $E|K$, the following are equivalent:
   (i) $[E : K(B)] < \infty$ for each transcendence basis B of $E|K$.

(ii) $\operatorname{trgd}(E|K) < \infty$ and $[E : K(B)] < \infty$ for each transcendence basis B of $E|K$.
(iii) There is a finite transcendence basis B of $E|K$ with $[E : K(B)] < \infty$.
(iv) There are finitely many $x_1, \dots, x_n \in E$ with $E = K(x_1, \dots, x_n)$.

7. Let $E|K$ be a field extension. If $E|K$ is purely transcendental, then K is algebraically closed in E.

21 The Hilbert Basis Theorem and the Nullstellensatz

21.1 Algebraic Geometry

An extremely important application of abstract algebra and an application central to all of mathematics is the subject of *algebraic geometry*. As the name suggests this is the branch of mathematics that uses the techniques of abstract algebra to study geometric problems. Classically, algebraic geometry involved the study of *algebraic curves*, which roughly are the sets of zeros of a polynomial or set of polynomials in several variables over a field. For example, in two variables a real algebraic plane curve is the set of zeros in $\mathbb{R}^2$ of a polynomial $p(x,y) \in \mathbb{R}[x,y]$. The common planar curves, such as parabolas and the other conic sections, are all plane algebraic curves. In actual practice, plane algebraic curves are usually considered over the complex numbers and are projectivized.

The algebraic theory that deals most directly with algebraic geometry is called *commutative algebra*. This is the study of commutative rings, ideals in commutative rings, and modules over commutative rings. A large portion of this book has dealt with commutative algebra.

Although we will not consider the geometric aspects of algebraic geometry in general, we will close the book by introducing some of the basic algebraic ideas that are crucial to the subject. These include the concept of an *algebraic variety* or *algebraic set* and its radical. We also state and prove two of the cornerstones of the theory as applied to commutative algebra—the Hilbert basis theorem and the nullstellensatz.

In this chapter, we also often consider a fixed field extension $C|K$ and the polynomial ring $K[x_1,\dots,x_n]$ of the n independent indeterminates $x_1,\dots,x_n$. Again, in this chapter, we often use letters $\mathfrak{a}, \mathfrak{b}, \mathfrak{m}, \mathfrak{p}, \mathfrak{P}, \mathfrak{A}, \mathfrak{Q}, \dots$ for ideals in rings.

21.2 Algebraic Varieties and Radicals

We first define the concept of an algebraic variety:

Definition 21.2.1. If $M \subset K[x_1,\dots,x_n]$, then we define

$$\mathcal{N}(M) = \{(\alpha_1,\dots,\alpha_n) \in C^n : f(\alpha_1,\dots,\alpha_n) = 0 \; \forall f \in M\}.$$

$\alpha = (\alpha_1,\dots,\alpha_n) \in \mathcal{N}(M)$ is called a *zero (Nullstelle) of M in C^n*, and $\mathcal{N}(M)$ is called the *zero set* of M in C^n. If we want to mention C, then we write $\mathcal{N}(M) = \mathcal{N}_C(M)$. A subset $V \subset C^n$ of the form $V = \mathcal{N}(M)$ for some $M \subset K[x_1,\dots,x_n]$ is called an *algebraic variety* or *(affine) algebraic set* of C^n over K, or just an *algebraic K-set of C^n*.

For any subset N of C^n, we can reverse the procedure and consider the set of polynomials, whose zero set is N.

https://doi.org/10.1515/9783111142524-021

Definition 21.2.2. Suppose that $N \subset C^n$. Then

$$I(N) = \{f \in K[x_1, \dots, x_n] : f(\alpha_1, \dots, \alpha_n) = 0 \; \forall (\alpha_1, \dots, \alpha_n) \in N\}.$$

Instead of $f \in I(N)$, we also say that f *vanishes on* N *(over* K*)*. If we want to mention K, then we write $I(N) = I_K(N)$.

What is important is that the set $I(N)$ forms an ideal. The proof is straightforward.

Theorem 21.2.3. *For any subset $N \subset C^n$, the set $I(N)$ is an ideal in $K[x_1, \dots, x_n]$; it is called the vanishing ideal of $N \subset C^n$ in $K[x_1, \dots, x_n]$.*

The following result examines the relationship between subsets in C^n and their vanishing ideals.

Theorem 21.2.4. *The following properties hold:*

(1) $M \subset M' \Rightarrow \mathcal{N}(M') \subset \mathcal{N}(M)$;
(2) *If $\mathfrak{a} = (M)$ is the ideal in $K[x_1, \dots, x_n]$ generated by M, then $\mathcal{N}(M) = \mathcal{N}(\mathfrak{a})$;*
(3) $N \subset N' \Rightarrow I(N') \subset I(N)$;
(4) *$M \subset I\mathcal{N}(M)$ for all $M \subset K[x_1, \dots, x_n]$;*
(5) *$N \subset \mathcal{N}I(N)$ for all $N \subset C^n$;*
(6) *If $(\mathfrak{a}_i)_{i \in I}$ is a family of ideals in $K[x_1, \dots, x_n]$, then $\bigcap_{i \in I} \mathcal{N}(\mathfrak{a}_i) = \mathcal{N}(\sum_{i \in I} \mathfrak{a}_i)$. Here $\sum_{i \in I} \mathfrak{a}_i$ is the ideal in $K[x_1, \dots, x_n]$, generated by the union $\bigcup_{i \in I} \mathfrak{a}_i$;*
(7) *If $\mathfrak{a}, \mathfrak{b}$ are ideals in $K[x_1, \dots, x_n]$, then $\mathcal{N}(\mathfrak{a}) \cup \mathcal{N}(\mathfrak{b}) = \mathcal{N}(\mathfrak{a}\mathfrak{b}) = \mathcal{N}(\mathfrak{a} \cap \mathfrak{b})$. Here $\mathfrak{a}\mathfrak{b}$ is the ideal in $K[x_1, \dots, x_n]$ generated by all products fg, where $f \in \mathfrak{a}$ and $g \in \mathfrak{b}$;*
(8) *$\mathcal{N}(M) = \mathcal{N}I\mathcal{N}(M)$ for all $M \subset K[x_1, \dots, x_n]$;*
(9) *$V = \mathcal{N}I(V)$ for all algebraic K-sets V;*
(10) *$I(N) = I\mathcal{N}I(N)$ for all $N \subset C^n$.*

Proof. The proofs are straightforward. Hence, we prove only (7), (8), and (9). The rest can be left as exercise for the reader.

Proof of (7): Since $\mathfrak{a}\mathfrak{b} \subset \mathfrak{a} \cap \mathfrak{b} \subset \mathfrak{a}, \mathfrak{b}$, we have, by (1), the inclusion

$$\mathcal{N}(\mathfrak{a}) \cup \mathcal{N}(\mathfrak{b}) \subset \mathcal{N}(\mathfrak{a} \cap \mathfrak{b}) \subset \mathcal{N}(\mathfrak{a}\mathfrak{b}).$$

Hence, we have to show that $\mathcal{N}(\mathfrak{a}\mathfrak{b}) \subset \mathcal{N}(\mathfrak{a}) \cup \mathcal{N}(\mathfrak{b})$.

Let $\alpha = (\alpha_1, \dots, \alpha_n) \in C^n$ be a zero of $\mathfrak{a}\mathfrak{b}$, but not a zero of $\mathfrak{a}$. Then there is an $f \in \mathfrak{a}$ with $f(\alpha) \neq 0$; hence, for all $g \in \mathfrak{b}$, we get $f(\alpha)g(\alpha) = (fg)(\alpha) = 0$. Thus, $g(\alpha) = 0$. Therefore, $\alpha \in \mathcal{N}(\mathfrak{b})$.

Proof of (8) and (9): Let $M \subset K[x_1, \dots, x_n]$. Then, on the one hand, $M \subset I\mathcal{N}(M)$ by (5), and further $\mathcal{N}I\mathcal{N}(M) \subset \mathcal{N}(M)$ by (1). On the other hand, $\mathcal{N}(M) \subset \mathcal{N}I\mathcal{N}(M)$ by (6). Therefore, $\mathcal{N}(M) = \mathcal{N}I\mathcal{N}(M)$ for all $M \subset K[x_1, \dots, x_n]$.

Now, the algebraic K-sets of C^n are precisely the sets of the form $V = \mathcal{N}(M)$. Hence, $V = \mathcal{N}I(V)$. ☐

We make the following agreement: if $\mathfrak{a}$ is an ideal in $K[x_1,\dots,x_n]$, then we write

$$\mathfrak{a} \lhd K[x_1,\dots,x_n].$$

If $\mathfrak{a} \lhd K[x_1,\dots,x_n]$, then we do not have $\mathfrak{a} = I\mathcal{N}(\mathfrak{a})$ in general. That is, $\mathfrak{a}$ is, in general, not equal to the vanishing ideal of its zero set in C^n. The reason for this is that not each ideal $\mathfrak{a}$ occurs as a vanishing ideal of some $N \subset C^n$. If $\mathfrak{a} = I(N)$, then we must have that $f^m \in \mathfrak{a}$ for $m \geq 1$ implies $f \in \mathfrak{a}$.

Hence, for instance, if $\mathfrak{a} = (x_1^2,\dots,x_n^2) \lhd K[x_1,\dots,x_n]$, then $\mathfrak{a}$ is not of the form $\mathfrak{a} = I(N)$ for some $N \subset C^n$. We now define the radical of an ideal:

Definition 21.2.5. Let R be a commutative ring and $\mathfrak{a} \lhd R$ an ideal in R. Then

$$\sqrt{\mathfrak{a}} = \{f \in R : f^m \in \mathfrak{a} \text{ for some } m \in \mathbb{N}\}$$

is an ideal in R. $\sqrt{\mathfrak{a}}$ is called the *radical of* $\mathfrak{a}$ (in R). $\mathfrak{a}$ is said to be *reduced* if $\sqrt{\mathfrak{a}} = \mathfrak{a}$.

We note that the $\sqrt{0}$ is called the *nil radical* of R; it contains exactly the *nilpotent elements* of R; that is, the elements $a \in R$ with $a^m = 0$ for some $m \in \mathbb{N}$.

Let $\mathfrak{a} \lhd R$ be an ideal in R and $\pi : R \to R/\mathfrak{a}$ the canonical mapping. Then $\sqrt{\mathfrak{a}}$ is exactly the preimage of the nil radical of $R/\mathfrak{a}$.

21.3 The Hilbert Basis Theorem

In this section, we show that if K is a field, then each ideal $\mathfrak{a} \lhd K[x_1,\dots,x_n]$ is finitely generated. This is the content of the Hilbert basis theorem. This has as an important consequence: any algebraic variety of C^n is the zero set of only finitely many polynomials.

The Hilbert basis theorem follows directly from the following Theorem 21.3.2. Before we state this theorem, we need a definition.

Definition 21.3.1. Let R be a commutative ring with an identity $1 \neq 0$. R is said to be *Noetherian* if each ideal in R is generated by finitely many elements; that is, each ideal in R is finitely generated.

Theorem 21.3.2. *Let R be a noetherian ring. Then the polynomial ring $R[x]$ over R is also noetherian.*

Proof. Let $0 \neq f_k \in R[x]$. We denote the degree of f_k with $\deg(f_k)$. Let $\mathfrak{a} \lhd R[x]$ be an ideal in $R[x]$. Assume that $\mathfrak{a}$ is not finitely generated. Then, particularly, $\mathfrak{a} \neq 0$. We construct a sequence of polynomials $f_k \in \mathfrak{a}$ such that the highest coefficients a_k generate an ideal in R, which is not finitely generated. This produces then a contradiction; hence, $\mathfrak{a}$ is in fact finitely generated. Choose $f_1 \in \mathfrak{a}, f_1 \neq 0$, so that $\deg(f_1) = n_1$ is minimal.

If $k \geq 1$, then choose $f_{k+1} \in \mathfrak{a}, f_{k+1} \notin (f_1, \dots, f_k)$ so that $\deg(f_{k+1}) = n_{k+1}$ is minimal for the polynomials in $\mathfrak{a} \setminus (f_1, \dots, f_k)$. This is possible, because we assume that $\mathfrak{a}$ is not finitely generated.

We have $n_k \leq n_{k+1}$ by construction. Furthermore, $(a_1, \dots, a_k) \subsetneq (a_1, \dots, a_k, a_{k+1})$.

To see this, assume that $(a_1, \dots, a_k) = (a_1, \dots, a_k, a_{k+1})$. Then $a_{k+1} \in (a_1, \dots, a_k)$. Hence, there are $b_i \in R$ with $a_{k+1} = \sum_{i=1}^{k} a_i b_i$. Let $g(x) = \sum_{i=1}^{k} b_i f_i(x) x^{n_{k+1}-n_i}$; hence, $g \in (f_1, \dots, f_k)$, and $g = a_{k+1} x^{n_{k+1}} + \cdots$.

Therefore, $\deg(f_{k+1} - g) < n_{k+1}$, and $f_{k+1} - g \notin (f_1, \dots, f_k)$, which contradicts the choice of f_{k+1}. This proves the claim. Hence, $(a_1, \dots, a_k) \subsetneq (a_1, \dots, a_k, a_{k+1})$, which contradicts the fact that R is Noetherian. Hence, $\mathfrak{a}$ is finitely generated. □

We now have the Hilbert basis theorem:

Theorem 21.3.3 (Hilbert basis theorem). *Let K be a field. Then any ideal $\mathfrak{a} \lhd K[x_1, \dots, x_n]$ is finitely generated; that is, $\mathfrak{a} = (f_1, \dots, f_m)$ for finitely many $f_1, \dots, f_m \in K[x_1, \dots, x_n]$.*

Corollary 21.3.4. *If $C|K$ is a field extension, then each algebraic K-set V of C^n is already the zero set of only finitely many polynomials $f_1, \dots, f_m \in K[x_1, \dots, x_n]$:*

$$V = \{(a_1, \dots, a_n) \in C^n : f_i(a_1, \dots, a_n) = 0 \text{ for } i = 1, \dots, m\}.$$

Furthermore, we write $V = \mathcal{N}(f_1, \dots, f_m)$.

21.4 The Nullstellensatz

Vanishing ideals of subsets of C^n are not necessarily reduced. For an arbitrary field C, the condition

$$f^m \in \mathfrak{a},\ m \geq 1 \Longrightarrow f \in \mathfrak{a}$$

is, in general, not sufficient for $\mathfrak{a} \lhd K[x_1, \dots, x_n]$ to be a vanishing ideal of a subset of C^n. For example, let $n \geq 2$, $K = C = \mathbb{R}$ and $\mathfrak{a} = (x_1^2 + \cdots + x_n^2) \lhd \mathbb{R}[x_1, \dots, x_n]$. $\mathfrak{a}$ is a prime ideal in $\mathbb{R}[x_1, \dots, x_n]$, because $x_1^2 + \cdots + x_n^2$ is a prime element in $\mathbb{R}[x_1, \dots, x_n]$. Hence, $\mathfrak{a}$ is reduced. But, on the other hand, $\mathcal{N}(\mathfrak{a}) = \{0\}$, and $I(\{0\}) = (x_1, \dots, x_n)$. Therefore, $\mathfrak{a}$ is not of the form $I(N)$ for some $N \subset C^n$. If this would be the case, then $\mathfrak{a} = I(N) = I\mathcal{N}I(N) = I\{0\} = (x_1, \dots, x_n)$, because of Theorem 21.2.4 (10), which gives a contradiction.

The nullstellensatz by Hilbert, which we give in two forms shows that if $\mathfrak{a}$ is reduced, that is, $\mathfrak{a} = \sqrt{\mathfrak{a}}$, then $I\mathcal{N}(\mathfrak{a}) = \mathfrak{a}$.

Theorem 21.4.1 (Hilbert's nullstellensatz, first form). *Let $C|K$ be a field extension with C algebraically closed. If $\mathfrak{a} \lhd K[x_1, \dots, x_n]$, then $I\mathcal{N}(\mathfrak{a}) = \sqrt{\mathfrak{a}}$. Moreover, if $\mathfrak{a}$ is reduced, that is, $\mathfrak{a} = \sqrt{\mathfrak{a}}$, then $I\mathcal{N}(\mathfrak{a}) = \mathfrak{a}$. Therefore, $\mathcal{N}$ defines a bijective map between the set of reduced ideals in $K[x_1, \dots, x_n]$ and the set of the algebraic K-sets in C^n, and I defines the inverse map.*

The proof follows from the following:

Theorem 21.4.2 (Hilbert's nullstellensatz, second form). *Let $C|K$ be a field extension with C algebraically closed. Let $\mathfrak{a} \lhd K[x_1,\dots,x_n]$ with $\mathfrak{a} \neq K[x_1,\dots,x_n]$. Then there exists an $\alpha = (\alpha_1,\dots,\alpha_n) \in C^n$ with $f(\alpha) = 0$ for all $f \in \mathfrak{a}$; that is, $\mathcal{N}_C(\mathfrak{a}) \neq \emptyset$.*

Proof. Since $\mathfrak{a} \neq K[x_1,\dots,x_n]$, there exists a maximal ideal $\mathfrak{m} \lhd K[x_1,\dots,x_n]$ with $\mathfrak{a} \subset \mathfrak{m}$. We consider the canonical map $\pi : K[x_1,\dots,x_n] \to K[x_1,\dots,x_n]/\mathfrak{m}$. Let $\beta_i = \pi(x_i)$ for $i = 1,\dots,n$. Then $K[x_1,\dots,x_n]/\mathfrak{m} = K[\beta_1,\dots,\beta_n] =: E$. Since $\mathfrak{m}$ is maximal, E is a field. Moreover, $E|K$ is algebraic by Corollary 20.3.11. Hence, there exists a K-homomorphism $\sigma : K[\beta_1,\dots,\beta_n] \to C$ (C is algebraically closed). Let $\alpha_i = \sigma(\beta_i)$. As a result we have $f(\alpha_1,\dots,\alpha_n) = 0$ for all $f \in \mathfrak{m}$. Since $\mathfrak{a} \subset \mathfrak{m}$ this holds also for all $f \in \mathfrak{a}$. Hence, we get a zero $(\alpha_1,\dots,\alpha_n)$ of $\mathfrak{a}$ in C^n. □

Proof of Theorem 21.4.1. Let $\mathfrak{a} \lhd K[x_1,\dots,x_n]$, and let $f \in I\mathcal{N}(\mathfrak{a})$. We have to show that $f^m \in \mathfrak{a}$ for some $m \in \mathbb{N}$. If $f = 0$, then there is nothing to show.

Now, let $f \neq 0$. We consider $K[x_1,\dots,x_n]$ as a subring of $K[x_1,\dots,x_n,x_{n+1}]$ of the $n+1$ independent indeterminates $x_1,\dots,x_n,x_{n+1}$. In $K[x_1,\dots,x_n,x_{n+1}]$, we consider the ideal $\bar{\mathfrak{a}} = (\mathfrak{a}, 1 - x_{n+1}f) \lhd K[x_1,\dots,x_n,x_{n+1}]$, generated by $\mathfrak{a}$ and $1 - x_{n+1}f$.

Case 1: $\bar{\mathfrak{a}} \neq K[x_1,\dots,x_n,x_{n+1}]$. Then $\bar{\mathfrak{a}}$ has a zero $(\beta_1,\dots,\beta_n,\beta_{n+1})$ in C^{n+1} by Theorem 21.2.4. Hence, for $(\beta_1,\dots,\beta_n,\beta_{n+1}) \in \mathcal{N}(\bar{\mathfrak{a}})$, we have the equations:

(1) $g(\beta_1,\dots,\beta_n) = 0$ for all $g \in \mathfrak{a}$, and

(2) $f(\beta_1,\dots,\beta_n)\beta_{n+1} = 1$.

From (1), we get $(\beta_1,\dots,\beta_n) \in \mathcal{N}(\mathfrak{a})$. In particular, $f(\beta_1,\dots,\beta_n) = 0$ for our $f \in I\mathcal{N}(\mathfrak{a})$. But this contradicts (2). Therefore, $\bar{\mathfrak{a}} \neq K[x_1,\dots,x_n,x_{n+1}]$ is not possible. Thus, we have

Case 2: $\bar{\mathfrak{a}} = K[x_1,\dots,x_n,x_{n+1}]$, that is, $1 \in \bar{\mathfrak{a}}$. Then there exists a relation of the form

$$1 = \sum_i h_i g_i + h(1 - x_{n+1}f) \quad \text{for some } g_i \in \mathfrak{a} \text{ and } h_i, h \in K[x_1,\dots,x_n,x_{n+1}].$$

The map given by $x_i \mapsto x_i$ for $1 \leq i \leq n$ and $x_{n+1} \mapsto \frac{1}{f}$ defines a homomorphism $\phi : K[x_1,\dots,x_n,x_{n+1}] \to K(x_1,\dots,x_n)$, the quotient field of $K[x_1,\dots,x_n]$. From (3), we get a relation $1 = \sum_i h_i(x_1,\dots,x_n,\frac{1}{f})g_i(x_1,\dots,x_n)$ in $K(x_1,\dots,x_n)$. If we multiply this with a suitable power f^m of f, we get $f^m = \sum_i \tilde{h}_i(x_1,\dots,x_n)g_i(x_1,\dots,x_n)$ for some polynomials $\tilde{h} \in K[x_1,\dots,x_n]$. Since $g_i \in \mathfrak{a}$, we get $f^m \in \mathfrak{a}$. □

21.5 Applications and Consequences of Hilbert's Theorems

Theorem 21.5.1. *Each nonempty set of algebraic K-sets in C^n contains a minimal element. In other words, for each descending chain*

$$V_1 \supset V_2 \supset \cdots \supset V_m \supset V_{m+1} \supset \cdots \tag{21.1}$$

of algebraic K-sets V_i in C^n, there exists an integer m such that $V_m = V_{m+1} = V_{m+2} = \cdots$, or equivalently, every strictly descending chain $V_1 \supsetneq V_2 \supsetneq \cdots$ of algebraic K-sets V_i in C^n is finite.

Proof. We apply the operator I; that is, we pass to the vanishing ideals. This gives an ascending chain of ideals

$$I(V_1) \subset I(V_2) \subset \cdots \subset I(V_m) \subset I(V_{m+1}) \subset \cdots. \tag{21.2}$$

The union of the $I(V_i)$ is an ideal in $K[x_1, \ldots, x_n]$, and hence, by Theorem 21.3.3, finitely generated. Therefore, there is an m with $I(V_m) = I(V_{m+1}) = I(V_{m+2}) = \cdots$.

Now we apply the operator $\mathcal{N}$ and get the desired result, because $V_i = \mathcal{N}I(V_i)$ by Theorem 21.2.4 (10). □

Definition 21.5.2. An algebraic K-set $V \neq \emptyset$ in C^n is called *irreducible* if it is not describable as a union $V = V_1 \cup V_2$ of two algebraic K-sets $V_i \neq \emptyset$ in C^n with $V_i \neq V$ for $i = 1, 2$. An irreducible algebraic K-set in C^n is also called a *K-variety* in C^n.

Theorem 21.5.3. *An algebraic K-set $V \neq \emptyset$ in C^n is irreducible if and only if its vanishing ideal $I_k(V) = I(V)$ is a prime ideal of $R = K[x_1, \ldots, x_n]$ with $I(V) \neq R$.*

Proof. (1) Let V be irreducible. Let $fg \in I(V)$. Then $V = \mathcal{N}I(V) \subset \mathcal{N}(fg) = \mathcal{N}(f) \cup \mathcal{N}(g)$; hence, $V = V_1 \cup V_2$ with the algebraic K-sets $V_1 = \mathcal{N}(f) \cap V$ and $V_2 = \mathcal{N}(g) \cap V$. Now V is irreducible; hence, $V = V_1$, or $V = V_2$, say $V = V_1$. Then $V \subset \mathcal{N}(f)$. Therefore, $f \in I\mathcal{N}(f) \subset I(V)$. Since $V \neq \emptyset$, we have further $1 \notin I(V)$; that is, $I(V) \neq R$.

(2) Let $I(V) \triangleleft R$ with $I(V) \neq R$ be a prime ideal. Let $V = V_1 \cup V_2$, $V_1 \neq V$, with algebraic K-sets V_i in C^n. First,

$$I(V) = I(V_1 \cup V_2) = I(V_1) \cap I(V_2) \supset I(V_1)I(V_2), \tag{$\star$}$$

where $I(V_1)I(V_2)$ is the ideal generated by all products fg with $f \in I(V_1)$, $g \in I(V_2)$. We have $I(V_1) \neq I(V)$, because otherwise $V_1 = \mathcal{N}I(V_1) = \mathcal{N}I(V) = V$ contradicting $V_1 \neq V$. Hence, there is a $f \in I(V_1)$ with $f \notin I(V)$. Now, $I(V) \neq R$ is a prime ideal; hence, necessarily $I(V_2) \subset I(V)$ by ($\star$). It follows that $V \subset V_2$. Therefore, V is irreducible. □

Note that the affine space K^n is, as the zero set of the zero polynomial 0, itself an algebraic K-set in K^n. If K is infinite, then $I(K^n) = \{0\}$. Hence, K^n is irreducible by Theorem 21.5.3. Moreover, if K is infinite, then K^n can not be written as a union of finitely many proper algebraic K-subsets. If K is finite, then K^n is not irreducible.

Furthermore, each algebraic K-set V in C^n is also an algebraic C-set in C^n. If V is an irreducible algebraic K-set in C^n, then—in general—it is not an irreducible algebraic C-set in C^n.

Theorem 21.5.4. *Let V be an algebraic K-set in C^n. Then V can be written as a finite union $V = V_1 \cup V_2 \cup \cdots \cup V_r$ of irreducible algebraic K-sets V_i in C^n. If here $V_i \not\subset V_k$ for all pairs*

(i,k) with $i \neq k$, then this presentation is unique, up to the ordering of the V_i, and then the V_i are called the irreducible K-components of V.

Proof. Let $\mathfrak{a}$ be the set of all algebraic K-sets in C^n, which can not be presented as a finite union of irreducible algebraic K-sets in C^n.

Assume that $\mathfrak{a} \neq \emptyset$. By Theorem 21.4.1, there is a minimal element V in $\mathfrak{a}$. This V is not irreducible, otherwise we have a presentation as desired. Hence, there exists a presentation $V = V_1 \cup V_2$ with algebraic K-sets V_i, which are strictly smaller than V. By definition, both V_1 and V_2 have a presentation as desired; hence, V also has one, which gives a contradiction. Hence, $\mathfrak{a} = \emptyset$.

Now suppose that $V = V_1 \cup \cdots \cup V_r = W_1 \cup \cdots \cup W_s$ are two presentations of the desired form. For each V_i, we have a presentation $V_i = (V_i \cap W_1) \cup \cdots \cup (V_i \cap W_s)$. Each $V_i \cap W_j$ is a K-algebraic set (see Theorem 21.2.4). Since V_i is irreducible, we get that there is a W_j with $V_i = V_i \cap W_j$, that is, $V_i \subset W_j$. Analogously, for this W_j, there is a V_k with $W_j \subset V_k$. Altogether, $V_i \subset W_j \subset V_k$. But $V_p \not\subset V_q$ if $p \neq q$. Hence, from $V_i \subset W_j \subset V_k$, we get $i = k$. Therefore, $V_i = W_j$; that means, for each V_i there is a W_j with $V_i = W_j$. Analogously, for each W_k, there is a V_l with $W_k = V_l$. This proves the theorem. □

Example 21.5.5. 1. Let $M = \{gf\} \subset \mathbb{R}[x,y]$ with $g(x) = x^2+y^2-1$ and $f(x) = x^2+y^2-2$. Then we have $\mathcal{N}(M) = V = V_1 \cup V_2$, where $V_1 = \mathcal{N}(g)$, and $V_2 = \mathcal{N}(f)$; V is not irreducible.
2. Let $M = \{f\} \subset \mathbb{R}[x,y]$ with $f(x,y) = xy-1$; f is irreducible in $\mathbb{R}[x,y]$. Therefore, the ideal (f) is a prime ideal in $\mathbb{R}[x,y]$. Hence, $V = \mathcal{N}(f)$ is irreducible.

Definition 21.5.6. Let V be an algebraic K-set in C^n. Then the residue class ring

$$K[V] = K[x_1,\dots,x_n]/I(V)$$

is called the *(affine) coordinate ring* of V.

$K[V]$ can be identified with the ring of all those functions $V \to C$, which are given by polynomials from $K[x_1,\dots,x_n]$. As a homomorphic image of $K[x_1,\dots,x_n]$, we get that $K[V]$ can be described in the form $K[V] = K[\alpha_1,\dots,\alpha_n]$; therefore, a K-algebra of the form $K[\alpha_1,\dots,\alpha_n]$ is often called an *affine K-algebra.* If the algebraic K-set V in C^n is irreducible—we can call V now an *(affine) K-variety in C^n*—then $K[V]$ is an integral domain with an identity, because $I(V)$ is then a prime ideal with $I(V) \neq R$ by Theorem 21.4.2. The quotient field $K(V) = \operatorname{Quot} K[V]$ is called the *field of rational functions on the K-variety V.*

We note the following:
1. If C is algebraically closed, then $V = C^n$ is a K-variety, and $K(V)$ is the field $K(x_1,\dots,x_n)$ of the rational functions in n variables over K.
2. Let the affine K-algebra $A = K[\alpha_1,\dots,\alpha_n]$ be an integral domain with an identity $1 \neq 0$. Then $A \cong K[x_1,\dots,x_n]/\mathfrak{p}$ for some prime ideal $\mathfrak{p} \neq K[x_1,\dots,x_n]$. Hence, if C is algebraically closed, then A is isomorphic to the coordinate ring of the K-variety $V = \mathcal{N}(\mathfrak{p})$ in C^n (see Hilbert's nullstellensatz, first form, Theorem 21.4.1).

3. If the affine K-algebra $A = K[a_1, \dots, a_n]$ is an integral domain with an identity $1 \neq 0$, then we define the transcendence degree $\operatorname{trgd}(A|K)$ to be the transcendence degree of the field extension $\operatorname{Quot}(A)|K$; that is, $\operatorname{trgd}(A|K) = \operatorname{trgd}(\operatorname{Quot}(A)|K)$, $\operatorname{Quot}(A)$ the quotient field of A.
   In this sense, $\operatorname{trgd}(K[x_1, \dots, x_n]|K) = n$. Since $\operatorname{Quot}(A) = K(a_1, \dots, a_n)$, we get $\operatorname{trgd}(A|K) \leq n$ by Noether's normalization theorem (Theorem 20.3.10).
4. An arbitrary affine K-algebra $K[a_1, \dots, a_n]$ is, as a homomorphic image of the polynomial ring $K[x_1, \dots, x_n]$, noetherian (see Theorem 21.2.4 and Theorem 21.2.3).

Example 21.5.7. Let $\omega_1, \omega_2 \in \mathbb{C}$ two elements which are linear independent over $\mathbb{R}$. An element $\omega = m_1\omega_1 + m_2\omega_2$ with $m_1, m_2 \in \mathbb{Z}$, is called a *period*. The periods describe an Abelian group $\Omega = \{m_1\omega_1 + m_2\omega_2 : m_1, m_2 \in \mathbb{Z}\} \cong \mathbb{Z} \oplus \mathbb{Z}$ and give a lattice in $\mathbb{C}$.

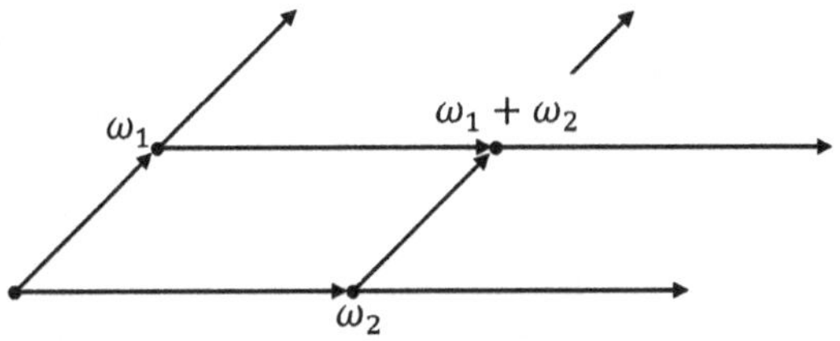

An elliptic function f (with respect to Ω) is a meromorphic function with period group Ω, that is, $f(z + w) = f(z)$ for all $z \in \mathbb{C}$. The Weierstrass $\wp$-function,

$$\wp(z) = \frac{1}{z^2} + \sum_{0 \neq w \in \Omega} \left(\frac{1}{(z - w)^2} - \frac{1}{w^2} \right),$$

is an elliptic function.

With $g_2 = 60 \sum_{0 \neq w \in \Omega} \frac{1}{w^4}$, and $g_3 = 140 \sum_{0 \neq w \in \Omega} \frac{1}{w^6}$, we get the differential equation $\wp'(z)^2 = 4\wp(z)^3 + g_2\wp(z) + g_3 = 0$. The set of elliptic functions is a field E, and each elliptic function is a rational function in $\wp$ and $\wp'$ (for details see, for instance, [44]).

The polynomial $f(t) = t^2 - 4s^3 + g_2 s + g_3 \in \mathbb{C}(s)[t]$ is irreducible over $\mathbb{C}(s)$. For the corresponding algebraic $\mathbb{C}(s)$-set V, we get $K(V) = \mathbb{C}(s)[t]/(t^2 - 4s^3 + g_2 s + g_3) \cong E$ with respect to $t \mapsto \wp'$, $s \mapsto \wp$.

21.6 Dimensions

From now we assume that C is algebraically closed.

Definition 21.6.1. (1) The *dimension* $\dim(V)$ of an algebraic K-set V in C^n is said to be the supremum of all integers m, for which there exists a strictly descending chain $V_0 \supsetneq V_1 \supsetneq \cdots \supsetneq V_m$ of K-varieties V_i in C^n with $V_i \subset V$ for all i.

(2) Let A be a commutative ring with an identity $1 \neq 0$. The *height* $h(\mathfrak{p})$ of a prime ideal $\mathfrak{p} \neq A$ of A is said to be the supremum of all integers m, for which there exists a strictly ascending chain $\mathfrak{p}_0 \subsetneq \mathfrak{p}_1 \subsetneq \cdots \subsetneq \mathfrak{p}_m = \mathfrak{p}$ of prime ideals $\mathfrak{p}_i$ of A with $\mathfrak{p}_i \neq A$. The *dimension* (Krull dimension) $\dim(A)$ of A is the supremum of the heights of all prime ideals $\neq A$ in A.

Theorem 21.6.2. *Let V be an algebraic K-set in C^n. Then* $\dim(V) = \dim(K[V])$.

Proof. By Theorem 21.2.4 and Theorem 21.4.2, we have a bijective map between the K-varieties W with $W \subset V$ and the prime ideals $\neq R = K[x_1,\dots,x_n]$ of R, which contain $I(V)$ (the bijective map reverses the inclusion). But these prime ideals correspond exactly with the prime ideals $\neq K[V]$ of $K[V] = K[x_1,\dots,x_n]/I(V)$, which gives the statement. □

Suppose that V is an algebraic K-set in C^n, and let $V_1,\dots,V_r$ the irreducible components of V. Then $\dim(V) = \max\{\dim(V_1),\dots,\dim(V_r)\}$, because if V is a K-variety with $V' \subset V$, Then, $V' = (V' \cap V_1) \cup \cdots \cup (V' \cap V_r)$. Hence, we may restrict ourselves on K-varieties V.

If we consider the special case of the K-variety $V = C^1 = C$ (recall that C is algebraically closed, and, hence, in particular, C is infinite). Then $K[V] = K[x]$, the polynomial ring $K[x]$ in one indeterminate x. Now, $K[x]$ is a principal ideal domain, and hence, each prime ideal $\neq K[x]$ is either a maximal ideal or the zero ideal $\{0\}$ of $K[x]$. The only K-varieties in $V = C$ are therefore V itself and the zero set of irreducible polynomials in $K[x]$. Hence, if $V = C$, then $\dim(V) = \dim K[V] = 1 = \operatorname{trgd}(K[V]|K)$.

Theorem 21.6.3. *Let $A = K[a_1,\dots,a_n]$ be an affine K-algebra, and let A be also an integral domain. Let $\{0\} = \mathfrak{p}_0 \subsetneq \mathfrak{p}_1 \subsetneq \cdots \subsetneq \mathfrak{p}_m$ be a maximal strictly ascending chain of prime ideals in A (such a chain exists since A is noetherian). Then $m =$* $\operatorname{trgd}(A|K) = \dim(A)$. *In other words;*

All maximal ideals of A have the same height, and this height is equal to the transcendence degree of A over K.

Corollary 21.6.4. *Let V be a K-variety in C^n. Then* $\dim(V) = \operatorname{trgd}(K[V]|K)$.

We prove Theorem 21.6.3 in several steps.

Lemma 21.6.5. *Let R be an unique factorization domain. Then each prime ideal $\mathfrak{p}$ with height $h(\mathfrak{p}) = 1$ is a principal ideal.*

Proof. $\mathfrak{p} \neq \{0\}$, since $h(\mathfrak{p}) = 1$. Hence, there is an $f \in \mathfrak{p}$, $f \neq 0$. Since R is an unique factorization domain, f has a decomposition $f = p_1 \cdots p_s$ with prime elements $p_i \in R$. Now, $\mathfrak{p}$ is a prime ideal; hence, some $p_i \in \mathfrak{p}$, because $f \in \mathfrak{p}$, say $p_1 \in \mathfrak{p}$. Then we have the chain $\{0\} \subsetneq (p_1) \subset \mathfrak{p}$, and (p_1) is a prime ideal of R. Since $h(\mathfrak{p}) = 1$, we get $(p_1) = \mathfrak{p}$. □

Lemma 21.6.6. *Let $R = K[y_1,\dots,y_r]$ be the polynomial ring of the r independent indeterminates $y_1,\dots,y_r$ over the field K (recall that R is a unique factorization domain). If $\mathfrak{p}$ is* *a prime ideal in R with height* $h(\mathfrak{p}) = 1$, *then the residue class ring* $\bar{R} = R/\mathfrak{p}$ *has transcendence degree* $r-1$ *over K.*

Proof. By Lemma 21.6.5, we have that $\mathfrak{p} = (p)$ for some nonconstant polynomial $p \in K[y_1,\dots,y_r]$. Let the indeterminate $y = y_r$ occur in p, that is, $\deg_y(p) \geq 1$, the degree in y. If f is a multiple of p, then also $\deg_y(f) \geq 1$. Hence, $\mathfrak{p} \cap K[y_1,\dots,y_r] \neq \{0\}$. Therefore, the residue class mapping $R \to \bar{R} = K[\bar{y}_1,\dots,\bar{y}_r]$ induces an isomorphism $K[y_1,\dots,y_{r-1}] \to K[\bar{y}_1,\dots,\bar{y}_{r-1}]$ of the subring $K[y_1,\dots,y_{r-1}]$; that is, $\bar{y}_1,\dots,\bar{y}_{r-1}$ are algebraically independent over K. On the other hand, $p(\bar{y}_1,\dots,\bar{y}_{r-1},\bar{y}_r) = 0$ is a nontrivial algebraic relation for $\bar{y}_r$ over $K(\bar{y}_1,\dots,\bar{y}_{r-1})$.

Hence, altogether $\operatorname{trgd}(\bar{R}|K) = \operatorname{trgd}(K(\bar{y}_1,\dots,\bar{y}_r)|K) = r-1$ by Theorem 20.3.9. □

Before we describe the last technical lemma, we need some preparatory theoretical material.

Let R, A be integral domains (with identity $1 \neq 0$), and let $A|R$ be a ring extension. We first consider only R.

(1) A subset $S \subset R \setminus \{0\}$ is called a multiplicative subset of R if $1 \in S$ for the identity 1 of R, and if $s,t \in S$, then also, $st \in S$. $(x,s) \sim (y,t) :\Leftrightarrow xt - ys = 0$ defines an equivalence relation on $M = R \times S$. Let $\frac{x}{s}$ be the equivalence class of (x,s) and $S^{-1}R$, the set of all equivalence classes. We call $\frac{x}{s}$ a fraction. If we add and multiply fractions as usual, we get that $S^{-1}R$ becomes an integral domain; it is called *the ring of fractions of R with respect to S*. If, in particular, $S = R \setminus \{0\}$, then $S^{-1}R = \operatorname{Quot}(R)$, the quotient field of R.

Now, back to the general situation. $i : R \to S^{-1}R$, $i(r) = \frac{r}{1}$, defines an embedding of R into $S^{-1}R$. Hence, we may consider R as a subring of $S^{-1}R$. For each $s \in S \subset R \setminus \{0\}$, we have that $i(s)$ is an unit in $S^{-1}R$. That is, $i(s)$ is invertible, and each element of $S^{-1}R$ has the form $i(s)^{-1}i(r)$ with $r \in R$, $s \in S$. Therefore, $S^{-1}R$ is uniquely determined up to isomorphisms, and we have the following universal property:

If $\phi : R \to R'$ is a ring homomorphism (of integral domains) with $\phi(s)$ invertible for each $s \in S$, then there exist exactly one ring homomorphism $\lambda : S^{-1}R \to R'$ with $\lambda \circ i = \phi$. If $\mathfrak{a} \triangleleft R$ is an ideal in $\mathfrak{a}$, then we write $S^{-1}\mathfrak{a}$ for the ideal in $S^{-1}R$, generated by $i(\mathfrak{a})$. $S^{-1}\mathfrak{a}$ is the set of all elements of the form $\frac{a}{s}$ with $a \in \mathfrak{a}$ and $s \in S$. Furthermore, $S^{-1}\mathfrak{a} = (1) \Leftrightarrow \mathfrak{a} \cap S \neq \emptyset$.

Vice versa; if $\mathfrak{A} \triangleleft S^{-1}R$ is an ideal in $S^{-1}R$, then we also denote the ideal $i^{-1}(\mathfrak{A}) \triangleleft R$ with $\mathfrak{A} \cap R$. An ideal $\mathfrak{a} \triangleleft R$ is of the form $\mathfrak{a} = i^{-1}(\mathfrak{A})$ if and only if there is no $s \in S$ such that its image in $R/\mathfrak{a}$ under the canonical map $R \to R/\mathfrak{a}$ is a proper zero divisor in $R/\mathfrak{a}$. Under the mapping $\mathfrak{P} \to \mathfrak{P} \cap R$ and $\mathfrak{p} \mapsto S^{-1}\mathfrak{p}$, the prime ideals in $S^{-1}R$ correspond exactly to the prime ideals in R, which do not contain an element of S.

We now identify R with $i(R)$:

(2) Now, let $\mathfrak{p} \triangleleft R$ be a prime ideal in R. Then $S = R \setminus \mathfrak{p}$ is multiplicative. In this case, we write $R_\mathfrak{p}$ instead of $S^{-1}R$, and call $R_\mathfrak{p}$ the *quotient ring* of R with respect to $\mathfrak{p}$, or the localization of R of $\mathfrak{p}$. Put $\mathfrak{m} = \mathfrak{p}R_\mathfrak{p} = S^{-1}\mathfrak{p}$. Then $1 \notin \mathfrak{m}$. Each element of $R_\mathfrak{p}/\mathfrak{m}$ is a unit in $R_\mathfrak{p}$ and vice versa. In other words, each ideal $\mathfrak{a} \neq (1)$ in $R_\mathfrak{p}$ is contained in $\mathfrak{m}$, or equivalently, $\mathfrak{m}$ is the only maximal ideal in $R_\mathfrak{p}$. A commutative ring with an identity

$1 \neq 0$, which has exactly one maximal ideal, is called a *local ring*. Hence, $R_\mathfrak{p}$ is a local ring. From part (1), we additionally get the prime ideals of the local ring $R_\mathfrak{p}$ correspond bijectively to the prime ideals of R, which are contained in $\mathfrak{p}$.

(3) Now we consider our ring extension $A|R$ as above. Let $\mathfrak{q}$ be a prime ideal in R.

Claim: If $\mathfrak{q}A \cap R = \mathfrak{q}$, then there exists a prime ideal $\mathfrak{Q} \lhd A$ with $\mathfrak{Q} \cap R = \mathfrak{q}$ (and vice versa).

Proof of the claim: If $S = R \setminus \mathfrak{q}$, then $\mathfrak{q}A \cap S = \emptyset$. Hence, $\mathfrak{q}S^{-1}A$ is a proper ideal in $S^{-1}A$, and hence contained in a maximal ideal $\mathfrak{m}$ in $S^{-1}A$. Here, $\mathfrak{q}S^{-1}A$ is the ideal in $S^{-1}A$, which is generated by $\mathfrak{q}$. Define $\mathfrak{Q} = \mathfrak{m} \cap A$; $\mathfrak{Q}$ is a prime ideal in A, and $\mathfrak{Q} \cap R = \mathfrak{q}$ by part (1), because $\mathfrak{Q} \cap S = \emptyset$, where $S = R \setminus \mathfrak{q}$.

(4) Now let $A|R$ be an integral extension (A, R integral domains as above). Assume that R is integrally closed in its quotient field K. Let $\mathfrak{P} \lhd A$ be a prime ideal in A and $\mathfrak{p} = \mathfrak{P} \cap R$.

Claim: If $\mathfrak{q} \lhd R$ is a prime ideal in A with $\mathfrak{q} \subset \mathfrak{p}$ then $\mathfrak{q}A_\mathfrak{p} \cap R = \mathfrak{q}$.

Proof of the claim: An arbitrary $\beta \in \mathfrak{q}A_\mathfrak{p}$ has the form $\beta = \frac{\alpha}{s}$ with $\alpha \in \mathfrak{q}A$, $\mathfrak{q}A$ (the ideal in A generated by $\mathfrak{q}$), and $s \in S = A \setminus \mathfrak{p}$. An integral equation for $\alpha \in \mathfrak{q}A$ over K is given a form $\alpha^n + a_{n-1}\alpha^{n-1} + \cdots + a_0 = 0$ with $a_i \in \mathfrak{q}$. This can be seen as follows: we have certainly a form $\alpha = b_1\alpha_1 + \cdots + b_m\alpha_m$ with $b_i \in \mathfrak{q}$ and $\alpha_i \in A$. The subring $A' = R[\alpha_1, \ldots, \alpha_m]$ is, as an R-module, finitely generated, and $\alpha A' \subset \mathfrak{q}A'$. Now, $a_i \in \mathfrak{q}$ follows with the same type of arguments as in the proof of Theorem 20.2.4.

Now, in addition, let $\beta \in R$. Then, for $s = \frac{\alpha}{\beta}$, we have an equation

$$s^n + \frac{a_{n-1}}{\beta}s^{n-1} + \cdots + \frac{a_0}{\beta^n} = 0$$

over K. But s is integral over R; hence, all $\frac{a_{n-1}}{\beta^i} \in R$.

We are now prepared to prove the last preliminary lemma, which we need for the proof of Theorem 21.6.3.

Lemma 21.6.7 (Krull's going up lemma). *Let $A|R$ be an integral ring extension of integral domains, and let R be integrally closed in its quotient field. Let $\mathfrak{p}$ and $\mathfrak{q}$ be prime ideals in R with $\mathfrak{q} \subset \mathfrak{p}$. Furthermore, let $\mathfrak{P}$ be a prime ideal in A with $\mathfrak{P} \cap R = \mathfrak{p}$. Then there exists a prime ideal $\mathfrak{Q}$ in A with $\mathfrak{Q} \cap R = \mathfrak{q}$, and $\mathfrak{Q} \subset \mathfrak{P}$.*

Proof. It is enough to show that there exists a prime ideal $\mathfrak{Q}$ in $A_\mathfrak{p}$ with $\mathfrak{Q} \cap R = \mathfrak{q}$. This can be seen from the preceding preparations. By part (1) and (2) such a $\mathfrak{Q}$ has the form $\mathfrak{Q} = \mathfrak{Q}'A_\mathfrak{p}$ with a prime ideal $\mathfrak{Q}'$ in A with $\mathfrak{Q}' \subset \mathfrak{P}$, and $\mathfrak{Q} \cap A = \mathfrak{Q}'$. It follows that $\mathfrak{q} = \mathfrak{Q}' \cap R \subset \mathfrak{P} \cap R = \mathfrak{p}$. And the existence of such a $\mathfrak{Q}$ follows from parts (3) and (4). □

Proof of Theorem 21.6.3. Let first be $m = 0$. Then $\{0\}$ is a maximal ideal in A; and hence, $A = K[\alpha_1, \ldots, \alpha_n]$ a field. By Corollary 20.3.11 then, $A|K$ is algebraic; therefore, $\operatorname{trgd}(A|K) = 0$. So, Theorem 21.3.3 holds for $m = 0$.

Now, let $m \geq 1$. We use Noether's normalization theorem. A has a polynomial ring $R = K[y_1, \ldots, y_r]$ of the r independent indeterminates $y_1, \ldots, y_r$ as a subring, and $A|R$ is an integral extension. As a polynomial ring over K, the ring R is a unique factorization domain, and hence, certainly, algebraically closed (in its quotient field).

Now, let

$$\{0\} = \mathfrak{P}_0 \subsetneq \mathfrak{P}_1 \subsetneq \cdots \subsetneq \mathfrak{P}_m \tag{21.3}$$

be a maximal strictly ascending chain of prime ideals in A. If we intersect with R, we get a chain

$$\{0\} = \mathfrak{p}_0 \subset \mathfrak{p}_1 \subset \cdots \subset \mathfrak{p}_m \tag{21.4}$$

of prime ideals $\mathfrak{p}_i = \mathfrak{P}_i \cap R$ of R. Since $A|R$ is integral, the chain (21.4) is also a strictly ascending chain. This follows from Krull's going up lemma (Lemma 21.6.7), because if $\mathfrak{p}_i = \mathfrak{p}_j$, then $\mathfrak{P}_i = \mathfrak{P}_j$. If $\mathfrak{P}_m$ is a maximal ideal in A, then also $\mathfrak{p}_m$ is a maximal ideal in R, because $A|R$ is integral (consider $A/\mathfrak{P}_m$ and use Theorem 20.2.19). If the chain (21.3) is maximal and strictly, then also the chain (2).

Now, let the chain (21.3) be maximal and strictly. If we pass to the residue class rings $\bar{A} = A/\mathfrak{P}_1$ and $\bar{R} = R/\mathfrak{p}_1$, then we get the chains of prime ideals

$$\{0\} = \bar{\mathfrak{P}}_1 \subset \bar{\mathfrak{P}}_2 \subset \cdots \subset \bar{\mathfrak{P}}_m \quad \text{and} \quad \{0\} = \bar{\mathfrak{p}}_1 \subset \bar{\mathfrak{p}}_2 \subset \cdots \subset \bar{\mathfrak{p}}_m$$

for the affine K-algebras $\bar{A}$ and $\bar{R}$, respectively, but with a 1 less length. By induction, we may assume that already $\operatorname{trgd}(\bar{A}|K) = m - 1 = \operatorname{trgd}(\bar{R}|K)$. On the other hand, by construction, we have $\operatorname{trgd}(A|K) = \operatorname{trgd}(R|K) = r$. Finally, to prove Theorem 21.3.3, we have to show that $r = m$. If we compare both equations, then $r = m$ follows if $\operatorname{trgd}(\bar{R}|K) = r - 1$. But this holds by Lemma 21.6.6. □

Theorem 21.6.8. *Let V be a K-variety in C^n. Then $\dim(V) = n-1$ if and only if $V = (f)$ for some irreducible polynomial $f \in K[x_1, \ldots, x_n]$.*

Proof. (1) Let V be a K-variety in C^n with $\dim(V) = n - 1$. The corresponding ideal (in the sense of Theorem 21.2.4) is by Theorem 21.4.2 a prime ideal $\mathfrak{p}$ in $K[x_1, \ldots, x_n]$. By Theorem 21.3.3 and Corollary 21.3.4, we get $h(\mathfrak{p}) = 1$ for the height of $\mathfrak{p}$, because $\dim(V) = n - 1$ (see also Theorem 21.3.2). Since $K[x_1, \ldots, x_n]$ is a unique factorization domain, we get that $\mathfrak{p} = (f)$ is a principal ideal by Lemma 21.6.5.

(2) Now let $f \in K[x_1, \ldots, x_n]$ be irreducible. We have to show that $V = \mathcal{N}(f)$ has dimension $n - 1$. For that, by Theorem 21.6.3, we have to show that the prime ideal $\mathfrak{p} = (f)$ has the height $h(\mathfrak{p}) = 1$. Assume that this is not the case. Then there exists a prime ideal $\mathfrak{q} \neq \mathfrak{p}$ with $\{0\} \neq \mathfrak{q} \subset \mathfrak{p}$. Choose $g \in \mathfrak{q}$, $g \neq 0$. Let $g = uf^{e_1}\pi_2^{e_2} \cdots \pi_r^{e_r}$ be its prime factorization in $K[x_1, \ldots, x_n]$. Now $g \in \mathfrak{q}$ and $f \notin \mathfrak{q}$, because $\mathfrak{q} \neq \mathfrak{p}$. Hence, there is a π_i in $\mathfrak{q} \subsetneq \mathfrak{p} = (f)$, which is impossible. Therefore $h(\mathfrak{p}) = 1$. □

21.7 Exercises

1. Let $A = K[a_1, \dots, a_n]$ and $C|K$ be a field extension with C algebraically closed. Show that there is a K-algebra homomorphism $K[a_1, \dots, a_n] \to C$.
2. Let $K[x_1, \dots, x_n]$ be the polynomial ring of the n independent indeterminates $x_1, \dots, x_n$ over the algebraically closed field K. The maximal ideals of $K[x_1, \dots, x_n]$ are exactly the ideals of the form $\mathfrak{m}(\alpha) = (x_1 - \alpha_1, x_2 - \alpha_2, \dots, x_n - \alpha_n)$ with $\alpha = (\alpha_1, \dots, \alpha_n) \in K^n$.
3. The nil radical $\sqrt{0}$ of $A = K[a_1, \dots, a_n]$ corresponds with the Jacobson radical of A, that is, the intersection of all maximal ideals of A.
4. Let R be a commutative ring with $1 \neq 0$. If each prime ideal of R is finitely generated, then R is noetherian.
5. Prove the theoretical preparations for Krull's going up lemma in detail.
6. Let $K[x_1, \dots, x_n]$ be the polynomial ring of the n independent indeterminates $x_1, \dots, x_n$. For each ideal $\mathfrak{a}$ of $K[x_1, \dots, x_n]$, there exists a natural number m with the following property: if $f \in K[x_1, \dots, x_n]$ vanishes on the zero set of $\mathfrak{a}$, then $f^m \in \mathfrak{a}$.
7. Let K be a field with char $K \neq 2$ and $a, b \in K^\star$. We consider the polynomial

$$f(x,y) = ax^2 + by^2 - 1 \in K[x,y]$$

as the polynomial ring of the independent indeterminates x and y. Let C be the algebraic closure of $K(x)$ and $\beta \in C$ with $f(x, \beta) = 0$. Show the following:
(i) f is irreducible over the algebraic closure C_0 of K (in C).
(ii) $\operatorname{trgd}(K(x,\beta)|K) = 1$, $[K(x,\beta) : K(x)] = 2$, and K is algebraically closed in $K(x,\beta)$.

22 Algebras and Group Representations

22.1 Group Representations

In Chapter 13, we spoke about group actions. These are homomorphisms from a group G into a set of permutations on a set S. The way a group G acts on a set S can often be used to study the structure of the group G, and, in Chapter 13, we used group actions to prove the important Sylow theorems.

In this chapter, we discuss a very important type of group action called a *group representation* or *linear representation*. This is a homomorphism of a group G into the set of linear transformations of a vector space V over a field K. It is a *finite-dimensional representation* if V is a finite dimensional vector space over K, and *infinite-dimensional* otherwise. For an n-dimensional representation, each element of the group G can be represented by an $(n \times n)$-matrix over K, and the group operation can be represented by matrix multiplication. As with general group actions, much information about the structure of the group G can be obtained from representations. In particular, in this chapter, we will present an important Burnside theorem, which shows that any finite group, whose order is divisible by only two primes, must be solvable.

Representations of groups are important in many areas of mathematics. Group representations allow many group-theoretic problems to be reduced to problems in linear algebra, which is well understood. They are also important in physics and the study of physical structure, because they describe how the symmetry group of a physical system affects the solutions of equations describing that system.

The theory of group representations can be divided into several areas depending on the kind of group being represented. The various areas can be quite different in detail, though the basic definitions and concepts are the same. The most important areas are:

(1) *The theory of finite group representations.* Group representations constitute a crucial tool in the study of finite groups. They also arise in applications of finite group theory to crystallography and to geometry.
(2) *Group representations of compact and locally compact groups.* Using integration theory and Haar measure, many of the results on representations of finite groups can be extended to infinite locally compact groups. The resulting theory is a central part of the area of mathematics called *harmonic analysis*. Pontryagin duality describes the theory for commutative groups as a generalized Fourier transform.
(3) *Representations of Lie Groups.* Lie groups are continuous groups with a differentiable structure. Most of the groups that arise in physics and chemistry are Lie groups, and their representation theory is important to the application of group theory in those fields.
(4) Linear algebraic groups are the analogues of Lie groups, but over more general fields than just the reals or complexes. Their representation theory is more complicated than that of Lie groups.

https://doi.org/10.1515/9783111142524-022

For this chapter, we will consider solely the representation theory of finite groups, and for the remainder of this chapter, when we say group, we mean finite group.

22.2 Representations and Modules

A group representation is a group action on a vector space that respects the vector space structure. In this section, we examine the basic definitions of group representations and the ties to general modules over rings, both commutative and non-commutative. The main reference for this chapter is the book entitled *Groups and Representations* by J. L. Alperin and R. B. Bell [1]. We follow the main lines of this book. As we mentioned in the previous section, throughout the remainder of the chapter, group refers to a finite group.

Let K be a field, and let G be a group action on a K-vector space V. We denote this action by gv for $g \in G$ and $v \in V$. The action is called *linear* if the following hold:

(1) $g(v + w) = gv + gw$ for all $g \in G$, and $v, w \in V$.
(2) $g(av) = a(gv)$ for all $g \in G$, $a \in K$, and $v \in V$.

Recall that group actions correspond to group homomorphisms into symmetric groups. For linear actions on a vector space V, we have a stronger result.

Theorem 22.2.1. *There is a bijective correspondence between the set of linear actions of a group G on a K-vector space V and the set of homomorphisms from G into $\mathrm{GL}(V)$, the group of all invertible linear transformations of V, which is called the general linear group over V.*

Proof. Suppose that $\rho : G \to \mathrm{GL}(V)$ is a homomorphism, then the action of G on V is defined by setting $gv = \rho(g)(v)$, and it is clear that this action is linear.

Conversely, if we have a linear action of G on V, then we can define a homomorphism $\rho : G \to \mathrm{GL}(V)$ by $\rho(g)v = gv$. These processes are mutually inverse, which gives the desired correspondence. □

Definition 22.2.2. A homomorphism $\rho : G \to \mathrm{GL}(V)$, where G is a group and V is a K-vector space called a *linear representation* or *group representation* of G in V.

From Theorem 22.2.1, it follows that the study of group representations is equivalent to the study of linear actions of groups. This area of study, with emphasis on finite groups and finite-dimensional vector spaces, has many applications to finite group theory.

The modern approach to the representation theory of finite groups involves another equivalent concept, namely that of finitely generated modules over group rings.

In Chapter 18, we considered R-modules over commutative rings R, and used this study to prove the fundamental theorem of finitely generated modules over principal ideal domains. In particular, we used the same study to prove the fundamental theorem of finitely generated Abelian groups. Here we must extend the concepts and allow R to be a general ring with identity.

Definition 22.2.3. Let R be a ring with identity 1, and let M be an Abelian group written additively. M is called *left R-module* if there is a map $R \times M \to M$ written as $(r, m) \mapsto rm$ such that the following hold:

(1) $1 \cdot m = m$;
(2) $r(m + n) = rm + rn$;
(3) $(r + s)m = rm + sm$;
(4) $r(sm) = (rs)m$;

for all $r, s \in R$ and $m, n \in M$.

We can similarly define the notion of a right R-module via a map from $M \times R$ to M sending (m, r) to mr, which satisfies the analogous properties to those above. If R is commutative, then every left module can in an obvious manner be given a right R-module structure; hence, it is not necessary in the commutative case to distinguish between left and right R-modules.

We always use the wording R-module to denote left R-module, unless otherwise specified.

Definition 22.2.4. An R-module M is *finitely generated* if every element of M can be written as an R-linear combination $m = r_1 m_1 + \cdots + r_k m_k$ for a finite subset $\{m_1, \ldots, m_k\}$ of M.

Finite minimal sets for a given module may have different numbers of elements. This is in contrast to the situation in free R-modules over a commutative ring R with identity, where any two finite bases have the same number of elements (Theorem 18.4.6).

In the following, we review the module theory that is necessary for the study of group representations. The facts we use are straightforward extensions of the respective facts for modules over commutative rings or for groups.

Definition 22.2.5. Let M be an R-module, and let N be a subgroup of M. Then N is an *R-submodule* (or just a submodule) if $rn \in N$ for every $r \in R$ and $n \in N$.

Example 22.2.6. The R-submodules of a ring R are exactly the left ideals of R (see Chapter 1). Every R-module M has at least two submodules, namely, M itself and the zero submodule $\{0\}$.

Definition 22.2.7. A *simple* R-module is an R-module $M \neq \{0\}$, which has only M and $\{0\}$ as submodules.

If N is a submodule of M, then we may construct the factor group M/N (recall that M is Abelian). We may give the factor group M/N an R-module structure by defining $r(m + N) = rm + N$ for every $r \in R$ and $m + N \in M/N$. We call M/N the *factor R-module*, or just factor module of M/N.

Definition 22.2.8. Let N_1, N_2 be submodules of an R-module M. Then we define the *module sum* $N_1 + N_2$ by

$$N_1 + N_2 = \{x + y \mid x \in N_1, y \in N_2\} \subset M.$$

The sum $N_1 + N_2$ and the intersection $N_1 \cap N_2$ are submodules of M. If $N_1 \cap N_2 = \{0\}$, then we call the sum $N_1 + N_2$ a *direct sum* and write $N_1 \oplus N_2$ instead of $N_1 + N_2$.

We say that a submodule N of M is a *direct summand* if there is some other submodule N' of M such that $M = N \oplus N'$. In general, we write kN or N^k to denote the direct sum

$$N \oplus N \oplus \cdots \oplus N$$

of k copies of N.

As for groups, we also have the external notion of a direct sum. If M and N are R-modules, then we give the Cartesian product $M \times N$ an R-module structure by setting $r(m,n) = (rm, rn)$, and we write $M \oplus N$ instead of $M \times N$.

The notions of internal and external direct sums can be extended to any finite number of submodules and modules, respectively.

Definition 22.2.9. A *composition series* of an R-module $M \neq \{0\}$ is a descending series

$$M = M_0 \supset M_1 \supset \cdots \supset M_k = \{0\}$$

of finitely many submodules M_i of M beginning with M and ending with $\{0\}$, where the inclusions are proper, and in which each successive factor module M_i/M_{i+1} is a simple module. We call the length of the composition series k.

Notice the following:

(1) A module need not have a composition series. For example, an infinite Abelian group, considered as a $\mathbb{Z}$-module, does not have a composition series (see Chapter 12).

(2) The analog of the Jordan–Hölder theorem for groups (see Theorem 12.3.3) holds for modules that have composition series.

Theorem 22.2.10 (Jordan–Hölder theorem for R-modules). *If an R-module $M \neq \{0\}$ has a composition series, then any two composition series are equivalent; that is, there exists a one-to-one correspondence between their respective factor modules. Hence, the factor modules are unique, and, in particular, the length must be the same.*

Therefore, we can speak in a well defined manner about the factor modules of a composition series. If an R-module M has a composition series, then each submodule N and each factor module M/N also has a composition series.

If the submodule N and the factor module M/N each have a composition series, then the module M also has one (see Chapter 13 for the respective proofs for groups).

Definition 22.2.11. Let M and N be R-modules, and let $\phi : M \to N$ be a group homomorphism. Then ϕ is an *R-module homomorphism* if $\phi(rm) = r\phi(m)$ for any $r \in R$ and $m \in M$.

As for all other structures, we define monomorphism, epimorphism, isomorphism, and automorphism of R-modules in analogy with the definition for groups.

Analogously, for groups, we have the following results:

Theorem 22.2.12 (First isomorphism theorem). *Let M and N be R-modules, and $\phi : M \to N$ an R-module homomorphism.*
(1) *The kernel* $\ker(\phi) = \{m \in M \mid \phi(m) = 0\}$ *of ϕ is a submodule of M.*
(2) *The image* $\mathrm{Im}(\phi) = \{n \in N \mid \phi(m) = n \text{ for some } m \in M\}$ *of ϕ is a submodule of N.*
(3) *The R-modules $M/\ker\phi$ and $\mathrm{Im}(\phi)$ are isomorphic via the map induced by ϕ.*

If the R-modules M and N are R-module isomorphic, then we write $M \cong N$.

Corollary 22.2.13. *An R-module homomorphism $\phi : M \to N$ is injective if and only if* $\ker(\phi) = \{0\}$.

Theorem 22.2.14 (Second isomorphism theorem). *Let N_1, N_2 be submodules of an R-module M. Then*

$$(N_1 + N_2)/N_2 \cong N_1/(N_1 \cap N_2).$$

Theorem 22.2.15 (Schur's lemma). *Let M and N be simple R-modules, and let $\phi : M \to N$ be a nonzero R-module homomorphism. Then ϕ is an R-module isomorphism.*

Proof. Since both M and N are simple, we must have either $\ker(\phi) = M$ or $\ker(\phi) = \{0\}$. If $\ker(\phi) = M$, then $\phi = 0$ the zero homomorphism. Hence, $\ker(\phi) = \{0\}$ and $\mathrm{Im}(\phi) = N$. Therefore, if $\phi \neq 0$, then ϕ is an R-module isomorphism. □

Group Rings and Modules over Group Rings

We now introduce the class of rings, whose modules we will study for group representations. They form the class of *group algebras.*

Definition 22.2.16. Let R be a ring and G a group. Then the *group ring* of G over R, denoted by RG, consists of all finite R-linear combinations of elements of G. This is the set of linear combinations of the form

$$\left\{\sum_{g \in G} \alpha_g g \mid \text{all } \alpha_g \in R\right\}.$$

For addition in RG, we take the rule

$$\sum_{g \in G} \alpha_g g + \sum_{g \in G} \beta_g g = \sum_{g \in G} (\alpha_g + \beta_g) g.$$

Multiplication in RG is defined by extending the multiplication in G:

$$\left(\sum_{g\in G}\alpha_g g\right)\left(\sum_{g\in G}\beta_g g\right)=\sum_{g\in G}\sum_{h\in G}\alpha_g\beta_h gh$$
$$=\sum_{x\in G}\left(\sum_{g\in G}(\alpha_g\beta_{g^{-1}x})\right)x.$$

The group ring RG has an identity element, which coincides with the identity element of G. We usually denote this by just 1.

From the viewpoint of abstract group theory, it is of interest to consider the case, where the underlying ring is an integral domain. In this connection, we mention the famous zero divisor conjecture by Higman and Kaplansky, which poses the question whether every group ring RG of a torsion-free group G over an integral domain R or over a field K has no zero divisors.

The conjecture has been proved only for a fairly restricted class of torsion-free groups.

In this chapter, we will primarily consider the case where $R = K$ is a field and the group G is finite, in which case the group ring KG is not only a ring, but also a finite-dimensional K-vector space having G as a basis. In this case, KG is called the *group algebra*.

In mathematics, in general, an *algebra* over a field K is a K-vector space with a bilinear product that makes it a ring. That is, an algebra over K is an algebraic structure A with both a ring structure and a K-vector space structure that are compatible. That is, $\alpha(ab) = (\alpha a)b = a(\alpha b)$ for any $\alpha \in K$ and $a, b \in A$. An algebra is finite-dimensional if it has finite dimension as K-vector space.

Example 22.2.17. (1) The matrix ring $M(n, K)$ is a finite-dimensional K-algebra for any natural number n.
(2) The group ring KG is a finite-dimensional K-algebra when the group G is finite.

Definition 22.2.18. A homomorphism of K-algebras is a ring homomorphism, which is also a K-linear transformation.

Modules over a group algebra KG can also be considered as K-vector spaces with $\alpha \in K$ acting as $\alpha \cdot 1 \in KG$.

Lemma 22.2.19. *If K is a field, and G is a finite group, then a KG-module is finitely generated if and only if it is finite-dimensional as a K-vector space.*

Proof. If V is generated as a KG-module by $\{v_1, \dots, v_k\}$, then V is generated as a K-vector space by $\{gv_1, \dots, gv_k\}$, and hence has finite dimension as a K-vector space. The converse is clear. □

We now describe the fundamental connections between modules over group algebras and group representation theory.

Theorem 22.2.20. *If K is a field and G is a finite group, then there is a one-to-one correspondence between finitely generated KG-modules and linear actions of G on finite-dimensional K-vector spaces V, and hence with the homomorphisms $\rho : G \to \mathrm{GL}(V)$ for finite dimensional K-vector spaces V.*

Proof. If V is a finitely generated KG-module, then $\dim K(V) < \infty$ by Lemma 22.2.19, and the map from $G \times V$ to V obtained by restricting the module structure map from $KG \times V$ to V is a linear action.

Conversely, let V be a finite-dimensional K-vector space, on which G acts linearly. Then we place a KG-module structure on V by defining

$$\left(\sum_{g\in G} a_g g\right)v = \sum_{g\in G} a_g(gv) \quad \text{for}$$

$$\sum_{g\in G} a_g g \in KG \quad \text{and} \quad v \in V.$$

The processes are inverses of each other. □

To define a KG-module structure on a K-vector space V, it suffices to stipulate the action of the elements of G on V. The action of arbitrary elements of KG on V is then defined by extending linearly.

As indicated for the remainder of this section, G will denote a finite group, and K will denote a field. All K-vector spaces will be finite dimensional, and all KG-modules will be finitely generated and hence of finite dimension as a K-vector space. Our attention will primarily be on KG-modules, although on occasion it will be convenient to work with the linear representation $\rho : G \to \mathrm{GL}(V)$ with $\rho(g) = gv$ for $g \in G$, $v \in V$ arising from a given KG-module V.

Example 22.2.21. (1) The field K can always be considered as a KG-module by defining $g\lambda = \lambda$ for all $g \in G$ and $\lambda \in K$. This module is called the *trivial module*.

(2) Let G act on the finite set $X = \{x_1, \dots, x_n\}$. Let KX be the set

$$\left\{\sum_{i=1}^{n} c_i x_i \mid c_i \in K, x_i \in X \text{ for } i = 1, \dots, n\right\}$$

of all formal sums of K-linear combinations of elements of X. This then has a K-vector space structure with basis X. On KX, we may define a KG-module in the following manner: If $g \in G$ and $\sum_{i=1}^{n} c_i x_i \in KX$, then

$$g\left(\sum_{i=1}^{n} c_i x_i\right) = \sum_{i=1}^{n} c_i(gx_i).$$

These modules are called the *permutation modules*.

(3) Let U, V be KG-modules. Then the (external) direct sum $U \oplus V$ has a KG-module structure given by

$$g(u, v) = (gu, gv).$$

(4) Let U, V be KG-modules, and let $\operatorname{Hom}_{KG}(U, V)$ be the set of all KG-module homomorphisms from U to V. For $\phi, \psi \in \operatorname{Hom}_{KG}(U, V)$ define $\phi + \psi \in \operatorname{Hom}_{KG}(U, V)$ by

$$(\phi + \psi)(u) = \phi(u) + \psi(u).$$

With this definition $\operatorname{Hom}_{KG}(U, V)$ is an Abelian group. Furthermore, $\operatorname{Hom}_{KG}(U.V)$ is a K-vector space with $(\lambda\phi)(u) = \lambda\phi(u)$ for $\lambda \in K$, $u \in U$ and $\phi \in \operatorname{Hom}_{KG}(U, V)$.

Note that this K-vector space has finite dimension. The K-vector space $\operatorname{Hom}_{KG}(U, V)$ also admits a natural KG-module structure. For $g \in G$ and $\phi \in \operatorname{Hom}_{KG}(U, V)$ then, we define

$$g\phi : U \to V \quad \text{by } (g\phi)(u) = g(\phi(g^{-1}(u))).$$

It is clear that $g\phi \in \operatorname{Hom}_{KG}(U, V)$.

For $g_1, g_2 \in G$, and $\phi \in \operatorname{Hom}_{KG}(U, V)$ then,

$$\begin{aligned}((g_1g_2)\phi)(u) &= g_1g_2\phi((g_1g_2)^{-1}u) = g_1(g_2\phi(g_2^{-1}(g_1^{-1}u))) \\ &= g_1(g_2\phi)(g_1^{-1}(u)) = (g_1(g_2(\phi))(u).\end{aligned}$$

Therefore, $(g_1g_2)\phi = g_1(g_2\phi)$. It follows that $\operatorname{Hom}_{KG}(U, V)$ has a KG-module structure.

G acts on $\operatorname{Hom}_{KG}(U, V)$, and we write $U^\star$ for $\operatorname{Hom}_{KG}(U, K)$, where K is the trivial module. $U^\star$ is called the *dual module* of U, and here we have $(g\phi)(u) = \phi(g^{-1}u)$.

Theorem 22.2.22 (Maschke's Theorem). *Let G be a finite group, and suppose that the characteristic of K is either 0 or co-prime to $|G|$; that is, $\gcd(\operatorname{char}(K), |G|) = 1$. If U is a KG-module and V is a KG-submodule of U, then V is a direct summand of U as KG-modules.*

Proof. U is, in particular, a finite-dimensional K-vector space, and V is a K-subspace. Any basis for V can be extended to a basis of U. Hence, there is some subspace W of U such that $U = V \oplus W$ as K-vector spaces. However, W may not be a KG-submodule of U. Let $\pi : U \to V$ be the projection of U onto V in terms of the vector space decomposition so that the map π is the unique linear transformation; that is, the identity on V and zero on W. We now define a linear transformation

$$\pi' : U \to U$$

by

$$\pi'(u) = \frac{1}{|G|} \sum_{g \in G} g\pi(g^{-1}u) \quad \text{for } u \in U.$$

Since $\operatorname{char}(K) = 0$, or $\gcd(\operatorname{char}(K), |G|) = 1$, it follows that $|G| \neq 0$ in K; hence, $\frac{1}{|G|}$ exists in K. Therefore, the definition of π' makes sense.

We have $gv \in V$ for any $g \in G$ and $v \in V$, because V is a KG-submodule of U. Therefore, the map π' maps U into V. moreover, since π is the identity on V, we have that $g\pi(g^{-1}v) = gg^{-1}(v) = v$ for any $g \in G$ and $v \in V$. Therefore, the restriction of π' to V is the identity. It also follows that $U = V \oplus \ker(\pi')$ as K-vector spaces. It remains to show that $\ker(\pi')$ is a KG-submodule of U.

To show this, it is sufficient to show that π' is a KG-module homomorphism; that is, we must show that $\pi'(xu) = x\pi'(u)$ for any $x \in G$ and $u \in U$. We have

$$\begin{aligned}\pi'(xu) &= \frac{1}{|G|}\sum_{g\in G} g\pi(g^{-1}xu)\\ &= \frac{1}{|G|}\sum_{g\in G} xx^{-1}g\pi(g^{-1}xu)\\ &= \frac{1}{|G|}x\Big(\sum_{g\in G} x^{-1}g\pi(g^{-1}xu)\Big).\end{aligned}$$

But as g varies through G further, $y = x^{-1}g$ varies through G for fixed $x \in G$.

Therefore,

$$\pi'(xu) = x\Big(\frac{1}{|G|}\sum_{y\in G} y\pi(y^{-1}u)\Big) = x\pi'(u)$$

as required. □

Definition 22.2.23. A module U is *semisimple* if it is a direct sum of simple modules. If $U = \{0\}$, then the sum is the empty sum.

Corollary 22.2.24. *Let G be a finite group and K a field. Suppose that either* $\operatorname{char}(K) = 0$ *or* $\operatorname{char}(K)$ *is relatively prime to* $|G|$. *Then every nonzero KG-module is semisimple.*

Proof. Let U be a nonzero KG-module. We use induction on $\dim_K(V)$. If U is simple, we are done. This includes the case where $\dim_K(V) = 1$. Suppose that $\dim_K(V) > 1$, and assume that U is not simple. Then U must have a nonzero proper KG-submodule V. By Maschke's theorem, we have $U = V \oplus W$ for some nonzero proper KG-submodule W of U. Then both V and W have dimension strictly less than $\dim_K(U)$. By the induction hypothesis, both V and W are semisimple; therefore, U is semisimple. □

We now present a version of Maschke's theorem for linear group representations $\rho : G \to \mathrm{GL}(V)$, where $\rho(g)(u) = gu$ for $g \in G$, $u \in U$, which arises from the given KG-module U. To formulate Maschke's result, we need some additional definitions and notation.

Definition 22.2.25. (1) A K-vector subspace V of U is a *G-invariant* subspace if $gv \in V$ for all $g \in G$ and $v \in V$.
(2) Let U be nonzero. A representation $\rho : G \to \mathrm{GL}(U)$ is *irreducible* if $\{0\}$ and U are the only G-invariant subspaces of U.
(3) Let U be nonzero. A representation $\rho : G \to \mathrm{GL}(U)$ is *fully reducible* if each G-invariant subspace V of U has a G-invariant complement W in U; that is, $U = V \oplus W$ as K-vector spaces.

Theorem 22.2.26 (Maschke's theorem). *Let G be a finite group and K a field. Suppose that either* $\mathrm{char}(K) = 0$ *or* $\mathrm{char}(K)$ *is relatively prime to $|G|$. Let U be a finite-dimensional K-vector space. Then each representation $\rho : G \to \mathrm{GL}(V)$ is fully reducible.*

Proof. By Theorem 22.2.1, we may consider U as a KG-module. Then the above version of Maschke's theorem follows from the proof for modules, because the KG-submodules of U together with the respective definitions for group representations represent the G-invariant subspaces of U. □

The theory of KG-modules, when $\mathrm{char}(K) = p > 0$ and p, divides $|G|$. In which case, arbitrary KG-modules need not be semisimple, and is called *modular representation theory*. The earliest work on modular representations was done by Dickson and many of the main developments were done by Brauer. More details and a good overview may be found in [1], [4], [5], and [18].

22.3 Semisimple Algebras and Wedderburn's Theorem

In this section, K will denote a field and all algebras will be finite dimensional K-algebras and, unless explicitly stated otherwise, will be algebras with an identity element. All modules and algebras are assumed to be finitely generated or equivalently finite-dimensional as K-vector spaces. All direct sums of modules will be assumed to be finite. Let A be an algebra. We are interested in semisimple A-modules, and want to determine conditions on A so that every A-module is semisimple.

Lemma 22.3.1. *Let M be an A-module. Then the following are equivalent:*
(1) *Any submodule of M is a direct summand of M.*
(2) *M is semisimple.*
(3) *M is a sum of simple submodules.*

Proof. The implication (1) $\Longrightarrow$ (2) follows in the same manner as Corollary 22.2.24. The implication (2) $\Longrightarrow$ (3) is direct.

Finally, we must show the implication (3) $\Longrightarrow$ (1). Suppose that (3) holds, and let N be a submodule of M. Let V also be a submodule of M; that is, maximal among all submodules of M that intersect N trivially. Such a submodule V exists by Zorn's lemma. We wish to show that $N+V = M$. Suppose that $N+V \neq M$ (certainly we have $N+V \subset M$). If every simple submodule of M were contained in $N+V$, then as M can be written as a sum of simple submodules, we would have $M \subset N+V$. This is not the case, since $N+V \neq M$. Hence, there is some simple submodule S of M that is not contained in $N+V$. Since $S \cap (N+V)$ is a proper submodule of the simple module S, we must have $S \cap (N+V) = \{0\}$. In particular, $S \cap V = \{0\}$, so we have $V \subset V+S$. Let $n \in N \cap (V+S)$. Then $n = s+v$ for some $v \in V$ and $s \in S$. This gives $s = n - v \in S \cap N + V$, and therefore $s = 0$. Hence, $n = v$, which forces n to be 0, because $N \cap V = \{0\}$. It follows that $N \cap (V+S) = \{0\}$, which contradicts the maximality of V. Hence, we now have $M = N+V$. Furthermore, since $N \cap V = \{0\}$, we get that the sum is direct and $M = N \oplus V$. Therefore, N is a direct summand of M, which proves the implication (3) $\Longrightarrow$ (1) completing the proof of the lemma. □

Lemma 22.3.2. *Submodules and factor modules of semisimple modules are also semisimple.*

Proof. Let M be a semisimple A-module. By the previous lemma and the isomorphism theorem for modules, we get that every submodule of M is isomorphic to a factor module of M. Therefore, it suffices to show that factor modules of M are semisimple. Let M/N be an arbitrary factor module, and let $\eta : M \to M/N$ with $m \mapsto m+N$ be the canonical map. Since M is semisimple, we have $M = S_1 + \cdots + S_n$ with $n \in \mathbb{N}$, and each S_i a simple module. Then $M/N = \eta(M) = \eta(S_1) + \cdots + \eta(S_n)$. But each $\eta(S_i)$ is isomorphic to a factor module of S_i, and hence each $\eta(S_i)$ is either $\{0\}$ or a simple module. Therefore, M/N is a sum of simple modules, and hence semisimple by Lemma 22.3.1. □

Definition 22.3.3. An algebra A is *semisimple* if all nonzero A-modules are semisimple.

Note that if G is a finite group, and either $\operatorname{char}(K) = 0$ or $\gcd(\operatorname{char}(K), |G|) = 1$, then KG is semisimple.

We now give some fundamental results on semisimple algebras.

Lemma 22.3.4. *The algebra A is semisimple if and only if the A-module A is semisimple.*

Proof. Suppose that the A-module A is semisimple, and let M be an A-module generated by $\{m_1, \ldots, m_r\}$.

Let A^r denote the direct sum of r copies of A; $(a_1, \ldots, a_r) \mapsto a_1 m_1 + \cdots + a_r m_r$ defines a map from A^r to M, which is an A-module epimorphism. Thus, M is isomorphic to a factor module of the semisimple module A^r, and hence semisimple by Lemma 22.3.2. It follows that A is a semisimple algebra.

The converse is clear. □

Theorem 22.3.5. *Let A be a semisimple algebra, and suppose that as an A-module, we have*

$$A \cong S_1 \oplus \cdots \oplus S_r, \quad r \in \mathbb{N},$$

where the S_i are simple submodules of A. Then any simple A-module is isomorphic to some S_i.

Proof. Let S be a simple A-module and $s \in S$ with $s \neq 0$. We define an A-module homomorphism $\phi : A \to S$ by $\phi(a) = as$ for $a \in A$. Since S is simple, the map ϕ is surjective. For each i, let be $\phi_i = \phi\ |_{S_i}$, the restriction of ϕ to S_i. If $\phi_i = 0$ for all i, then we would have $\phi = 0$. Hence, ϕ_i is nonzero for some i, and it follows from Schur's lemma that $\phi_i : S_i \to S_i$ is an isomorphism for such an i. □

Theorem 22.3.6. *Suppose that A is a semisimple algebra, and let $S_1, \dots, S_r$ be a collection of simple A-modules such that every simple A-module is isomorphic with exactly one S_i. Let M be an A-module, and let*

$$M \cong m_1 S_1 + \cdots + m_r S_r$$

for some integers $m_i \in \mathbb{N} \cup \{0\}$. Then the m_i are uniquely determined.

Proof. There is a composition series of $m_1 S_1 \oplus \cdots \oplus m_r S_r$ having $m_1 + \cdots + m_r$ terms, in which S_i appears m_i times as a composition factor. The result then follows from the Jordan-Hölder theorem for modules (Theorem 22.2.10). □

Whenever the modules $S_1, \dots, S_r$ are stated as $m_1 S_1 + \cdots + m_r S$, as in the previous theorem, we will say that the S_i are the distinct simple A-modules. The S_i are nonisomorphic.

We want to classify all semisimple algebras. We start by showing the semi-simplicity of a certain class of algebras, and then showing that all semisimple algebras fall in this class. We will introduce this class in steps.

Let D be a finite-dimensional K-algebra. Then for any $n \in \mathbb{N}$, the set $M(n, D)$ of $(n \times n)$-matrices with entries in D is a finite-dimensional K-algebra of dimension $n^2 \dim_K(D)$. Algebras of this form are called matrix algebras over D.

For $1 \leq i, j \leq n$, and $a \in D$, let $E_{ij}(a)$ be the matrix, whose only nonzero entry is equal to a, and occurs in the (i, j)-th position.

Let D^n be the set of column vectors of length n with entries from D, then D^n forms an $M(n, D)$-module under matrix multiplication.

Definition 22.3.7. An algebra D is a *division algebra* or *skew field* if the nonzero elements of D form a group. Equivalently, it is a ring, where every nonzero element has a multiplicative inverse. It is exactly the definition of a field without requiring commutativity.

Any field K is a division algebra over itself, but there may be division algebras that are noncommutative. If the interest is on the ring structure of D, one often speaks about *division rings* (see Chapter 7).

Theorem 22.3.8. *Let D be a division algebra and $n \in \mathbb{N}$. Then any simple $M(n, D)$-module is isomorphic to D^n, and $M(n, D)$ is an $M(n, D)$-module isomorphic to the direct sum of n copies of D^n. In particular, $M(n, D)$ is a semisimple algebra.*

Proof. A nonzero submodule of D^n must contain some nonzero vector, which must have a nonzero entry x in the j-th place for some j. This x is invertible in D.

By premultiplying this vector by $E_{jj}(x^{-1})$, we see that the submodule contains the j-th canonical basis vector. By premultiplying this basis vector by appropriate permutation matrices, we get that the submodule contains every canonical basis vector, and hence contains every vector.

It follows that D^n is the only nonzero $M(n,D)$-submodule of D^n, and hence D^n is simple. Now for each $1 \le k \le n$, let C_k be the submodule of $M(n,D)$ consisting of those matrices, whose only nonzero entries appear in the k-th column. Then we have

$$M(n,D) \cong \bigoplus_{k=1}^{n} C_k$$

as $M(n,D)$-modules. But each C_k is isomorphic as an $M(n,D)$-module to D^n.

It follows that $M(n,D)$ is a semisimple algebra by Lemma 22.3.4, and then D^n is the unique simple $M(n,D)$-module by Theorem 22.3.5. □

Definition 22.3.9. A nonzero algebra is *simple* if its only (two-sided) ideals (as a ring) are itself and the zero ideal.

Lemma 22.3.10. *Simple algebras are semisimple.*

Proof. Let A be a simple algebra, and let Σ be the sum of all simple submodules of A. Let S be a simple submodule of A, and let $a \in A$. Then the map $\phi : S \to Sa$, given by $s \mapsto sa$, is a module epimorphism. Therefore, Sa is simple, or $Sa = \{0\}$. In either case, we have $Sa \subset \Sigma$ for any submodule S and any $a \in A$.

It follows that Σ is a right ideal in A, and hence that Sa is a two-sided ideal. However, A is simple, and $\Sigma \neq \{0\}$, so we must have $\Sigma = A$. Therefore, A is the sum of simple A-modules, and from Lemmas 22.3.1 and 22.3.4, it follows that A is a semisimple algebra. □

Theorem 22.3.11. *Let D be a division algebra, and let $n \in \mathbb{N}$. Then $M(n,D)$ is a simple algebra.*

Proof. Let $M \in M(n,D)$ with $M \neq \{0\}$. We must show that the principal two-sided ideal J of $M(n,D)$ generated by M is equal to $M(n,D)$.

It suffices to show that J contains each $E_{ij}(1)$, since these matrices generate $M(n,D)$ as an $M(n,D)$-module. Since $M \neq \{0\}$, there exists some $1 \le r, s \le n$ such that the (r,s)-entry of M is nonzero. We call this entry x. By calculation, we have

$$E_{ss}(1) = E_{sr}(x^{-1})ME_{ss}(1) \in J.$$

Now let $1 \le i,j \le n$, and let w, w' be the permutation matrices corresponding to the transpositions (i,s) and (s,j), respectively. Then $E_{ij}(1) = wE_{ss}(1)w' \in J$. □

Let $B_1, \dots, B_r$ be algebras. The external direct sum $B = B_1 \oplus B_2 \oplus \cdots \oplus B_r$ is the algebra, whose underlying set is the Cartesian product, and whose addition, multiplication, and scalar multiplication are defined componentwise.

If M is a B_i-module for some i, then M has a B-module structure given by

$$(b_1, \dots, b_r)m = b_i m.$$

If M is simple (respectively semisimple) as a B_i-module, then M is also simple (respectively semisimple) as a B-module. For each i, the set of elements of B, whose only nonzero entry is in the ith component of B, is an ideal in B, and this ideal is B-module isomorphic to B_i.

Now suppose that B is an algebra having ideals $B_1, \dots, B_r$ such that, as vector spaces, B is the direct sum of the B_i. Then B is isomorphic to the external direct sum $B_1 \oplus \cdots \oplus B_r$ by the map

$$b = b_1 + \cdots + b_r \mapsto (b_1, \dots, b_r).$$

The algebra B is the internal direct sum as algebras of the B_i. This can be seen as follows. If $i \neq j$ and $b_i \in B_i$, $b_j \in B_j$, then we must have $b_i b_j \in B_i \cap B_j = \{0\}$, since B_i and B_j are ideals. Therefore, the product in B of $b_1 + \cdots + b_r$ and $b_1' + \cdots b_r'$ is just $b_1 b_1' + \cdots + b_r b_r'$.

Lemma 22.3.12. *Let $B = B_1 \oplus \cdots \oplus B_r$ be a direct sum of algebras. Then the (two-sided) ideals of B are precisely the sets of the form $J_1 \oplus \cdots \oplus J_r$, where J_i is a (two-sided) ideal of B_i for each i.*

Proof. Let J be a (two-sided) ideal of B, and let $J_i = J \cap B_i$ for each i. Certainly, $\bigoplus_{i=1}^r J_i \subset J$.

Let $b \in J$, then $b = b_1 + \cdots + b_r$ with $b_i \in B_i$ for each i. For some i, consider $e_i = (0, \dots, 0, 1, 0, \dots, 0)$; that is, the element of B, whose only nonzero entry is the identity element of B_i. Then $b = be_i \in J \cap B_i = J_i$. Therefore, $b \in \bigoplus_{i=1}^r J_i$, which shows that $J = J_1 \oplus \cdots \oplus J_r$.

The converse is clear. □

Theorem 22.3.13. *Let $r \in \mathbb{N}$. For each $1 \le i \le r$, let D_i be a division algebra over K. Let $n_i \in \mathbb{N}$, and let $B_i = M(n_i, D_i)$. Let B be the external direct sum of the B_i. Then B is a semisimple algebra having exactly r isomorphism classes of simple modules and exactly 2^r (two-sided) ideals, namely, every sum of the form $\bigoplus_{j \in J} B_j$, where J is a subset of $\{1, \dots, r\}$.*

Proof. For each i, we write $B_i = C_{i1} \oplus \cdots \oplus C_{in}$ using Theorem 22.3.8, where the C_{i_j} are mutually isomorphic B_i-modules. As we saw above, each C_{ij} is also simple as a B-module. Therefore, as B-modules, we have $B \cong \bigoplus_{i,j} C_{ij}$, and hence B is a semisimple algebra by Lemma 22.3.4. From Theorem 22.3.5, we get that any simple B-module is isomorphic to some C_{ij}, but $C_{ij} \cong C_{kl}$ if and only if $i = k$. Hence, there are exactly r isomorphisms of simple B-modules. The final statement is a straightforward consequence of Theorem 22.3.11 and Lemma 22.3.12. □

We saw that a direct sum of matrix algebras over a division algebras is semisimple. We now start to show that the converse is also true; that is, any semisimple algebra is isomorphic to a direct sum of matrix algebras over division algebras. This is *Wedderburn's theorem.*

Definition 22.3.14. If M is an A-module, then let $\mathrm{End}_A(M) = \mathrm{Hom}_A(M,M)$ denote the set of all A-module endomorphisms of M. In a more general context, we have seen that $\mathrm{End}_A(M)$ has the structure of an A-module via

$$(\phi + \psi)(m) = \phi(m) + \psi(m)$$
$$(\lambda\phi)(m) = \phi(\lambda m)$$

for all $\phi, \psi \in \mathrm{End}_A(M), \lambda \in A$, and $m \in M$. This composition of mappings gives a multiplication in $\mathrm{End}_A(M)$, and hence $\mathrm{End}_A(M)$ is a K-algebra, called the *endomorphism algebra* of M.

Definition 22.3.15. The *opposite algebra* of B, denoted B^{op}, is the set B together with the usual addition and scalar multiplication, but with the opposite multiplication, that is, the multiplication rule of B reversed.

Given $a, b \in B$, we use ab to denote their product in B, and $a \cdot b$ to denote their product in B^{op}. Hence, $a \cdot b = ba$. We certainly have $(B^{op})^{op} = B$. If B is a division algebra, then so is B^{op}. The opposite of a direct sum of algebras is the direct sum of the opposite algebras, because the multiplication in the direct sum is defined componentwise.

Endomorphism algebras and opposite algebras are closely related.

Lemma 22.3.16. *Let B be an algebra. Then $B^{op} \cong \mathrm{End}_B(B)$.*

Proof. Let $\phi \in \mathrm{End}_B(B)$, and let $a = \phi(1)$. Then $\phi(b) = b\phi(1) = ba$ for any $b \in B$; hence, ϕ is equal to the automorphism ψ_a, given by right multiplication of a. Therefore, $\mathrm{End}_B(B) = \{\psi_a : a \in B\}$; hence, $\mathrm{End}_B(B)$ and B are in one-to-one correspondence. To finish the proof, we must show that $\psi_a\psi_b = \psi_{a\cdot b}$ for any $a, b \in B$.

Let $a, b \in B$. Then $\psi_a\psi_b(x) = \psi_a(xb) = xba = \psi_{ba}(x) = \psi_{a\cdot b}(x)$, as required. □

Lemma 22.3.17. *Let $S_1, \ldots, Sr$ be the r distinct simple A-modules of Theorem 22.3.6. For each i, let U_i be a direct sum of copies of S_i, and let $U = U_1 \oplus \cdots \oplus U_r$. Then*

$$\mathrm{End}_A(U) \cong \mathrm{End}_A(U_1) \oplus \cdots \oplus \mathrm{End}_A(U_r).$$

Proof. Let $\phi \in \mathrm{End}_A(U)$. Fix some i. Then every composition factor of U_i is isomorphic to S_i. Therefore, by the Jordan–Hölder theorem for modules (Theorem 22.3.10), we see that the same is true for $\phi(U_i)$, since $\phi(U_i)$ is isomorphic to a quotient of U_i. Assume that $\phi(U_i)$ is not contained in U_i. Then the image of $\phi(U_i)$ in U/U_i under the canonical map is a nonzero submodule, having S_i as a composition factor. However, the composition factors of U/U_i are exactly those S_j for $j \neq i$. This gives a contradiction. It follows that

$\phi(U_i) \subset U_i$, and a submodule of U/U_i cannot have S_i as a composition factor. For each i, we can define $\phi_i = \phi_{|_{U_i}}$, and we have $\phi_i \in \mathrm{End}_A(U_i)$. In this way, we define a map

$$\Gamma : \mathrm{End}_A(U) \mapsto \mathrm{End}_A(U_1) \oplus \cdots \oplus \mathrm{End}_A(U_r)$$

by setting

$$\Gamma(\phi) = (\phi_1 \ldots, \phi_r) \in \mathrm{End}_A(U_1) \oplus \cdots \oplus \mathrm{End}_A(U_r).$$

It is straightforward that Γ is an A-module monomorphism.

Now let $(\phi_1, \ldots, \phi_r) \in \mathrm{End}_A(U_1) \oplus \cdots \oplus \mathrm{End}_A(U_r)$. We define $\hat{\phi} \in \mathrm{End}_A(U)$ as follows: Given $x \in U$ with $x = x_1 + \cdots + x_r$, and $x_i \in U_i$ for each i, then

$$\hat{\phi}(x) = \phi_1(x_1) + \cdots + \phi_r(x_r).$$

We then have $(\phi_1, \ldots, \phi_r) = \Gamma(\hat{\phi})$, which shows that Γ is surjective, and hence an isomorphism. □

Lemma 22.3.18. *If S is a simple A-module, then* $\mathrm{End}_A(nS) \cong M(n, \mathrm{End}_A(S))$ *for $n \in \mathbb{N}$.*

Proof. We regard the elements of nS as being column vectors of length n with entries from S. Let $\Phi = (\phi_{ij}) \in M(n, \mathrm{End}_A(S))$. We now define the map

$$\Gamma(\Phi) : nS \to nS$$

by

$$\begin{aligned}\Gamma(\Phi)\begin{pmatrix} s_1 \\ \vdots \\ s_n \end{pmatrix} &= \begin{pmatrix} \phi_{11} & \cdots & \phi_{in} \\ \vdots & & \vdots \\ \phi_{n1} & \cdots & \phi_{nn} \end{pmatrix}\begin{pmatrix} s_1 \\ \vdots \\ s_n \end{pmatrix} \\ &= \begin{pmatrix} \phi_{11}(s_1) + \cdots + \phi_{1n}(s_n) \\ \vdots \\ \phi_{n1}(s_1) + \cdots + \phi_{nn}(s_n) \end{pmatrix}.\end{aligned}$$

We write $\vec{s} = \begin{pmatrix} s_1 \\ \vdots \\ s_n \end{pmatrix} \in nS$. Then

$$\Gamma(\Phi(a\vec{s} + \vec{t})) = a\Gamma(\Phi)(\vec{s}) + \Gamma(\Phi)(\vec{t})$$

for any $a \in A$ and $\vec{s}, \vec{t} \in nS$, because each ϕ_{ij} is an A-module homomorphism. Therefore, $\Gamma(\Phi) \in \mathrm{End}_A(nS)$, and we easily obtain that

$$\Gamma : M(n, (\mathrm{End}_A(S))) \to \mathrm{End}_A(nS)$$

by

$$\Phi \mapsto \Gamma(\Phi)$$

is an algebra monomorphism.

Now let $\psi \in \mathrm{End}_A(nS)$. For each $1 \le i, j \le n$, we define $\psi_{ij} : S \to S$ implicitly by

$$\psi\begin{pmatrix} s \\ 0 \\ \vdots \\ 0 \end{pmatrix} = \begin{pmatrix} \psi_{11}(s) \\ \vdots \\ \psi_{n1}(s) \end{pmatrix}, \dots, \psi\begin{pmatrix} 0 \\ 0 \\ \vdots \\ s \end{pmatrix} = \begin{pmatrix} \psi_{1n}(s) \\ \vdots \\ \psi_{nn}(s) \end{pmatrix}.$$

We get that each $\psi_{ij} \in \mathrm{End}_A(S)$. Now let $\Psi = (\psi_{ij}) \in M(n, \mathrm{End}_A(S))$. Then $\Gamma(\Psi) = \psi$, showing that Γ is also surjective, and hence an isomorphism. □

If S is a simple A-module, then $\mathrm{End}_A(S)$ is a division algebra by Schur's lemma (Theorem 22.2.15). If the ground field K is algebraically closed, then more specific results can be stated about the structure of $\mathrm{End}_A(S)$.

Lemma 22.3.19. *Suppose that K is algebraically closed, and let S be a simple A-module. Then* $\mathrm{End}_A(S) \cong K$.

Proof. Let $\phi \in \mathrm{End}_A(S)$. Consider ϕ as an invertible K-linear map of the finite-dimensional K-vector space S onto itself. Since K is algebraically closed, ϕ has a nonzero eigenvalue $\lambda_\phi \in K$.

If I is the identity element of $\mathrm{End}_a(S)$, then $(\phi - \lambda_\phi I) \in \mathrm{End}_A(S)$ has a nonzero kernel, and therefore is not invertible. From this, it follows that $\phi = \lambda_\phi I$, since $\mathrm{End}_A(S)$ is a division algebra. The map $\phi \mapsto \lambda_\phi$ is then an isomorphism from $\mathrm{End}_A(S)$ to K. □

Lemma 22.3.20. *Let B be an algebra. Then $(M(n,B))^{op} \cong M(n, B^{op})$ for any $n \in \mathbb{N}$.*

Proof. Define the map $\psi : (M(n,B))^{op} \to M(n, B^{op})$ by $\psi(X) = X^t$, where X^t is the transpose of the matrix X. This map is bijective.

Let $X = (x_{ij})$ and $Y = (y_{ij})$ be elements of $(M(n,B))^{op}$. Then for any i and j we have

$$\begin{aligned}(\psi(X)\psi(Y))_{ij} &= \sum_{k=1}^{n} \psi(X)_{ij} \cdot \psi(Y)_{kj} = \sum_{k=1}^{n} (X^t)_{ik} \cdot (Y^t)_{kj} \\ &= \sum_{k=1}^{n} X_{ki} \cdot Y_{jk} = \sum_{k=1}^{n} Y_{jk} X_{ki} = (YX)_{ji} \\ &= ((YX)^t)_{ij} = ((X \cdot Y)^t)_{ij} = \psi(X \cdot Y)_{ij}.\end{aligned}$$

Therefore, $\psi(X \cdot Y) = \psi(X)\psi(Y)$, and then ψ is an algebra homomorphism, and since it is bijective also an algebra isomorphism. □

We are now at the point of stating Wedderburn's theorem.

Theorem 22.3.21 (Wedderburn). *The algebra A is semisimple if and only if it is isomorphic to a direct sum of matrix algebras over division algebras.*

Proof. Suppose that the algebra A is semisimple. Then A is of the form $A = U_1 \oplus \cdots \oplus U_r$, where each U_i is the direct sum of n_i copies of a simple A-module S_i, and no two of the distinct S_i are isomorphic.

We have $A^{op} \cong \mathrm{End}_A(A)$ by Lemma 22.3.16, and $A^{op} \cong \mathrm{End}_A(U_1) \oplus \cdots \oplus \mathrm{End}_A(U_r)$ by Lemma 22.3.17. Therefore, $A^{op} \cong \mathrm{End}_A(n_1 S_1) \oplus \cdots \oplus \mathrm{End}_A(n_r S_r)$, and then by Lemma 22.3.16, $A^{op} \cong M(n_1, \mathrm{End}_A(S_1)) \oplus \cdots \oplus M(n_r, \mathrm{End}_A(S_r))$. Now, from Lemma 22.3.18,

$$\begin{aligned} A &\cong M(n_1, \mathrm{End}_A(S_1)) \oplus \cdots \oplus M(n_r, \mathrm{End}_a(S_r))^{op} \\ &\cong M(n_1, \mathrm{End}_A(S_1))^{op} \oplus \cdots \oplus M(n_r, \mathrm{End}_a(S_r))^{op} \\ &\cong M(n_1, \mathrm{End}_A(S_1))^{op} \oplus \cdots \oplus M(n_r, \mathrm{End}_A(S_r)^{op})^{op}. \end{aligned}$$

Since the endomorphism algebra of a simple module is a division algebra, and the opposite algebra of a division algebra is also a division algebra, it follows that a semisimple algebra is isomorphic to a direct sum of matrix algebras over division algebras. The converse is a direct consequence of Theorem 22.3.13. □

Theorem 22.3.22. *The algebra A is simple if and only if it is isomorphic to a matrix algebra over a division ring.*

Proof. Suppose that A is a simple algebra. Then by Lemma 22.3.10, A is semisimple; hence, by Theorem 22.3.21, A is isomorphic to a direct sum of R matrix algebras over division algebras. From Theorem 22.3.13, we have that A has exactly 2^r ideals. However, A is simple, and hence has only 2 ideals. Therefore, $r = 1$, and any simple algebra is isomorphic to a matrix algebra over a division algebra. The converse follows from Theorem 22.3.11. □

We see that an algebra is semisimple if and only if it is a direct sum of simple algebras. This affirms the consistency of the choice of terminology.

Theorem 22.3.23. *Suppose that the field K is algebraically closed. Then any semisimple algebra is isomorphic to a direct sum of matrix algebras over K.*

Proof. This follows directly from Lemma 22.3.19 and Theorem 22.3.21. □

22.4 Ordinary Representations, Characters and Character Theory

In this section, we look at a concept, the character of a representation, which gives more information than one might expect at first glance. Throughout this section, we will be concerned with the case, where the ground field K is $\mathbb{C}$, the field of complex numbers. In this case, representation theory of groups is called *ordinary representation theory*. Recall that $\mathbb{C}$ has characteristic 0 and is algebraically closed. For this section, G will denote a finite group, and all $\mathbb{C}G$-modules are finitely generated, or equivalently have finite dimension as $\mathbb{C}$-vector spaces. From Theorem 22.3.21, we see that every nonzero $\mathbb{C}G$-modules is semisimple for any group G. It follows, from Wedderburn's theorem, that we have very specific information about the nature of the group algebra $\mathbb{C}G$.

Theorem 22.4.1. *There exists some $r \in \mathbb{N}$ and some $f_1, \dots, f_r \in \mathbb{N}$ such that*

$$\mathbb{C}G = M(f_1, \mathbb{C}) \oplus \cdots \oplus M(f_r, \mathbb{C})$$

as $\mathbb{C}$-algebras. Furthermore, there are exactly r isomorphism classes of simple $\mathbb{C}G$-modules, and if we let $S_1, \dots, S_r$ be representations of these r classes, then we can order the S_i so that

$$\mathbb{C}G \cong f_1 S_1 \oplus \cdots \oplus f_r S_r$$

as $\mathbb{C}G$-modules, where $\dim_{\mathbb{C}} S_i = f_i$ for each i. Any $\mathbb{C}G$-module can be written uniquely in the form $a_1 S_1 \oplus \cdots \oplus a_r S_r$ where all $a_i \in \mathbb{N} \cup \{0\}$.

Proof. The theorem follows from our results on the classification of simple and semisimple algebras. The first statement follows from Corollary 22.2.24 and Theorem 22.3.23. The second statement follows from Theorems 22.3.8 and 22.3.13, where we take S_i as the space of column vectors of length f_i with the canonical module structure over the ith summand $M(f_i, \mathbb{C})$.

The final statement follows from Theorem 22.3.6. □

Definition 22.4.2. The $\mathbb{C}$-dimensions $f_1, \dots, f_r$ of the r simple $\mathbb{C}G$-modules are called the *degrees* of the representations of G.

The trivial $\mathbb{C}G$-module $\mathbb{C}$ is one-dimensional, and hence simple. Therefore, G will always have at least one representation of degree 1. By convention, we let $f_1 = 1$. The sizes of the degrees are determined by the order of the group G.

Corollary 22.4.3. *We have*

$$\sum_{i=1}^{r} f_i^2 = |G|.$$

Proof. Theorem 22.4.1 gives

$$|G| = \dim_{\mathbb{C}}(\mathbb{C}G) = \dim\left(\bigoplus_{i=1}^{r} M(f_i, \mathbb{C})\right)$$
$$= \sum_{i=1}^{r} \dim_{\mathbb{C}} M(f_i, \mathbb{C}) = \sum_{i=1}^{r} f_i^2.$$

□

We note that the degrees of G divide $|G|$. We do not need this fact. For a proof see the appendix in the book [1].

Theorem 22.4.4. *The number r of simple G-modules is equal to the number of conjugacy classes of G.*

Proof. Let Z be the center of $\mathbb{C}G$; that is, the subalgebra of $\mathbb{C}G$ consisting of all elements that commute with every element of $\mathbb{C}G$. From Theorem 22.4.1, it follows that Z is isomorphic to the center of $M(f_1, \mathbb{C}) \oplus \cdots \oplus M(f_r, \mathbb{C})$, and therefore is isomorphic to the direct sum of the centers of the $M(f_i, \mathbb{C})$. It is straightforward that the center of $M(f_i, \mathbb{C})$ is equal to the set of diagonal matrices

$$\{aI : I \text{ is the identity matrix in } M(f_i, \mathbb{C}), a \in \mathbb{C}\}.$$

Hence, the center of $M(f_i, \mathbb{C})$ is isomorphic to $\mathbb{C}$, and therefore $Z \cong \mathbb{C}^r$, which implies that $\dim_{\mathbb{C}}(Z) = r$.

We now consider an element $\sum_{g \in G} \lambda_g G$ of Z. For any $h \in G$, we have

$$\Big(\sum_{g \in G} \lambda_g G\Big) h = h\Big(\sum_{g \in G} \lambda_g g\Big),$$

which leads to

$$\sum_{g \in G} \lambda_g g = \sum_{g \in G} \lambda_g h^{-1} g h = \sum_{g \in G} \lambda_{hgh^{-1}} g.$$

It follows that we must have $\lambda_g = \lambda_{hgh^{-1}}$ for all $g, h \in G$.

It also follows then that the coefficients of elements of the center Z are constant on conjugacy classes of G, and that a basis for Z is the set of *class sums*, which are the sums of the form $\sum_{g \in C} g$, where C is a conjugacy class of G. Thus, $\dim_{\mathbb{C}} Z$ is equal to the number of conjugacy classes of G. □

Characters and Character Theory

We now define and study the characters of an ordinary representation.

Definition 22.4.5. If U is a $\mathbb{C}G$-module, then each $g \in G$ defines an invertible linear transformation of U via $u \mapsto gu$ for $u \in U$. The *character* of U is the function $\chi_U : G \to \mathbb{C}$ defined by $\chi_U(g)$, the trace of the linear transformation of U defined by g.

We note that for any representation U, we have $\chi_U(1) = \dim_{\mathbb{C}}(U)$, since the identity element of G induces the identity transformation of U. Furthermore, if $\rho : G \to \mathrm{GL}(U)$ is the representation corresponding to U, then $\chi_U(g)$ is just the trace of the map $\rho(g)$. Thus, isomorphic $\mathbb{C}G$-modules have equal characters.

If $g, h \in G$, then the linear transformations of U, defined by g and hgh^{-1}, have the same trace. These linear transformations are called *similar*. Therefore, any character is constant on each conjugacy class of G; that is, the value of the character on any two conjugate elements is the same.

Example 22.4.6. Let $U = \mathbb{C}G$ and $g \in G$. By considering the matrix of the linear transformation defined by g with respect to the basis G of $\mathbb{C}G$, we get that $\chi_U(g)$ is equal to the number of elements $x \in G$, for which $gx = x$. Therefore, we have $\chi_U(1) = |G|$ and $\chi_U(g) = 0$ for every $g \in G$ with $g \neq 1$. This character is called the *regular character* of G.

The theory of characters was introduced by Frobenius. In connection with number theory, he defined characters as being functions from G to $\mathbb{C}$ satisfying certain properties. However, it turned out that his characters were exactly the trace functions of finitely generated $\mathbb{C}G$-modules. In what follows, we describe the properties of characters.

We first consider the characters of the r simple $\mathbb{C}G$-modules. We denote these by $\chi_1, \ldots, \chi_r$. These are called the *irreducible characters* of G.

Whenever we have that $S_1, \ldots, S_r$ are the distinct (up to isomorphism) $\mathbb{C}G$-modules, we order them so that $\chi_{S_i} = \chi_i$ for each i. Because $S_1 = \{1\}$ for the trivial representation, we let χ_1 be the character of the trivial representation, and call χ_1 the *principal character* of G. We then have $\chi_1(g) = 1$ for all $g \in G$.

Definition 22.4.7. A character of a one-dimensional representation $\mathbb{C}G$-module is called a *linear character*.

Since one-dimensional modules are simple, we get that all linear characters are irreducible. Let χ be the linear character arising from the $\mathbb{C}G$-module U, and let $g, h \in G$. Since U is one-dimensional for any $u \in U$, we have $gu = \chi(g)u$, and $hu = \chi(h)u$. Then $\chi(gh)u = (gh)u = \chi(g)\chi(h)u$. Hence, χ is a homomorphism from G to the multiplicative group $\mathbb{C}^\star = \mathbb{C} \setminus \{0\}$. On the other hand, given a homomorphism $\phi : G \to \mathbb{C}^\star$, we can define a one-dimensional $\mathbb{C}G$-module U by $gu = \phi(g)u$ for $g \in G$ and $u \in U$. Therefore, $\chi_U = \phi$. It follows that the linear characters of G are precisely the group of homomorphisms from G to $\mathbb{C}^\star$.

Theorem 22.4.8. *Let U be a $\mathbb{C}G$-module, and let $\rho : G \to \mathrm{GL}(V)$ be the representation corresponding to U. Let $g \in G$ be of order n. Then the following hold:*
(i) *$\rho(g)$ is diagonalizable.*
(ii) *$\chi_U(g)$ equals the sum (with multiplicities) of the eigenvalues of $\rho(g)$.*
(iii) *$\chi_U(g)$ is the sum of the $\chi_U(1)$th roots of unity.*
(iv) *$\chi_U(g^{-1}) = \overline{\chi_U(g)}$ the complex conjugate of $\chi_U(g)$.*
(v) *$|\chi_U(g)| \leq \chi_U(1)$.*
(vi) *The set $\{x \in G \mid \chi_U(x) = \chi_U(1)\}$ is a normal subgroup of G.*

Proof. Since $g^n = 1$, we get that $\rho(g)$ is a zero of the polynomial $X^n - 1$. However, $X^n - 1$ splits into distinct linear factors in $\mathbb{C}[X]$, and so it follows that the minimal polynomial of $\rho(g)$ does also. Hence, $\rho(g)$ is diagonalizable by way of proving (i). From this, we have that the trace of $\rho(g)$ is the sum (with multiplicities) of the eigenvalues proving (ii). The eigenvalues are precisely the zeros of the minimal polynomial of $\rho(g)$, which divides $X^n - 1$. Consequently, these roots are nth roots of unity, which proves (iii), since $\chi_U(1) = \dim_{\mathbb{C}}(U)$. Each eigenvector of $\rho(g)$ is also an eigenvector for $\rho(g^{-1})$ with the eigenvalue for $\rho(g^{-1})$ being the inverse of the eigenvalue for $\rho(g)$. Since the eigenvalues are roots of unity, it follows that $\chi_U(g^{-1}) = \overline{\chi_U(g)}$. From this we obtain (iv).

Now (v) follows directly from (iii). We have already seen that $\chi_U(g)$ is the sum of its $\chi_U(1)$ eigenvalues, each of which is a root of unity. If the sum is equal to $\chi_U(1)$, then it follows that each of these eigenvalues must be 1, in which case $\rho(g)$ must be the identity map. Conversely, if $\rho(g)$ is the identity map, then $\chi_U(g) = \dim_{\mathbb{C}}(U) = \chi_U(1)$. Therefore, $\{x \in G : \chi_U(x) = \chi_U(1)\} = \ker(\rho)$, and hence is a normal subgroup of G. □

Suppose that χ and ψ are characters of G. We define new functions $\chi + \psi$ and $\chi\psi$ from G to $\mathbb{C}$ by $(\chi + \psi)(g) = \chi(g) + \psi(g)$ and $(\chi\psi)(g) = \chi(g)\psi(g)$ for $g \in G$. These new functions are not a priori characters themselves. Given a scalar $\lambda \in \mathbb{C}$, define a new function $\lambda\chi : G \to \mathbb{C}$ by $(\lambda\chi)(g) = \lambda\chi(g)$. Consequently, we can view the characters of G as elements of a $\mathbb{C}$-vector space of functions from G to $\mathbb{C}$.

Theorem 22.4.9. *The irreducible characters of G are, as functions from G to $\mathbb{C}$, linearly independent over $\mathbb{C}$.*

Proof. We have $\mathbb{C}G \cong M(f_1, \mathbb{C}) \oplus \cdots \oplus M(f_r, \mathbb{C})$ by Theorem 22.4.1. Let $S_1, \ldots, S_r$ be the distinct simple $\mathbb{C}G$-modules. For each i, let e_i be the identity element of $M(f_i, \mathbb{C})$. We fix some i.

Recall that $\chi_i(g)$ is the trace of the linear transformation on S_i defined by $g \in G$. The linear transformation on S_i, given by e_i, is the identity. Hence, $\chi_i(e_i) = \dim_{\mathbb{C}}(S_i) = f_i$. Moreover, if $j \neq i$, then the linear transformation on S_j given by e_i is the zero map, and hence $\chi_j(e_i) = 0$ for $j \neq i$. Now suppose that $\lambda_1, \ldots, \lambda_r \in \mathbb{C}$ such that $\sum_{j=1}^r \lambda_j\chi_j = 0$. From above, we see that $0 = \sum_{j=1}^r \lambda_j\chi_j(e_i) = \lambda_i f_i$ for each i. It follows that $\lambda_i = 0$ for all i; therefore, the characters are linearly independent. □

Lemma 22.4.10. $\chi_{U\oplus V} = \chi_U + \chi_V$ *for any $\mathbb{C}G$-modules U and V.*

Proof. By considering a $\mathbb{C}$-basis for $U \oplus V$, whose first $\dim_{\mathbb{C}}(U)$ elements form a $\mathbb{C}$-basis for $U \oplus \{0\}$, and whose remaining elements form a $\mathbb{C}$-basis for $\{0\} \oplus V$, we get that $\chi_{U\oplus V}(g) = \chi_U(g) + \chi_V(g)$ for any $g \in G$. □

Theorem 22.4.11. *If $S_1, \ldots, S_r$ are the distinct (up to isomorphism), simple $\mathbb{C}G$-modules, then the character of the $\mathbb{C}G$-module $a_1S_1 \oplus \cdots \oplus a_rS_r$ with $a_i \in \mathbb{N} \cup \{0\}$ is $a_1\chi_1 + \cdots + a_r\chi_r$. Consequently, two $\mathbb{C}G$-modules are isomorphic if and only if their characters are equal.*

Proof. The first statement follows directly from Lemma 22.4.10. Now, suppose that $\chi_U = \chi_V$ for some $\mathbb{C}G$-modules U and V.

Since $\mathbb{C}G$ is semisimple, we can write $U \cong a_1S_1 \oplus \cdots \oplus a_rS_r$ and $V \cong b_1S_1 \oplus \cdots \oplus b_rS_r$ with $a_i, b_i \in \mathbb{N} \cup \{0\}$. By taking characters, we have

$$0 = \chi_U - \chi_V = \sum_{i=1}^{r} (a_i - b_i)\chi_i.$$

By Theorem 22.4.9, this forces $a_i = b_i$ for all i, and therefore $U \cong V$. □

Definition 22.4.12. A *class function* on G is a function from G to $\mathbb{C}$, whose value within any conjugacy class is constant.

For example, characters of $\mathbb{C}G$-modules are class functions.

The set of all class functions on G forms a $\mathbb{C}$-vector space of dimension r, where r is the number of conjugacy classes within G. An obvious basis for this vector space is the set of class functions on G that have the value 1 on a single conjugacy class, and 0 on all other conjugacy classes.

Theorem 22.4.13. *The irreducible characters for G form a basis for the $\mathbb{C}$-vector space of class functions on G.*

Proof. By Theorem 22.4.9, the irreducible characters of G are linearly independent elements of the space of class functions. Their number equals the number of conjugacy classes of G by Theorem 22.4.4, and this number is equal to the dimension of the space of class functions. □

Definition 22.4.14. If α, β are class function of G, then their *inner product* is the complex number

$$\langle \alpha, \beta \rangle = \frac{1}{|G|} \sum_{g \in G} \alpha(g)\overline{\beta(g)}.$$

This inner product is a traditional complex inner product on the space of class function. Therefore, we have the following properties:

(1) $\langle \alpha, \alpha \rangle \geq 0$, and $\langle \alpha, \alpha \rangle = 0$, if and only if $\alpha = 0$;
(2) $\langle \alpha, \beta \rangle = \overline{\langle \beta, \alpha \rangle}$;
(3) $\langle \lambda\alpha, \beta \rangle = \lambda\langle \alpha, \beta \rangle$ for all $\lambda \in \mathbb{C}$;
(4) $\langle \alpha_1 + \alpha_2, \beta \rangle = \langle \alpha_1, \beta \rangle + \langle \alpha_2, \beta \rangle$.

From these basic properties we further have
(5) $\langle \alpha, \lambda\beta \rangle = \overline{\lambda}\langle \alpha, \beta \rangle$,
(6) $\langle \alpha, \beta_1 + \beta_2 \rangle = \langle \alpha, \beta_1 \rangle + \langle \alpha, \beta_2 \rangle$,

for all class functions $\alpha_1, \beta_1, \alpha_2, \beta_2$, and all $\lambda \in \mathbb{C}$.

Definition 22.4.15. If U is a $\mathbb{C}G$-module, then

$$U^G = \{u \in U : gu = u \text{ for all } g \in G\}.$$

Lemma 22.4.16. *If U is a $\mathbb{C}G$-module, then*

$$\dim_{\mathbb{C}}(U^G) = \frac{1}{|G|}\sum_{g\in G}\chi_U(g).$$

Proof. Let $a = \frac{1}{|G|}\sum_{g\in G} g \in \mathbb{C}G$. Clearly, $ga = a$ for any $g \in G$, and hence $a^2 = a$. If T is a linear transformation of U, defined by a, then T must satisfy the equation $X^2 - X = 0$, and consequently, T is diagonalizable. It follows that the only eigenvalues of T are 0 and 1. Let $U_1 \subset U$ be the eigenspace of T corresponding to the eigenvalue 1. If $u \in U_1$, then $gu = gau = au = u$ for any $g \in G$. Therefore, $u \in U^G$. Conversely, suppose that $u \in U^G$. Then

$$|G|au = \Big(\sum_{g\in G} g\Big)u = \sum_{g\in G} gu = \sum_{g\in G} u = |G|u,$$

and hence $a \in U_1$. It follows that $U^G = U_1$. However, the trace of T is equal to the dimension of U_1, and then the result follows from the linearity of the trace map. □

Theorem 22.4.17. *We have* $\langle \chi_U, \chi_V\rangle = \dim_{\mathbb{C}}(\mathrm{Hom}_{\mathbb{C}G}(U, V))$ *for any* $\mathbb{C}G$*-modules* U, V.

Recall that $\mathrm{Hom}_{\mathbb{C}G}(U, V)$ is an $\mathbb{C}$-vector space with $(\phi + \psi)(u) = \phi(u) + \psi(u)$, and $(\lambda\phi)(u) = \lambda\phi(u)$ for any $\lambda \in \mathbb{C}$, $u \in U$ and $\phi, \psi \in \mathrm{Hom}_{\mathbb{C}G}(U, V)$.

Proof. We observe that $\mathrm{Hom}_{\mathbb{C}G}(U, V)$ is a subspace of the $\mathbb{C}G$-module $\mathrm{Hom}_{\mathbb{C}G}(U, V)$. If $\phi \in \mathrm{Hom}_{\mathbb{C}G}(U, V)$ and $g \in G$, then $(g\phi)(u) = g\phi(g^{-1}u) = gg^{-1}\phi(u) = \phi(u)$ for any $u \in U$. Hence, $g\phi = \phi$ for all $g \in G$. This implies that $\phi \in \mathrm{Hom}_{\mathbb{C}G}(U, V)^G$. By reversing the elements, we get $\mathrm{Hom}_{\mathbb{C}G}(U, V) = \mathrm{Hom}_{\mathbb{C}G}(U, V)^G$.

Therefore,

$$\begin{aligned}\dim_{\mathbb{C}}(\mathrm{Hom}_{\mathbb{C}G}(U, V)) &= \dim_{\mathbb{C}}(\mathrm{Hom}_{\mathbb{C}G}(U, V)^G)\\ &= \frac{1}{|G|}\sum_{g\in G}\chi_{\mathrm{Hom}_{\mathbb{C}G}(U,V)}(g)\\ &= \frac{1}{|G|}\sum_{g\in G}\overline{\chi_U(g)}\chi_V(g)\\ &= \langle \chi_V, \chi_U\rangle\end{aligned}$$

by Lemma 22.4.16, and part (iii) of Theorem 22.4.8.

This implies that

$$\langle \chi_U, \chi_V\rangle = \overline{\langle \chi_V, \chi_U\rangle} = \langle \chi_V, \chi_U\rangle = \dim(\mathrm{Hom}_{\mathbb{C}G}(U, V)),$$

since we know that $\langle \chi_V, \chi_U\rangle$ is real. □

The Character Table and Orthogonality Relations

We have seen that the number of conjugacy classes r in a finite group G is the same as the number of irreducible characters. Furthermore, the set of irreducible characters form a basis for the space of class functions on G. If $\chi_1, \dots, \chi_r$ are the set of irreducible characters, and $g_1, \dots, g_r$ are a complete set of conjugacy class representatives, then the $r \times r$-matrix $\chi = (\chi_i(g_j))$ is called the *character table* for G.

We close this section by showing that the rows and columns of the character table are orthogonal vectors relative to the defined inner product. These results are called the *orthogonality relations*. As a consequence of these relations, we obtain the fact that the irreducible characters form an orthonormal basis for the space of characters. There is great deal of other information that can be obtained from the character table. We refer to the book by Alperin and Bell [1] for further discussion.

Theorem 22.4.18 (First orthogonality relation). *Let $\chi_1, \dots, \chi_r$ be the set of irreducible characters of G. Then*

$$\frac{1}{|G|}\sum_{g \in G} \chi_i(g)\overline{\chi_j}(g) = \begin{cases} 0, & \text{if } i \neq j, \\ 1, & \text{if } i = j. \end{cases}$$

In other words, the irreducible characters form an orthonormal set with respect to the defined inner product.

Proof. Let $S_1, \dots, S_r$ be the distinct simple $\mathbb{C}G$-modules that go with the irreducible characters. From the previous theorem, we have

$$\langle \chi_i, \chi_j \rangle = \dim_{\mathbb{C}}(\operatorname{Hom}_{\mathbb{C}G}(S_i, S_j))$$

for any i, j. We further have $\operatorname{Hom}_{\mathbb{C}G}(S_i, S_i) \cong \mathbb{C}$, and by Schur's lemma $\operatorname{Hom}_{\mathbb{C}G}(S_i, S_j) = 0$ for $i \neq j$, proving the theorem. □

Corollary 22.4.19. *The set of irreducible characters form an orthonormal basis for the vector space of class functions.*

Proof. The irreducible characters form a basis for the space of characters, and from the orthogonality result they are an orthonormal set relative to the inner product. □

The second orthogonality relation says that the columns of the character table are also a set of orthogonal vectors. That is, the irreducible characters of a set of conjugacy class representatives also forms an orthogonal set with respect to the defined inner product.

Theorem 22.4.20 (Second orthogonality relation). *Let $\chi_1, \dots, \chi_r$ be the set of irreducible characters of G, and suppose that $g_1, \dots, g_r$ are a set of conjugacy class representatives, and $k_1, \dots, k_r$ are the orders of the conjugacy classes. Then for any $1 \leq i, j \leq r$, we have*

$$\sum_{s=1}^{r} \chi_s(g_i)\overline{\chi_s(g_j)} = \begin{cases} 0, & \text{if } i \neq j, \\ \frac{|G|}{k_i}, & \text{if } i = j. \end{cases}$$

Proof. Let $\chi = (\chi_i(g_j))_{1 \leq i,j \leq r}$ be the character table for G, and let K be the $r \times r$ diagonal matrix with the set $\{k_1, \ldots, k_r\}$ as its main diagonal. Then we have $(\chi K)_{i,j} = \chi_i(g_j)k_j$ for any i, j. Then

$$(\chi K \overline{\chi}^t)_{ij} = \sum_{\ell=1}^{r} k_\ell \chi_i(g_\ell)\overline{\chi_j}(g_\ell) = \sum_{g \in G} \chi_i(g)\overline{\chi_j(g)},$$

but this equals $= |G|\langle \chi_i, \chi_j \rangle$ by the first orthogonality relation.

Hence, $\chi K \overline{\chi}^t = |G| I$, where I is the identity matrix. It follows that for any i, j, we have

$$|G| = \sum_{\ell=1}^{r} k_j \overline{\chi_\ell(g_j)} \chi_\ell(g_j),$$

and

$$0 = \sum_{\ell=1}^{r} k_j \overline{\chi_\ell(g_j)} \chi_\ell(g_i) \quad \text{for } i \neq j,$$

completing the proof. □

As mentioned before, more information about character tables and their consequences can be found in [1].

22.5 Burnside's Theorem

We conclude this chapter by presenting a very important result in finite group theory, whose proof uses representation theory. This is *Burnside's Theorem*, which asserts that any group of order $p^a q^b$ with p, q distinct primes must be solvable. Burnside's result was important in the proof of the famous *Feit–Thompson theorem*, which asserted that any group of odd order must be solvable. This was crucial in the classification of finite simple groups.

Recall that a group G is *solvable* if it has a normal series with Abelian factors. Solvable groups play a crucial role in the proof of the insolvability of the quintic polynomial, and we discussed solvable groups in detail in Chapter 12. For the proof, we need the following two facts about solvable groups:

1. If a group G has a normal solvable subgroup N with G/N solvable, then G is solvable (Theorem 12.1.3).
2. Any finite group of prime power order is solvable (Theorem 12.1.8).

We start with several lemmas that depend on representation theory.

Let G be a finite group, and suppose it has r irreducible representations $\chi_1, \chi_2, \dots, \chi_r$ of respective degrees $m_1, m_2, \dots, m_r$. Suppose the respective orders of the r conjugacy classes are $h_1, h_2, \dots, h_r$. The statements in the lemmas depend on some mild facts on algebraic integers. An algebraic integer is a complex number, which is a zero of a monic integral polynomial. Here we just need the following two facts:

1. The set of algebraic integers forms a subring of $\mathbb{C}$.
2. If an algebraic integer is a rational number, then it is an ordinary integer.

For more information about algebraic integers see Chapter 21.

Lemma 22.5.1. *Let χ be a character of G. The value $\chi(g)$ for any $g \in G$ is an algebraic integer.*

Proof. For any $g \in G$, the value $\chi(g)$ is a sum of roots of unity. However, any root of unity satisfies a monic integral polynomial $X^n - 1 = 0$, and hence is an algebraic integer. Since the algebraic integers form a ring, any sum of roots of unity is an algebraic integer. □

Lemma 22.5.2. *Let χ be an irreducible character of G. Let $g \in G$ and $C_G(g)$ the centralizer of g in G. Then*

$$\frac{|G : C_G(g)|}{\chi(1)}\chi(g)$$

is an algebraic integer.

Proof. Let S be the simple $\mathbb{C}G$-module having character χ.

Let $g \in G$, and let C be the conjugacy class of g in G. By Theorem 13.2.1, we have $|C| = |C : C_G(g)|$.

Let $\alpha \in \mathbb{C}$, $\alpha = \sum_{x \in K} x$, be the class sum of K. We consider the map $\varphi : S \to S$, $\varphi(s) = \alpha s$, for $s \in S$. From Theorem 22.4.4 and its proof, we get that α is in the center of $\mathbb{C}G$.

This gives $\varphi \in \operatorname{End}_{\mathbb{C}G}(S)$, and there exists a $\lambda \in \mathbb{C}$ with $\alpha s = \lambda s$ for all $s \in S$ by Schur's lemma. We obtain

$$\lambda\chi(1) = \sum_{x \in C} \chi(x) = |C|\chi(g) = |G : C_G(g)|\chi(g)$$

by taking traces. Therefore,

$$\lambda = \frac{|G : C_G(g)|}{\chi(1)} = \chi(g).$$

Let $\tau : \mathbb{C}G \to \mathbb{C}G$, $\tau(z) = z\alpha$ for $z \in \mathbb{C}G$. We get $\tau \in \operatorname{End}_{\mathbb{C}G}(\mathbb{C}G)$ by the proof of Lemma 22.3.6. Since S is a simple $\mathbb{C}G$-module. Therefore, we may consider S as a submodule of $\mathbb{C}G$, and for $0 \neq s \in S \subset \mathbb{C}G$, we have $\tau(s) = s\alpha = \alpha s = \lambda s$, since α is a central element.

Therefore, λ is an eigenvalue of τ, and so $\det(\lambda I - A) = 0$, where I is the identity matrix, and A the matrix of τ with respect to the $\mathbb{C}$-basis G for $\mathbb{C}G$. Each entry of A is either 0 or 1, which means that, in particular, $f(X) = \det(XI - A)$ is a monic polynomial in X with integer coefficients. Since $f(\lambda) = 0$, we get that λ is an algebraic integer. □

Lemma 22.5.3. *Let χ be an irreducible character of G. Then $\chi(1)$ divides $|G|$.*

Proof. Let $g_1, g_2, \ldots, g_r$ be a set of representatives of the conjugacy classes of G. We know that

$$\frac{|G : C_G(g_i)|\chi(g_i)}{\chi(1)} \quad \text{and} \quad \overline{\chi(g_i)} = \chi(g_i^{-1})$$

are algebraic integers. By the first orthogonality relation

$$\frac{|G|}{\chi(1)} = \frac{1}{\chi(1)} \sum_{i=1}^{r} |G : C_G(g_i)| \chi(g_i)\overline{\chi(g_i)} = \sum_{i=1}^{r} |G : C_G(g_i)| \frac{\chi(g_i)}{\chi(1)} \overline{\chi}(g_i),$$

which is an algebraic integer, and hence an ordinary integer. □

Lemma 22.5.4. *Let G be a character of G, $g \in G$ and $\gamma = \frac{\chi(g)}{\chi(1)}$.*
If γ is a nonzero algebraic integer, then $|\gamma| = 1$.

Proof. From Theorem 22.4.8, we know that $|\gamma| \leq 1$.

Suppose that $0 < |\gamma| < 1$, and assume that γ is an algebraic integer.

Now, γ is an average of complex roots of unity. The same will be true for all $\sigma(\gamma)$ with $\sigma \in \mathrm{Aut}(K \mid \mathbb{Q}) =: H$, where K is the splitting field of the minimal polynomial of γ over $\mathbb{Q}$.

In particular, $|\sigma(\gamma)| \leq 1$ for all $\sigma \in H$. Hence, $p := |\prod_{\sigma \in H} \sigma(\gamma)| < 1$.

On the other hand, $p \in \mathbb{Z}$ by Theorems 7.3.12 and 16.5.1 (recall that γ is a zero of a irreducible, monic polynomial with integer coefficients, see Theorem 4.4.3).

This implies $p = 0$, and therefore the constant term of the minimal polynomial of γ over $\mathbb{Q}$ must be zero, which gives a contradiction.

Hence, γ cannot be an algebraic integer. □

Theorem 22.5.5. *If G has a conjugacy class of nontrivial prime power order, then G is not simple.*

Proof. Suppose that G is simple and that the conjugacy class of $1 \neq g \in G$ has order p^n with p a prime number, and $n \in \mathbb{N}$. From the second orthogonality relation, we get

$$0 = \frac{0}{p} = \frac{1}{p} \sum_{i=1}^{r} \chi_i(g)\chi_i(1) = \frac{1}{p} + \frac{1}{p} \sum_{i=2}^{r} \chi_i(g)\chi_i(1),$$

where $\chi_1, \chi_2, \ldots, \chi_r$ are the irreducible characters of G (recall that χ_1 is the principal character).

Since $-\frac{1}{p}$ is not an algebraic integer, it follows that $\frac{\chi_i(g)\chi_i(1)}{p}$ is not an algebraic integer for some $2 \le i \le r$. As $\chi_i(g)$ is an algebraic integer, this implies that $p \nmid \chi_i(1)$, and $\chi_i(g) \neq 0$.

Now $|G : C_G(g)| = p^n$ is relatively prime to $\chi_i(1)$.

Therefore,

$$a\big|G : C_G(g)\big| + b\chi_i(1) = 1$$

for some $a, b \in \mathbb{Z}$ (see Theorem 3.1.9).

Thus,

$$\frac{\chi_i(g)}{\chi_i(1)} = \frac{a|G : C_G(g)|\chi_i(g)}{\chi_i(1)} + b\chi_i(g),$$

which is an algebraic integer, and therefore $|\chi_i(x)| = \chi_i(1)$.

Consequently,

$$g \in Z_i = \{x \in G : \big|\chi_i(x)\big| = \chi_i(1)\}.$$

We show that Z_i is a subgroup of G. □

First of all, if $g \in Z_i$, then $g^{-1} \in Z_i$. From Theorem 22.4.8, we also get $|\chi_i(g)| = \chi_i(1)$ if and only if g has exactly one eigenvalue. If $g \in Z_i$, let this eigenvalue be $\lambda(g)$, so that, if U is the $\mathbb{C}G$-module corresponding to χ_i, then we have $gu = \lambda(g)u$ for all $u \in U$. We now see that for $g, h \in Z_i$, then $(gh)u = \lambda(g)\lambda(h)u$ for all $u \in U$. Hence, $\chi_i(gh) = \chi_i(1)\lambda(g)\lambda(h)$, and thus $|\chi_i(gh)| = \chi_i(1)$, which gives $gh \in Z_i$. Therefore, Z_i is a subgroup of G.

Now, let $K_i = \{x \in G : \chi_i(x) = \chi_i(1)\}$. K_i is a normal subgroup of G, and also in Z_i.

We now want to show that

$$Z_i/K_i = Z(G/K_i),$$

the center of G/K_i. If $\rho : G \to \mathrm{GL}(U)$ is the representation corresponding to χ_i, then for any $g \in Z_i$, the matrix of $\rho(g)$ (with respect to any $\mathbb{C}$-basis of U) will be scalar, and hence $\rho(g) \in Z(\rho(G))$. Since $\rho(G) \cong G/K_i$, it follows that Z_i/K_i is a subgroup of $Z(G/K_i)$. Now, we apply that χ_i is irreducible. If $gK_i \in Z(G/K_i)$, then $\rho(g)$ commutes with $\rho(x)$ for every $x \in G$. Consequently, the map defined by $u \mapsto gu$, $u \in U$, is a $\mathbb{C}G$-endomorphism of U. But U is simple, so we have $\mathrm{End}_{\mathbb{C}G}(U) \cong \mathbb{C}$ by Schur's lemma.

Therefore, there is a complex root of unity μ such that $gu = \mu u$ for all $u \in U$. We now have $\chi_i(g) = \chi_i(1)$, and hence $g \in Z_i$. Therefore, $Z_i/K_i = Z(G/K_i)$.

Consequently, if G is non-Abelian and simple, then $Z_i = \{1\}$. But this gives a contradiction.

Theorem 22.5.6 (Burnside's Theorem). *If $|G| = p^a q^b$, where p and q are prime numbers and $a, b \in \mathbb{N}$, then G is solvable.*

Proof. We use induction on $a + b$. If $a + b = 1$, then G has a prime order, and hence is solvable. We now assume that $a + b \geq 2$, and that any group of order $p^r q^s$, $r, s \in \mathbb{N}$, is solvable whenever $r + s < a + b$.

First of all, if the center $Z(G)$ is nontrivial, then G is solvable, because $Z(G)$ is solvable and $G/Z(G)$ is solvable by the inductive hypothesis.

Now, let $Z(G) = \{1\}$.

Then we may take $h_1 = 1$ for the conjugacy class of 1.

By the class equation (see Theorem 13.2.2), we then have

$$p^a q^b = |G| = 1 + h_2 + h_3 + \cdots + h_r.$$

It follows that pq cannot divide each $h_2, h_3, \ldots, h_r$. Hence, h_i is a prime power of either p or q for some $i \geq 2$. If h_i is a nontrivial prime power, then from Theorem 22.5.5 it follows that G is not simple.

If $h_i = 1$ for some $i \geq 2$, then G has at least two representations into $\mathbb{C}$. The number of these representations is given by the Abelianizations, which is given by $|G : G'|$, where G' is the commutator subgroup of G. Then $|G : G'| > 1$, and since G' is non-Abelian, G' is a proper normal subgroup. Hence, G is not simple. So, in any case, G is not simple. Therefore, G contains a proper normal subgroup N. Since $|N|||G|$, we have $|N| = p^{a_1} q^{b_1}$ with $a_1 + b_1 < a + b$, since N is a proper subgroup.

By the inductive hypothesis, N is solvable. Furthermore, $|G/N|$ also divides $|G|$. So, for the same reason, G/N is solvable. Therefore, both N and G/N are solvable, so G is solvable by Theorem 12.1.3. □

22.6 Exercises

1. Let K be a field, and let G be a finite group. Let U and V be KG-modules having the same dimension n, and let $\rho : G \to \mathrm{GL}(U)$ and $\tau : G \to \mathrm{GL}(V)$ be the corresponding representations.
   By fixing K-bases for U and V, consider ρ and τ as homomorphisms from G to $\mathrm{GL}(n, K)$. Show that U and V are KG-module isomorphic if and only if there exists some $M \in \mathrm{GL}(n, K)$ such that $\rho(g)M = M\tau(g)$ for every $g \in G$.
2. Let K be a field, and let G be a finite group. Let $x = \sum_{g \in G} g \in KG$.
   (i) Show that the subspace Kx of KG is the unique submodule of KG, that is, isomorphic to the trivial module.
   (ii) Let $\epsilon : KG \to K$ be the KG-module epimorphism defined by $\epsilon(g) = 1$ for all $g \in G$.
   Show that $\ker(\epsilon)$ is the unique KG-submodule of KG, whose quotient is isomorphic to the trivial module. This kernel is called the augmentation ideal of KG.

(iii) Suppose that $\operatorname{char}(K) = p$, with p dividing $|G|$. Show that $KG \subset \ker(\epsilon)$, the augmentation ideal of KG. Show that $\ker(\epsilon)$ is not a direct summand of KG, and hence that the KG-module KG is not semisimple.

3. Show that the converse of Corollary 22.2.24 is true.
4. Let U be a finite-dimensional K-vector space and let G be a finite group with fully reducible representation $\rho : G \to \mathrm{GL}(U)$. Show that ρ gives a direct decomposition

$$U = V_1 \oplus \cdots \oplus V_k$$

of U with all V_i, $i = 1, \ldots, k$, irreducible G-invariant subspaces of U.
5. Show that A is a simple A-module if and only if A is a division algebra.
6. Let $n \in \mathbb{N}$, and let $T_n(K)$ be the algebra of upper triangular $n \times n$ matrices over K.
   (i) Show that the set $V_n(K)$ of column vectors of K of length n is a $T_n(K)$-module that has a unique composition series, in which every simple $T_n(K)$-module appears exactly once as a composition factor.
   (ii) Show that the $T_n(K)$-module $T_n(K)$ is isomorphic to the direct sum of all nonzero submodules of $V_n(K)$.
7. Let U be an A-module, let $n \in \mathbb{N}$, and let U^n be the set of column vectors of length n with entries from U, considered in the obvious way as an $M(n, A)$-module. Show that U is a simple A-module if and only if U^n is a simple $M(n, A)$-module.
8. Let χ be an irreducible character of G. Let λ be any $|G|$th root of unity. Show that the set $\{x \in G : \chi(x) = \lambda\chi(1)\}$ is a normal subgroup of G.
9. Prove that the set of algebraic integers forms a subring of $\mathbb{C}$.
10. Prove that if an algebraic integer is rational, then it is an ordinary integer.
11. Prove that G is simple if and only if the only irreducible character χ_i, for which $\chi_i(g) = \chi_i(1)$ for some $1 \neq g \in G$ is the principal character χ_1.

23 Algebraic Cryptography

23.1 Basic Algebraic Cryptography

23.1.1 Cryptosystems Tied to Abelian Groups

Cryptography refers to the science of sending and receiving coded messages. Coding and hidden ciphering is an old endeavor used by governments and military, and between private individuals from ancient times. Recently, it has become even more prominent because of the necessity of sending secure and private information, such as credit card numbers and passwords, over essentially open communication systems.

Traditionally, cryptography deals with devising and implementing secret codes or cryptosystems. Cryptoanalysis is the science of breaking cryptosystems while cryptology refers to the whole field of cryptography plus cryptoanalysis.

A cryptosystem or code is an algorithm to change a plain message, called the plain text message, into a coded message, called the ciphertext message. In general, both the plaintext message (uncoded message) and the ciphertext message (coded message) are written in some N-letter alphabet which is usually the same for both plaintext and code. The method of coding, or the encoding algorithm, is then a transformation of the N letters. The most common way to perform this transformation is to consider the N letters as N integers modulo N and then apply a number theoretical function to them. Therefore, many encoding algorithms use modular arithmetic and hence cryptography is tied to number theory and Abelian groups.

Modern cryptography is usually separated into classical cryptography, called symmetric key cryptography, and public key cryptography. In the former, both the encoding and decoding algorithms are supposedly known only to the sender and receiver, usually referred to as Bob and Alice. In the latter, the encryption method is public knowledge but only the receiver knows how to decode.

The message that one wants to send is written in plaintext and then converted into code. The coded message is written in ciphertext. The plaintext message and the ciphertext message are written in some alphabets that are usually the same. The process of putting the plaintext into code is called *enciphering* or *encryption* while the reverse process is called *deciphering* or *decryption*.

Encryption algorithms break the plaintext and ciphertext message into message units. These are single letters or, more generally, k-vectors of letters. The transformations are done in these message units and the encryption algorithm is a mapping from the set of plaintext message units to the set of ciphertext message units.

Putting this into a mathematical formulation, we let $\mathcal{P}$ to be the set of all plaintext message units and $\mathcal{C}$ be the set of all ciphertext message units. The encryption algorithm is then the application of an injective map $f: \mathcal{P} \to \mathcal{C}$. The map f is the *encryption map*. The left inverse map $g: \mathcal{C} \to \mathcal{P}$ is the *decryption* or *deciphering map*. The collection $\{\mathcal{P}, \mathcal{C}, f, g\}$ is called a *basic cryptosystem*.

https://doi.org/10.1515/9783111142524-023

We may place this in a more general context. We call this wider model a *(general) cryptosystem*, indexed by a set $\mathcal{K}$, called the *key space*. Formally, a cryptosystem is a tuple $(\mathcal{P}, \mathcal{C}, \mathcal{K}, \mathcal{E}, \mathcal{D})$ where $\mathcal{P}$ is the set of plaintext message units, called the *plaintext space*, $\mathcal{C}$ is the set of ciphertext message units, called the *ciphertext space*, the elements $k \in \mathcal{K}$ are called *keys*, $\mathcal{E}$ is a set of injective maps $f_k : \mathcal{P} \to \mathcal{C}$ indexed by the key space. This is called the *set of encryption maps*. Hence, for each $k \in K$, there is an injective map $f_k : \mathcal{P} \to \mathcal{C}$. The set $\mathcal{D}$ consists of maps $g_k : \mathcal{C} \to \mathcal{P}$, also indexed by the key space. This is called the *set of decryption maps*.

The central property of a cryptosystem is that, for each $k \in K$, there exists a corresponding key $k' \in \mathcal{K}$ and a decryption map $g_{k'} : \mathcal{C} \to \mathcal{P}$ such that $g_{k'}$ is the left inverse of f_k. In our previous language this means that for each $k \in \mathcal{K}$ we have a basic cryptosystem $\{\mathcal{P}, \mathcal{C}, f_k, g_{k'}\}$ with k the *encryption key* and k' the *decryption key*.

Using this model, we can easily distinguish symmetric from asymmetric cryptosystems. In a symmetric key cryptosystem, if the encryption key k is given, it is easy to find the corresponding decryption key k'. In fact, most of the time we have $k = k'$. In an asymmetric or public key cryptosystem, even if the encryption key k is known, it is infeasible to find or to compute the corresponding decryption key k'.

In the following, we describe some cryptosystems and start with the symmetric key cryptosystems.

23.1.1.1 Permutation Cipher

The simplest type of encryption algorithm is a permutation cipher. Here, the letters of the plaintext alphabet are permuted and the plaintext message is sent in the permuted letters. A very straightforward example of a permutation encryption algorithm is a shift algorithm.

Here, we consider the plaintext alphabet as the integers $0, 1, \ldots, N-1 \pmod{N}$. We choose a fixed integer k and the encryption algorithm is

$$f(m) \equiv (m + k) \pmod{N}.$$

This is often known as a Caesar code after Julius Caesar who supposedly invented it. Any permutation encryption algorithm is very simple to attack using statistical analysis. Polyalphabetic ciphers are an attempt to thwart statistical attacks. One variation of the basic Caesar code is the following where message units are k-vectors. It is actually a type of polyalphabetic cipher called a Vigenère code. In this code, message units are considered as t-vectors of integers modulo N from an N letter alphabet. Let $(b_1, \ldots, b_t)$ be a fixed t-vector in $\mathbb{Z}_n^t$. This Vigenère code then takes a message unit $(a_1, \ldots, a_t)$ to $(a_1 + b_1, \ldots, a_t + b_t) \pmod{N}$. For a long period of time polyalphabetic ciphers where considered unbreakable. In 1920, the Friedmann test was developed. Given a sequence of letters of length m representing a Vigenère encrypted cipher text, the Friedmann test calculates the length t of the key word $(b_1, \ldots, b_t)$, see for instance [66]. A statistical analysis then allows to break the Vigenère code.

23.1.1.2 One-Time Pad

We now describe the one-time pad which has perfect security. Here, let $\mathcal{P}$ be the set of plaintext messages, $\mathcal{C}$ the set of ciphertext messages, and $\mathcal{K}$ the set of keys for a cryptosystem $\mathcal{E}$. Then $\mathcal{E}$ has perfect security if for any given plaintext message $\mathcal{P}$ and corresponding ciphertext message C we have that the conditional probability $\text{Prob}(P|C)$ of determining the plaintext message P, given knowledge of the ciphertext message C, is exactly the same as the absolute probability $\text{Prob}(P)$ of determining the plaintext P.

Definition 23.1.1. Suppose the sets $\mathcal{P}$ of plaintext messages, $\mathcal{C}$ of ciphertext messages and $\mathcal{K}$ of keys are all given by elements of $\{0,1\}^n$. That is, plaintext messages, ciphertext messages and keys are all random bit strings of fixed length n. For a given $k \in \mathcal{K}$ the encryption function is given by $F_k(p) = p \oplus k$ for $p \in \mathcal{P}$. Here, $\oplus$ denotes the XOR operation on each pair of corresponding bits. This is simply the operation on bits $\{0,1\}$, that is, addition modulo 2. We assume that the distribution on all three sets is the uniform distribution and a key k is only used once. The resulting cryptosystem is called a *one-time pad.*

Shannon, see [98], proved that the one-time pad, under the assumptions provided in the definition, is perfectly secure, as long as the keys are randomly chosen and used only once.

Theorem 23.1.2. *A one-time pad has perfect security if the keys are randomly chosen from the uniform distribution of keys and a key is used only once.*

Although the one-time pad is theoretically secure there are many problems with its practical use because of the assumptions described above. For these reasons the one-time pad, while important theoretically, is not used to a great extent in encryption. However, a stream cipher is a method to attempt to mimic the important properties of the one-time pad. A *stream cipher* is a symmetric key cipher where plaintext characters are combined with a pseudo-random key generator called a *key stream*. In a stream cipher the plaintext characters are encrypted one at a time and the encryption of successive characters varies during the encryption.

Stream ciphers require sequences of pseudo-random digits. These are sequences that behave as if they are random. Here we will discuss a procedure to generate pseudo-random sequences and hence stream cipher key generation. First we need the concept of a linear congruence generator. For a given natural number n we denote by $\mathbb{Z}_n$ the ring of integers modulo n. Elements of $\mathbb{Z}_n$ are residue classes of integers modulo n. If a is an integer, we will denote the corresponding residue class in $\mathbb{Z}_n$ by $\overline{a}$.

Definition 23.1.3. Let $n \in \mathbb{N}$ and $\overline{a}, \overline{b} \in \mathbb{Z}_n$. A bijective map $f\colon \mathbb{Z}_n \to \mathbb{Z}_n$ given by $x \mapsto \overline{a}x + \overline{b}$ is called a *bilinear congruence generator*.

Notice that the map $x \mapsto \overline{a}x + \overline{b}$ is bijective if and only if $\gcd(a,b) = 1$. If we choose a large modulus n, linear congruence generators are used to generate pseudo-random integers. In using a linear congruence generator $f\colon x \mapsto \overline{a}x + \overline{b}$ the integers a, b should be chosen such that the function g has no fixed point in $\mathbb{Z}_n$. Then $\overline{b} \neq 0$ for otherwise $\overline{0}$ is a fixed point. Hence, let $\overline{b} \neq \overline{0}$. If $\overline{a} = \overline{1}$, then f has no fixed point but then the function is just a linear shift which is insecure. Therefore, let $\overline{a} \neq \overline{1}$. Then f has a fixed point in $\mathbb{Z}_n$ if $\gcd(a-1, n) = 1$ because then there exists a $d \in \mathbb{Z}$ with $(\overline{d(a-1)}) = \overline{1}$ and then $x = -\overline{db}$ is a fixed point in $\mathbb{Z}_n$. Therefore, altogether for a linear congruence generator we should choose a and b such that $\gcd(a-1, n) > 1$, $\overline{a} \neq 1$, and $\overline{b} \neq \overline{0}$.

Using the idea of a linear congruence generator, we now give a procedure for the generation of a stream cipher.

1. Choose a seed $s \in \mathbb{Z}$ by key agreement or as a random number.
2. Let $n \in \mathbb{N}$, $a, b \in \mathbb{Z}$ and $f: \mathbb{Z}_n \to \mathbb{Z}_n$, $x \mapsto \overline{a}x + \overline{b}$ be a linear congruence generator. Define the sequences $x_0 = \overline{s}, \overline{x}_1 \equiv f(\overline{x}_0) \pmod{n}, \overline{x}_2 \equiv f(\overline{x}_1) \pmod{n}, \ldots$.
3. Transform the sequence of plaintext units into a sequence of residue classes $\overline{m}_0, \overline{m}_1, \ldots$ in $\mathbb{Z}_n$.
4. Encrypt the m_i into $\overline{c}_i = \overline{m}_i + \overline{x}_i \in \mathbb{Z}_n$. The secret key is $\overline{s} \in \mathbb{Z}_n$.

We give the following remarks.

1. The integer n should be very large and the residue classes should occur with the same probability. Further the function f should not have a fixed point. To accomplish this we must choose f and $\overline{s} \in \mathbb{Z}_n$ such that the period length $\overline{x}_0, \overline{x}_1, \ldots$ is as large as possible. Best would be the maximal length n.
2. If we know sufficiently many plain text units which follow each other and we know the linear congruence generator used then we may calculate $\overline{s}$.

Theorem 23.1.4 (Maximal period length for $n \geq 2$). *Let $n \in \mathbb{N}$ with $n = 2^m$, $m \geq 1$, and let $a, b \in \mathbb{Z}$ such that $f: \mathbb{Z}_n \to \mathbb{Z}_n$, $x \mapsto \overline{a}x + \overline{b}$ is a linear congruence generator. Further let $s \in \{0, 1, \ldots, n-1\}$ be given, $\overline{x}_0 = \overline{s}, \overline{x}_1 = f(\overline{x}_0), \ldots$. Then the sequence $\overline{x}_0, \overline{x}_1, \ldots$ is periodic with the maximal period length $n = 2^m$ if and only if the following holds:*

(1) *a is odd.*

(2) *If $m \geq 2$ then $a \equiv 1 \pmod{4}$.*

(3) *b is odd.*

Proof. We show that (1), (2), and (3) hold if the period length is maximal. First we must have $\gcd(a, n) = 1$ since f is a linear congruence generator. Further f has no fixed point because the period length is maximal. We show that $a \equiv 3 \pmod{4}$ is not possible if $m \geq 2$. Suppose that $a \equiv 3 \pmod{4}$ and $m \geq 2$. Suppose that $a \equiv 3 \pmod{4}$ and $m \geq 2$. Then $a + 1 \equiv 0 \pmod{4}$ and it follows that

$$(1 + a + a^2 + \cdots + a^{2i-1}) = (1 + a^2 + a^4 + \cdots + a^{2i-2})(1 + a) \equiv 0 \pmod{4}. \qquad (*)$$

We now consider

$$\overline{x}_{i+1} - \overline{x}_i = f(\overline{x}_i) - f(\overline{x}_{i-1}) = (\overline{a}\,\overline{x}_i + \overline{b}) - (\overline{a}\,\overline{x}_{i-1} + \overline{b}) = \overline{a}(\overline{x}_i - \overline{x}_{i-1})$$

for $i \geq 1$. It then follows recursively that

$$\begin{aligned}\overline{x}_k - \overline{x}_0 &= (\overline{x}_k - \overline{x}_{k-1}) + (\overline{x}_{k-1} - \overline{x}_{k-2}) + \cdots + (\overline{x}_1 - \overline{x}_0)\\ &= \overline{a}^{k-1}(\overline{x}_1 - \overline{x}_0) + \overline{a}^{k-2}(\overline{x}_1 - \overline{x}_1) + \cdots + (\overline{x}_1 - \overline{x}_0)\\ &= (\overline{x}_1 - \overline{x}_0)(\overline{1} + \overline{a} + \overline{a}^2 + \cdots + \overline{a}^{k-1})\end{aligned}$$

for $k \geq 1$. Therefore

$$x_{2i} \equiv x_0 \pmod 4 \quad \text{and} \quad x_{2i+1} \equiv x_1 \pmod 4$$

from relation $(*)$ above.

Hence half of the elements in the sequence have the same residue class as x_0 modulo 4 and the other half the same as x_1 modulo 4 which gives a contradiction to the maximality of the period length. Therefore $a \equiv 1 \pmod 4$ if $m \geq 2$. To show (3) notice that in a sequence with maximal period length the residue class $\overline{0}$ must occur.

Hence, without loss of generality, we may assume that $\overline{x}_0 = \overline{0}$. Then $\overline{x}_1 = \overline{b}$ and recursively we have

$$\overline{x}_k = (\overline{1} + \overline{a} + \cdots + \overline{a}^{k-1})\overline{b}$$

for $k \geq 1$ since $\overline{x}_0 = \overline{0}$ and $\overline{x}_1 = \overline{b}$. All elements in the sequence are multiples of $\overline{b}$. There is an $\overline{x}_i = \overline{1}$ and therefore $\overline{b}$ is invertible in $\mathbb{Z}_n$ and hence b is odd.

Now, assume that (1), (2), and (3) are satisfied. The theorem follows directly if $n = 2$ since then if $\overline{x}_0 = \overline{0}$ we have $\overline{x}_1 = \overline{1}$ and if $\overline{x}_0 = \overline{1}$ we have $\overline{x}_1 = \overline{0}$. Now suppose that $m \geq 2$, so that $n \geq 4$. We show that we may obtain the maximal length $n = 2^m$ for $\overline{x}_0 = \overline{0}$ which proves the theorem.

Let $\overline{x}_0 = \overline{0}$. Then as before we obtain recursively $\overline{x}_k = (\overline{1} + \overline{a} + \cdots + \overline{a}^{k-1})\overline{b}$ for $k \geq 1$. Since $\overline{b}$ is odd we have $\overline{x}_k = \overline{0}$ if and only if $(\overline{1} + \overline{a} + \ldots \overline{a}^{k-1}) = \overline{0}$ in $\mathbb{Z}_n$.

We write $k = 2^r t$ with $r \geq 0$ and t odd. Then

$$\begin{aligned}\overline{0} &= (\overline{1} + \ldots \overline{a}^{k-1})\\ &= \left(\overline{1} + \overline{a} + \cdots + \overline{a}^{2^r-1}\right)\left(\overline{1} + \overline{a}^{2^r} + (\overline{a}^{2^r})^2 + \cdots + (\overline{a}^{2^r})^{t-1}\right).\end{aligned}$$

The second factor is congruent to 1 modulo 2 and hence $2^m | (1 + a + \cdots + a^{k-1})$ if and only if $2^m | (1 + a + \cdots + a^{2^r-1})$. The integer $1 + a + \cdots + a^{2^r-1}$ is divisible by 2^r since it is the sum of 2^r odd numbers but not divisible by 2^{r+1}. It follows that $r \geq m$ if and only if $2^m | k$. Therefore $\overline{x}_k = \overline{0}$ occurs for $k \geq 1$ for the first time when $k = n = 2^m$. □

We now describe some of the current public key cryptosystems. We start with the RSA cryptosystem named after L. Rivest, A. Shamir, and L. Adleman.

23.1.1.3 RSA Cryptosystem

Alice chooses two distinct primes p and q and computes the product $n = pq$; n must be chosen large enough. For the Euler φ-function we have

$$\begin{aligned}\varphi(n) &= \left|\{a \in \mathbb{N} \mid 1 \le a \le n \text{ and } \gcd(a,n) = 1\}\right| \\ &= pq - p - q + 1 \\ &= (p-1)(q-1).\end{aligned}$$

Now Alice computes two numbers $e, s \ge 3$ such that $es \equiv 1 \pmod{\varphi(n)}$. The number s should be large; otherwise, the private key (n, s) is insecure due to an attack by Wiener, see [104]. Assume that the plaintext message is given by an integer $x \in \{0, 1, \ldots, n-1\}$. The public key is the pair (n, e), and the encryption is done by $x \mapsto x^e \pmod{n}$. Alice decrypts by $y \mapsto y^s \pmod{n}$.

Now, let $y = x^e \pmod{n}$. If $es = 1 + (p-1)k$, then

$$\begin{aligned}y^s &\equiv x^{es} \\ &\equiv x \cdot \left(x^{p-1}\right)^k \\ &= \begin{cases} 0 & \text{if } p|x \\ x \cdot 1^k & \text{otherwise} \end{cases} \\ &\equiv x \pmod{p}\end{aligned}$$

by Fermat's Little Theorem.

Analogously, $y^s \equiv x \pmod{q}$. In other words, both p and q divide $y^s - x$. Since p and q are coprime, $n = pq$ divides $y^s - x$, and hence we have $y^s \equiv x \pmod{n}$. Especially, $x \equiv y^s \pmod{n}$ if $x \in \{0, 1, \ldots, n-1\}$. Thus, every encrypted message is decrypted correctly.

The security certificate of the RSA cryptosystem is based on the assumption that the factorization into prime factors is difficult for large numbers. It is not really known how difficult the factorization problem really is. It is possible that there exists an easy solution to the factorization problem that is not yet known. At the present time we can say that the factorization problem is in the complexity class NP.

Recall that a mathematical problem Π belongs to NP if there exists a polynomial time algorithm which can prove if a general solution is correct or not. The factorization problem for an integer $n \ge 1$ is in NP because it can be checked with the division algorithm if a general divisor is or is not a divisor of n. If the input value is n then we have to make $\mathcal{O}(2^n)$ tests. We now discuss the ElGamal encryption.

23.1.1.4 ElGamal Encryption

The basic scheme for an ElGamal encryption system is the following. Each user chooses a common large prime p and a generator g for the cyclic group $\mathbb{Z}_p^* = \mathbb{Z}_p \setminus \{\bar{0}\}$, the unit group within $\mathbb{Z}_p$. Given a large prime p there is a fixed efficiently invertible procedure to encrypt plaintext into residue classes within $\mathbb{Z}_p^*$. For each message transmission the user's public key is (p, g, A) where $A = g^a$ for some integer a.

The encryption works as follows. Suppose that Bob wants to send a message to Alice. Alice's public key is (p, g, A) as above. The message is m, and as above, is encrypted in some workable efficient manner within $\mathbb{Z}_p^*$, that is, the message is encrypted in a manner known to all users as an integer in $\{0, 1, \ldots, p-1\}$. Bob now randomly chooses an integer b and computes $B = g^b$. He now sends to Alice (B, mC) where $C = g^{ab}$. To decrypt, Alice first uses B to determine the common shared key C. Since $B = g^b$, and she knows $A = g^a$, she knows $C = g^{ab}$ and the modulus p. Hence, she can compute the inverse $g^{(-ab)}$ to obtain the message m.

The security certificate of the ElGamal cryptosystem is based on the difficulty of the Computational Diffie–Hellman problem (CDH) for $\mathbb{Z}_p^*$: given a prime p, a generator g of $\mathbb{Z}_p^*$, g^a modulo p and g^b modulo p, determine g^{ab} modulo p. Certainly, the CDH can be formulated for each cyclic group $G = \langle g \rangle$: the CDH is the problem to find g^{ab} for two elements g^a and g^b. At present, the only known solution of CDH is to solve the discrete logarithm problem (DLP): for $G = \langle g \rangle$ being a cyclic group and $h \in G$, find $a \in \mathbb{Z}$ such that $h = g^a$.

The DLP appears to be very hard for large orders $|G|$ of G. Solving the DLP for $\mathbb{Z}_p^*$ breaks the ElGamal cryptosystem, as does solving the CDH. It is not known whether the CDH can be solved without solving the DLP. The ElGamal encryption becomes the basis for elliptic curve cryptography which we discuss briefly.

23.1.1.5 Elliptic Curve Cryptography

A very powerful approach which has wide ranging applications in cryptography is to use elliptic curves. If K is a finite field of characteristic not equal to 2 or 3 then an elliptic curve over K (in Weierstrass form) is the locus of points $(x, y) \in K \times K$ satisfying the equation $y^2 = x^3 + ax + b$ with $4a^3 + 27b^2 \neq 0$. We denote by $\mathcal{O}$ a single point at infinity and let

$$E(K) = \{(x, y) \in K \times K : y^2 = x^3 + ax + b\} \cup \{\mathcal{O}\}.$$

The important thing about elliptic curves from the viewpoint of cryptography is that a group structure can be placed on $E(K)$. In particular, we define the operation $+$ on $E(K)$ by:

1. $\mathcal{O} + P = P$ for any point $P \in E(K)$.
2. If $P = (x, y)$, then $-P = (x, -y)$ and $-\mathcal{O} = \mathcal{O}$.
3. $P + (-P) = \mathcal{O}$ for any point $P \in E(K)$.
4. If $P_1 = (x_1, y_1)$ and $P_2 = (x_2, y_2)$ such that $P_1 \neq -P_2$, then $P_1 + P_2 = (x_3, y_3)$ with $x_3 = m^2 - (x_1 + x_2)$ and $y_3 = -m(x_3 - x_1) - y_1$ where $m = \frac{y_2 - y_1}{x_2 - x_1}$ if $x_2 \neq x_1$ and $m = \frac{3x_1^2 + a}{2y_1}$ if $x_2 = x_1$.

This operation has a very nice geometric interpretation if $K = \mathbb{R}$. It is known as the *chord and tangent method.* If $P_1 \neq P_2$ are two points on the curve then the line through P_1 and P_2 intersects the curve at another point P_3. If we reflect P_3 through the x-axis we get $P_1 + P_2$. If $P_1 = P_2$ we take the tangent line at P_1. With this operation $E(K)$ becomes an Abelian group. A very detailed proof can be found in [6]. The structure of the group can be worked out.

Theorem 23.1.5. *If K is a finite field of order p^k then the group $E(K)$ is either cyclic or is isomorphic to $\mathbb{Z}_{m_1} \times \mathbb{Z}_{m_2}$ with $m_1|m_2$ and $m_1|(p^k - 1)$.*

A proof of this result is given in [60].

We now consider the case $K = \mathbb{Z}_p$, $p \geq 5$, and write $\overline{a}$ and $\overline{b}$ instead of a and b in $\mathbb{Z}_p$ for the residue classes. Let $f(x) = x^3 + \overline{a}x + \overline{b}$. We have p elements x in $\mathbb{Z}_p$. If $f(x) = \overline{0}$, then we have exactly one point $(x, \overline{0})$ in $E(\mathbb{Z}_p)$. If $f(x)$ is a nontrivial square in $\mathbb{Z}_p$, especially $f(x)^{\frac{p-1}{2}} = \overline{1}$, then for x there are two points (x, y) and $(x, -y)$ in $E(\mathbb{Z}_p)$. If $f(x)$ is not a square in $\mathbb{Z}_p$, then for x there is no point in $E(\mathbb{Z}_p)$. Finally we have to add 1 for the element $\mathcal{O}$. Hence, $|E(\mathbb{Z}_p)| = 1 + s + 2t$ where s is the number of x with $f(x) = \overline{0}$ and t is the number of nontrivial squares in $\mathbb{Z}_p$.

We now give a version of Hasse's Theorem for $\mathbb{Z}_p$, $p \geq 5$.

Theorem 23.1.6 (Hasse's Theorem). *Let $I = [p+1-2\sqrt{p}, p+1+2\sqrt{p}] \cap \mathbb{N}$. Then there exists for each $k \in I$ at least one elliptic curve with $|E(\mathbb{Z}_p)| = k$.*

A proof is given in the book [61].

In [66] there are described efficient probabilistic algorithms to calculate points on $E(\mathbb{Z}_p) \setminus \{0\}$ and to construct an injective, efficiently invertible map $\mathcal{M} \to E(\mathbb{E}_p) \setminus \{0\}$, $p \geq 5$ prime, where $\mathcal{M}$ is the set of plain text units. Using these, we may describe the elliptic curve public key system for $E(\mathbb{Z}_p)$ as follows:

(1) Choose a large prime $p \geq 5$ and $a, b \in \mathbb{Z}_p$ such that $y^2 = x^3 + ax + b$ is an elliptic curve.
(2) Choose an injective efficiently invertible (on the image) map $\rho: \mathcal{M} \to E(\mathbb{Z}_p) \setminus \{0\}$, where $\mathcal{M}$ is the set of plain text units.
(3) Choose a point $P \in E(\mathbb{Z}_p) \setminus \{0\}$.
(4) Choose a secret integer $d \in \mathbb{Z}$ and calculate $dP \in E(\mathbb{Z}_p)$.

The public key is (P, dP) and the elliptic curve itself. The secret key is d.

For encryption, let $m \in \mathcal{M}$ be a plain text message unit. Calculate $Q = \rho(m)$. Choose a random integer k and define $c = (kP, Q + k(dP)) \in \mathcal{C}$, where $\mathcal{C}$ is the set of cipher text units. This is the encrypted message unit.

For decryption, let $C = (c_1, c_2) \in \mathcal{C}$ be a ciphertext unit. Calculate $Q = c_2 - dc_1$ and $m = \rho^{-1}(Q)$, the preimage of Q. Recall that $Q \in E(\mathbb{Z}_p) \setminus \{0\}$ if $Q = \rho(m)$ and $(c_1, c_2) = (kP, Q+k(dP))$. The elliptic curve public key cryptosystem provides a valid cryptosystem: if $(c_1, c_2) = (kP, Q+k(dP))$, then $c_2 - dc_1 = Q = \rho(m)$. The security certificate of the elliptic curve public key cryptosystem is also based on the difficulty of the Computational Diffie–Hellman problem for $E(\mathbb{Z}_p)$. For this, care should be taken that the discrete logarithm problem in $E(\mathbb{Z}_p)$ is difficult. Elliptic curve public key cryptosystems are at present the most important commutative alternatives to the use of the RSA algorithm. There are several reasons for that. They are more efficient in many cases than RSA and keys in elliptic curve systems are much smaller than keys in RSA.

23.1.2 Cryptographic Protocols

Besides secure confidential message transmission there are many other tasks that are important in cryptography, both symmetric key and public key. Although it is not entirely precise, we say that a cryptographic task is where one or more people must communicate with some degree of secrecy. The set of algorithms and procedures needed to accomplish a cryptographic task is called *cryptographic protocol*. A cryptosystem is just one type of a cryptographic protocol. More formally, suppose that several parties want to manage a cryptographic task. Then they must communicate with each other and cooperate. Hence, each party must follow certain rules and implement a certain algorithm that they agreed upon.

We now discuss some cryptographic tasks that we will occasionally refer to in this book but many more can be found in detail in the book [66].

23.1.2.1 Secret Sharing

Given a secret S, a (t,n)-secret sharing threshold scheme is a cryptographical primitive in which a secret is split into pieces (shares) and distributed among a collection of n participants $\{p_1,\dots,p_n\}$ so that any group of t or more participants, with $t \leq n$, can recover the secret. Meanwhile, any group of $t-1$ or fewer participants cannot recover the secret. Shamir solved the secret sharing problem in a very simple but beautiful manner using polynomial interpolation.

The general idea in a Shamir (t,n)-secret sharing threshold scheme is the following. Let K be any field and $(x_1,y_1),\dots,(x_n,y_n)$ be n points in K^2 with pairwise distinct x_i. A polynomial $p(x)$ over K interpolates these points if $p(x_i) = y_i$ for $i = 1,\dots,n$. The polynomial $p(x)$ is called the *interpolating polynomial* for the given points. The crucial theoretical result is that for any n points $(x_1,y_1),\dots,(x_n,y_n)$ with distinct x_i there always exists a unique interpolating polynomial of degree $\leq n-1$.

We now present a more explicit version of the Shamir scheme using the finite field $K = GF(q)$ where $q = p^k$ with $k \geq 1$ and p is a large prime. Let S be the secret. The dealer generates a polynomial $p(x)$ of degree at most $t-1$ over K where q is much larger than n as follows:

$$p(x) = a_0 + a_1x + \cdots + a_{t-1}x^{t-1}$$

where $a_0 = S$ is the secret and $a_1, \dots, a_{t-1} \in K$. The dealer chooses pairwise distinct $x_i \in K \setminus \{0\}$, $i = 1, \dots, n$, which are stored in a public area. The dealer calculates $y_i = p(x_i)$, $i = 1, \dots, n$, and distributes to the n participants via a secure channel so that each participant p_i gets one share y_i.

For the secret recovery we use the Lagrange interpolation. We can construct the Lagrange interpolating polynomial with respect to $(x_1, y_1), \dots, (x_n, y_n)$, all $x_i \in K \setminus \{0\}$ pairwise distinct, as

$$p(x) = \sum_{i=1}^{t} y_i l_i(x)$$

where $l_i(x) = \prod_{j=1, j \neq i} \frac{x_i - x_j}{x_i - x_j}$. Clearly, $p(x)$ is a polynomial of degree at most $t-1$. In particular, the secret a_0 will be

$$a_0 = p(0) = \sum_{i=1}^{t} y_i \prod_{j=i, j \neq i}^{t} \frac{-x_j}{x_i - x_j}.$$

This scheme is perfect in the sense that for $t-1$ participants any secret $S \in K$ is equally likely.

We now describe a geometric alternative scheme which depends on the closest vector theorem. Let W be a real inner product space and V be a subspace of finite dimension t. Suppose that $\vec{w} \in W$ and $\{\vec{e}_1, \dots, \vec{e}_t\}$ is an orthonormal basis of V. Note that, given any basis for the subspace V, the Gram–Schmidt orthonormalization procedure can be used to find an orthonormal basis for V. Suppose that $\vec{w} \in W$ is not in V. Then the unique vector $\vec{w}^* \in V$ closest to $\vec{w}$ is given by

$$\vec{w}^* = \langle \vec{w}, \vec{e}_1 \rangle \vec{e}_1 + \dots + \langle \vec{w}, \vec{e}_t \rangle \vec{e}_t$$

where $\langle \, , \, \rangle$ is the inner product on W.

We now describe the secret sharing scheme. We start with an inner product space W of dimension m and an access control group of size n. We assume that m is much greater than n. Within W there is a hidden subspace V of dimension $t < n$. The secret to be shared is given as an element in this hidden subspace, that is, the secret $\vec{v} \in V$, a vector in V. The dealer distributes two vectors $\vec{v}_i$ and $\vec{w}$ where $\vec{v}_i \in V$ and $\vec{w}$ is a vector in $W \setminus V$, and let $\vec{v} \in V$ be the vector closest to $\vec{w}$.

In general, the vector $\vec{w}$ can be given publically. The set $\{\vec{v}_1, \dots, \vec{v}_n\}$ has the property that any subset of size t is linearly independent. Hence, any subset of size t determines a basis for V. Suppose t valid users get together. They can determine an orthonormal basis of V. Since $\vec{w}$ is given, they can determine $\vec{v}$ by the closest vector theorem and recover the secret. Given a subset of size less than t, the given vectors generate a subspace of V of dimension less than t and hence in W there are infinitely many extensions of subspaces of dimension t. This implies that determining V with less than t elements of a basis has zero probability.

23.1.2.2 Key Exchange and Key Transport

In a key exchange two people, usually called Bob and Alice, exchange a secret shared key to be used in some encryption. In a key transport one party transports to another a secret key that is to be used. We briefly describe the Diffie–Hellman key exchange protocol.

Bob and Alice choose a large prime p and a generator g of the cyclic multiplicative group $\mathbb{Z}_p^*$. The element g is public to all. Alice chooses an a with $1 < a < p-1$. Her public information or public key is g^a given modulo q. This is open to all. Her private information or (secret) private key is a. Bob chooses a b with $1 < b < p-1$. His public information or public key is g^b given modulo p. This is open to all. His private information or (secret) private key is g^b.

Communication: The secret sharing key is g^{ab}. This can be computed easily by both Bob and Alice using their private keys. Alice knows her private key a and the value g^a is public from Bob. Hence, she can compute $g^{ab} = (g^b)^a$. The analogous situation holds for Bob. The security certificate of the Diffie–Hellman key exchange protocol is again the Computational Diffie–Hellman problem for $\mathbb{Z}_p^*$.

23.1.2.3 Authentication Protocols and Zero-Knowledge Proof Protocols

There are two more important cryptographic protocols which are discussed in detail in [66] and also to some extent in Chapter 24 on noncommutative group based cryptography: the authentication protocols and the zero-knowledge proof protocols.

When a confidential message is transmitted there are several aspects that must be verified. First, there must be a verification to the receiver that the sender is who he claims to be. Secondly, there must be a verification to the sender that the receiver is also who he claims to be. Next there should be a verification that the message has not been altered in any way. Finally, there should be in many message transmissions some form of undeniability, that is a procedure that makes it impossible for the sender that he did not send the message. All of these verifications are handled by an *authentication protocol*. In Section 24.5 we discuss a password-authentication protocol using combinatorial group theory.

Now, a *zero-knowledge proof protocol* is a method by which one party (the prover) can prove to another party (the verifier) that a given statement is true while the prover avoids conveying any additional information apart from the fact that the statement is indeed true. The essence of zero-knowledge proofs is that it is trivial to prove that one possesses knowledge of certain information by simply revealing it; the challenge is to prove such possession without revealing the information itself or any additional information. For a classical prototype of a zero-knowledge proof we mention the Ali Baba cave problem with a magic secret door, see [97].

Exercises

1. Let $F: \mathbb{Z}_{24} \to \mathbb{Z}_{24}$ be given by $x \mapsto \overline{5}x + 3$. Calculate the period length for $\overline{x}_0 = \overline{0}$.
2. We use the standard allocation $A = 01, B = 02, \ldots, Z = 26$. Calculate the plaintext number M for the plaintext message 'Louisa is born on Christmas Day.'
3. Distribute the secret 42 using the Shamir secret sharing scheme evenly among three people such that any two can put together the secret.
4. The company Ruin Invest has two directors, seven department managers, and 87 further employees. A valuable customer file is protected by a secret key. Develop a procedure of the information about the key among the following groups of authorized people:
   (1) both directors,
   (2) one director and all seven department managers together, and
   (3) one director, at least four department managers, and also at least 11 employees.
5. Given are prime numbers p and q with $q < p$ and $n = pq$. For an RSA cryptosystem assume that $p - q$ is very small. Show that n can be factorized using the following procedure:
   (1) Let $t \in \mathbb{N}$ be the smallest number with $t \geq \sqrt{n}$.
   (2) If $t^2 - n$ is a square, that is, $t^2 - n = s^2$ for some $s \in \mathbb{N}$, then $p = t + s$ and $q = t - s$ provides the factorization.
   (3) Otherwise take the next integer $t \geq \sqrt{n}$ and go back to (2).
   Use the procedure to factorize $n = 9898828507$.
6. Let $(n, e) = (2047, 179)$ be the public RSA key. A plaintext alphabet has the 26 letters $A, B, \ldots, Z$ and the empty sign $\emptyset$ between words. The plaintext message c with $\emptyset$ between words will be subdivided into double blocks with $\emptyset$ at the end, if necessary. By the assignment $A \mapsto 00, B \mapsto 01, \ldots, Z \mapsto 25, \emptyset \mapsto 26$ each double block gives a block with 4 digits. We consider the four digit numbers as residue classes modulo 2047. Encryption with the public key $(2047, 179)$ gives the ciphertext message $\overline{1054}$, $\overline{92}$, $\overline{1141}$, $\overline{1571}$, $\overline{92}$, $\overline{832}$ in the form of residue classes modulo 2047.
   (a) Break the encryption by factoring 2047.
   (b) Why is the number 2047 besides the small size, a particularly unfavorable choice?
   It is possible to break the encryption without factoring 2047?
7. Alice and Bob agree on the following public key cryptosystem:
   (1) Alice chooses $a, b \in \mathbb{Z}$ with $ab \neq 1$ and calculates $M = ab - 1$. Then Alice chooses two integers a', b' and calculates $e = a'M + a$ and $d = b'M + b$. She then calculates $n = \frac{ed-1}{M}$.
   (2) Alice publishes the pair (n, e). The secret key is d.
   (3) Bob wants to send a message $m \in \{0, 1, \ldots, n-1\}$ to Alice.
   He calculates $c \equiv em \pmod{n}$ and sends c to Alice.
   (4) She decrypts the message by calculating cd modulo n.
   Show that this is a valid cryptosystem, that is, Alice gets the message.

8. Show that breaking the ElGamal encryption scheme and breaking the Diffie–Hellman key exchange protocol are equally difficult.
9. (a) Let $K = \mathbb{Z}_5$ and $y^2 = x^3 + x$. This equation defines an elliptic curve over $\mathbb{Z}_5$. Show that $E(\mathbb{Z}_5) \cong \mathbb{Z}_2 \times \mathbb{Z}_2$.
   (b) Let $K = \mathbb{Z}_{11}$ and $y^2 = x^3 + x + \overline{6}$ be a curve over $\mathbb{Z}_{11}$. Show that $y^2 = x^3 + x + \overline{6}$ is an elliptic curve over $\mathbb{Z}_{11}$ and that $E(\mathbb{Z}_{11})$ is cyclic of order 13.
10. Determine all possible groups $E(\mathbb{Z}_5)$ for elliptic curves over $\mathbb{Z}_5$. Give all possible orders for a group $E(\mathbb{Z}_5)$.

24 Non-Commutative Group Based Cryptography

24.1 Group Based Methods

The public key cryptosystems and public key exchange protocols that we have discussed, such as the RSA algorithm, or the Diffie–Hellman, ElGamal and elliptic curve methods, are number theory based, and thus depend on the structure of Abelian groups. As computing machinery has gotten stronger, and computational techniques have become more sophisticated and improved, there have been successful attacks on both RSA and Diffie–Hellman for smaller and specialized parameters (RSA and Diffie–Hellman moduli). Furthermore, there exist quantum algorithms that specifically break both RSA and Diffie–Hellman. As a consequence, when and if a workable quantum computer will be realized, these cryptographic methods will have to be altered.

Because of these attacks there is a feeling that these number theoretic techniques are theoretically susceptible to attack. Somehow the relatively simple structure of Abelian groups opens up the possibility of weaknesses in cryptographic protocols. As a result there has been an active line of research to develop cryptosystems and key exchange protocols using noncommutative cryptographic platforms which is called *noncommutative algebraic cryptography*. Since most of the cryptographic platforms are groups this is also known as *group based cryptography*.

The main sources for non-Abelian groups are combinatorial group theory and linear group theory, that is matrix groups. Braid group cryptography where encryption is done within the classical braid groups, is one prominent example. The one-way functions in braid group systems are based on the difficulty of solving group theoretic decision problems such as the conjugacy problem and conjugator search problem. Recall that a one-way function is a function which is easy to implement but very hard to invert. Although braid group cryptography had initial spectacular success, various potential attacks have been identified. Borovik, Myasnikov, Shpilrain, see [70], and others have studied the statistical aspects of these attacks and have identified what is termed *black holes* in the platform groups, the outsides of which present cryptographic problems.

The extension of the cryptographic ideas to noncommutative platforms involves the following ideas:

1. general algebraic techniques for developing cryptosystems;
2. potential algebraic platforms (specific groups, rings, etc.) for implementing the techniques; and
3. cryptanalysis and security analysis of the resulting systems.

The basic idea in using combinatorial group theory for cryptography is that elements of groups can be expressed as words in some alphabet. If there is an easy method to rewrite group elements in terms of these words, and further the technique used in this rewriting process can be supplied by a secret key, then a cryptosystem can be created.

https://doi.org/10.1515/9783111142524-024

In Section 14.7 we discussed group presentations and fundamental group decision problems. Given a group G there exists a presentation $G = \langle X; R\rangle$ and vice versa. We recall that the three fundamental group decision problems by Dehn, that is, the word problem, the conjugacy problem, and the isomorphism problem, have negative answers in general but have simple and elegant solutions for finitely generated free groups.

These three problems are only the basic decision problems and other algorithmic problems concerning presentations can be considered. The conjugacy problem asks to algorithmically determine if two elements given in terms of the generators are conjugate. The *conjugator search problem* asks: given a group presentation for G and two elements g_1, g_2 in G that are known to be conjugate, to determine algorithmically a conjugator, that is an element h such that $h^{-1}g_1h = g_2$. It is known, as with the conjugacy problem itself, that the conjugator search problem is undecidable in general.

There are several other group theoretical decision problems. We just mention two. For a subgroup H of a group G, where H has generating set $\{x_1, \ldots, x_n\} \subset H$, the *membership problem* asks whether a given element $g \in G$ lies in H, and the *constructional membership problem* asks whether a given element $g \in G$ lies in H, and if so, how to express g as a word in the generators $x_1, \ldots, x_n$. Michailova, see [38], showed that in general the constructional membership problem is undecidable for infinite matrix groups, also see [39].

The second is the *root extraction problem* in a group G. Given an element $g \in G$, and a number $k \in \mathbb{N}$, find an $h \in G$ such that $h^k = g$. Many cryptosystems such as authentication schemes and digital signatures are based on the root extraction problem. We mention that the root extraction problem is solvable in free groups.

The computational difficulty of solving various group decision problems will play the role of a hard problem used to construct a one-way function in several non-Abelian group based cryptosystems.

The book [93] by Myasnikov, Shpilrain and Ushakov has discussions of the complexity of many of these group decision problems.

If a cryptographic protocol is based on an algebraic object, e. g., group, ring, lattice, or finite field, then this object is called the *(cryptographic) platform*. In group based cryptography this is then a *platform group* for the cryptographic protocol. The security of the cryptographic protocol is then dependent upon the difficulty, computational or theoretic, of solving a group theoretic problem within the platform group.

To be a reasonable platform group for a group based cryptographic protocol, a group G must possess certain properties that make the protocol both efficient to implement and secure.

We assume that the group G has a finite presentation

$$G = \langle X; R\rangle = \langle x_1, \ldots, x_n; r_1 = \cdots = r_m = 1\rangle$$

and that the protocol security is based on a group theoretic problem that we denote by $\mathcal{P}$. The first necessity is that there is an efficient way to uniquely represent and then multiply the elements of G. In most cases this requires a *normal form* for elements $g \in G$, that is, a unique representation in terms of the generators $\{x_1, \ldots, x_n\}$. In particular, reduced words provide normal forms for elements of free groups. Normal forms provide an effective method of disguising group elements. Without this, one can determine a secret key simply by inspection of group elements. The existence of a normal form in a group implies solvable word problem, which is also essential for these protocols. For $g \in G$ we will denote its normal form, in terms of the set of generators X, by $NF_X(g)$.

To be useful in cryptography, given $g \in G$, expressed as a word in $x_1, \ldots, x_n$, the process of moving between the word and the unique normal form must be efficiently computable. Usually we require at most polynomial time in the input length of g.

In addition to the platform group having normal forms, ideally, it would also exhibit exponential growth. That is, the growth function for G, $\gamma : \mathbb{N} \to \mathbb{R}$, defined by $\gamma(n) = \#\{w \in G : l(w) \leq n\}$, has an exponential growth rate, also see [93]. In the definition $l(W)$ stands for the minimal number of letters needed to express W as a word in $x_1, \ldots, x_n$. Exponential growth is a necessity that ensures that the group will provide a large key space.

Further, the normal form must exhibit *good diffusion* in determining the normal forms of products. This means that in finding the normal forms of products it is computationally difficult to rediscover the factors, that is if we know $NF_X(g_1g_2)$ it is computationally difficult to discover g_1, g_2 or $NF_X(g_1)$, $NF_X(g_2)$.

Other necessities for a platform group depend on the particular protocol. If the security is based on the group problem $\mathcal{P}$, such as the word problem or conjugacy problem, we have to assume that in G, the solution to $\mathcal{P}$ is computationally hard (NP-hard) or unsolvable. However, what we really want is *generically hard*, that is, hard on most inputs. The solution to $\mathcal{P}$ might be unsolvable but have polynomial average case complexity. In this case, if care is not taken in choosing the inputs, the solution to $\mathcal{P}$ is easy and the cryptographic protocol is broken. This does not eliminate a group G as a possible platform group but indicates that one must take great care in choosing cryptographic inputs.

Among the first attempts to use non-Abelian groups as platforms for public key cryptosystems were the schemes [62] by Anshel, Anshel and Goldfeld, and the schemes [85] by Ko, Lee et al. The first protocol was developed by I. Anshel, M. Anshel and D. Goldfeld. The original version of the Ko–Lee protocol was published by K. H. Ko, S. J. Lee, J. H. Han, J. Kang and C. Park. We will refer to the second protocol as Ko–Lee. Both sets of authors, at about the same time, proposed using non-Abelian groups and combinatorial group theory for public key exchange.

The Anshel–Anshel–Goldfeld and Ko–Lee methods can be considered as group theoretic analogs of the number theory based Diffie–Hellman method. The basic underlying idea is the following. If G is a group and $g, h \in G$ we let g^h denote the conjugate of g by h, that is $g^h = h^{-1}gh$. The simple observation is that $(g^{h_1})^{h_2} = g^{h_1h_2}$. Therefore writing conjugation in this exponential manner behaves like ordinary exponentiation. From this straightforward idea one can almost exactly mimic the Diffie–Hellman protocol, now within a non-Abelian group.

In Section 24.8, we examine the Ko–Lee and Anshel–Anshel–Goldfeld protocols. Both sets of developers originally suggested using braid groups as the basic and most appropriate group theoretic platform. Here, we just give a presentation for the braid group B_n, $n \geq 3$, in the form

$$B_n = \langle \sigma_1, \dots, \sigma_{n-1}; [\sigma_i, \sigma_j] = 1 \text{ if } |i-j| > 1, \sigma_{i+1}\sigma_i\sigma_{i+1} = \sigma_i\sigma_{i+1}\sigma_i \text{ for } i = 1, \dots, n-1\rangle$$

which is now called the Artin presentation. We remark that there are several possibilities for normal forms for elements of B_n, see [24].

We describe both protocols in a most general context, that is, with a general platform group. This platform group must have a finite presentation with efficiently computable normal forms, exponential growth, and good diffusion in determining the normal form of products. For the following Ko–Lee protocol and the Anshel–Anshel–Goldfeld protocols, the platform group must also contain an abundant collection of subgroups that commute elementwise and that can be efficiently described.

24.2 Initial Group Theoretic Cryptosystems—The Magnus Method

One of the earliest descriptions of using a non-Abelian group in cryptography appeared in a paper by Magnus in the early 1970's, see [89]. This was what is now called a *free group cryptosystem*. The seminal idea of using the difficulty of group theory decision problems in infinite non-Abelian groups as one-way functions in cryptography was first developed by Magyarik and Wagner in 1985. Neither of these two methods proved successful as workable encryption methods yet their introduction ushered in a subsequent complete theory and other ideas. In this section we describe Magnus' idea and in the next subsection the Wagner–Magyarik method.

In [89], Magnus studied rational representations of Fuchsian groups and non-parabolic subgroups of the classical modular group M. Recall that $M = \mathrm{PSL}(2, \mathbb{Z})$. That is, M consists of the 2×2 projective integral matrices

$$M = \left\{ \pm \begin{pmatrix} a & b \\ c & d \end{pmatrix} : ad - bc = 1, a, b, c, d \in \mathbb{Z} \right\}.$$

Equivalently, M can be considered as the set of integral linear fractional transformations with determinant 1:

$$z' = \frac{az+b}{cz+d} \quad \text{with } ad - bc = 1 \text{ and } a, b, c, d \in \mathbb{Z}.$$

Theorem 24.2.1 (B. H. Neumann). *The matrices*

$$\pm \begin{pmatrix} 1 & 1 \\ 1 & 2 \end{pmatrix}, \quad \pm \begin{pmatrix} 1+4t^2 & 2t \\ 2t & 1 \end{pmatrix}, \quad t = 1, 2, 3, \dots,$$

freely generate a free subgroup F of infinite index in M. Further, distinct elements of F have distinct first columns (up to sign). The group F is of infinite rank.

Proof. Without loss of generality we first work in the homogenous modular group

$$\Gamma = \left\{ \begin{pmatrix} a & b \\ c & d \end{pmatrix} : a, b, c, d \in \mathbb{Z}, ad - bc = 1 = \mathrm{SL}(2, \mathbb{Z}) \right\}.$$

B. H. Neumann, see [40], constructed infinitely many subgroups N of Γ with the following properties:

(i) N contains the matrix $T = \left(\begin{smallmatrix} 0 & -1 \\ 1 & 0 \end{smallmatrix}\right)$.

(ii) Let a and c be any pair of coprime integers. Then N contains exactly one matrix in which the first column consists of the ordered pair (a, c).

We remark that Neumann showed that such an N has properties (i) and (ii) if it contains T and has exactly all the elements U^n, $n = 0, \pm 1, \pm 2, \ldots$, as right coset representatives in Γ where $U = \left(\begin{smallmatrix} 1 & 1 \\ 0 & 1 \end{smallmatrix}\right)$.

To prove Theorem 24.2.1 we do not need the whole procedure, also not the additional remark (for the complete construction see [40]). We just pick up the single procedure for the special group given in Theorem 24.2.1. We consider the bijective map $f\colon \mathbb{Z} \to \mathbb{Z}$ given by $f(f(n)) = n$, $f(0) = 0$, $f(-1) = -1$, and for any positive integer k we have $f(2k) = 2k$, $f(6k-1) = -3k-1$, $f(6k-3) = -3k$, $f(6k-5) = 1-3k$.

We define the subgroup N generated by the elements

$$\gamma_n = \begin{pmatrix} n & -1 - nf(n) \\ 1 & -f(n) \end{pmatrix}.$$

We now consider N as a subgroup of the modular group M and use the Reidemeister–Schreier method in combination with Tietze transformations, see Chapter 14. We see that N is generated by the elements

$$\gamma_{-1} \quad \text{and} \quad \gamma_{2k}, \quad k = 1, 2, 3, \ldots$$

with the defining relations

$$\gamma_{-1}^2 = \gamma_{2k}^2, \quad k = 1, 2, 3, \ldots.$$

This shows that the elements $A = \gamma_0^{-1}\gamma_{-1}$ and $B_{2k} = \gamma_{2k}\gamma_0^{-1}$, $k = 1, 2, 3, \ldots$, freely generate a free subgroup F of infinite rank in N using the Reidemeister–Schreier method. This, in fact, also follows if we consider F acting on the upper half plane.

We have $A = \pm\left(\begin{smallmatrix} 1 & 1 \\ 1 & 2 \end{smallmatrix}\right)$ and $B_{2k} = \pm\left(\begin{smallmatrix} 1+4k^2 & 2k \\ 2k & 1 \end{smallmatrix}\right)$, $k = 1, 2, 3, \ldots$. The group F does not contain any power of U^t, $t \in \mathbb{Z} \setminus \{0\}$. In fact, all the elements U^n, $n = 0, \pm 1, \pm 2, \pm 3, \ldots$, are right coset representatives of F in Γ because f is bijective. If $C = \pm\left(\begin{smallmatrix} a & b \\ c & d \end{smallmatrix}\right)$ is any element of F, then $C \neq 0$ because no power of U is in F and the elements $CU^t \in M$, $t \in \mathbb{Z}$, have the same first column as C, up to the sign, and if t runs through the integers, we get all elements of M with the same first column. This we can see as follows.

Let $D = \pm\begin{pmatrix} a & g \\ c & h \end{pmatrix}$ be any element of M with the same first column. Then

$$1 = ad - bc = ah - gc$$

from the determinant. It follows that $a(d-h) = c(b-g)$. Since $\gcd(a,c) = 1$ we get $c|(d-h)$, that is, there exists a $t \in \mathbb{Z}$ with $ct = d-h$, and therefore $h = d-ct$. We get with this that $ad - bc = a(d-ct) - gc$, that is, $g = b - at$.

Hence, $D = \pm\begin{pmatrix} a & b-at \\ c & d-ct \end{pmatrix}$. Now consider $CU^{-t} \in M$, $t \in \mathbb{Z}$, then

$$CU^{-t} = \pm\begin{pmatrix} a & b \\ c & d \end{pmatrix}\begin{pmatrix} 1 & -t \\ 0 & 1 \end{pmatrix} = \pm\begin{pmatrix} a & b-at \\ c & d-ct \end{pmatrix}.$$

This shows that distinct elements of F have distinct first columns, up to sign. □

Magnus, see [89], had the idea to use this for cryptographic protocols. Since the entries in the generating matrices are positive we can do the following.

Choose a set $T_1, \ldots, T_n$ of projective matrices from the set above with n large enough to encode a desired plaintext alphabet $\mathcal{A}$. Any message would be encoded by a word $w(T_1, \ldots, T_n)$ with nonnegative exponents. This represents an element g of F. The two elements in the first column determine w and therefore g. Receiving w then determines the message uniquely. Pure free cryptography as Magnus proposed is subject to many attacks. We will discuss this further in Section 24.3.

24.2.1 The Wagner–Magyarik Method

The idea of using the difficulty of group theory decision problems in devising hard one-way functions for cryptographic purposes was first developed by Magyarik and Wagner in 1985, see [103]. They devised a public key protocol based on the difficulty of the solution to the word problem. Although this was a seminal idea, their basic cryptosystem was really unworkable and not secure in the form they presented.

Wagner and Magyarik outlined a conceptual public key cryptosystem based on the hardness of the word problem for finitely presented groups. At the same time, they gave a specific example of such a system. Gonzalez Vasco and Steinwandt, see [78], proved that their approach is vulnerable to so-called reaction attacks. In particular, for the proposed instance it is possible to retrieve the private key just by watching the performance of a legitimate recipient.

The general scheme of the Wagner and Magyarik public key cryptosystem is as follows. Let X be a finite set of generators, and let R and S be finite sets of relators on X. Consider the two groups G and G_0 with presentations

$$G = \langle X; R \rangle \quad \text{and} \quad G_0 = \langle X; R \cup S \rangle.$$

The group G_0 is then a homomorphic image of G. We assume first that G has a hard word problem so that the word problem in G is not solvable in polynomial time. We next assume that the homomorphic image G_0 has a word problem solvable in polynomial time, that is an easy word problem.

Choose two words w_0 and w_1 which are not equivalent in G_0 (and hence not equivalent in G since G_0 is a homomorphic image of G). The public key is the presentation $\langle X; R\rangle$ and the chosen words w_0 and w_1. To encrypt a single bit $\in \{0,1\}$, pick w_i and transform it into a ciphertext word w by repeatedly and randomly applying Tietze transformations to the presentation $\langle X; R\rangle$. To decrypt a word w, run the algorithm for the word problem of G_0 in order to decide which of $w_i w^{-1}$ is equivalent to the empty word for the presentation $\langle X; R \cup S\rangle$. The private key is the set S. As pointed out by González Vasco and Steinwandt, this is not sufficient and Wagner and Magyarik are not clear on this point. The public key should be a deterministic polynomial-time algorithm for the word problem of $G_0 = \langle X; R \cup S\rangle$. Just knowing S does not automatically and explicitly give us an efficient algorithm (even if such an algorithm exists).

Although the Wagner–Magyarik protocol was not workable as a public key system, the idea opened the door for using similar types of encryption involving group theoretic decision problems.

24.3 Free Group Cryptosystems

The simplest example of a non-Abelian group based cryptosystem is perhaps a *free group cryptosystem*. This can be described in the following manner.

Consider a free group F on free generators $x_1, \dots, x_r$. Then each element g in F has a unique expression as a reduced word $w(x_1, \dots, x_r)$. Let $w_1, \dots, w_k$, where each $w_i = w_i(x_1, \dots, x_r)$, be a set of words in the generators $x_1, \dots, x_r$ of the free group F. At the most basic level, to construct a cryptosystem, suppose that we have a plaintext alphabet $\mathcal{A}$. For example, suppose $\mathcal{A} = \{a, b, \dots\}$ are the symbols needed to construct meaningful messages in English. To encrypt, use a substitution ciphertext

$$\mathcal{A} \mapsto \{w_1, \dots, w_k\}$$

given by $a \mapsto w_1$, $b \mapsto w_2, \dots$. Then, for a word $w(a, b, \dots)$ in the plaintext alphabet, form the free group word $w(w_1, w_2, \dots)$. This represents an element g in F. Send out g as the secret message.

In order to implement this scheme we need a concrete representation of g and then for decryption a way to rewrite g back in terms of $w_1, \dots, w_k$. This concrete representation is the idea behind *homomorphic cryptosystems*.

The decryption algorithm in a free group cryptosystem then depends on the *Reidemeister–Schreier rewriting process*, see Section 14.4. Let F be a free group on $\{x_1, \dots, x_n\}$. The Reidemeister–Schreier process allows one to construct a set of generators $w_1, \dots, w_k$ for H by using a Schreier transversal. Further, given the Schreier transversal from which the set of generators for H was constructed, the *Reidemeister–Schreier rewriting process* allows us to algorithmically rewrite an element of H. Given such an element expressed as a word $w = w(x_1, \ldots, x_r)$ in the generators of F this algorithm rewrites w as a word $w^\star(w_1, \ldots, w_k)$ in the generators of H.

Pure free group cryptosystems are subject to various attacks and can be broken often easily. However, a public key free group cryptosystem using a free group representation in the modular group was developed by Baumslag, Fine and Xu, see [67] and [68]. The most successful attacks on free group cryptosystems are called *length based attacks*. The general idea in a length based attack is that an attacker multiplies a word in ciphertext by a generator to get a shorter word which then could possibly be decoded. We refer to [76] for more on length based attacks.

Baumslag, Fine and Xu in [67] described the following general encryption scheme using free group cryptography. A further enhancement was discussed in the paper [68].

We start with a finitely presented group

$$G = \langle X; R \rangle,$$

where $X = \{x_1, \ldots, x_n\}$, and a faithful representation

$$\rho : G \mapsto \overline{G}.$$

$\overline{G}$ can be any one of several different kinds of objects; linear group, permutation group, power series ring, etc.

We assume that there is an algorithm to re-express an element of $\rho(G)$ in $\overline{G}$ in terms of the generators of G. That is if $g = w(x_1, \ldots, x_n) \in G$, where w is a word in these generators and we are given $\rho(g) \in \overline{G}$, we can algorithmically find g and its expression as the word $w(x_1, \ldots, x_n)$.

Once we have G, we assume that we have two free subgroups K, H with

$$H \subset K \subset G.$$

We assume that we have fixed Schreier transversals for K in G and for H in K both of which are held in secret by the communicating parties Bob and Alice. Now based on the fixed Schreier transversals we have sets of Schreier generators constructed from the Reidemeister–Schreier process for K and for H:

$$k_1, \ldots, k_m, \ldots \quad \text{for } K$$

and

$$h_1, \ldots, h_t, \ldots \quad \text{for } H.$$

Notice that the generators for K will be given as words in $x_1, \ldots, x_n$, the generators of G, while the generators for H will be given as words in the generators $k_1, k_2, \ldots$ for K.

We note further that H and K may coincide and that H and K need not in general be free but only have a unique set of normal forms so that the representation of an element in terms of the given Schreier generators is unique.

We will encode within H, or more precisely within $\rho(H)$. We assume that the number of generators for H is larger than the set of characters within our plaintext alphabet. Let $\mathcal{A} = \{a, b, c, \ldots\}$ be our plaintext alphabet. At the simplest level we choose a starting point i, within the generators of H, and encode

$$a \mapsto h_i, \quad b \mapsto h_{i+1}, \ldots, \qquad \text{etc.}$$

Suppose that Bob wants to communicate the message $w(a, b, c, \ldots)$ to Alice where w is a word in the plaintext alphabet. Recall that both Bob and Alice know the various Schreier transversals which are kept secret between them. Bob then encodes $w(h_i, h_{i+1}, \ldots)$ and computes the element $w(\rho(h_i), \rho(h_{i+1}), \ldots)$ in $\overline{G}$ which he sends to Alice. This is sent as a matrix if $\overline{G}$ is a linear group or as a permutation if $\overline{G}$ is a permutation group and so on.

Alice uses the algorithm for $\overline{G}$ relative to G to rewrite $w(\rho(h_i), \rho(h_{i+1}), \ldots)$ as a word $w^\star(x_1, \ldots, x_n)$ in the generators of G. She then uses the Schreier transversal for K in G to rewrite using the Reidemeister–Schreier process $w^\star$ as a word $w^{\star\star}(k_1, \ldots, k_s)$ in the generators of K. Since K is free or has unique normal forms this expression for the element of K is unique. Once she has the word written in the generators of K she uses the transversal for H in K to rewrite again, using the Reidemeister–Schreier process, in terms of the generators for H. She then has a word $w^{\star\star\star}(h_i, h_{i+1}, \ldots)$ and using the allocation $h_i \mapsto a$, $h_{i+1} \mapsto b, \ldots$ decodes the message.

In an actual implementation an additional *random noise factor* is added. This is explained in more detail below.

We now describe an implementation of this process using for the base group G the classical modular group $M = \mathrm{PSL}(2, \mathbb{Z})$. Further, this implementation uses a polyalphabetic cipher which is secure. This was introduced originally in [67] and [68].

The system in the modular group M works as follows. A list of finitely generated free subgroups $H_1, \ldots, H_m$ of M is public and presented by their systems of generators (presented as matrices). In a full practical implementation it is assumed that m is large. For each H_i we have a Schreier transversal

$$h_{1,i}, \ldots, h_{t(i),i}$$

and a corresponding ordered set of generators

$$w_{1,i}, \ldots, w_{m(i),i}$$

constructed from the Schreier transversal by the Reidemeister–Schreier process. It is assumed that each $m(i) \gg l$ where l is the size of the plaintext alphabet, that is, each subgroup has many more generators than the size of the plaintext alphabet. Although

Bob and Alice know these subgroups in terms of free group generators what is made public are generating systems given in terms of matrices.

The subgroups on this list and their corresponding Schreier transversals can be chosen in a variety of ways. For example the commutator subgroup of the modular group is free of rank 2 and some of the subgroups H_i can be determined from homomorphisms of this subgroup onto a set of finite groups.

Suppose that Bob wants to send a message to Alice. Bob first chooses three integers (m, q, t) where m is the choice of the subgroup H_m, q is the choice of the starting point among the generators of H_m for the substitution of the plaintext alphabet, and t is the choice of the size of the message unit.

We clarify the meanings of q and t. Once Bob chooses m, to further clarify the meaning of q, he makes the substitution

$$a \mapsto w_{m,q}, \quad b \mapsto w_{m,q+1}, \dots.$$

Again the assumption is that $m(i) \gg l$ so that starting almost anywhere in the sequence of generators of H_m will allow this substitution. The message unit size t is the number of coded letters that Bob will place into each coded integral matrix.

Once Bob has chosen (m, q, t) he takes his plaintext message $w(a, b, \dots)$ and groups blocks of t letters. He then makes the given substitution above to form the corresponding matrices in the modular group:

$$T_1, \dots, T_s.$$

We now introduce a *random noise factor*. After forming $T_1, \dots, T_s$ Bob then multiplies on the right each T_i by a random matrix in M say R_{T_i} (different for each T_i). The only restriction on this random matrix R_{T_i} is that there is no free cancellation in forming the product $T_i R_{T_i}$. This can be easily checked and ensures that the freely reduced form for $T_i R_{T_i}$ is just the concatenation of the expressions for T_i and R_{T_i}. Next he sends Alice the integral key (m, q, t) by some public key method (RSA, Anshel–Goldfeld, etc.). He then sends the message as s random matrices

$$T_1 R_{T_1}, \quad T_2 R_{T_2}, \dots, \quad T_s R_{T_s}.$$

Hence what is actually being sent out are not elements of the chosen subgroup H_m but rather elements of random right cosets of H_m in M. The purpose of sending coset elements is two-fold. The first is to hinder any geometric attack by masking the subgroup. The second is that it makes the resulting words in the modular group generators longer—effectively hindering a brute force attack.

To decode the message Alice first uses public key decryption to obtain the integral keys (m, q, t). She then knows the subgroup H_m, the ciphertext substitution from the generators of H_m and how many letters t each matrix encodes. She next uses the algorithms described in Section 24.2 to express each $T_i R_{T_i}$ in terms of the free group generators of M

say $w_{T_i}(y_1, \ldots, y_n)$. She has knowledge of the Schreier transversal, which is held secretly by Bob and Alice, so now uses the Reidemeister–Schreier rewriting process to start expressing this freely reduced word in terms of the generators of H_m. The Reidemeister–Schreier rewriting is done letter by letter from left to right. Hence when she reaches t of the free generators she stops. Notice that the string that she is rewriting is longer than what she needs to rewrite in order to decode as a result of the random matrix R_{T_i}. This is due to the fact that she is actually rewriting not an element of the subgroup but an element in a right coset. This presents a further difficulty to an attacker. Since these are random right cosets it makes it difficult to pick up statistical patterns in the generators even if more than one message is intercepted. In practice the subgroups should be changed with each message.

The initial key (m, q, t) is changed frequently. Hence as mentioned above this method becomes a type of polyalphabetic cipher which is difficult to decode.

24.4 Non-Abelian Digital Signature Procedure

We present a digital signature procedure based on non-Abelian groups developed by Ko, Lee et al., see [84]. In describing this protocol we must first introduce additional group theoretic decision problems. In Section 14.7 we discussed the three basic group decision problems for a finitely presented group G: the *word problem*, the *conjugacy problem*, and the *isomorphism problem*. Recall that in a finitely presented group G the *conjugacy problem* asks if there exists an algorithm to decide whether or not arbitrary words u and v in the generators of G are conjugate? That is, is there an $x \in G$ such that $x^{-1}ux = v$? To distinguish this from certain other decision problems using conjugacy we call this the *decision conjugacy problem*. For a finitely presented group G the *conjugator search problem* is the following. Given $u, v \in G$ that we know to be conjugate is there an algorithm to find $z \in G$ satisfying $z^{-1}uz = v$?

In the following we use the notation u^z for $z^{-1}uz$.

Let G be a non-Abelian group in which the conjugator search problem is infeasible and the decision conjugacy problem is solvable. Let $\{0,1\}^*$ be the set of all $0, 1$ sequences and let $h : \{0,1\}^* \to G$ be a hash function. Recall that a (cryptographic) hash function is a deterministic function $h: S \to \{0,1\}^n$, which returns for each arbitrary block of data, called a message, a fixed size of bit strings. It should have the property that a change in the data will change the hash value.

An ideal hash function has the following properties:

(i) It is easy to compute the hash value for any given message.
(ii) It is infeasible to find a message that has a given hash value (preimage resistant).
(iii) It is infeasible to modify a message.
(iv) It is infeasible to find two different messages with the same hash (collision resistant).

With these ideas here is the Ko–Lee digital signature scheme.

- *Key Generation:* Alice wants to sign and send a message, m, to Bob. Alice begins by choosing two conjugate elements $u, v \in G$ with conjugator a. The conjugate pair (u, v) is public information while the conjugator a is Alice's secret key.
- *Signature Generation:* Alice chooses arbitrary $b \in G$, and computes $\alpha = u^b$ and $y = h(m\alpha)$. Then a signature σ on the message m is the triple (α, β, γ) where $\beta = y^b$ and $\gamma = y^{a^{-1}b}$. She sends this to Bob for verification and acceptance.
- *Verification:* Upon receiving the signature, Bob checks whether or not the following hold:
  (1) There exists $c_1 \in G$ such that $u = \alpha^{c_1}$.
  (2) There exist $c_2, c_3 \in G$ such that $y = \beta^{c_2}$ and $y = \gamma^{c_3}$.
  (3) There exists $c_4 \in G$ such that $uy = (\alpha\beta)^{c_4}$.
  (4) There exists $c_5 \in G$ such that $vy = (\alpha\gamma)^{c_5}$.
  Bob accepts the signature if and only if conditions (1)–(4) hold.

The security of this scheme lies in the assumption that, given a pair of conjugate elements $u, v \in G$, finding elements α, β, γ such that (1)–(4) above hold is infeasible. If the conjugator a can be found, then $(\alpha, \beta, \gamma) = (u^b, y^b, y^{a^{-1}b})$ satisfy properties (1)–(4) for any $b \in G$. Hence the conjugacy search problem has to be infeasible.

24.5 Password Authentication Using Combinatorial Group Theory

Closely related to digital signatures is the problem of *secure password authentication*. With the increased use of online credit card transactions there is at present more than ever a need for secure password identification. For many online purchases, this is being carried out by a *challenge response system* accompanying the password. In the simplest systems this takes the form of secondary password questions such as the user's mother's maiden name or place of birth. There are inherent difficulties with these types of challenge response systems. First of all there is the trivial problem of the users remembering their responses. More critical is the problem that this type of information for many people is readily available and easily found or guessed by would-be attackers or eavesdroppers.

Challenge response systems are also subject to man-in-the-middle attacks and replay attacks. In this section we present an alternative method for challenge response password authentication using combinatorial group theory. In particular this method depends upon the difficulty of solving the word problem within a given finitely presented group without knowing the presentation and the difficulty of solving systems of equations within free groups. This latter problem has been proved to be NP-hard.

These group theoretic techniques have several major advantages over other challenge response systems. We will call the password presenter, the *prover*, and the password presentee, the *verifier*. The methods we present can be used for *two-way authentication*, that is to both verify the prover to the verifier and to verify the verifier to the prover. To each user in conjunction with a standard password there will be assigned a finitely presented group with a solvable word problem. We call this the *challenge group*. This will be done randomly by the group randomizer system and will be held in secret by the prover and the verifier.

Cryptographically, we assume the adversary can steal the encrypted form of the group theoretic responses. Probabilistically this does not present a problem. Each challenge response set of questions forms a virtual one time key pad as we will explain. Therefore the adversary must steal three things: the original password, the challenge group and the group randomizer. Hence there is almost total security in the challenge response system.

Further there is an infinite supply of finitely presented groups to use as challenge groups and an infinite supply of challenge response questions that never have to be duplicated. We will explain these in the section on this protocol's security. Finally the method is symmetric between the verifier and the prover, so while the verifier verifies the prover's password simultaneously the prover verifies that he or she is dealing with the verifier.

The theoretical security of the system is provided by several results in asymptotic group theory which we discuss in Section 24.6. In particular, a result of Lysenok and Myasnikov, see [91], implies that stealing the challenge group is NP-hard while a result of Jitsukawa, see [81], says that the asymptotic density of using homomorphisms to attack the group randomizer protocol is zero.

The whole password protocol depends upon the *group randomizer system*. This is a computer program that can handle several elementary tasks involving finitely presented groups. The scope of the particular group randomizer system will depend on the type of login protocol or cryptographic protocol desired. At the most basic level the group randomizer system has the ability to do the following things:

1. To recognize a finite presentation of a finitely presented group with a solvable word problem and manipulate arbitrary words in the alphabet of generators according to the rewriting rules of the presentation. In particular, if the group has a normal form for each element, the group randomizer can rewrite an arbitrary word in the generators in terms of its group normal form.
2. Given a finite presentation of a group with a solvable word problem, to recognize whether two free group words have the same value in the given group when considered in terms of the given generators of the group.
3. To randomly generate free group words on an alphabet of any finite size.
4. To recognize and store sets of free group words $w_1, \dots, w_k$ on an alphabet $x_1, \dots, x_n$ and rewrite words $w(w_1, \dots, w_k)$ as the corresponding word in $x_1, \dots, x_n$.
5. Given a free group of finite rank on $x_1, \dots, x_n$ and a set of words $w_1, \dots, w_k$ on an alphabet $x_1, \dots, x_n$, to solve the membership problem in F relative to $H = \langle w_1, \dots, w_k \rangle$, the subgroup of F generated by $w_1, \dots, w_k$.
6. Given a stored finitely presented group or a stored set of free group words, the randomizer can accept a random free group word and rewrite it as a normal form in

the finitely presented group in the former case or as a word in the ambient free group in the latter case.

We now present several variations on secure password authentication using the group randomizer. First we give an overall outline of the protocol.

24.5.1 General Outline of the Authentication Protocol

This is a symmetric key cryptographic authentication protocol. Both the prover and verifier use a single private key to both encrypt and decrypt within the authentication process. At the first step the prover and verifier must communicate directly, either face-to-face or by a public key method, to set the private shared secret. This is the model now used for most password/password back-up schemes. We assume that both the prover and verifier have a group randomizer system. For security analysis we assume that an adversary or eavesdropper has access to the encrypted form of the transmission but is passive in that the adversary will not change any transmissions.

1. The prover and verifier communicate directly to set up a common shared secret (P, G) where P is a standard password and G is a challenge group. Each prover's challenge group is unique to that prover. The challenge group is a finitely presented group with a solvable word problem and satisfying the strong generic free group property which we discuss in Section 24.6. The password is chosen by the prover while the challenge group is randomly chosen by the group randomizer system.
2. The prover presents the password to the verifier. The group randomizer of the verifier presents a group theoretic "question" concerning the challenge group G to the prover. The assumption is that this "question" is difficult in the sense that it is infeasible to answer it if the group G is unknown. The question is then answered by the group randomizer. This is repeated a finite number of times. If the answers are correct, the prover (and the password) is verified.
3. The protocol is then repeated from the viewpoint of the prover, authenticating the verifier to the prover.

24.5.2 Free Subgroup Method

We assume that both the prover and the verifier has a group randomizer. Each prover has a standard password. Suppose that F is a free group on $\{x_1, \dots, x_n\}$. The prover's password is linked to a finitely generated subgroup of a free group given as words in the generators, that is, the prover's password is linked to $w_1, \dots, w_k$ where each w_i is a word in $x_1, \dots, x_n$. The group $G = \langle w_1, \dots, w_k \rangle$ is called the *challenge group*. In general we have $k \neq n$. The prover does not need to know the generators. The randomizer can randomly choose words from this subgroup and then freely reduce them. The prover has the challenge group or subgroup also stored in its randomizer.

The prover submits his or her standard password to the prover. This activates the verifier's randomizer to the prover's set of words. The verifier now submits a random free group word on $y_1, \ldots, y_k$ to the prover's randomizer say $w(y_1, \ldots, y_k)$. The prover's randomizer treats this as $w(w_1, \ldots, w_k)$ and then reduces it in terms of the free group generators $x_1, \ldots, x_n$ and rewrites it as $w^{\star}(x_1, \ldots, x_n)$. The verifier checks that this is correct, that is, $w(w_1, \ldots, w_k) = w^{\star}(x_1, \ldots, x_n)$ on the free group on $x_1, \ldots, x_n$. If it is, the verifier continues and does this three (or some other finite number) of times. There is one proviso. The verifier submits a word to the prover only once, so that a submitted word can never be reused. The prover's randomizer will recognize if it has (this is a verification to the prover of the verifier).

To verify that the verifier is legitimate, the process is repeated from the prover's randomizer to the verifier.

An attacker only has access to the transmitted words. Given a series of free group words there is essentially zero probability of reconstructing the subgroup. To prevent an attacker using an already used word to gain access, the group randomizer system allows a free group word, submitted as a challenge word, to be used only once. If an attacker gets access to the verifier and submits an already submitted word or vice versa from the prover, this will red flag the attempt. We also suggest that if there is a previously used word, indicating perhaps an attack, the group randomizer should change the prover's group. The beauty of this system is that this can be done extremely easily; change several of the words for example. Essentially this presents an essential one-time key pad each time the prover presents the password. The map $y_i \to w_i$ is a homomorphism and an attacker can manipulate various equations in an attempt to solve. Presumably, if there are enough equations, the words $w_1, \ldots, w_k$ can be discovered. However, in Section 24.6 we present a security proof based on several results in asymptotic group theory showing that this cannot happen with asymptotic density one.

We suggest a noise/diffusion enhancement. The provers challenge group generator words $w_1, \ldots, w_k$ are indexed. With each use the randomizer applies a random permutation ϕ on $\{1, \ldots, k\}$ to scramble the indices. These permutations are coded and stored both in the prover's randomizer and the verifier's one. This prevents a length based attack by an eavesdropper since discovering, for example, what w_{37} is, is of no use since it will be indexed differently for the next use. The coded permutation is sent as part of the challenge.

24.5.3 General Finitely Presented Group Method

This is essentially the same method, however, rather than working with an ambient free group we work with a given finitely presented group with a solvable word problem. Let $G = \langle X; R \rangle$ be the group. As before we assume that both the prover and the verifier has a group randomizer. Each prover has a standard password. Suppose that $X = \{x_1, \ldots, x_n\}$

and F is a free group on $\{x_1, \ldots, x_n\}$. The prover's password is linked to a finitely generated subgroup of G again given as words in the generators X, that is, the prover's password is linked to $w_1, \ldots, w_k$ where each w_i is a word in $x_1, \ldots, x_n$. As before, we let $k \neq n$. The randomizer can randomly choose words from this subgroup and then reduce them via the finite presentation. The verifier has the group and subgroup also stored in its randomizer.

The remainder of the procedure is exactly the same as in the free group case. The prover submits his or her standard password to the verifier. This activates the verifier's randomizer to the prover's set of words. The verifier now submits a random free group word on $y_1, \ldots, y_k$ to the prover's randomizer, say, $w(y_1, \ldots, y_k)$. The prover's randomizer treats this as $w(w_1, \ldots, w_k)$ and rewrites it as $w^\star(x_1, \ldots, x_n)$. The verifier checks that this is correct, that is, $w(w_1, \ldots, w_k) = w^\star(x_1, \ldots, x_n)$, however, this time in the group G. If it is, the verifier continues and does this three (or some other finite number) of times. There is one proviso. The verifier submits a word to the prover only once so that a submitted word can never be reused. The prover's randomizer will recognize if it has (this is a verification to the prover of the verifier).

To verify that the verifier is legitimate, the process is repeated from the prover's randomizer to the verifier.

As in the free group method, an attacker only has access to the transmitted words. Given a series of group words there is zero probability of reconstructing the group, however, as in the free group method a given challenge response word is to be used only once.

24.6 The Strong Generic Free Group Property

Part of the theoretical security of the group randomizer protocols depends on the *strong generic free group property* and *asymptotic density*. Asymptotic density is a general method to compute densities and/or probabilities on infinite discrete sets where each individual outcome is tacitly assumed to be equally likely. The origin of asymptotic density lie in the attempt to compute probabilities on the whole set of integers where each integer is considered equally likely. The method can also be used where some probability distribution is assumed on the elements. It has been effectively applied to determining densities within infinite finitely generated groups where random elements are considered as being generated from random walks on the Cayley graph of the group. The paper [70] by Borovik, Myasnikov and Shpilrain provides a good general description of the probability method in group theory. Let $\mathcal{P}$ be a group property and let G be a finitely generated group. We want to determine the measure of the set of elements which satisfy $\mathcal{P}$. For each positive integer n let B_n denote the n-ball in G. Let $|B_n|$ denote the actual size of B_n (which is an integer since G is finitely generated) or the measure of $|B_n|$ if a distribution has been placed on the elements of G. Let S be the set of elements in G satisfying $\mathcal{P}$. The asymptotic density of S is then

$$\lim_{n\to\infty} \frac{|S \cap B_n|}{|B_n|}$$

provided this limit exists. We say that the property $\mathcal{P}$ is *generic* if the asymptotic density of the set S of elements satisfying $\mathcal{P}$ equals 1.

This concept can be easily extended to properties of finitely generated subgroups. We consider the asymptotic density of finite sets of elements that generate subgroups that have a considered property. For example, to say that a group has the generic free group property we mean that

$$\lim_{m,n\to\infty} \frac{|S_m \cap B_{m,n}|}{|B_{m,n}|} = 1$$

where S_m is the collection of finite sets of elements of size m that generate a free subgroup while $B_{m,n}$ are all the m-element subsets within the n-ball. We refer to the paper [70] and the book [93] for terminology and further definitions.

We say that a group G has the *generic free group property* if a finitely generated subgroup is generically a free group. For example, a result by Epstein, see [25], says that the group $\mathrm{GL}(n, \mathbb{R})$ satisfies the generic free group property. A group G has the *strong generic free group property* if given randomly chosen elements $g_1, \ldots, g_n$ in G then generically they are a free basis for the free subgroup they generate. Jitsukawa, see [81], proved that free groups have the strong generic free group property. That is, given k random elements $w_1, \ldots, w_k$ in the free group on $y_1, \ldots, y_n$, then with asymptotic density one the elements $w_1, \ldots, w_k$ are a free basis for the subgroup they generate. We compare this with the Nielsen–Schreier theorem that says that $w_1, \ldots, w_k$ generate a free group. In the context of the group randomizer protocols, the strong generic free group property implies that if $v_1(y_1, \ldots, y_m), \ldots, v_k(y_1, \ldots, y_m)$ have already been presented as challenge words then the probability is approximately zero that a new challenge word $v(y_1, \ldots, y_m)$ lies in the subgroup generated by $v_1, \ldots, v_k$, and hence a homomorphism attack is nullified.

The strong generic free group property has been extended to many classes of groups including surface groups by Fine, Myasnikov and Rosenberger, see [29]. Let us mention some further results. Gilman, Myasnikov and Osin, see [77], showed that torsion-free hyperbolic groups have the generic free group property. Myasnikov and Ushakov, see [94], showed that pure braid groups P_n with $n \geq 3$ also have the strong generic free group property. We will show that all Fuchsian groups of finite co-volume and all braid groups B_n with $n \geq 3$ have the strong generic free group property.

Extremely useful in proving that a group has the generic or strong generic free group property is the following, see Exercise 6.

Theorem 24.6.1. *Let G be a group and N a normal subgroup. If the quotient G/N satisfies the strong generic free group property then G also satisfies the strong generic free group property.*

Corollary 24.6.2. *Any orientable surface group*

$$\left\langle a_1,\ldots,a_g,b_1,\ldots,b_g;\prod_{i=1}^{g}[a_i,b_i]=1\right\rangle$$

of genus $g \geq 2$ and any nonorientable surface group

$$\langle a_1,\ldots,a_g;a_1^2\cdots a_g^2=1\rangle$$

of genus $g \geq 4$ satisfies the strong generic subgroup property.

In general, asymptotic density is not independent of finite generating systems. Indeed, it is possible for a group property to be generic with respect to one finite generating system and negligible with respect to another, see [32]. We call a group property $\mathcal{P}$ *suitable* for a finitely generated group G if it is preserved under isomorphisms and its asymptotic density is independent of finite generating systems and *supersuitable* for G if its suitable both for G and all subgroups of finite index in G. It is clear that the strong generic free group property is suitable in any group G which has a non-Abelian free quotient.

Corollary 24.6.3. *The strong generic free group property is suitable in any finitely generated group G which has a non-Abelian free quotient.*

We remark that in a strong generic free group the conjugacy problem and the root extraction problem are generic problems.

In [23] it was shown that there is an interesting connection between the strong generic free group property of a group G and its subgroups of finite index. The main result of that paper is that a finitely generated group which has a non-Abelian free quotient satisfies the strong generic free group property if and only if each subgroup of finite index satisfies the strong generic free group property. As a consequence of this and Theorem 24.6.4, it follows that many important classes of groups, such as finitely generated Fuchsian groups with finite co-volume and the braid groups B_n for $n \geq 3$ satisfy the strong generic free group property.

Theorem 24.6.4 (Inheritance Theorem). *Let G be a finitely generated group and $H \subset G$ a subgroup of finite index $[G : H] = n < \infty$. Let $\mathcal{P}$ be the strong generic free group property. Then:*

1. *If $\mathcal{P}$ is a suitable and generic property in H then it is also suitable and generic in G.*
2. *If $\mathcal{P}$ is a suitable and generic property in G then it is also suitable and generic in H.*

Proof. Let X be a finite generating system for G. As X is finite, it follows that H is finitely generated, and H has finite index in G. Let Y be a finite generating system for H. Let $\mathcal{P}$ be the strong generic free group property and suppose that $\mathcal{P}$ is a suitable and generic property in H. Let S_m be the collection of m element subsets that generate a free subgroup of G.

Let $B_k(G)$ be the ball of radius k in the Cayley graph of G (with respect to X). Since H is a subgroup of finite index n in G, there exists a complete system of representatives $a_1, \ldots, a_n \in G$ for the left cosets of H in G. We consider the elements of H as vertices in the Cayley graph of G. Let $B_k(H)$ be the set of vertices in $B_k(G)$ which belong to H. For all i let $a_iB_k(H)$ denote the displaced $B_k(H)$ around the representative a_i in the Cayley graph of G, that is the set of all elements of the form a_ih, where $h \in H$ is of length $\leq k$. Define $B'_k(H) = \bigcup_{i=1}^n a_iB_k(H)$ as the (disjoint) union of these $B_k(H)$. We have $|B'_k(H)| = n \cdot |B_k(H)|$, since the cosets a_iH and also the $a_iB_k(H)$ with them are pairwise disjoint. Let $t \in \mathbb{N}$ be the length of the longest geodesic in the Cayley graph of G from the identity element 1 to one of the representatives a_i. With this t we have

$$B'_{k-t}(H) \subset B_k(G) \subset B'_{k+t}(H). \tag{1}$$

Now let $B_{m,k}(G)$ and $a_iB_{m,k}(H)$ be the collection of m element subsets within $B_k(G)$ and $a_iB_k(H)$, respectively, for $i = 1, \ldots, m$. Let A be any m-element subset within $B_k(G)$. Then A splits into the disjoint union $\bigcup_{i=1}^n A_i$ of m_i-element subsets A_i within $a_iB_{k+t}(H)$ for $i = 1, \ldots, n$ and we have $0 \leq m_i \leq m$ for all i (some of the m_i may be zero).

In this sense, if we define $B'_{m,k}(H) = \bigcup_{i=1}^n a_iB_{m_i,k}(H)$, $m = m_1 + \cdots + m_n$, then we get the inclusions

$$B'_{m,k-t}(H) \subset B_{m,k}(G) \subset B'_{m,k+t}(H). \tag{1'}$$

Here, we consider a disjoint union $\bigcup_{i=1}^n A_i$ of m_i-element subsets A_i in $a_iB_{k-t}(H)$ with $m_1 + \cdots + m_n = m$ as an m-element subset $B_k(G)$. If A is a free generating system for a free subgroup of G, then each A_i is a free generating system for a free subgroup of G. Then intersecting with S_m leads to

$$S_m \cap B'_{m,k-t}(H) \subset S_m \cap B_{m,k}(G) \subset S_m \cap B'_{m,k+t}(H). \tag{2}$$

On the other hand, if some $A_i \subset a_iB_k(H)$ contains a subset which generates a free subgroup of G, then also $A_j = a_ja_i^{-1}A_i \subset a_jB_k(H)$ contains a subset which generates a free subgroup of G. More concretely, if A_i freely generates a free subgroup of rank p, then $\langle A_j \rangle$ has a p generating system which contains a basis for a free subgroup of rank at least $p - 1$.

This shows that for k large enough the sets $S_m \cap a_iB_{m,k}(H)$ are of the same order of magnitude in m. Applying this we get approximately the equality

$$\frac{|S_m \cap B'_{m,k}(H)|}{|B'_{m,k}(H)|} = \frac{|S_m \cap B_{m,k}(H)|}{|B_{m,k}(H)|} \tag{3}$$

for k and m large enough.

Assume that $\mathcal{P}$ holds and is suitable in H. Then there exists a constant integer $s > 0$ such that the length of each $y \in Y$ written as a word in X is less then s. Therefore the fraction on the right hand side of (3) converges to 1 as $k \to \infty$ and $m \to \infty$. Therefore

$$\lim_{m,k\to\infty} \frac{|S_m \cap B_{m,k}(G)|}{|B_{m,k}(G)|} = 1$$

from the inclusions (1') and (2), completing the proof of (1). The proof for (2) follows in an entirely analogous manner. □

As mentioned above if a finitely generated group G has a non-Abelian free quotient then the strong generic free group property holds in G and is suitable. Therefore we have the following corollary.

Corollary 24.6.5. *Let G be a finitely generated group and $H \subset G$ a subgroup of finite index. Assume that both G and H have non-Abelian free quotients. Then G has the strong generic free group property if and only if H has the strong generic free group property.*

We now show the strong generic free group property for braid groups.

Theorem 24.6.6. *The braid group B_n, $n \geq 3$, has the strong generic free group property.*

Proof. Denote by $\sigma_{i,i+1}$ the transposition $(i, i+1)$ in the symmetric group S_n. The map $\sigma_i \mapsto \sigma_{i,i+1}$, $i = 1, \ldots, n-1$, defines a canonical epimorphism $\pi: B_n \to S_n$. The kernel of σ is a subgroup of index $n!$ in B_n, called the pure braid group PB_n. The group PB_n, $n \geq 3$, maps onto the group PB_3, and the group PB_3 is isomorphic to $F_2 \times \mathbb{Z}$, where F_2 is the free group of rank 2. Hence, PB_n, $n \geq 3$, maps onto F_2. Now, the result follows from Corollary 24.6.3 and the Inheritance Theorem 24.6.4. □

Corollary 24.6.7. *The root extraction problem is a generic problem in B_n, $n \geq 3$.*

We now describe an authentication scheme based on the root extraction problem as given in [88]. Let B_n, $n \geq 3$, be the braid group generated by $\sigma_1, \ldots, \sigma_n$ with n even. Write LB_n for the braid group generated by $\sigma_1, \ldots, \sigma_{\frac{n}{2}-1}$ and UB_n for the group generated by $\sigma_{\frac{n}{2}+1}, \ldots, \sigma_n$.

Alice chooses two integers $r, s \geq 2$, and two elements $a \in LB_n$ and $c \in B_n$. Then $B_n, LB_n, UB_n, X = a^r c a^s, c, r, s$ are public and a is secret. The authentication is as follows. Bob chooses an element $b \in UB_n$, and sends to Alice $Y = b^r c b^s$. Alice computes $Z = a^r Y a^s$ and sends it to Bob. Finally, Bob verifies that $Z = b^r X b^s$. The security is based on finding a root x in B_n when x^m, $m \geq 2$, is given.

In the protocol a secret braid x is chosen at random, and the braid $y = x^m$ is made public. Hence, we are dealing with braids for which an mth root is known to exist. This means generically we may find the mth root of y very fast. The interest of braid groups for cryptography has decreased due to the appearance of algorithms which solve, for instance, the conjugacy problem and the root extraction problem, fast in the generic case. The main problem with the cryptographic protocols based on braid groups turns out to be the key generation.

Public and secret keys are so far chosen at random, and this implies often that the protocols are insecure against algorithms which have generically a fast complexity. The importance and the future of braid groups cryptography depends on finding a suitable key generation procedure, or in popular words, in finding so-called suitable black holes. Another promising possibility is to look for nongeneric properties of braid groups which could be used for cryptographic protocols.

24.6.1 Security Analysis of the Group Randomizer Protocols

In order to analyze the security of the group randomizer password protocols, we make the security assumption that an adversary has access to the coded group theoretic responses. The strength of the proposed protocol include that an attacker must steal three things: the original password, the group randomizer and the challenge group. There is no access without all three. This immediately nullifies middleman attacks. If the adversary pretends to be the verifier to obtain the group words the attack is thwarted by the facts that the prover can verify the verifier and further if the attacker just transmits from the middle, nothing can be stolen since each time through a new challenge word must be used. Further, the group randomizer has an infinite supply of both subgroups and challenge responses that are done randomly. In addition, since a challenge word can be used only once the protocol nullifies replay attacks. Since challenge responses are machine to machine there is essentially zero probability of an incorrect response. The protocol shuts down with an incorrect response and hence repeat attacks are harmless.

These are in distinction to answer-driven challenge–response systems where a prover often forgets or misspells a response. In these systems a prover is usually permitted several opportunities to answer making it susceptible to both man-in-the-middle and repeat attacks.

There are two theoretical attacks that must be dealt with. Relative to these the security of the system, and hence a security proof for the protocol, is provided by several results in asymptotic group theory.

The most straightforward attack is for the adversary to collect enough challenge words and responses. This provides a system of equations in a free group (or a finitely presented group)

$$y_{i_1} \cdots y_{i_t} = w_i(x_1, \ldots, x_n), \quad i = 1, \ldots, m.$$

An adversary can then break the protocol by solving the system

$$z_i = w_i(x_1, \ldots, x_n)$$

to obtain the challenge group.

However, a result by Lysenok and Myasnikov, see [91], shows that solving such systems of equations in free groups (and in most finitely presented groups) is NP-hard. Hence this method of attack is impractical in most cases.

A second method of attack is based on the following. The mapping $y_i \to w_i$ is a homomorphism. If a challenge word appears in the subgroup generated by previous challenge words then an attacker can use this to answer a challenge without ever solving for the challenge group. However, the probability of succeeding with this approach is essentially zero due to Jitsukawa's result mentioned in the previous section. Each challenge word lies in a free group which has the strong generic free group property. Hence as explained in the previous section the probability is essentially zero that a new challenge word is in the subgroup generated by previous challenge words.

24.6.2 Implementation of a Group Randomizer System Protocol

The actual implementation of a workable group randomizer system protocol involves several choices of parameters and subprograms. These include the following choices.

1. The choice of the rank of the ambient free group in the group randomizer systems A and B.
2. An enhancement program which takes randomly chosen words $w_1, \ldots, w_k$ in a free group F and finds a new set of words $v_1, \ldots, v_k$ generating the same subgroup for which the words formed in $v_1, \ldots, v_k$ have a great deal of free cancellation. This involves Nielsen transformations, see Section 14.3.
3. The choice of parameter sizes for the lengths of the randomly chosen words. In an actual implementation all words in the generators will have lengths between a and b where a and b are to be determined. All words used as test logins will have lengths between c and d with c and d to be determined.
   The determination of the optimal values of a, b, c, d are being studied.
4. The implementation of a coded permutation system on $\{1, \ldots, k\}$ where k is the rank of the challenge group and which can be sent with each challenge word.
5. The development of an automatic reset protocol for the challenge group. In an ideal situation this can be done without actually communicating the changes between verifier and prover. That is, each randomizer system does the same protocol automatically when reset is called for.

24.7 A Secret Sharing Scheme Using Combinatorial Group Theory

Recall that the secret sharing problem is the following. We have a secret K and a group of n participants. This group is called the *access control group*. A *dealer* allocates shares to each participant under given conditions. If a sufficient number of participants combine their shares then the secret can be recovered. If $t \leq n$ then an (t, n)*-threshold scheme* is the one with n total participants and in which any t participants can combine their shares to recover the secret but not fewer than t. The number t is called the *threshold.* The scheme is called a *secure secret sharing scheme* if, given fewer shares than the threshold, there is no chance to recover the secret.

Panagopoulos, see [96], devised a secret sharing scheme based on the word problem in finitely presented groups. It is an (t, n)-threshold scheme and its main advantage over many other secret sharing schemes is that it does not require the secret message to be determined before each individual person receives his share of the secret. For this scheme it is assumed that the secret is given in the form of a binary sequence. The scheme is as follows.

1. A finitely presented group $G = \langle x_1, x_2, \ldots, x_k; r_1 = \cdots = r_m = 1\rangle$ is chosen. It is assumed that the word problem is solvable for the presentation and that $m = \binom{n}{t-1}$.
2. Let $A_1, \ldots, A_m$ be an enumeration of the subsets of $\{1, \ldots, n\}$ with $t - 1$ elements. Define n subsets $R_1, \ldots, R_n$ of $\{r_1, \ldots, r_m\}$ such that $r_j \in R_i$ if and only if $i \notin A_j$ for $i = 1, \ldots, n$ and $j = 1, \ldots, m$. Then for every $j \in \{1, \ldots, m\}$, the word r_j is not contained in exactly $t - 1$ of the subsets $R_1, \ldots, R_n$. It follows that r_j is contained in any union of t of them, whereas if we take any $t-1$ of the sets $R_1, \ldots, R_n$, there exists an index j such that r_j is not contained in their union.
3. Distribute to each of the n persons one of the sets $R_1, \ldots, R_n$. The set $\{x_1, \ldots, x_k\}$ is known to all participants.
4. If the binary sequence to be distributed is $a_1, \ldots, a_k$, construct and distribute a sequence of elements $w_1, \ldots, w_k$ of G such that we have $w_i = 1$ in G if and only if $a_i = 1$ for $i = 1, \ldots, k$. The word w_i must involve most of the relations $r_1 = 1, \ldots, r_m = 1$ if $w_i = 1$. Furthermore, all of the relations must be used at some point in the construction of some element.

Then any t of the n persons can obtain the sequence $a_1, \ldots, a_k$ by taking the union of the subsets of the relations of G that they possess. Thus they obtain the presentation $G = \langle x_1, x_2, \ldots, x_k; r_1 = \cdots = r_m = 1\rangle$ and can solve the word problem $w_i = 1$ in G for $i = 1, \ldots, k$. A collection of fewer than t persons cannot decode the message correctly, since the union of fewer than t of the sets $R_1, \ldots, R_n$ contains some but not all of the relations $r_1, \ldots, r_m$.

Such a collection leads to a group presentation

$$\widetilde{G} = \langle x_1, x_2, \ldots, x_k; r_{j_1} = \cdots = r_{j_p} = 1\rangle$$

with $p < m$ and $G \neq \widetilde{G}$, where $w_i = 1$ in G is, in general, not equivalent to $w_i = 1$ in $\widetilde{G}$.

Notice that the secret sequence to be shared is not needed until the final step. It is possible for someone to distribute the sets $R_1, \ldots, R_m$ and decide at a later time what the sequence $a_1, \ldots, a_k$ would be. In that way the scheme can also be used so that t of the n persons can verify the authenticity of the message. In particular, the binary sequence in Step 4 may contain a predetermined subsequence (signature) along with the actual message. Then any t persons may check whether this predetermined sequence is contained in the encoded message and thus validate it.

In the paper by Panagopoulos, see [96], he also describes some methods for attacking this scheme and makes some suggestions for possible group presentation types to use.

Moldenhauer [90] proposed a modification of Panagopoulos' (t, n)-threshold scheme using Nielsen transformations. We need the following.

Theorem 24.7.1. *Let $T_1, T_2, \ldots$ be a countable number of matrices of the form*

$$T_j = \begin{pmatrix} -r_j & -1 + r_j^2 \\ 1 & -r_j \end{pmatrix}$$

where r_j are integers and $r_{j+1} - r_j \geq 3$, $r_1 \geq 2$. Then the $T_1, T_2, \ldots$ form a basis of a free group of countable rank.

Proof. The isometric circle of T_j is given by $|z - r_j| = 1$ and that of T_j^{-1} is given by $|z + r_j| = 1$. The respective isometric disks

$$K(T_1), K(T_1^{-1}), K(T_2), K(T_2^{-1}), \ldots$$

are pairwise disjoint because of the restriction on r_j. Let F be the group generated by $\{T_1, T_2, \ldots\}$. Clearly, F is a subgroup of $\mathrm{SL}(2, \mathbb{Z})$. Let $S_k \cdots S_1$ be a reduced word in F. Each S_i is a T_j or T_j^{-1}. It may happen that $S_{i+1} = S_i$. Suppose p lies outside every isometric disk $K(T_j), K(T_j^{-1}), j = 1, 2, \ldots$. Such a P exists because F is a subgroup of the $\mathrm{SL}(2, \mathbb{Z})$. Then $S_1(P)$ lies inside $K(S_1^{-1})$. Since $S_1(P)$ lies outside $K(S_2)$, this is true even if $S_1 = S_2$, it is seen that $S_2 S_1(P)$ is inside $K(S_2^{-1})$. We conclude that $Q = S_k \cdots S_1(P)$ is inside $K(S_k^{-1})$. Hence, $S_k \cdots S_1 \neq 1 (= E_2)$. This shows that F is free on $\{T_1, T_2, \ldots\}$. □

We now describe the modified (t, n)-threshold scheme. We write N_{r_j} instead of T_j and choose a large number m of the form $m = 2^n$, $n \geq 64$. This allows us to use the idea of linear congruence generators (modulo m) to get a stream cipher. The dealer performs the following to distribute the secret among n participants:

1. Start with a set $(x_1, N_{r_1}), \ldots, (x_m, N_{r_m})$, where $x_1, \ldots, x_m$ are the generators of the free group $F(x_1, \ldots, x_m)$ and $N_{r_1}, \ldots, N_{r_m}$ are matrices in $\mathrm{SL}(2, \mathbb{Z})$ of the form

   $$N_{r_i} = \begin{pmatrix} -r_i & -1 + r_1^2 \\ 1 & -r_i \end{pmatrix}$$

   satisfying $r_1 \geq 2$ and $r_{i+1} \geq r_i + 3$ (more generally, any free generating set for a free subgroup in $\mathrm{SL}(2, \mathbb{Q})$). The secret is a rational number

   $$\sum_{i=1}^{m} \frac{1}{|\operatorname{tr}(N_{r_i})|} = \sum_{i=1}^{m} \frac{1}{2|r_i|}.$$

2. Apply a sequence of Nielsen transformations to the set of pairs above to obtain a new set

$$(v_1, M_1), \ldots, (v_m, M_m).$$

3. Distribute subsets in Panagopoulos' scheme. To recover the secret, perform the following:
4. Take a union of their shares. In the case that t participants gather, they are able to recover the set $(v_1, M_1, \ldots, v_m, M_m)$.
5. Apply a sequence of Nielsen transformations to the obtained set of pairs in order to Nielsen-reduce the first components and obtain the set $x_1, \ldots, x_m$ in the first components. As a result, in the second components, we get the original matrices $N_{r_1}, \ldots, N_{r_m}$. Compute the sum $\sum_{i=1}^{m} \frac{1}{|\mathrm{tr}(N_i)|}$.

Kotov, Panteleev and Ushakov, see [87], analyzed this secret sharing protocol. They could reduce it to a system of polynomial equations over the free group $F(\{x_1, \ldots, x_m\} \cup \{a_1, \ldots, a_{m-1}\})$ where x_i stands for an unknown matrix N_{r_i} and a_i stands for the matrix M_i. Replacing x_i with an unknown matrix N_{r_i} and a_i with M_i and performing matrix multiplication, we obtain a system of polynomial equations which can further fed to any computer algebra system that can solve polynomial equations, for instance CoCoA.

The solution of the systems provides the original matrices $M_1, \ldots, M_m$. The attack reconstructs the original data generated by the dealer and does not depend on the function of $M_1, \ldots, M_m$ used to calculate the shared secret. It seems unlikely that their attack is successful if $m \geq 2^{64}$. If so, for chosen matrices $N_{r_1}, \ldots, N_{r_m}$ we still may collect in each round m new matrices from the countably many and/or may use the stream cipher for a one-time pad.

Moreover, increasing the length of keys, the number of Nielsen transformations increases the sizes of polynomials and seems to be successful countermeasure against their attack. Another possibility to repel such attacks is to change the tactic and to work with more general matrices $N_{r_1}, N_{r_2}, \ldots$ which form a free generating set of a free subgroup in $\mathrm{SL}(k, \mathbb{R})$, $k \geq 2$.

24.8 Ko–Lee and Anshel–Anshel–Goldfeld Protocols

All of the non-Abelian group based protocols depend on the difficulty of solving certain group decision problems and group theoretical computational problems. Recall that the conjugacy problem, also called the decision conjugacy problem, for a group G, or more precisely for a group presentation for G, is the following: given $g, h \in G$, determine algorithmically if they are conjugate.

The conjugacy problem is unsolvable in general, that is, there exist group presentations for which there does not exist an algorithm that solves the conjugacy problem.

Hence a solution to the conjugacy problem is usually associated with a particular class of group presentations. For example, the conjugacy problem is solvable in free groups and in torsion-free hyperbolic groups.

Relevant to the Ko–Lee protocol is the conjugator search problem. This is, given a group presentation for G, and two elements g_1, g_2 in G, that are known to be conjugate, to determine algorithmically a conjugator, that is, an element $h \in G$ with $g_1 = hg_2h^{-1}$. It is known, as with the decision conjugacy problem, that the conjugator search problem is undecidable in general.

24.8.1 The Ko–Lee Protocol

Ko, Lee et al., see [85], developed a public key exchange system that is a direct translation of the Diffie–Hellman protocol to a non-Abelian group theoretic setting. Its security is based on the difficulty of the conjugacy problem. We assume that the platform group has nice unique normal forms that are easy to compute for a given group element but hard to recover the individual group elements under group multiplication.

Recall from Section 24.1 that by this we mean that if $G = \langle X; R\rangle$ is a finite presentation for the group G and $g \in G$ then there is a unique expression $NF_X(g)$ called a normal form as a word in the generators X. Further, given any $g \in G$ it is computationally easy to find $NF_X(g)$. On the other hand, given $g_1, g_2 \in G$ and given the normal form $NF_X(g_1g_2)$, it is computationally difficult to recover g_1 and g_2. We say that there is good diffusion in terms of normal forms in forming products.

In any group G and for $g, h \in G$ the notation g^h indicates the conjugate of g by h, that is, $g^h = h^{-1}gh$. What is important for both the Ko–Lee and Anshel–Anshel–Goldfeld protocols is that relative to this notation, group conjugation behaves exactly as ordinary exponentiation. That is for groups elements $g, h_1, h_2 \in G$ we have $(g^{h_1})^{h_2} = g^{h_1h_2}$. That this is true is a straightforward computation

$$(g^{h_1})^{h_2} = h_2^{-1}g^{h_1}h_2 = h_2^{-1}h_1^{-1}gh_1h_2 = (h_1h_2)^{-1}g(h_1h_2) = g^{h_1h_2}.$$

With this observation, the Ko–Lee protocol exactly mimics, using group conjugation, the traditional Diffie–Hellman protocol. We first start with a platform group G satisfying the necessary requirements on normal forms. We assume further that the platform group G has a collection of large (noncyclic) subgroups that commute elementwise. That is, if A, B are two of these subgroups and $a \in A$ and $b \in B$, then $ab = ba$. It is not necessary that the subgroups themselves be Abelian.

Alice and Bob choose a pair of these commuting subgroups A and B of the platform group G. A is Alice's subgroup while Bob's subgroup is B and these are secret. By assumption each element of A commutes with each element of B. Further, it is not assumed that A and/or B are themselves Abelian. Now the method completely mimics the classical Diffie–Hellman technique.

There is a public element $g \in G$, Alice chooses a random secret element $a \in A$ and makes public g^a, the conjugate of g by a.

Bob chooses a random secret element $b \in B$ and makes public g^b the conjugate of g by b. The secret shared key is g^{ab}. Notice that $ab = ba$ since the subgroups commute. It follows then that

$$(g^a)^b = g^{ab} = g^{ba} = (g^b)^a$$

just as if these were ordinary exponents.

It follows, as in the number theoretic based Diffie–Hellman protocol, that both Bob and Alice can determine the common secret. Alice knows her secret key a and Bob's public key g^b. Hence she knows $(g^b)^a = g^{ba}$. Bob knows his secret key b and g^a is public. Hence Bob knows $(g^a)^b = g^{ab}$. However, as explained $g^{ab} = g^{ba}$. The difficulty is in that of the decision conjugacy problem.

It is known that both the decision conjugacy problem and the conjugator search problem are undecidable in general. However, there are groups where both are solvable but hard, that is the problems are solvable but are not solvable in polynomial time. These groups then become the target platform groups for the Ko–Lee protocol. Ko and Lee in their initial work suggest the use of the braid groups. We will discuss braid group cryptography later in this chapter.

We now summarize the formal setup for the Ko–Lee Key Exchange Protocol. After this we will show how to use the ElGamal method to construct a public key encryption system from this.

24.8.1.1 Ko–Lee Preparation

1. We start with a platform group G. We assume that G has a finite presentation with efficiently computable normal forms that have good diffusion. Further the group G must have a large collection of subgroups that commute elementwise.
2. We choose an element $g \in G$.
3. We assume that Alice wants to share a common key with Bob. Alice and Bob choose subgroups A and B that elementwise commute. A is Alice's subgroup and B is Bob's subgroup. These subgroups are kept secret and known only to Bob and Alice, respectively.

Ko–Lee Key Exchange

1. Alice randomly chooses an $a \in A$. This element a will be her secret key. Her public key is (g, g^a) where $g^a = a^{-1}ga$ is the conjugate of g by her secret key a. All public information and communication is done in terms of the normal forms of these elements.
2. Bob randomly chooses an element $b \in B$. This element b will be his secret key. His public key is (g, g^b) where $g^b = b^{-1}gb$ is the conjugate of g by his secret key b. As

with Alice all public information and communication is done in terms of the normal forms of these elements.

3. The secret shared key is g^{ab}.

24.8.1.2 ElGamal Encryption Using the Ko–Lee Protocol

As with the standard Diffie–Hellman key exchange protocol using number theory, the Ko–Lee protocol can be changed to an encryption system via the ElGamal method. There are several different variants of noncommutative ElGamal systems. At the simplest level we assume that we have a group G appropriate for the Ko–Lee key exchange and that Alice and Bob want to communicate secretly. The element $g \in G$ is public and Alice and Bob, respectively, have chosen their appropriate commuting subgroups A and B. Bob has made public g^b for $b \in B$ in normal form and Alice has made public g^a for $a \in A$ also in normal form. The secret shared key is then g^{ab}. We assume that Alice wants to send an encrypted message to Bob and further we assume the encrypted message can be encoded as $h \in G$, that is as an element of the group G. Alice then sends to Bob the normal form of hg^{ab}. Bob can determine the common shared secret g^{ab}. He then multiplies hg^{ab} by $(g^{ab})^{-1}$ to obtain the secret h.

As with the number theoretic based public key cryptosystems, the Ko–Lee method can be used to provide methods for other protocols, especially authentication and digital signature protocols.

24.8.2 The Anshel–Anshel–Goldfeld Protocol

We now describe another non-Abelian group-based public key exchange protocol. It is somewhat similar to the Ko–Lee protocol and was developed at approximately the same time. This is the Anshel–Anshel–Goldfeld public key exchange protocol.

As in the Ko–Lee protocol we start with a group G given by a finite presentation $G = \langle X; R \rangle$. We further assume as before that there are efficiently computable normal forms relative to the presentation $\langle X; R \rangle$. The Ko–Lee protocol required two large commuting subgroups. For communication, the Anshel–Anshel–Goldfeld protocol requires a choice of subgroups of G, but they need not commute. While the difficulty of the decision conjugacy problem provides the security for the Ko–Lee method, it is the difficulty of the conjugator search problem that provides the hard problem, and hence the security, in the Anshel–Anshel–Goldfeld protocol.

Once we have our platform group G, we assume that Alice and Bob want to obtain a common shared secret or a common shared secret key. We assume that this secret key can be expressed as a group element $g \in G$. The first step is for Alice and Bob to choose random finitely generated subgroups of G by giving a set of generators for each,

$$A = \{a_1, \ldots, a_n\}, \quad B = \{b_1, \ldots, b_m\},$$

and make them public. The subgroup A is Alice's subgroup while the subgroup B is Bob's subgroup.

Alice chooses a secret group word $a = w(a_1, \dots, a_n)$ in her subgroup while Bob chooses a secret group word $b = v(b_1, \dots, b_m)$ in his subgroup. As before, for an element $g \in G$ we denote by $NF_X(g)$ the normal form for g. Alice knows her secret word a and knows the generators b_i of Bob's subgroup. She can then form the conjugates of the generators of Bob's subgroup B by her secret element $a \in A$. That is, she can compute $b_i^a = a^{-1}b_ia$ for each b_i. She then makes public the normal forms of these conjugates

$$NF_X(b_i^a), \quad i = 1, \dots, m.$$

Bob does the analogous thing. He knows his secret word b and the generators a_i, $i = 1, \dots, n$ of Alice's subgroup A and hence can compute the conjugates $a_i^b = b^{-1}a_ib$ for $i = 1, \dots, n$. He then makes public the normal forms of the conjugates

$$NF_X(a_j^b), \quad j = 1, \dots, n.$$

The common shared secret is the commutator

$$[a, b] = a^{-1}b^{-1}ab = a^{-1}a^b = (b^a)^{-1}b.$$

Notice that this is known for both Alice and Bob. Alice knows $a^b = b^{-1}ab$ since she knows a in terms of generators a_i of her subgroup and she knows the conjugates by b, since Bob has made the conjugates of the generators of A by b public. That is, Alice knows $a = w(a_1, \dots, a_n)$ and $a^b = b^{-1}ab = w(b^{-1}a_1b, \dots, b^{-1}a_nb) = w(a_1^b, \dots, a_n^b)$. Since Alice knows a^b, she knows

$$[a, b] = a^{-1}b^{-1}ab = a^{-1}a^b.$$

In an analogous manner Bob knows $[a, b] = (b^a)^{-1}b$, since he knows his secret element b in terms of the generators $b_j, j = 1, \dots, m$, of his subgroup B and Alice has made public the conjugates of each of his generators by her secret element a. Hence $b = v(b_1, \dots, b_m)$ so that $b^a = v(b_1^a, \dots, b_m^a)$ and this is known to Bob. Since Bob knows b^a and b, he knows

$$[a, b] = a^{-1}b^{-1}ab = a^bb = (b^{-1})^ab = (b^a)^{-1}b.$$

Notice that in this system there is no requirement that the chosen subgroups A and B commute.

An attacker would have to know the corresponding *conjugator*, that is the element that conjugates each of the generators, that is, the conjugator search problem: Given elements g, h in a group G, where it is known that $g^k = k^{-1}gk = h$, determine the conjugator k. It is known that this problem is undecidable in general, that is, there are groups where the conjugator cannot be determined algorithmically. On the other hand there are groups where the conjugator search problem is solvable but difficult, that is, the complexity of solving the conjugator search problem is hard. Such groups become the ideal platform groups for the Anshel–Anshel–Goldfeld protocol.

The security in this system is then in the computational difficulty of the conjugator search problem. Anshel, Anshel, Goldfeld suggested, as did Ko, Lee et al., the braid groups, B_n, as potential platforms. The braid groups are a class of infinite, finitely presented groups that arise in many different contexts. The braid group B_n has a standard presentation with $n - 1$ generators.

The necessary parameters that must be decided in using the braid groups as platforms for either the Ko–Lee protocol or the Anshel–Anshel–Goldfeld protocol are then the number of generators of the braid groups used and the number of generators for the chosen subgroups. For example B_{200}, the braid group on 200 strands with 12 or more generators in the chosen subgroups might be used. It has been shown that the larger the number of strands, the harder it is to attack the protocol. The suggested use of the braid groups by both Anshel, Anshel and Goldfeld and Ko and Lee led to the development of *braid group cryptography*. There have been various attacks on the braid group cryptosystems.

We now summarize the formal setup for the Anshel–Anshel–Goldfeld Key Exchange Protocol. After this we will show how to use the ElGamal method to construct a public key encryption system from this.

24.8.2.1 Anshel–Anshel–Goldfeld Preparation

1. We start with a platform group G. We assume that G has a finite presentation with efficiently computable normal forms that have good diffusion. Further, there is a large collection of efficiently computable subgroups.
2. We assume that Alice wants to share a common key with Bob. Alice and Bob choose random finitely generated subgroups of G by giving a set of generators for each,

$$A = \{a_1, \ldots, a_n\}, \quad B = \{b_1, \ldots, b_m\},$$

   and make them public. The subgroup A is Alice's subgroup while the subgroup B is Bob's subgroup.

24.8.2.2 Anshel–Anshel–Goldfeld Key Exchange

1. Alice chooses a secret group word $a = w(a_1, \ldots, a_n)$ in her subgroup. Alice knows her secret word a and knows the generators b_i of Bob's subgroup. She can then form the conjugates of the generators of Bob's subgroup B by her secret element $a \in A$. That is, she can compute $b_i^a = a^{-1} b_i a$ for each b_i. She then makes public the normal forms of these conjugates

$$NF_X(b_i^a), \quad i = 1, \ldots, m.$$

2. Bob chooses a secret group word $b = w(b_1, \ldots, b_m)$ in his subgroup. Bob knows his secret word b and knows the generators a_i of Alice's subgroup. He can then form the conjugates of the generators of Alice's subgroup A by his secret element $b \in B$. That is, he can compute $a_i^b = b^{-1}a_ib$ for each a_i. He then makes public the normal forms of these conjugates

$$NF_X(a_i^b), \quad i = 1, \ldots, m.$$

3. The secret shared key is the commutator

$$[a,b] = a^{-1}b^{-1}ab = a^{-1}a^b = (b^a)^{-1}b.$$

24.8.2.3 ElGamal Encryption using the Anshel–Anshel–Goldfeld Protocol

As with all public key exchange protocols, the Anshel–Anshel–Goldfeld key exchange can be developed into a cryptosystem by the ElGamal method. This works essentially in the same manner as for the Ko–Lee protocol. We assume that we have a group G appropriate for the Anshel–Anshel–Goldfeld key exchange and that Alice and Bob want to communicate secretly.

Alice and Bob, respectively, have chosen their appropriate subgroups A and B whose generators have been made public. Bob has made public the conjugates of the generators of A by his secret element $b \in B$ in normal form and Alice has made public the conjugates of the generators of B by her secret element $a \in A$, also in normal form. The secret shared key is then the commutator $[a,b]$.

We assume that Alice wants to send an encrypted message to Bob, and further we assume that the encrypted message can be encoded as $h \in G$, that is, as an element of the group G. Alice then sends to Bob the normal form of $h[a,b]$. Bob can determine the common shared key $[a,b]$. He then multiplies $h[a,b]$ by $[a,b]^{-1}$ to obtain the secret h.

Exercises

1. Bob has a backup authentication security system as described in Section 24.5. His basic words are $w_1 = x_1^{-1}x_2^2x_3^{-2}$, $w_2 = x_1^5x_2^3$, and $w_3 = x_2^5x_1^3x_2^{-2}x_3^4$. The bank sends him $w = y_1^2y_3^3y_1$. What must the group randomizer send back?
2. Let $M = \mathrm{PSL}(2,\mathbb{Z})$ be the modular group. Let $\mathcal{A} = \{a,b,c,d,e,f,g\}$ be a 7 letter plaintext alphabet. Choose a free subgroup of the modular group to encrypt these.
   (a) Using your basic encryption and message units of size 3, what would be the encryption matrices for the message *abbdceffgcba*?
   (b) Using your basic encryption and the algorithm given in Problem 1, what is the plaintext message for $\left(\begin{smallmatrix} 8 & 5 \\ 5 & 3 \end{smallmatrix}\right)$ and $\left(\begin{smallmatrix} 7 & 9 \\ 3 & 4 \end{smallmatrix}\right)$?
3. The following protocol is based on the factorization search problem which is: Given two subgroups A, B of a group G and $w \in G$, to find $a \in A$, $b \in B$ with $w = ab$. This

protocol is described in [93]. For this problem you must show and explain that the protocol works.
The requirements for the protocol are as follows: a public group and two public subgroups A, B that commute elementwise. Alice randomly chooses two private elements $a_1 \in A$ and $b_1 \in B$ and sends $a_1 b_1$ to Bob. Bob does the same choosing $a_2 \in A$ and $b_2 \in B$ and sends $a_2 b_2$ to Alice.
The common shared secret is $K = a_2 a_1 b_1 b_2$.
4. Prove Epstein's theorem: Given a random finitely generated subgroup of $\mathrm{GL}(n, \mathbb{R})$, with probability 1 it is a free group. The probability is standard measure on $\mathbb{R}^{n^2}$.
Hint: Given a finite set of matrices in $\mathrm{GL}(n, \mathbb{R})$, think what a relation between them would mean algebraically on the coefficients and where this would place the matrices topologically.
5. Let $G = H_1 * \cdots * H_n$ with $n \geq 2$ be the free product of finitely many nontrivial groups. Suppose that $H_1 \geq 3$ if $n = 2$. Show that G has the strong generic free group property.
6. Let G be a group and N be a normal subgroup. Show: If the quotient G/N satisfies the strong generic free group property then G also satisfies the strong generic free group property.
7. Show that a group with a generating set X is an epimorphic image of $F(X)$. Moreover, every map $X \to G$ with G a group can be extended to a unique homomorphism $f: F(X) \to G$.
8. Let F be a free group on $\{x_1, \ldots, x_n\}$. Show that each conjugation $x_i \mapsto g x_i g^{-1}$ with $g \in F$ can be written as a sequence of elementary Nielsen transformations.
9. Let F be a free group on $\{x_1, \ldots, x_n\}$. Show that the automorphism group $\mathrm{Aut}(F)$ is generated by the elementary Nielsen transformations (N1) and (N2).
10. Let PB_n stand for the pure braid group, $n \geq 3$. Using the Reidemeister–Schreier method, show that this group has a presentation with generators

$$A_{ij} = \sigma_{j-1}\sigma_{j-2}\cdots\sigma_{i+1}\sigma_i^2\sigma_{i+1}^{-1}\cdots\sigma_{j-2}^{-1}\sigma_{j-1}^{-1}$$

where $1 \leq i < j \leq n$ and relations

$$A_{rs}A_{ij}A_{rs}^{-1} = \begin{cases} A_{ij} & \text{if } s < i \text{ or } j > r, \\ A_{is}A_{ij}A_{is} & \text{if } i < j = r < s, \\ A_{ij}^{-1}A_{ir}^{-1}A_{ij}A_{ir}A_{ij} & \text{if } i < r < j = s, \\ A_{is}^{-1}A_{ir}^{-1}A_{is}A_{ir}A_{ij}A_{ir}^{-1}A_{is}^{-1}A_{ir}A_{is} & \text{if } i < r, j, s. \end{cases}$$

11. Show that the pure braid group PB_3 is isomorphic to the direct product $F_2 \times \mathbb{Z}$.
12. Let F_n be the free group of rank n on the free generating system $X = \{x_1, \ldots, x_n\}$ and let $\beta \in \mathrm{Aut}(F_n)$. Show that $\beta \in B_n$ if and only if β satisfies the following two conditions:
(1) $\beta(x_i)$ is conjugate to another generator.
(2) $\beta(x_1 \cdots x_n) = x_1 \cdots x_n$.

13. Let G be B_{20}, the braid group on 19 generators $\sigma_1, \dots, \sigma_{19}$. Let A be the subgroup generated by $\sigma_1, \dots, \sigma_5$ and B the subgroup generated by $\sigma_{16}, \dots, \sigma_{19}$.
Let $g = \sigma_7^3 \sigma_{12} \sigma_3^{-1} \sigma_5^{-2} \sigma_{10}$, $a = \sigma_2^4 \sigma_3^2 \sigma_1$, and $b = \sigma_{17}^4 \sigma_{18}^{-1} \sigma_{17}$.
   (a) What is the secret shared key using the Ko–Lee protocol?
   (b) What is the secret shared key using the Anshel–Anshel–Goldfeld protocol?

Bibliography

General Abstract Algebra

[1] J. L. Alperin and R. B. Bell, *Groups and Representations*, Springer-Verlag, 1995.
[2] M. Artin, *Algebra*, Prentice-Hall, 1991.
[3] C. Curtis and I. Reiner, *Representation Theory of Finite Groups and Associative Algebras*, Wiley Interscience, 1966.
[4] C. Curtis and I. Reiner, *Methods of Representation Theory I*, Wiley Interscience, 1982.
[5] C. Curtis and I. Reiner, *Methods of Representation Theory II*, Wiley Interscience, 1986.
[6] V. Diekert, M. Kufleitner, G. Rosenberger, and U. Hertrampf, *Discrete Algebraic Methods*, De Gruyter, 2016.
[7] B. Fine and G. Rosenberger, *The Fundamental Theorem of Algebra*, Springer-Verlag, 2000.
[8] J. Fraleigh, *A First Course in Abstract Algebra*, 7th ed., Addison-Wesley, 2003.
[9] E. G. Hafner, *Lineare Algebra*, Wiley-VCH, 2018.
[10] P. R. Halmos, *Naive Set Theory*, Springer-Verlag, 1998.
[11] I. Herstein, *Topics in Algebra*, Blaisdell, 1964.
[12] M. Kreuzer and S. Robiano, *Computational Commutative Algebra I and II*, Springer-Verlag, 1999.
[13] S. Lang, *Algebra*, Addison-Wesley, 1965.
[14] S. MacLane and G. Birkhoff, *Algebra*, Macmillan, 1967.
[15] N. McCoy, *Introduction to Modern Algebra*, Allyn and Bacon, 1960.
[16] N. McCoy, *The Theory of Rings*, Macmillan, 1964.
[17] G. Stroth, *Algebra. Einführung in die Galoistheorie*, De Gruyter, 1998.
[18] A. Zimmermann, *Representation Theory*, Spinger-Verlag, 2014.

Group Theory and Related Topics

[19] G. Baumslag, *Topics in Combinatorial Group Theory*, Birkhäuser, 1993.
[20] O. Bogopolski, *Introduction to Group Theory*, European Mathematical Society, 2008.
[21] T. Camps, V. Große Rebel, and G. Rosenberger, *Einführung in die kombinatorische und die geometrische Gruppentheorie*, Heldermann Verlag, 2008.
[22] T. Camps, S. Kühling, and G. Rosenberger, *Einführung in die mengenteoretische und die algebraische Topologie*, Heldermann Verlag, 2006.
[23] C. Carstensen, B. Fine, and G. Rosenberger, *On asymptotic densities and generic properties in finitely generated groups*, Groups Complex. Cryptol., **2**, 212–225, 2010.
[24] P. Dehornoy, *Braids and Self-Distributivity*, Birkhäuser, 2000.
[25] D. B. A. Epstein, *Almost all subgroups of Lie groups are free*, J. Algebra, **19**, 261–262, 1971.
[26] B. Fine, A. Moldenhauer, G. Rosenberger, and L. Wienke, *Topics in Infinite Group Theory*, De Gruyter, 2021.
[27] B. Fine, A. Moldenhauer, G. Rosenberger, A. Schürenberg, and L. Wienke, *Geometry and Discrete Mathematics: A Selection of Highlights*, 2nd ed., De Gruyter, 2022.
[28] B. Fine and G. Rosenberger, *Algebraic Generalizations of Discrete Groups*, Marcel Dekker, 2001.
[29] B. Fine, A. Myasnikov, and G. Rosenberger, *Generic subgroups of amalgams*, Groups Complex. Cryptol., **1**, 51–61, 2009.
[30] D. Gorenstein, *Finite Simple Groups. An Introduction to Their Classification*, Plenum Press, 1982.
[31] D. Johnson, *Presentations of Groups*, Cambridge University Press, 1990.
[32] I. Kapovich, I. Kaimonovich, and P. Schupp. *The Subadditive Ergodic Theorem and generic stretching factors for free group automorphisms*, Isr. J. Math., **157**, 1–46, 2007.

https://doi.org/10.1515/9783111142524-025

[33] S. Katok, *Fuchsian Groups*, Univ. of Chicago Press, 1992.

[34] G. Kern-Isberner and G. Rosenberger, *A note on numbers of the form* x^2+Ny^2, Arch. Math., **43**, 148–155, 1986.

[35] R. C. Lyndon, *Groups and Geometry*, LMS Lecture Note Series **101**, Cambridge University Press, 1985.

[36] R. C. Lyndon and P. Schupp, *Combinatorial Group Theory*, Springer-Verlag, 1977.

[37] W. Magnus, A. Karrass, and D. Solitar, *Combinatorial Group Theory*, Wiley, 1966.

[38] K. A. Mihailova, *The occurence problem for direct products of groups*, Dokl. Akad. Nauk SSSR, **119**, 1103–1105, 1958.

[39] C. F. Miller, *On Group-Theoretic Decision Problems and Their Classification*, Princeton University Press, 1971.

[40] B. H. Neumann, *Über ein gruppentheoretisch-arithmetisches Problem*, Sitz.ber. Preuss. Akad. Wiss. Phys. Math. Kl., 429–444, 1933.

[41] D. J. S. Robinson, *A Course in the Theory of Groups*, Springer-Verlag, 1982.

[42] J. Rotman, *Group Theory*, 3rd ed., Wm. C. Brown, 1988.

[43] J. Rotman, *An Introduction to the Theory of Groups*, Springer-Verlag, 1999.

Number Theory

[44] L. Ahlfors, *Introduction to Complex Analysis*, Springer-Verlag, 1968.

[45] T. M. Apostol, *Introduction to Analytic Number Theory*, Springer-Verlag, 1976.

[46] A. Baker, *Transcendental Number Theory*, Cambridge University Press, 1975.

[47] H. Cohn, *A Classical Invitation to Algebraic Numbers and Class Fields*, Springer-Verlag, 1978.

[48] L. E. Dickson, *History of the Theory of Numbers*, Chelsea, 1950.

[49] B. Fine, *A note on the two-square theorem*, Can. Math. Bull., **20**, 93–94, 1977.

[50] B. Fine, *Sums of squares rings*, Can. J. Math., **29**, 155–160, 1977.

[51] B. Fine, *The Algebraic Theory of the Bianchi Groups*, Marcel Dekker, 1989.

[52] B. Fine, A. Gaglione, A. Moldenhauer, G. Rosenberger, and D. Spellman, *Algebra and Number Theory: A Selection of Highlights*, De Gruyter, 2017.

[53] B. Fine and G. Rosenberger, *Number Theory: An Introduction via the Distribution of Primes*, 2nd ed., Birkhäuser, 2016.

[54] G. H. Hardy and E. M. Wright, *An Introduction to the Theory of Numbers*, 5th ed., Clarendon Press, 1979.

[55] E. Landau, *Elementary Number Theory*, Chelsea, 1958.

[56] M. Newman, *Integral Matrices*, Academic Press, 1972.

[57] I. Niven and H. S. Zuckerman, *The Theory of Numbers*, 4th ed., John Wiley, 1980.

[58] O. Ore, *Number Theory and its History*, McGraw-Hill, 1949.

[59] H. Pollard and H. Diamond, *The Theory of Algebraic Numbers*, Carus Mathematical Monographs **9**, Math. Assoc. of America, 1975.

[60] J. H. Silverman, *The Arithmethic of Elliptic Curves*, Springer-Verlag, 1986.

[61] W. C. Waterhouse, *Elliptic Curves: Number Theory and Cryptography*, Chapman and Hall, 2003.

Cryptography

[62] I. Anshel, M. Anshel, and D. Goldfeld, *An algebraic method for public key cryptography*, Math. Res. Lett., **6**, 287–291, 1999.

[63] G. Baumslag, Y. Brjukhov, B. Fine, and G. Rosenberger, *Some cryptoprimitives for noncommutative algebraic cryptography*, in Aspects of Infinite Groups, 26–44, World Scientific Press, 2009.

[64] G. Baumslag, Y. Brjukhov, B. Fine, and D. Troeger, Challenge response password security using combinatorial group theory, Groups Complex. Cryptol., **2**, 67–81, 2010.
[65] G. Baumslag, T. Camps, B. Fine, G. Rosenberger, and X. Xu, *Designing key transport protocols using combinatorial group theory*, Contemp. Math., **418**, 35–43, 2006.
[66] G. Baumslag, B. Fine, M. Kreuzer, and G. Rosenberger, *A Course in Mathematical Cryptography*, De Gruyter, 2015.
[67] G. Baumslag, B. Fine, and X. Xu, *Cryptosystems using linear groups*, Appl. Algebra Eng. Commun. Comput., **17**, 205–217, 2006.
[68] G. Baumslag, B. Fine, and X. Xu, A proposed public key cryptosystem using the modular group, Contemp. Math., **421**, 35–44, 2007.
[69] J. Birman, *Braids, Links and Mapping Class Groups*, Annals of Math Studies, **82**, Princeton University Press, 1975.
[70] A. V. Borovik, A. G. Myasnikov, and V. Shpilrain, *Measuring sets in infinite groups*, in Computational and Statistical Group Theory, Contemp. Math., **298**, 21–42, 2002.
[71] J. A. Buchmann, *Introduction to Cryptography*, Springer 2004.
[72] T. Camps, *Surface braid groups as platform groups and applications in cryptography*, Ph. D. thesis, Universität Dortmund, 2009.
[73] R. E. Crandall and C. Pomerance, *Prime Numbers. A Computational Perspective*, 2nd ed., Springer-Verlag, 2005.
[74] P. Dehornoy, *Braid-based cryptography*, Contemp. Math., **360**, 5–34, 2004.
[75] B. Eick and D. Kahrobaei, *Polycyclic groups: A new platform for cryptology?*, arXiv:math/0411077, 1–7, 2004.
[76] D. Garber, *Braid group cryptography*, World Scientific Review Volume, arXiv:0711.3941, 2008.
[77] R. Gilman, A. G. Myasnikov, and D. Osin, *Exponentially generic subsets of groups*, Ill. J. Math., **54**, 371–388, 2010.
[78] M. I. Gonzalez Vasco and R. Steinwandt, *Group Theoretic Cryptography*, Chapman & Hall, 2015.
[79] D. Grigoriev and I. Ponomarenko, *Homomorphic public-key cryptosystems over groups and rings*, Quad. Mat., 2005.
[80] P. Hoffman, *Archimedes' Revenge*, W. W. Norton & Company, 1988.
[81] T. Jitsuwaka, Malnormal subgroups of free groups, Contemp. Math., **298**, 83–96, 2002.
[82] D. Kahrobaei and B. Khan, *A non-commutative generalization of the El-Gamal key exchange using polycyclic groups*, in Proceedings of IEEE, 1–5, 2006.
[83] I. Kapovich and A. Myasnikov, *Stallings foldings and subgroups of free groups*, J. Algebra, **248**, 608–668, 2002.
[84] K. H. Ko, D. Choi, M. Cho, and J. Lee, *New signature scheme using conjugacy problem*, IACR Cryptology ePrint Archive, **168**, 1–13, 2002.
[85] K. H. Ko, S. J. Lee, J. H. Cheon, J. H. Han, J. S. Kang, and C. Park, *New public-key cryptosystems using Braid groups*, in Advances in Cryptography, Proceedings of Crypto 2000, Lecture Notes in Computer Science, **1880**, 166–183, 2000.
[86] N. Koblitz, *Algebraic Methods of Cryptography*, Springer, 1998.
[87] M. Kotov, D. Panteleev, and A. Ushakov, *Analysis of the secret schemes based on Nielsen transformations*, Groups Complex. Cryptol., **10**, 1–8, 2018.
[88] S. Lal and A. Chaturvedi, *Authentication schemes using braid groups*, arXiv:cs/0507066, 2005.
[89] W. Magnus, *Rational Representations of Fuchsian Groups and Non-Parabolic Subgroups of the Modular Group*, Nachrichten der Akad. Göttingen, 179–189, 1973.
[90] A. Moldenhauer, *Cryptographic protocols based on inner product spaces and group theory with a special focus on the use of Nielsen transformations*, Ph. D. thesis, University of Hamburg, 2016.
[91] I. G. Lysenok and A. G. Myasnikov, *A polynomial bound on solutions of quadratic equations in free groups*, Proc. Steklov Inst. Math., **274**, 136–173, 2011.

[92] A. G. Myasnikov, V. Shpilrain, and A. Ushakov, *A practical attack on some braid group based cryptographic protocols*, in CRYPTO 2005, Lecture Notes in Computer Science, **3621**, 86–96, 2005.
[93] A. G. Myasnikov, V. Shpilrain, and A. Ushakov, *Group-Based Cryptography*, Advanced Courses in Mathematics, CRM, Barcelona, 2007.
[94] A. D. Myasnikov and A. Ushakov, *Length based attack and braid groups: Cryptanalysis of Anshel–Anshel–Goldfeld key exchange protocol*, Lect. Notes Comput. Sci., **4450**, 76–88, 2007.
[95] G. Petrides, *Cryptoanalysis of the public key cryptosystem based on the word problem on the Grigorchuk groups*, in Cryptography and Coding, Lecture Notes in Computer Science, **2898**, 234–244, 2003.
[96] D. Panagopoulos, *A secret sharing scheme using groups*, arXiv:1009.0026, 2010.
[97] J. J. Quisquarter, L. C. Guillou, and T. A. Bersom, *How to explain zero-knowledge protocols to your children*, in Advances in Cryptology – CRYPTO' 89 Proceedings, Lecture Notes in Computer Science, **435**, 628–631, 1990.
[98] C. E. Shannon, *Communication theory of secrecy systems*, Bell Syst. Tech. J., **28**, 656–715, 1949.
[99] V. Shpilrain and A. Ushakov, *The conjugacy search problem in public key cryptography; unnecessary and insufficient*, Appl. Algebra Eng. Commun. Comput., **17**, 285–289, 2006.
[100] V. Shpilrain and A. Zapata, *Using the subgroup memberhsip problem in public key cryptography*, Contemp. Math., **418**, 169–179, 2006.
[101] R. Steinwandt, *Loopholes in two public key cryptosystems using the modular groups*, preprint, University of Karlsruhe, 2000.
[102] R. Stinson, *Cryptography; Theory and Practice*, Chapman and Hall, 2002.
[103] N. R. Wagner and M. R. Magyarik, *A public-key cryptosystem based on the word problem*, in Advances in Cryptology, 19–36, 1985.
[104] M. J. Wiener, *Cryptoanalysis of short RSA secret exponents*, IEEE Trans. Inf. Theory, **36**, 553–558, 1990.
[105] X. Xu, *Cryptography and infinite group theory*, Ph. D. thesis, CUNY, 2006.
[106] A. Yamamura, *Public key cryptosystems using the modular group*, in Public Key Cryptography, Lecture Notes in Computer Sciences, **1431**, 203–216, 1998.

Index

https://doi.org/10.1515/9783111142524-026

www.ingramcontent.com/pod-product-compliance
Lightning Source LLC
Chambersburg PA
CBHW080649220326
41598CB00033B/5152
* 9 7 8 3 1 1 1 1 3 9 5 1 7 *